ECOLOGY OF DUNES, SALT MARSH AND SHINGLE

Oystercatcher (*Haematopus ostralegus*). (Drawn by P.R. Hobson.)

ECOLOGY OF DUNES, SALT MARSH AND SHINGLE

J.R. Packham

Emeritus Professor of Ecology,
University of Wolverhampton, UK

and

A.J. Willis

Emeritus Professor of Botany
University of Sheffield, UK

The assistance of Dr N.J.W. Clipson (part-author of Chapter 3 and section 10.3) and of Dr R.A. Stuttard (part-author of sections 2.7 and 2.8) is gratefully acknowledged.

CHAPMAN & HALL
London · Weinheim · New York · Tokyo · Melbourne · Madras

Published by Chapman & Hall, 2–6 Boundary Row, London SE1 8HN, UK

Chapman & Hall, 2–6 Boundary Row, London SE1 8HN, UK

Chapman & Hall GmbH, Pappelallee 3, 69469 Weinheim, Germany

Chapman & Hall USA, 115 Fifth Avenue, New York, NY 10003, USA

Chapman & Hall Japan, ITP-Japan, Kyowa Building, 3F, 2-2-1 Hirakawacho, Chiyoda-ku, Tokyo 102, Japan

Chapman & Hall Australia, 102 Dodds Street, South Melbourne, Victoria 3205, Australia

Chapman & Hall India, R. Seshadri, 32 Second Main Road, CIT East, Madras 600 035, India

First edition 1997

Typeset in 10/12pt Palatino by Saxon Graphics Ltd, Derby

Printed at the University Press, Cambridge

ISBN 0 412 57980 4

A catalogue record for this book is available from the British Library

Library of Congress Catalog Card Number: 97-65570

Printed on permanent acid-free text paper, manufactured in accordance with ANSI/NISO Z39.48-1992 and ANSI/NISO Z39.48-1984 (Permanence of Paper).

Dedicated to the memory of Bill Carter (1946–1993)
and Derek Ranwell (1924–1988)

To all those who strive to protect, maintain and conserve the ever-changing shorelines of the World

To every thing there is a season, and a time to every purpose under the heaven:
A time to be born, and a time to die; a time to plant, and a time to pluck up that which is planted.

Ecclesiastes 3, v.1-2

CONTENTS

PREFACE

This book is concerned with coastal communities – and particularly the vegetation – developed upon granular deposits, and is intended mainly for use by undergraduates in biological and environmental sciences and by all with an interest in conservation and coastal management. It seeks to describe and interpret the complex interactions between organisms and the winds, tides and other water movements which shape the silt, sand and shingle in coastal ecosystems. Such interactions illuminate many problems of zonation and succession and necessarily involve both animals and plants; ecophysiological studies of halophytes, hydrophytes and xerophytes are particularly important. Grazing and nutrient requirements of sand dune and salt marsh plants have long attracted attention; such work now has an added relevance as grazing, mowing and fertilizer treatments are used more extensively for the management of vegetation. Many salt marsh systems developed under grazing pressures that remained relatively constant for decades or even centuries; too little grazing of historically grazed marshes is potentially as harmful for the wildlife interest as too much.

Control and protection of such coastal features as dune systems, salt marshes, shingle ridges and offshore barrier islands can be effective only if the physical processes involved in substrate movement and deposition are understood. There is also a pressing need for further studies of the movement of water and solutes in particulate deposits, a matter of practical importance in the Netherlands where the dunes are extensively used as water catchments.

Coastal ecosystems are usually very distinctive and well defined; the processes by which they are maintained or changed continue to attract the attention of ecologists around the world. This has resulted in a large and rapidly expanding literature from which we have drawn; certain topics, e.g. animal ecophysiology, however, are treated only briefly but references are given. We have tried to convey the authentic 'feel' of ecological processes in coastal ecosystems by describing and discussing (Chapters 1, 4, 6 and 8) a number of sand dune, salt marsh, shingle and barrier island sites, largely from Europe and North America.

Examples have been taken on a world-wide basis because many of the principles have a universal application, but the treatment of British systems is more detailed, major references for many of the important coastal systems being given opposite maps showing their location (Figures 4.14, 6.5 and 8.1). Key ideas and concept words are emphasized in **bold type** when first explained, and references to definitions are printed in bold in the index. The organisms most commonly referred to in the text are listed after the index. Where possible, nomenclature for vascular plants follows that of Stace (1991).

Much of the early work on coastal systems, particularly of Europe and North America, was initiated by a small number of notable individuals. In the second half of the 20th century, however, the International Biological Programme (which made major contributions to the study of energy flow and nutrient cycling), the first two European symposia on the ecology of coastal ecosystems (Norwich 1977, Haamstede 1983) and symposia elsewhere, greatly stimulated studies in this field. In recent years a number of investigations and surveys concerned with conservation have yielded further useful information.

Scientific studies of coastal ecosystems are likely to promote their wise use, and to provide the data necessary to secure their permanent protection as naturally evolving systems. There is, moreover, an urgent need to learn as much as possible about coastal dynamics: even a modest rise in sea level resulting from climatic change will produce major alterations, especially of salt marshes which are under threat from erosion, pollution and changes of land use. Current work on managed retreat and the use of vegetation in sea defence ('soft engineering') is providing further impetus for ecological and geomorphological studies, which should in turn enable coastal ecosystems to be more effectively utilized for recreation and conservation. The European Union for Coastal Conservation, established at the Leiden congress in 1987 and now with a permanent secretariat, provides, along with a number of other conservation bodies, a strong focus for future work.

We are especially indebted to Dr A.J. Davy, Prof. C.H. Gimingham, and Dr M.C.F. Proctor for their most valuable comments on the manuscript, and to Dr N.J.W. Clipson and Dr R.A. Stuttard for their specialist assistance. Dr M.C.F. Proctor in addition made available many informative photographs. We are also indebted to Dr K.E. Carpenter, Dr E.V.J. Cohn, Mrs J.M. Croft, Dr J.P. Doody, Dr D.J.L. Harding, Dr O.L. Gilbert, Mr P.R. Hobson, Dr. T.J. Hocking, Prof. D.J. Read, Prof. R.L. Jefferies, Dr A. Neal, Mr S.J. Pittman, Dr R.E. Randall, Mr. A. Spiers, Prof. I.C. Trueman, Dr C.A. Walmsley, Dr A.J. Wheeler and numerous other colleagues for information and discussion. We are grateful for technical assistance from both Universities and in particular to Glyn Woods and Jayne Young.

We thank all those individuals and organizations who have granted use of copyright material and gratefully acknowledge the consistent help and encouragement provided by Helen Sharples, Dr R.C.J. Carling, Dr C. Earle and Martin Tribe of Chapman & Hall. Finally, we thank our wives for their patience and support.

J.R.P.
A.J.W.
March 1997

INTRODUCTION AND PRIMARY CONCEPTS 1

'All is flux, nothing is stationary'
Heracleitus

1.1 AT THE EDGE OF THE SEA

This book is concerned with sand-dune, shingle beach and salt-marsh ecosystems based upon relatively unconsolidated granular deposits which frequently rest upon solid rock or, much more rarely, on peat. The particles from which these deposits are formed result from the unending battle between the sea and the land, and the constant supply of estuarine silt and clay. Until, after millenia, consolidated into sedimentary rock, these deposits often remain exposed to the action of the wind and the sea being, in terms of geological time, never stable for long. Such stability as they acquire, however, results in part from the plant communities which develop upon them: there is a constant interaction between the substrate, the physical forces tending to move it and the organisms living upon and within it. Of much greater stability are hard-rock cliffed coasts. Rocky shores are characteristic of **high-energy coasts** with strong wave action; they support a substantial and distinctive flora and fauna but these shores and their communities are not considered here. Mangroves, very important on tropical coasts, are included, but only very briefly.

1.1.1 A LOCAL EXAMPLE

Dunes, salt marsh and shingle, although in some places well separated, may occur close together, as in the river valley which opens into Three Cliffs Bay on the Gower Peninsula, South Wales, UK (Figure 1.1). Sand dunes occur on the Carboniferous Limestone hills on both flanks of the valley, Pennard Burrows on the eastern side being very extensive. Marram grass (*Ammophila arenaria*), red fescue (*Festuca rubra*), sand sedge (*Carex arenaria*) and a host of forbs including a number of nitrogen-fixing legumes and *Geranium sanguineum* – a calcicole favoured by the high calcium carbonate of the substrate – are among the numerous species found here. On grey dunes the purplish–black hips of burnet rose (*Rosa pimpinellifolia*) are a striking feature in autumn. Present also are many lichens and a range of bryophytes, of which *Brachythecium albicans*, *Homalothecium lutescens*, *Hypnum cupressiforme* and *Tortula ruralis* ssp. *ruraliformis* are characteristic. Pennard Pill meanders gently through a wide salt marsh established on a mixture of sand and finer material. The flora here is more limited; it includes *Juncus maritimus*, the grasses *Puccinellia maritima* and *F. rubra*, and a number of dicotyledonous species, several of which occur in a salt marsh/sand dune ecotone (see Figures 1.10 and 1.11). *Beta vulgaris* ssp. *maritima*, *Glaucium flavum* (yellow horned poppy) and *Rumex crispus* are among the shingle species on the pebble ridge forming the barrier at the margin of the bay. Autumn lady's-tresses (*Spiranthes spiralis*) is a nationally rarer plant whose flowering spikes can be found in September towards the top of the pebble ridge, where there is a matrix of finer material between the pebbles. Flowering in this species is very variable; there were hundreds all down the valley in 1981, whereas in 1987 it was hardly to be seen (D.J.L. Harding, personal communication).

Pennard Pill valley itself is an excellent example of the continual change – on a variety of scales – to which coastal habitats are subject. In previous centuries the castle and the former village of Pennard were overwhelmed by sand blown inland; the 14th century was a particularly stormy period (Gillham, 1977). The river has sanded up substantially in the last two centuries; in the 18th century ships used to moor beneath Pennard Castle. In recent years trampling pressure has risen considerably, although broadwalks have helped to reduce the damage. The site is also a meeting point where birds of woodland, grassland, sea and shore can all be seen, and where the invertebrate population of the stream changes from species typical of freshwater to those which tolerate much higher salinities approaching the sea.

1.1.2 THE OBJECTS OF STUDY

At the species and community level, ecologists are interested in **pattern and process** within coastal ecosystems such as these. They seek to discover how plants and animals colonize initially hostile environments at the edge of the sea, and how various factors govern the subsequent processes of succession and cyclic change. Studies of particular communities often involve the inter-relationships and roles of the component species, whereas function at the ecosystem level involves such processes as the flow of energy and the cycling of nutrients. Salt marshes in particular are frequently characterized by large-scale import and export of organic material through tidal action, so investigations of nutrient budgets involving energy flow, as well as geomorphology, frequently involve cooperation with physical geographers.

The description and classification of coastal vegetation have been pursued vigorously in recent years, but there is also a major emphasis on mechanism and function. The ultimate aim of some long-term studies is to identify the structure and processes characteristic of particular coastal ecosystems, and to be able to predict the likely outcome of disturbances by storms, changes in sea level or human inter-

Figure 1.1 Pennard Pill meandering through the salt marsh at Three Cliffs Bay, Gower Peninsula, South Wales. Steeply dipping Carboniferous Limestone on the adjacent hills is covered by sand dunes. These also cover alluvial deposits further up the valley, which has been over-deepened by glacial scouring. The river discharges into the Bay to the west of the pebble ridge bounding the southern margin of the salt marsh. (Photograph by John R. Packham.)

ference. Such investigations are very relevant to coastal protection, where the use of physical and computer modelling may avoid costly errors. Knowledge of natural systems also makes it possible to engineer suitable new environments for coastal organisms when creating vast new projects such as the Dutch Delta scheme, and to predict the outcome of policies of 'planned retreat' in which sea walls are broken down or allowed to deteriorate.

Much of the interest of coastal systems derives from the constant change to which they are subject. Sand dunes, shingle banks and salt marshes – the result of processes which have occurred in the past – are also potentially the source of components of land forms which will develop in the future. Many of the species present, such as *Salix repens* (creeping willow) which may start as part of the cover of a damp slack and end as a 'hedgehog' projecting from a dune, show considerable tolerance to changes in the environment that they help to shape. Others do not; such organisms are often very limited in terms of the niches which they can exploit.

Studies of the **natural flux in granular substrates** are intrinsic to an understanding of the geomorphology of coastal dunes, shingle systems and salt marshes described in section 1.2. Silt and clay travel in suspension in water, sand is carried by turbulent water and swept by the wind, and shingle is moved by wave action. Particle size gradings within these very different materials, and their modes of movement, are considered in section 1.3. The chapter then goes on to examine the diverse roles played by the species of a salt marsh/sand dune ecotone, the importance of nutrient cycling and energy flow in coastal ecosystems, succession, zonation and species strategies.

1.2 GEOMORPHOLOGY OF COASTAL DUNES, SHINGLE SYSTEMS AND SALT MARSHES

The main types of dune system found in Great Britain, in some cases associated with salt marshes, are shown in Figure 1.2. The distribution and areas of British coastal dunes, shingle systems and salt marshes are shown in Figures 6.5, 8.1 and 4.14. Quigley (1991) provides a guide to the sand dunes of Ireland; there is a also a sand dune inventory of Europe (Doody, 1991). In a wider context Géhu (1985) provides maps showing the coastal dunes and shingle banks of Europe, which has extensive coarse sedimentary coasts consisting primarily of sand, gravel and shingle. Dune coasts, which are widely represented, are replaced by gravel and shingle ridges where sea currents are very strong or the sand supply insufficient. Extensive new dune systems are not building up at present. Indeed, many dune fronts are being eroded and show a cliff-like face to the sea, frontal sand having been blown inland without replacement. Contemporary frontal erosion of Atlantic dunes is sometimes caused by natural factors such as changes in current direction, but may be the result of man's interference such as the removal of offshore shingle by dredging.

Extensive dunes, with well-marked crests and depressions, develop mainly on low-lying coasts perpendicular to the prevailing wind and with an adequate supply of sand. Dune development as a result of the fixation of blown sand is considered in Chapter 6, where the important role of dune-fixing grasses, especially *Ammophila arenaria*, is emphasized. In Europe apart from the Mediterranean, many of the principal dune systems face dominant west winds, as on the coasts of Denmark, Holland, the Cotentin Peninsula (northern France), central and south-western France, and the western Iberian coast. In Britain, the important dune systems of the Hebridean coasts, of north-east England, and of Ainsdale, Aberffraw, Newborough Warren, Harlech, Ynyslas, Tenby, Whiteford Burrows, Oxwich, Kenfig and Braunton Burrows are on the west coast, but there are others – including the Sands of Forvie, St Cyrus, Tentsmuir, Gibraltar Point, Holkham and Winterton – on the east coast. The form of European coastal dunes (Figure 1.2; see also Figures 6.18–6.21) ranges from flat narrow sand ridges to major systems whose crests, which may reach 30–40 m in

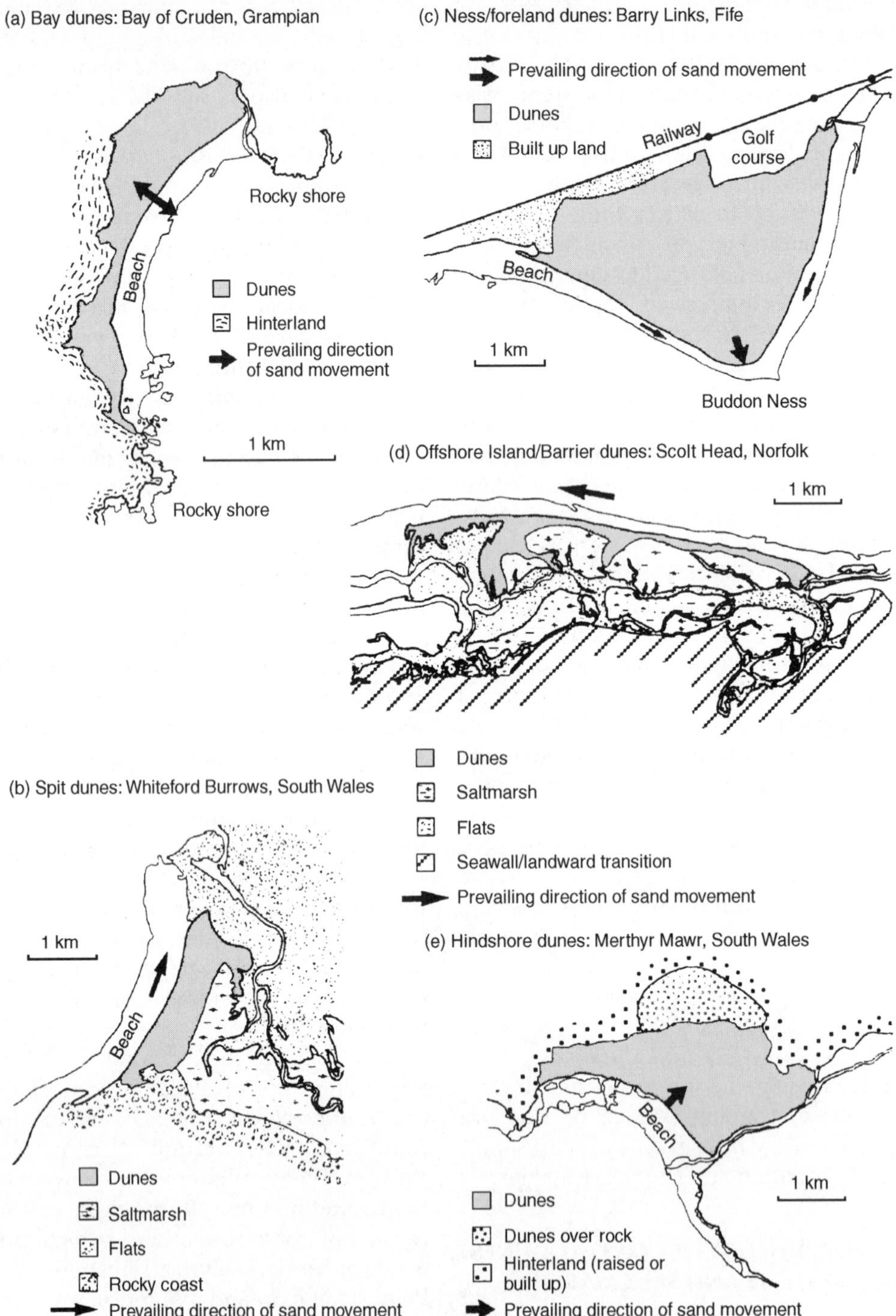

Figure 1.2 Major types of dune systems in Great Britain. (From Doody, 1989c, after Ranwell and Boar, 1986.)

height, alternate with depressions (**slacks**; '**pannes**' in French). The most spectacular dunes have developed where prevailing winds have deposited sand on clifftops 50 or more metres high as on the Gower Peninsula, South Wales, and in continental Europe on the Boulonnais coast, north-western coast of the Cotentin Peninsula, south-western coast of Portugal and the Mediterranean dunes at Leucate on the Golfe du Lion, southern France.

1.2.1 DETACHED BEACHES

Granular deposits frequently give rise to **detached beach** forms including spits, **cuspate forelands** (in which the beach forms a roughly triangular projection from the coast), tombolos and barrier islands (Pethick, 1984). The transport of sediment by shore drift commonly results in fringing beaches. Where the coast is indented this process results, dependent on the size of particles concerned, in the formation of **sand** and **shingle spits** and **bars** (Holmes, 1965; Randall, 1989). Where a coast turns in at the entrance to an estuary or bay, sediment transported by beach drift and longshore currents is carried on so that some of the coarser material drops into the deeper water beyond. What starts as a shoal gradually forms an embankment that continues to grow out from the shore until limited by waves or currents from another direction. Storm waves roll and throw material over to their sheltered sides so that **spits**, coastal ridges terminating in open water, tend to migrate landwards often becoming curved. The tendency of oblique waves to swing round the end of spits (becoming refracted), in places where the shore slopes rapidly into deeper water, also encourages the development of **hooks**. Spits frequently lengthen by the addition of successive hooks. Spurn Head, Yorkshire, now under threat from erosion, is an excellent example of a curved spit. Similar southward extensions of sand and shingle spits have resulted in the deflection of the River Yare in Norfolk, and of the River Alde in Suffolk, the latter giving rise to the shingle formation of Orfordness.

Bars or **barrier beaches** often extend from one headland to another, each enclosing a marsh or, if the bay inside receives streams from the mainland, a shore-line lake such as Slapton Ley, south Devon, UK. More commonly, however, outflow occurs via a deep narrow channel often kept open by vigorous tidal scour. Figure 1.3 shows two very large and smoothly curved bay-mouth bars surmounted by sand dunes and enclosing broad lagoons along the south-east coast of the Baltic. A **tombolo** is a bar connecting an island

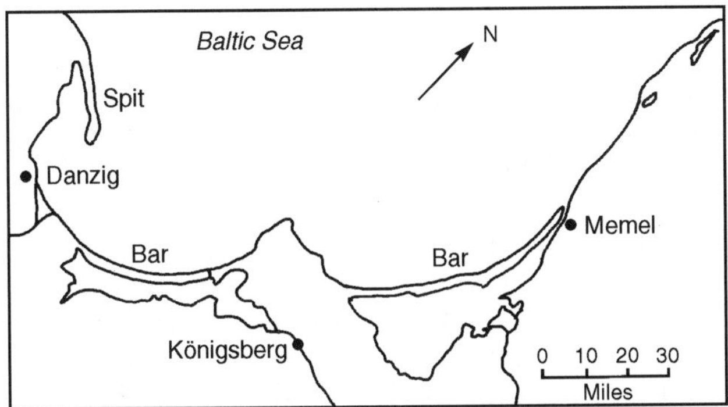

Figure 1.3 A spit north of Gdansk (Danzig) and two smoothly curved bay-mouth bars enclosing the broad coastal lagoons of Vistula and Kursk along the south-east Baltic coast of Poland and Lithuania (Königsberg and Memel are now called Kalingrad and Klaipeda respectively). (After Holmes, 1965.)

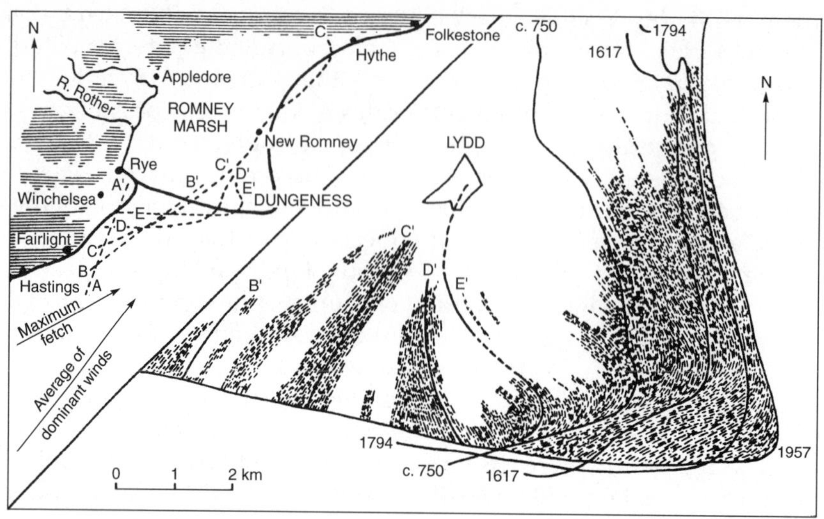

Figure 1.4 Map of the cuspate foreland of Dungeness showing the pattern and sequence of the shingle ridges. The scale is about six times that of the regional map on the left. The latter shows Romney Marsh and (horizontal shading) the higher ground bordering it on the north and west. It also illustrates the Lewis (1932) hypothesis of the development of Dungeness (section 8.4). (From Holmes, 1965, after W.V. Lewis and W.G.V. Balchin.)

to the mainland or another island; the eastern end of Chesil Beach, Dorset, which reaches the Isle of Portland, is a notable English example. Dunes and shingle ridges often arise on **banner-banks** (trailing spits) developed in the lee of offshore islands. Figure 1.4 shows the vast complex of parallel shingle ridges developed at Dungeness. Such structures are rare, others being found at Somme and Talbert (Géhu, 1985).

In contrast, **salt marshes** develop on low-energy coasts on a variety of substrates, including sand, silt and peat, behind spits and barrier islands, in estuaries and in shallow bays. **Creeks** and **salt pans**, the most characteristic topographic features of salt marshes, are well seen in the salt marsh at Blakeney Point (Figure 1.5), while Figure 1.6 shows the sub-habitats of a typical British salt marsh. The processes whereby sediments are transported to salt-marsh sites, deposited and stabilized, are complex (Pethick, 1984). The topic is introduced in section 1.3 and discussed at greater length in section 4.2.

1.2.2 BARRIER ISLANDS

Some 13% of the world's coastlines are guarded by offshore bars and barrier islands, which provide many striking examples of dune, shingle beach and salt marsh habitats. The longest stretches occur along the eastern seaboard and gulf coast of North America (Sackett, 1983), where their combined length totals 2800 miles (4500 km). Such islands (Figure 1.7) possess very low profiles and are much longer than they are wide; although the longest may be well over a kilometre wide, many are very narrow indeed.

Although technically a linear shingle storm beach for much of its length, Chesil Beach (see Figure 8.12), which runs for 29 km from its western junction with the mainland until reaching a height of 13 m at the Isle of Portland, functions in much the same way as a barrier island such as Scolt Head Island, Norfolk (Figure 1.2(d)). It has a long smooth margin facing the English Channel and a marshy tidal lagoon, the Fleet, between its shoreward margin and the indented Dorset coast.

Figure 1.5 Distribution of creeks and salt pans in a salt marsh at Blakeney Point, Norfolk. (From Pethick, 1984.)

Barriers seem to develop on coasts with a low tidal range and relatively high wave energy; there are three main explanations for their existence (Pethick, 1984). They may represent long-shore spits which have been broken through by storms. Alternatively they may be the results of the post-glacial sea-level transgression which swept sediments towards the present-day coastline (cf. Figure 10.2(d)). Hoyt (1967), however, maintained that barrier islands arise from the drowning of sand dune or beach berm features, forming a lagoon behind them. (A **berm** is the flat-topped ridge or bar on the landward side of a beach.) All three explanations may be true in particular cases.

'Wash-over' is a process important in maintaining the morphology of barrier islands as they move towards the mainland in a high-energy environment. Sediments are carried, usually by storm waves, from the seaward

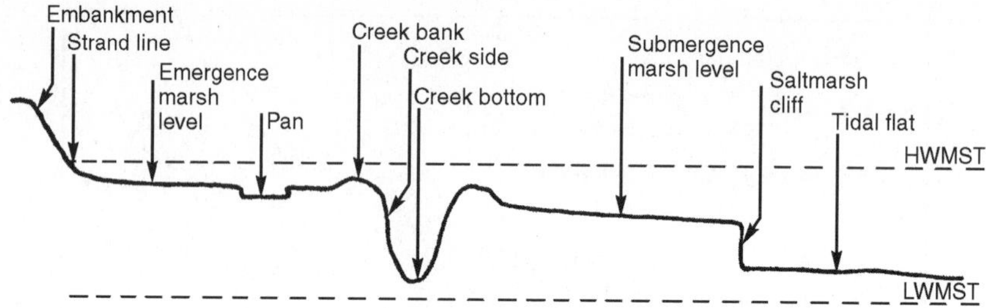

Figure 1.6 Zonation and distribution of sub-habitats across the profile of a typical salt marsh. (From Ranwell, 1972.)

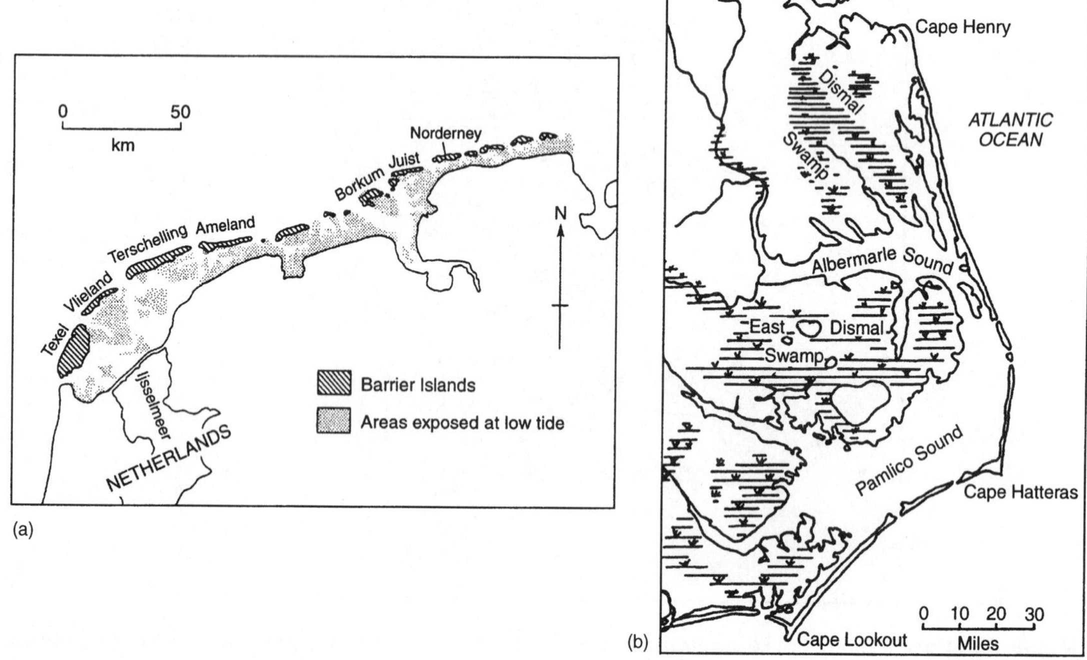

Figure 1.7 Barrier islands on the coasts of (a) the Netherlands and (b) the USA. ((a) from Pethick, 1984; (b) from Holmes, 1965.)

face to the landward side of a barrier island. In low latitude coasts, wash-over events caused by hurricanes can shift large volumes of sand across the crests of the islands, forming extensive wash-over fans. Ultimately such catastrophic events can transport a barrier island onto the marshes or mangroves of the coast. The distinctive landform known as a **chenier plain** results from the translation of barrier islands onto inner marshland by wash-over mechanisms (Pethick, 1984).

1.3 SUBSTRATE MATERIAL AND MOVEMENT: WINDS, TIDES AND PLANTS

1.3.1 SUBSTRATE PARTICLE SIZE

Particle size greatly influences soil aeration and drainage. The percentage pore space in soils whose particles are of similar shape and packing is identical irrespective of the size of particle. In ordinary soils, however, pore space varies from a little over 50% in stiff clays down to 25–30% in coarse sands of uniform texture. This is because the weight of the small particles of clay and silt in finer-grained soils is insufficient to overcome the cohesive forces and move the particles into the arrangement giving the minimum pore space, as occurs in sands.

In coastal systems the size of the particles largely controls the mechanisms by which they are transported and the texture of the soils to which they give rise. Shingle, the largest material, is shifted by wave action; the nature of the vegetation which manages to establish on it is strongly influenced by any finer particles which may form a matrix between its pebbles (section 8.2). Sand, which is moved by turbulent water and by wind, is naturally well drained; although sandy soils characteristically have lower percentage pore space than clays or silts, their capillaries are of relatively large diameter. Silt and clay, on the other hand, are usually moved in suspension in water. When initially deposited they have capillaries so small that drainage is greatly impeded, and in most muddy salt marshes an anaerobic layer blackened by sulphide comes very close to the surface.

There are a number of methods of grading the **soil particle sizes** determined by analyses made by sieving, sedimentation, the use of hydrometers, or more advanced methods (in many cases aggregates of fine materials are previously dispersed by chemical reagents). The International grading scheme classified **coarse sand** as particles of 2.0–0.2 mm equivalent diameter, **fine sand** as 0.2–0.02 mm, **silt** as 0.02–0.002 mm, and **clay** as below 0.002 mm median diameter. In the system now used by the British Soil Survey, **silt particles** are taken as 0.05–0.002 mm, largely because particles within this range have a similar effect in impeding drainage. The particle size analyses shown in Figure 2.7 can be interpreted on either of these systems.

The general term **shingle** is commonly applied to material between 2 mm and 200 mm equivalent diameter (section 8.1). On the Udden–Wentworth scale – now in almost universal use among sedimentologists – material of diameter >1 mm is described as **very coarse sand**, **granules** (2–4 mm), **pebbles** (4–64 mm), **cobbles** (64–256 mm), or **boulders** (>256 mm). This scale – a geometric series in which the fine class intervals are separated by a factor of two – facilitates the plotting of grain sizes in very varied samples on a logarithmic basis (Allen, 1985). The origin, composition and properties of the soils of coastal ecosystems are reviewed in section 2.5, which is cross-referenced to edaphic topics elsewhere.

1.3.2 SUBSTRATE MOVEMENT

Winds, tides and plants all influence the dynamic processes which shape sand dunes. The initial source of sand of maritime dune systems is the water-deposited sand of beaches. When the foreshore is exposed long enough at low tide for sand to dry, much sand can be blown inland and large dunes built. Dune systems are not always enlarging, however; they often go through cycles when they erode for long periods of time. At extreme high tide the plants nearest the sea may occasionally be inundated by salt water, to which species such as sand couch *Elytrigia juncea* (= *Elymus farctus* = *Agropyron junceiforme*) are tolerant, and they are often subject to saline spray.

Bagnold (1941), in his classic discussion of the physics of blown sand, showed that the movement of unfixed sand is dependent on the wind velocity gradient close to the sand surface (section 6.1); studies such as that of

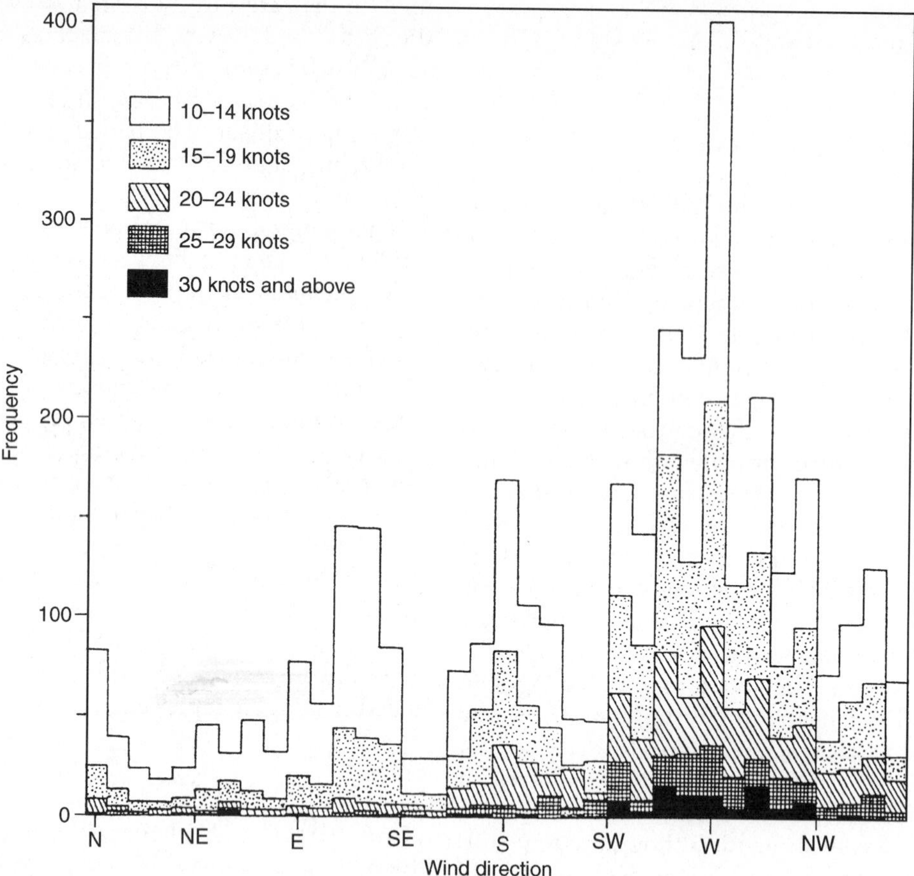

Figure 1.8 The direction and speed of winds at Chivenor, North Devon, UK. The frequency of winds of different velocity is shown against direction in a composite histogram. Frequency is expressed as number of winds recorded at four fixed hours daily during the period June 1949 to May 1954. (From Willis *et al.*, 1959a; courtesy *Journal of Ecology*.)

the example described below are founded upon his work. Wind direction at any particular site varies at different times, and a wind rose is often used to represent the sum of wind activity over a period. In their investigations of sand movement at Braunton Burrows, North Devon, where annual resultant sand movement is from 4° south of west, Willis *et al.* (1959a) used wind data from the weather station at Chivenor some 4 km to the east (Figure 1.8). A comparative measure of the sand-carrying capacity of winds from different directions was provided by use of the expression $An(v-V_t)^3$, where A is a constant associated with the volume of sand, n is the number of occasions on which wind is recorded in a given direction with speed v in knots (1 knot = c. 0.5 m s^{-1}), and V_t is the threshold wind velocity for sand movement (about 10 knots for the mean particle size of many British sand dune systems). Calculations showed that, over a 5-year period, 41.8% of relative sand movement occurred during the stormy winter months of December–February when the movement was from 2° 30′ south of west. In June–August, relative sand movement was only 14.2% and was from 11° south of west.

The presence of any barrier along the driftline, such as half-empty casks or other jetsam, impedes windflow and may cause its speed to drop below the threshold required for sand movement so that a small dune may build up. Vegetation is much more effective in this respect: Figure 1.9 shows the pattern of windflow over the marram grass of an isolated but typical embryo dune at Braunton Burrows. The barrier formed by the taller grass at the top of the dune has the effect of heightening the zone of low-velocity air near the ground, thus greatly increasing the chance that sand will be deposited on the inland side of the barrier. Coarse sand is dropped first; fine sand has a greater chance of being blown to the inland side of the dune system.

The vortex shown in the lee of the tallest tussock (Figure 1.9) caused a scouring in a direction opposite to the main windflow so that sand was thrown backwards and built up around the marram shoots. Similar but smaller eddies were detected behind the other tussocks of the dune whose wind shadow extended approximately 10 m beyond the dense vegetation (but see section 6.1). The extensive deposit of sand in the wind shadow, which is not directly protected or stabilized by the vegetation, is very sensitive to changes of wind direction. In extreme cases plants may be permanently buried by accreting sand; this can occur when sand from blowouts is deposited in dune slacks. Often, however, a rough balance develops in which burial of the vegetation caused by an excessive supply of sand leads to decreased wind resistance at the dune surface and less sand is deposited, thus giving the plants an opportunity of growing to the surface again.

1.3.3 ORIGIN, TRANSPORT AND EROSION OF WATER-BORNE SEDIMENTS

Salt marshes are typically developed on **low-energy coasts** and many of them, particularly in eastern and southern England, build up on

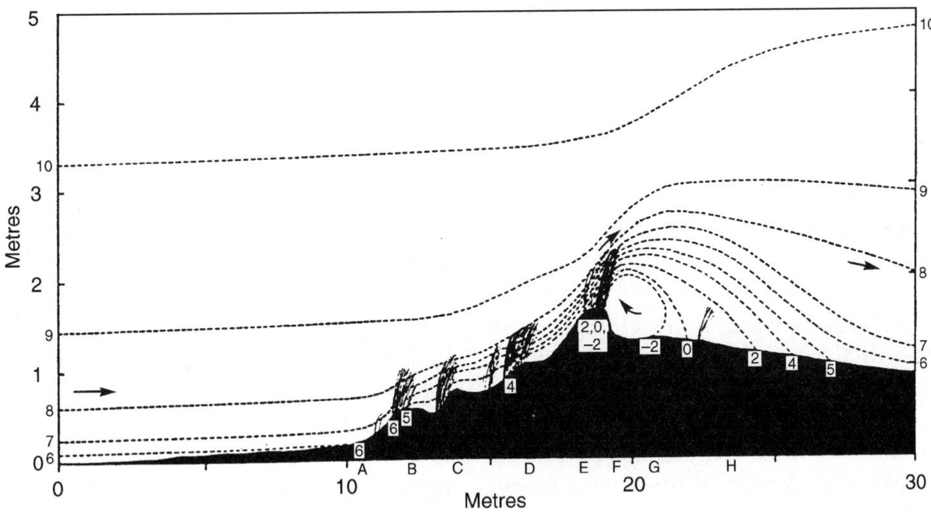

Figure 1.9 Wind flow over an embryo dune. Measurements of wind velocity were made at positions A–I at four known heights above the sand surface. From the vertical gradients the heights above the sand surface at which the wind reached the horizontal velocity of 0, 2, 4, 5, 6, 7, 8, 9 and 10 m s^{-1} were calculated and are shown as far as is practical by broken lines on the diagram. In the central region of the transect a reversal of the direction of air movement was noted, and is represented as a negative velocity (–2 m s^{-1}). The height of the sand surface along the transect is shown; the position and height of the marram shoots together with the density of their distribution are indicated. (From Willis *et al.*, 1959b; courtesy *Journal of Ecology*.)

fine-grained sediments or muds whose behaviour is considered by McLusky (1989). Tidal mudflats and salt marshes receive sedimentary material from rivers or the sea bed; in other cases material is locally reworked or washed in from land surrounding the estuary. In many North European estuaries most sediment is carried into the estuary by inflowing bottom currents of salt water. Whatever their source, estuarine and other tidally influenced sediments tend to accrete to a level at which flowering plants can colonize them, thus further slowing water currents and causing greater deposition (Figure 4.2).

Deposition of sediments is controlled by the size of their particles and the speed of the current. Pebbles 1 cm in diameter are eroded by currents with speeds exceeding 150 cm s^{-1}, transported by those of 90–150 cm s^{-1}, and deposited at speeds below 90 cm s^{-1}. Erosion of fine sand with particles 10^2 μm (0.1 mm) in diameter occurs at speeds exceeding 30 cm s^{-1}, while deposition takes place at less than 15 cm s^{-1}. The very small particles of silts and clays are deposited at even lower speeds. However, once they have **consolidated**, and thus considerably reduced their water content, silts and clays behave as if they were formed of larger-sized particles so that much faster currents are required to cause erosion.

Settling velocities of sediments are related to particle diameter and with small particles are governed by **Stokes' law**. With spherical quartz particles in water at 16°C this can be simplified to $V = 8100\ d^2$, where V = settling velocity in cm s^{-1} and d = diameter of the grain (cm). Sands and coarser materials settle rapidly in water and any sediment coarser than 15 μm (0.015 mm) settles within one tidal cycle. Particles less than 4 μm in diameter settle very slowly and cannot do so within a single tidal cycle. The presence of a few parts of salt per thousand causes 'salt flocculation' to occur, clay particles adhering to each other to form aggregates which fall more rapidly. Constant turbulence delays settling and causes turbidity which greatly reduces the rate of photosynthesis in submerged aquatic plants.

Technically, short-term **deposition** is distinguished from **accretion**, the difference between total deposition and erosion over a given time interval. The complex mechanisms involved in these processes are reviewed by Pethick (1984) and Pye and French (1993b). **Stabilization** of salt-marsh sediments – in which microalgae play an important role – is considered in section 4.2, as is the question of **subsidence** resulting from auto-compaction of sediments after accretion.

1.3.4 TIDAL CYCLES

Tides and currents often cause material to move towards or away from the foreshore, thus partially controlling the availability of sand to dune systems. Only plants of embryo dunes are likely to be directly influenced by tidal inundation, however. Nonetheless, salt marsh and mangrove associations (mangals) are regularly flooded by salt water: often twice daily at the bottom of the marsh (though there is commonly a respite during neap tides, see Figure 4.5) and rather seldom at the top. The length of time a plant is submerged in a year is directly related to the height at which it occurs on the shore and if this is known a **submergence/exposure ratio** can be calculated with the aid of tide tables. Tidal cycles are considered in more detail in section 4.2; each has a duration of one lunar month (c. 29 days) and includes two periods of **neap tides** in which deviations from mean sea level are least, and another two of **spring tides** in which low tides are much lower and high tides much higher than those of the neaps. Periods of greatest predicted tidal range occur in March and September, the times at which the very top of the marsh is most likely to be flooded by the sea.

Species regularly immersed in sea water have cell sap with high osmotic potential and many, such as glasswort (*Salicornia*), are succulent. Flooding restricts soil aeration and prevents gas exchange by salt-marsh plants as long as they are submerged. *Spartina* (cord-grass) is an example of a plant which has such well-developed aerenchyma (tissue with large intercellular air spaces) that oxygen produced in photosynthesis of the shoot system can reach the roots – a 'snorkel' system effective enough to cause the oxidation of reduced iron and produce a rusty halo around roots in anaerobic soils.

1.4 ENVIRONMENTAL VARIATION IN A COASTAL ECOTONE

In many wide expanses of salt marsh, sand dune or shingle ridge, variations in the vegetation are either gradual or apparently not clearly related to directional changes along environmental gradients. In some situations, however, the impact of the environment and the nature of the vegetation alter from metre to metre. These microcosms of coastal vegetation pose, in a very direct manner, many of the questions which apply to much larger areas. Frequently, such sites are **ecotones**, zones intermediate between well-defined types of ecosystem such as sand dune and salt marsh; they may be only a few metres across or much broader.

The setting of the salt marsh/sand dune ecotone discussed below is described in section 1.1. It is situated in the flood plain of the River Pennard, which meanders through a wide expanse of salt marsh (in which *Puccinellia maritima* is prominent) before discharging through a gap in the pebble ridge into Three Cliffs Bay, Gower Peninsula, South Wales. Seaward from Pennard Castle, the top of the ridge shows as a yellowish-green undulation edged by the litter of the drift line (which shows up clearly in the profile) and surrounded by the deeper green of the salt marsh, frequently flecked with grey silt brought down by the river and carried back and up by inundating tides.

Quadrat data for a transect across the ecotone were classified by indicator species analysis (ISA). The first part of this computer analysis is a stand ordination in which the stands are placed in order along an axis of variation (whose nature the experimenter has subsequently to determine) according to the species they contain. The point along this axis which represents the mean of the ordination scores for all the stands present is used to separate the stands into Groups 0 and Group 1 (Figure 1.10). ISA also selects indicator species which can, if necessary, be used to 'key out' the groups to which stands belong without further recourse to a computer. This process is then repeated to continue the sub-division of Groups 0 and 1. The original description of the method (Hill, Bunce and Shaw, 1975), evaluates both its advantages and limitations.

When classified in this way the stands of the transect across the low ridge shown in Figure 1.11 divide rather sharply into two groups whose distribution is closely related to topography. The indicator species for Group 1 – *Elytrigia juncea*, *Honckenya peploides* and *Poa pratensis* – are all sand-dune plants, while those for Group 0 are characteristic of salt marshes, though thrift (*Armeria maritima*) has a wide ecological amplitude and is often found on dunes and sea cliffs. In terms of cover values the grasses are the most important family present. *Puccinellia maritima* is a very strong indicator for Group 0 and occurs in lower and damper areas than does *Festuca rubra*.

When we follow the stand division within the transect a little further it is of interest that the stands of Group 01, for which the indicator species are sea aster (*Aster tripolium*) and sea arrow-grass (*Triglochin maritimum*), generally occur in areas even damper and less well drained than those occupied by Group 00. It should, incidentally, be noted that, with ISA, indicator species are not necessarily confined to a group, but are statistically preferential to it. Though *Triglochin* occurs in 14 out of 21 Group 01 stands and in none of the 19 Group 00 stands, the equivalent figures for *Aster* are 21 and 9, for example.

The only indicator for Group 10 is *Elytrigia juncea*, a rhizomatous pioneer characteristic of embryo dunes that is more tolerant of salinity

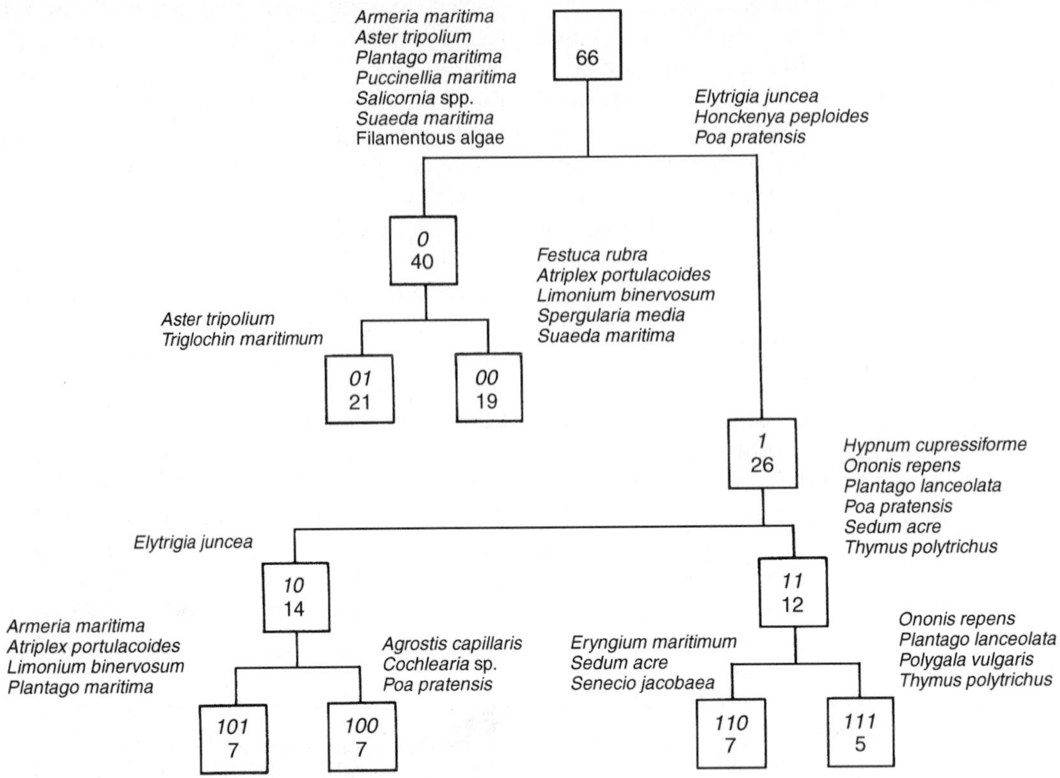

Figure 1.10 Indicator species analysis (Hill, Bunce and Shaw, 1975) of data from 66 quadrats (each 0.33 m × 0.33 m), from a transect across a salt marsh/sand dune ecotone in the Pennard Valley, Gower Peninsula, South Wales. The upper figure in each box refers to the ISA group; the lower figure gives the number of quadrats allocated to the group. The position of Group 01 has been transposed with Group 00, and that of Group 101 with Group 100, giving a sequence which more closely fits that shown in Figure 1.11. (Unpublished data of J.R. Packham and I.C. Trueman.)

than the larger *Ammophila arenaria*, which occurs higher on the ridge and is the dominant species of mobile or yellow dunes in larger maritime dune systems. When Group 10 is itself subdivided, the Group 101 indicators – *Armeria maritima* (thrift), *Atriplex* (= *Halimione*) *portulacoides* (sea purslane), *Limonium binervosum* (rock sea lavender) and *Plantago maritima* (sea plantain), all of which can withstand repeated inundation by saline water – reflect the gradual nature of the change from salt marsh to sand dune. Plant litter along the drift line in the area occupied by Group 10 often tends to kill or weaken the plants beneath it. Consequently, in spring, areas of bare sand soon to be penetrated by shoots of sea purslane and sand couch-grass are not uncommon here.

The six species selected as indicators for Group 11 are relatively intolerant of immersion and raised salinity. The same is true of several other species of the ridge top which are not listed as indicators or shown in Figure 1.11 (*Pilosella officinarum*, *Hypochaeris radicata*, *Luzula campestris*, *Plantago coronopus*, *Polygala vulgaris*, *Ranunculus bulbosus*, *Rhinanthus minor*, *Senecio jacobaea*, *Taraxacum* sp.). Sea couch-grass *Elytrigia atherica* (= *Elymus pycnanthus* = *Agropyron pungens*), sea holly *Eryngium maritimum* and Babington's orache *Atriplex glabriuscula*, are three species growing on the ridge

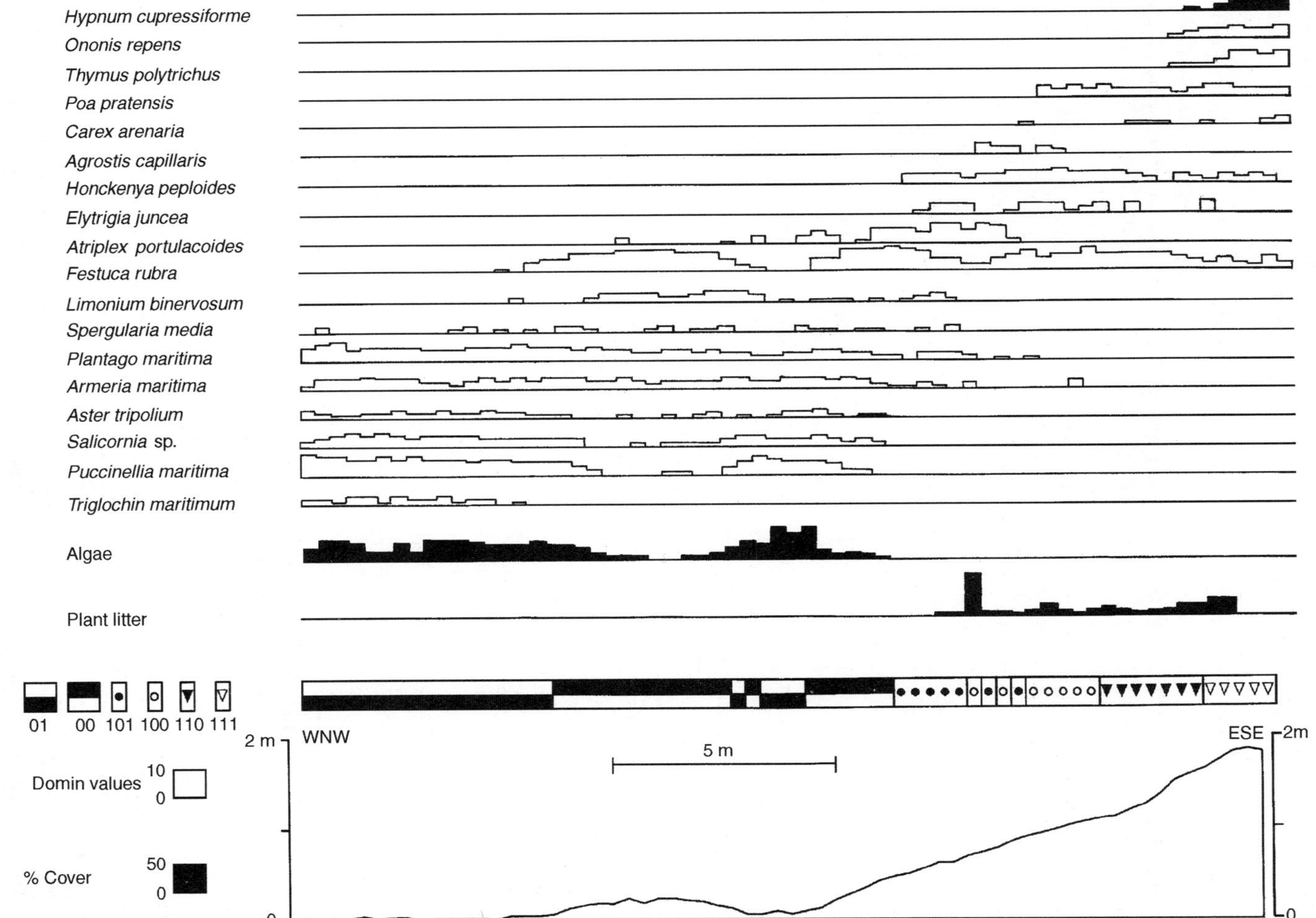

Figure 1.11 Profile of the Pennard Valley transect analysed in Figure 1.10. Cover is shown as a percentage (*Hypnum cupressiforme*, algae and plant litter) or a Domin value. The stands are allocated to the six main groups derived from the indicator species analysis. (Unpublished data of J.R. Packham and I.C. Trueman.)

which are more salt tolerant and may occur in areas flooded by extremely high tides.

Variation in soil particle size across the transect from the WNW, where silt is a conspicuous feature though much sand is present, to the ESE, where the ridge consists entirely of sand, is highly significant. Coastal plants are frequently subject to unfavourable conditions associated with drought and drainage, so the properties of the soils on which they grow are of particular importance. Sand, with its large grain size and frequently low amounts of associated colloids, is often wind-borne. Even when stabilized by vegetation there is always the chance that gales will transform a small gap in the cover of a dune into a major blowout.

Surface sand on the ridge summit has in most places been stabilized for a some time and has acquired a considerable organic content. Bryophytes are common here with *Brachythecium albicans*, *Hypnum cupressiforme*, *Lophocolea* sp., *Rhytidiadelphus squarrosus* and *Tortula ruralis* ssp. *ruraliformis* occurring in the transect. Lichens such as *Cladonia rangiformis* and *Peltigera rufescens* help to stabilize the maturing dune. Those in which the non-fungal partner belongs to the Cyanobacteria are frequently important in fixing atmospheric nitrogen, which may later become available to the flowering plants.

The vegetation of the transect was recorded in late September, and is thus fixed in time as well as space. In nature the spatial and temporal mosaics formed by the organisms and their environment are in constant flux, particularly with regard to the seasons. Seedlings of salt-marsh plants are abundant in spring, while winter annuals develop on sand dunes from late autumn onwards. Birds and other animals dependent on coastal ecosystems show similar seasonal differences in activity.

Animals frequently graze and trample coastal vegetation. *Salicornia* often germinates in the footprints of horses ridden over the Pennard salt marsh, while rabbits graze many coastal plants differentially. The smaller herbivores, and also the decomposers which break down dead plant and animal tissues and waste products, play even more important roles in processes such as nutrient cycling and energy flow.

Observations based on a small area of coast suggest hypotheses which can be tested only by further systematic observation or by experiment. Is *Puccinellia maritima* more tolerant of waterlogging and low soil oxygen tensions than *Festuca rubra*? Can *Bolboschoenus maritimus* withstand higher salinities than *Phragmites australis*? Theories concerning such questions are best developed or refuted by means of ecophysiological investigations. As in ecology generally, comprehensive studies of coastal ecosystems involve the repeated records and stimulus produced by field observations followed by experimental work, often in the laboratory, and *vice versa*.

1.5 GRAZING, DECOMPOSITION AND RENEWAL

Numerous modern studies, including several of salt marshes, have attempted to investigate and quantify fluxes of energy and materials into, within, and from complete ecosystems. As an **ecosystem** involves all the organisms in a community, such as that of a sand dune or a salt marsh, together with the abiotic environment – particularly its chemical and physical features – in which the animals and plants are living, this is a demanding approach.

Relationships involving the fundamental ecosystem processes of energy flow and nutrient cycling can be understood more easily when the various organisms are grouped according to their modes of nutrition. Figure 1.12 shows the three subsystems: plant, herbivore and decomposition, which result from this division. **Primary production** is carried out by the plant subsystem, which uses only a small fraction of the light energy reaching the Earth's surface to fix the organic carbon which subsequently drives both the **grazing chain** of the herbivore subsystem and the **decomposer** or **detritus chain** of the decomposition subsystem. The concept of **trophic levels** is frequently used to indicate the number of feeding links between the original solar energy and a particular organism. Primary producers (T1) are autotrophs which use much of the energy trapped in photosynthesis (or in certain bacteria, chemosynthesis), for their metabolism. The remainder is available for exploitation by heterotrophs, which use it to build up the **biomass** (living material) of their own bodies in the process of **secondary production**.

Herbivores, which subsist on living autotrophs ranging from unicellular algae to grasses and trees, form the second trophic level (T2) and may themselves be devoured or parasitized by carnivores (T3), which are in turn eaten by top carnivores (T4). A **food chain** is a simple progression through the trophic levels as in the following sand dune example:

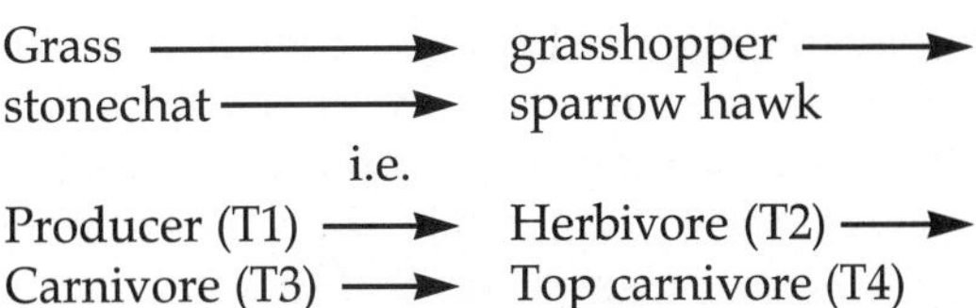
Grass ⟶ grasshopper ⟶
stonechat ⟶ sparrow hawk
i.e.
Producer (T1) ⟶ Herbivore (T2) ⟶
Carnivore (T3) ⟶ Top carnivore (T4)

There are also other grazing chains, such as that given below, in which the producer does not occur in the dune system so that import of both energy and nutrients is involved.

Diatom → *Calanus* (a copepod crustacean) → sand eel → herring gull → fox

Such chains form part of much more complex food webs, which also include decomposer chains involving animals and microbes that derive their nourishment from dead tissues. The amphipod *Corophium volutator*, which consumes both organic debris and small algae, is one of many salt-marsh animals which form links in both grazing and decomposer chains. Grazing by rabbits, sheep and cattle is an

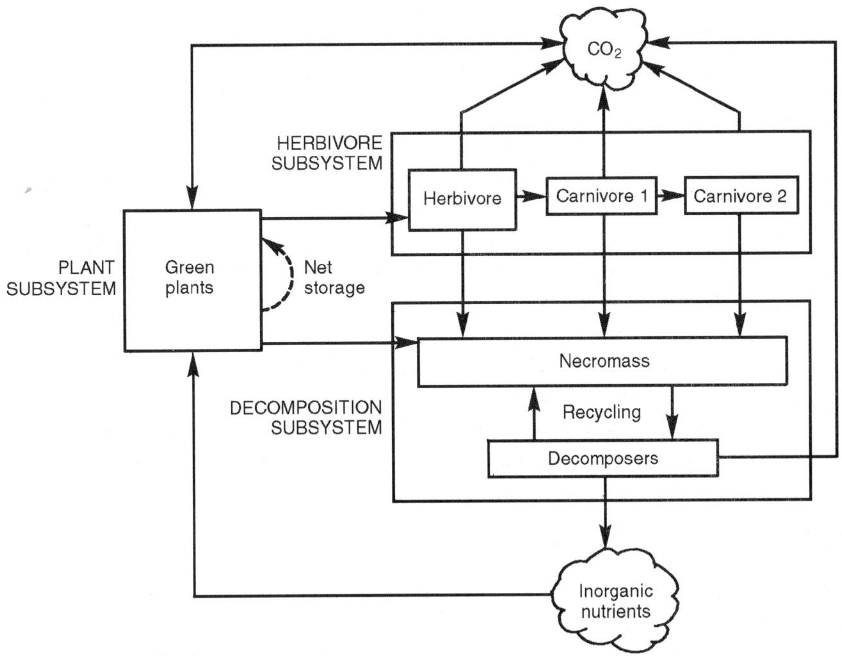

Figure 1.12 Generalized model of a terrestrial ecosystem showing the three subsystems. Arrows indicate major transfers between organic matter pools (rectangles) and to and from inorganic pools ('clouds'). (From Packham *et al.*, 1992.)

important influence on vegetation, frequently enhancing biodiversity in dune systems and salt marshes (sections 6.6 and 9.3). Excessive grazing pressure, sometimes exerted by migratory geese on salt-marshes and mudflats, may lead to degeneration of the plant cover (section 5.4).

1.5.1 ENERGY FLOW

The International Biological Programme (IBP), which ran from 1964–1974, took as an important theoretical base the concept of energy flow through the various trophic levels of the ecosystem put forward in a posthumous paper by Lindeman (1942), whose views have largely withstood the test of time though there are some situations where their elegant simplicity needs modification.

One of these concerns the trophic level to which particular species of varied diet should be allocated. Food chains are simple linear relationships; **food webs** are much more complicated because many organisms have a very varied diet and may themselves be subject to predation by several species. The diet of humans, for example, may include all the trophic levels including the decomposers; badgers have a similarly varied diet. To allocate such omnivores to an appropriate place in the trophic system demands a knowledge of their average diet. In constructing energy flow diagrams, attempts are made to partition the various energy transfers among the trophic levels; the food web shown in Figure 5.8 illustrates the difficulties associated with this allocation. In salt marshes (Pomeroy and Wiegert, 1981), grasslands and woodlands the **decomposers** are responsible for a much **higher proportion of the total energy flow** than are members of the herbivore subsystem. In the open sea the reverse applies; more energy flows through the grazing chain than through the decomposer or detritus chain.

There is considerable loss of energy at every link in a food chain, since even if digestion is 100% efficient, the energy released in respiration is not available to the next trophic level. The efficiency of energy transfer between successive trophic levels varies considerably and is often less than the figure of 10% commonly quoted. Whatever the value, the storage of energy as **secondary production** in heterotroph tissues is necessarily far less than primary production.

1.5.2 NUTRIENT CYCLING

Cycling of nutrients results from the activities of organisms such as those shown in Figure 1.13, a salt marsh food web, and geochemical processes, e.g. weathering of rocks, volcanic action and transport by wind and water. In sand dunes the main repositories for mineral nutrients are the green plants and the soil, particularly the colloidal organic matter. The sea has its influence in that most of the calcium in young dunes originates from the broken shells of marine molluscs; the pH of dune soils usually decreases away from the sea owing to loss of calcium carbonate as a result of prolonged leaching. Within many terrestrial ecosystems the major **nutrient pathways** are from the soil to green plants and back again (Figure 1.14). Such cycles are largely closed and self-maintained, with usually minor inputs from rocks or the atmosphere (with the notable exception of carbon and, to a lesser extent, nitrogen), and with losses as wind- or water-borne material or in harvests. Wind-borne losses are often higher in sand dunes than in many other ecosystems, while in salt marshes the ionic balance is strongly influenced by the sea and tidal mechanisms may result in very considerable import or export of litter containing nutrients; indeed particulate matter derived from the plant litter of salt marshes frequently affords an important food source for young fish in shallow waters.

Inorganic plant nutrients are absorbed from the soil, or the surrounding water in the case of salt-marsh algae, and incorporated, largely as organic compounds, into the bodies of autotrophs. Nutrients are returned to the soil

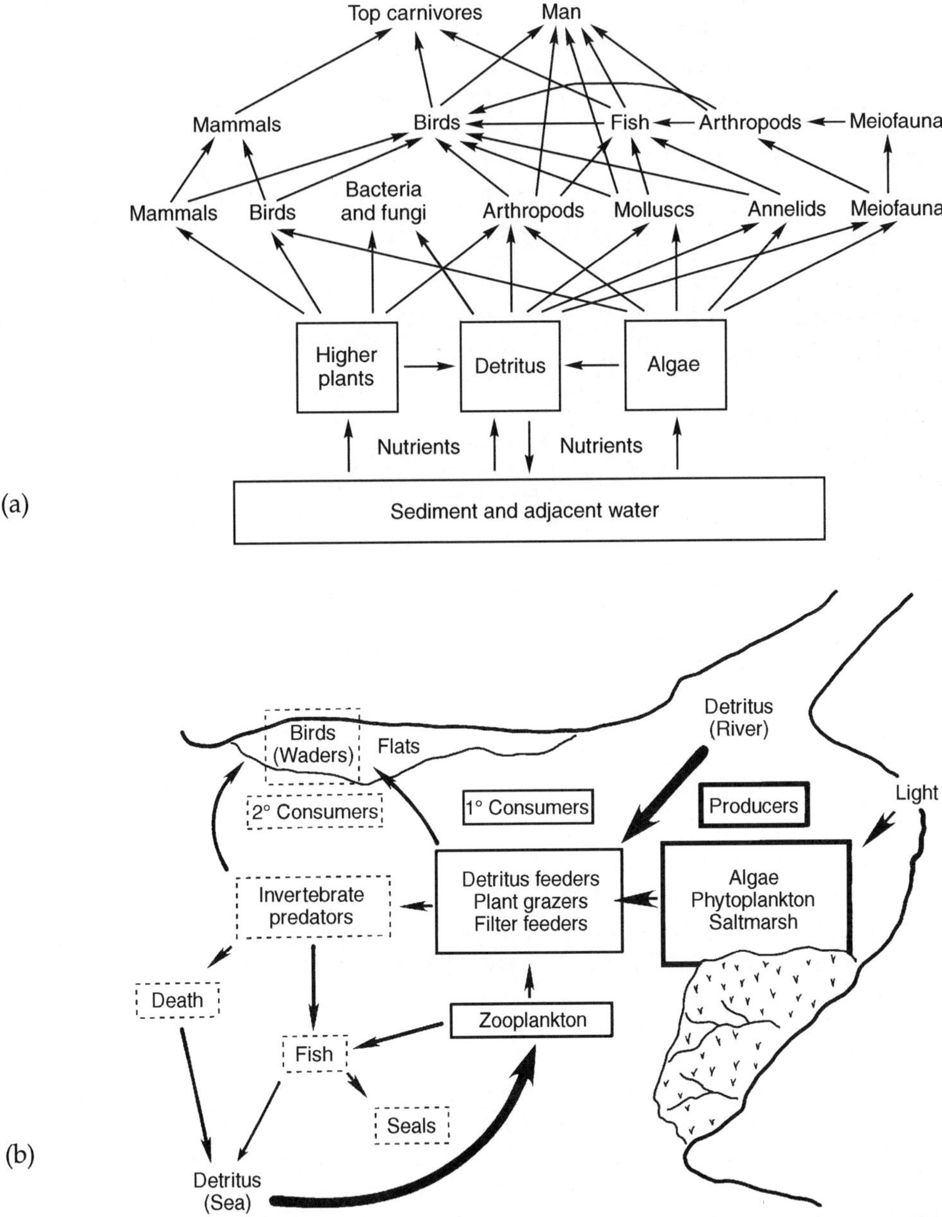

Figure 1.13 Trophic relationships of salt-marsh organisms. (a) Simplified food web on a salt marsh. (From Long and Mason, 1983.) The smaller invertebrates can have a high productivity and for this reason are frequently graded according to size. The **meiofauna** (meiobenthos) consists of those animals able to pass through a 0.5-mm (500-μm) sieve, but which are retained by finer sieves (usually 62 μm). The **macrofauna** (macrobenthos) is retained by a 500-μm sieve, while the **microfauna** (microbenthos) is able to pass a 62-μm sieve. (See also Green, 1968, Chapter 13.)
(b) Simplified diagram showing the main trophic pathways within an estuary and the important contribution made by salt-marsh producers. (From Doody, 1992; courtesy Cambridge University Press.)

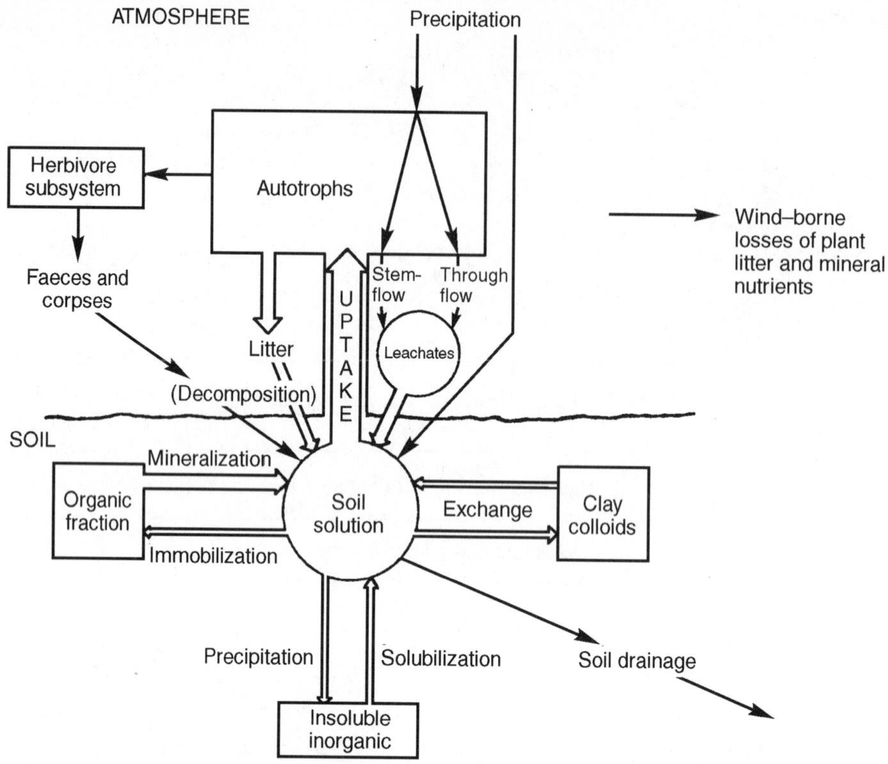

Figure 1.14 Major pathways of nutrient transfer in a terrestrial ecosystem such as a sand dune. In salt marshes and mangals there is often considerable import or export of litter and other components via tidal mechanisms. (Adapted from Packham and Harding, 1982.)

(or to the sea) as leachates and exudates and in the form of **necromass** (dead organic matter) especially litter. Nutrients are released from dead tissues by leaching and other abiotic processes, including fire, and by the catabolic activities of decomposers, the resultant release of inorganic nutrients from their organic, **immobilized** combinations being referred to as **mineralization**. Despite their relatively small total biomass, heterotrophs are particularly important in facilitating the effective operation of nutrient cycles.

1.6 PATTERNS IN SPACE AND TIME

Sand dune, salt marsh and shingle communities normally arise on bare substrate and when mature are often eroded, creating bare areas which are frequently re-colonized. They are thus very suitable habitats in which to study processes of vegetation change (Burrows, 1990), of which **succession** (non-seasonal directional change in the types and numbers of organisms present in a particular habitat over a period of time), is of major importance. They also illustrate **zonation**, the segregation of different species and communities in space. Viewed from above, many salt marsh and sand dune systems are seen to have belts of vegetation running parallel to the coast. These are sometimes interpreted as exhibiting **developmental zonation** in which the vegetation grades from that dominated by pioneer species verging on the sea to climax communities, frequently characterized by highly competitive mesophytes, at the landward margin where they merge imperceptibly into normal terrestrial vegetation. Such a model best fits salt marshes with a relatively uniform slope extending from the lowest region of the shore occupied by flowering plants to the topmost limit of the equinoctial spring tides. The succession that has occurred at this latter point is, on this concept, taken to correspond with the vegetational zones now present, the pioneer plants now growing at nearest the sea having been the first to colonize. Evidence to support the concept of developmental zonation has been presented by a number of authors including Eilers (1979), who investigated the salt marshes of Nehalem Bay, North West Oregon, USA and referred to the phenomenon as **successional zonation**. Here, creeks of the high marsh eroded through the sediments of earlier marshes, revealing the remains of low marsh species, notably the bulbaceous roots of *Spartina alterniflora* underlain by fine-grained sand as in active mudflat, at considerable depths.

Of the three major models of succession outlined below (there are also others), none by itself can account for the complete range of floristic replacements found in nature. In the **facilitation** model, species replacement is assisted by environmental changes brought about by organisms in earlier stages in the succession, whereas in the **inhibition model** it is prevented by the present occupants of the site, perhaps through heavy shading or allelopathic mechanisms, and replacement will occur only when these are damaged or killed. The species which colonizes the site first thus gains a major advantage; succession on this model often involves the replacement of short-lived species by long-lived ones. *Spartina* is a good example of a plant which once established inhibits the growth of other species at its level on salt marshes, while *Hippophae rhamnoides* acts similarly on many dunes. The **tolerance** (= **competitive hierarchy**) **model** does not depend on the initial presence of early successional species. Species which occupy a site early have little or no influence on recruitment of other species, which grow to maturity despite their presence (Burrows, 1990). Any species can start the succession but those which establish first are replaced by others that are more tolerant (competitively superior) and usually longer-lived. The mode of operation of the three models is shown by Archibold (1995, Figure 1.7).

The **facilitation** or **relay floristics** model of succession was developed by Clements (1916, 1936), who recognized a number of distinct phases in the process. It begins with migration of available species to an open area produced by some major disturbance of the environment, such as the large-scale blowout of a sand dune, the emergence of a fresh area from the sea as a result of isostatic readjustment, or the development of a volcanic island such as Surtsey in 1963. Once established these organisms both compete among themselves and react with the habitat. On sand dunes, for example, this leads to the creation of reserves of soil nitrogen by *Ononis repens* (restharrow) and other legumes common in the grey dunes. Competition and reaction do not follow each other, they occur simultaneously. In the final process the community was considered to reach a **climax** at which it was in equilibrium with its environment.

This is called the facilitation model because of the importance given to site modification as a driving force, a feature now seen to have been over-emphasized by Clements. The term relay floristics derives from the concept that a site undergoing succession is occupied by one set of species after another, with each set further ameliorating the conditions present. This model, though subject to criticism from the first, e.g. Gleason (1917, 1926), gave a dynamic aspect to plant ecology and is still of value today, although Miles (1979) expressed a widely held view in commenting on the lack of a single convincing and all-embracing model of succession, and many apparently similar successions may result in more than one end-point. Above all, there is now a general realization that complete stabilization never comes; vegetation is always changing and in coastal areas such change is often rapid.

The developmental phases described above constitute a **primary sere** in which climax vegetation – that Clements supposed to be perfectly adapted to its environment – develops on an area of bare substrate. Salt marshes, whose pioneer vegetation is regularly submerged by sea water, show **haloseres**. The complete succession occurring on a sand dune is called a **psammosere** (from the Greek, *psammos* = sand). Vegetation may undergo succession after being damaged but not destroyed. **Secondary succession** occurs after coastal vegetation has been damaged by fire or by excessive trampling and grazing. Washover processes in which dunes and salt marshes are covered in storm-driven sediment may completely destroy the existing vegetation, but in other cases it grows through a thin layer of fresh sediment.

The simplistic picture of developmental zonation given earlier has been criticized by Davy and Costa (1992), partly because in many British salt marshes the zone nearest the sea consists of *Spartina anglica*, a species which did not exist when many of them began to develop. The situation is further complicated by the topographical features, including creeks and salt pans, characteristic of the salt marsh subhabitats shown in Figure 1.6. Careful survey of apparently uniform slopes often reveals very small scale wave-like undulations which cause sufficient variation in drainage to influence plant distribution (Figure 4.13). Shrubs commonly occur at the top of salt marshes and in the oligohaline (low salinity) marshes of the Fal estuary, Cornwall, ungrazed salt marsh has succeeded to tidal woodland without major isostatic change (section 4.1). As a result of poor drainage the trees tend to be short-lived, and are often undermined by flooding.

Sand dune systems have hummocky profiles rather than gradual slopes. At the edge of the sea is the **drift line** (= strand zone) and the low **embryo dunes**, with the frequently tall **mobile** or **yellow dunes** behind. Next come the **semi-consolidated dunes** followed by the **mature** (= **grey**) **dunes** with maritime sward or dune heath on the inland side of the system (Figures 2.10 and 6.18). Areas where the sand has been eroded below the winter water level are called **slacks**; they commonly occur amidst the semi-consolidated and grey dunes. Similar depressions which do not become flooded are known as dry slacks or **lows**. This is the situation in a typical system, but some are eroded or lack embryo dunes, having mobile dunes facing directly down the beach, at least after the autumn gales. Dune systems change constantly but may be stabilized on the inland side by woodland, often coniferous and planted.

No single model can account for all successions but the facilitation model put forward by Clements fits – at least partially – a number of cases including some sand dune and salt marsh systems arising *de novo*. In other cases a secondary succession starts with a mature soil and a wide variety of plant propagules left after the former vegetation has been virtually destroyed. In such successions most of the plant species are either present from the outset as buried seeds, bulbs, rootstocks, rhizomes, etc. or invade shortly afterwards. Soil **seed banks** have been demonstrated in many coastal ecosystems: 110 taxa were germinated from soils of a barrier island off the coast of Florida (Looney and Gibson, 1995). This **initial floristic composition** factor (Egler, 1954) is an essential feature of the **inhibitional model** of succession, where the dominant species of plants at any one stage prevent the successful growth of other species, often by restricting available light. This is well seen in many 'old field' successions which develop when arable fields are left uncultivated. Although almost all the species found in the complete succession are present as propagules from the outset, annual weeds, perennial grasses and herbs, shrubs and finally trees, follow each other as the dominant species, inhibiting the growth of the other forms as they do so. Such changes in the composition of vegetation result largely from different rates of growth, reproduction and survival of species present from the start. When grey dunes are devastated by a blowout, *Ammophila arenaria* often has a new lease of life; fragments of almost moribund 'relict' marram develop vigorously, sometimes initiating – often with *Carex arenaria* – fresh dunes on which other species present as propagules from an earlier stage may be restricted to a fairly small area for several years.

Ammophila dunes frequently build up in places previously occupied by dune slack; the **cyclic sub-seral succession** described by Ranwell (1960a) is driven by alternating phases of erosion and accretion operating in particular areas. Figure 1.15 illustrates the way in which this process occurs at Newborough Warren, Anglesey, and is further discussed in section 6.1. Figure 1.16 illustrates the successional processes leading to climax forest in coastal dunes of the Pacific Northwest. These have been less modified by humans than European dunes, but the introduction of *Ammophila arenaria* has caused the major changes described in section 7.1. In contrast, long-term change from calcareous fixed-dune grassland to acid grassland or dune heath results from edaphic processes associated with leaching and hence is primarily driven by an allogenic (external environmental) mechanism.

As already noted, Miles (1987) considers, with justification, that no succession occurs in which any one of the generally accepted models operates alone, a view supported by Castellanos, Figueroa and Davy (1994) on the basis of investigations of the coastal salt marshes of Odiel (Huelva, south-west Spain).

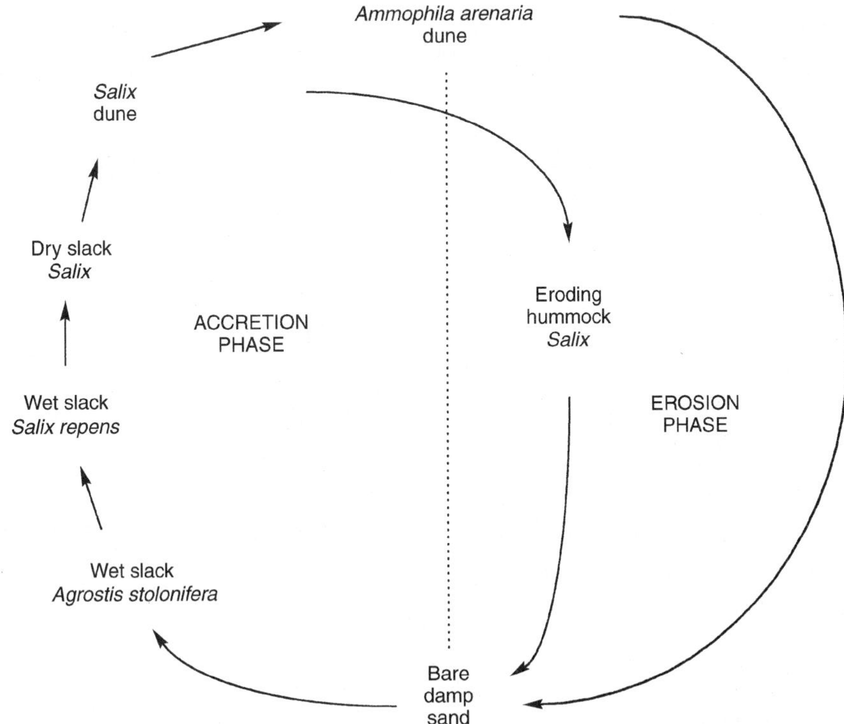

Figure 1.15 Cyclic sub-seral succession in the mobile dune-slack complex at Newborough Warren, Anglesey. The *Agrostis stolonifera* associes often develops in crescent-shaped areas behind advancing parabolic dunes. *Salix repens* is polymorphic; it grows poorly or is absent from very wet slacks, but flowers and seeds vigorously in places where the sand is drier and accretion rates of up to 40 cm per year occur. (After Ranwell, 1960a.)

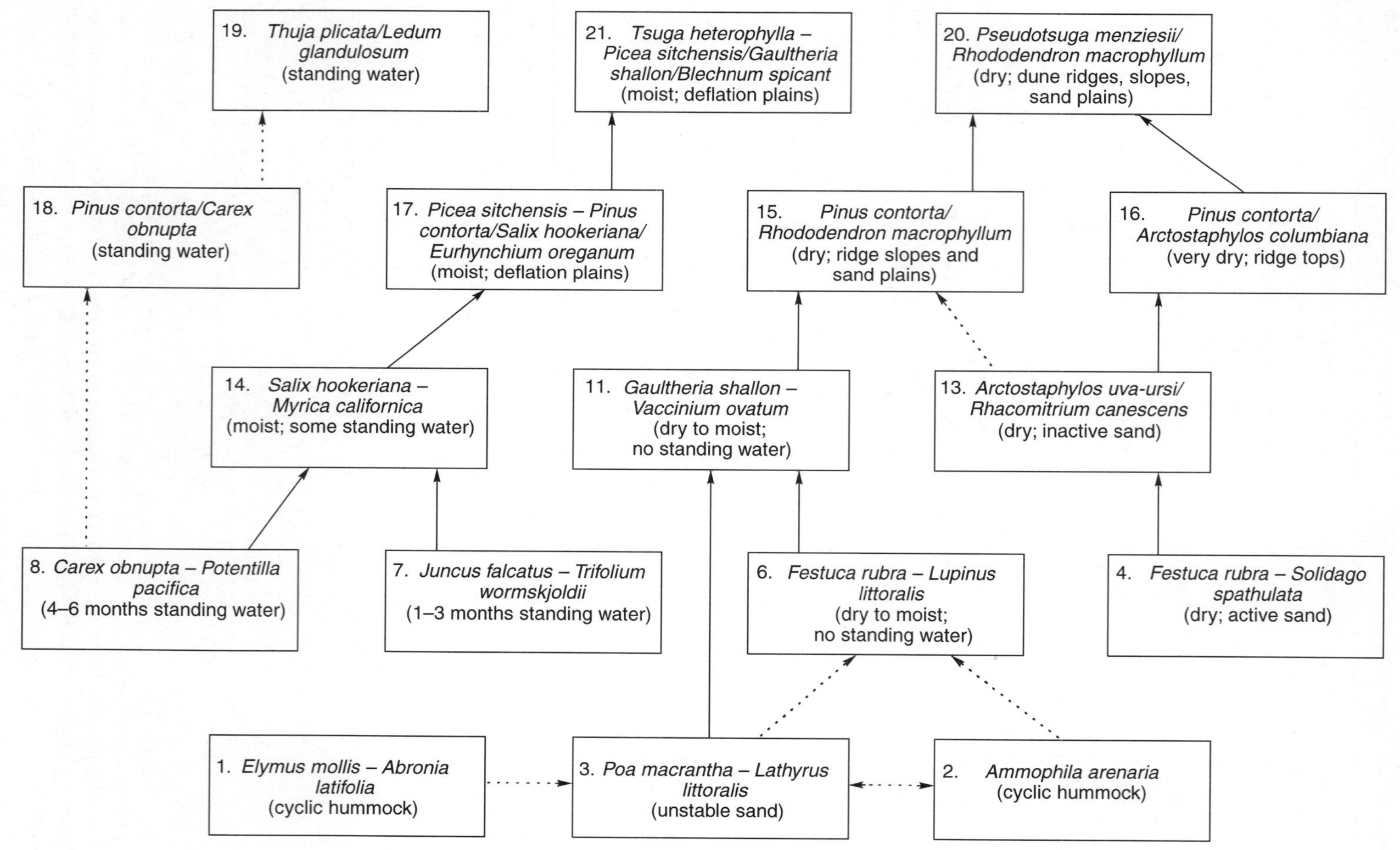

Figure 1.16 Successional relationships of plant communities of the coastal sand dunes, Pacific Northwest, USA. Four communities have been omitted; these either result from stabilization plantings or are found only on sand dunes south of Bandon, Oregon. Solid lines show the commonest successional pathways. Species of the same life form are separated by a dash (-); those of different life forms by a solidus (/). (Adapted from Wiedemann, 1984.)

Construction of a raised dyke divided a uniform area of low-lying sediment into two lagoons, both of which were colonized by isolated clones of *Spartina maritima*. These locally enhanced accretion and formed domed tussocks, a process termed **nucleation**. Only in the better-drained of the two lagoons were the central, higher areas of the *S. maritima* tussocks invaded by *Sarcocornia perennis* (= *Arthrocnemum perenne*), though its seed is freely available in both lagoons. The sprawling, dense canopy and superficial, relatively impermeable root system of this species rapidly suppressed the remaining *Spartina* tillers, leaving only a ring around the edge of the tussock. Soils in the *Spartina*-dominated areas in both lagoons remained highly reducing, whereas the upper 10 cm of soil beneath *Sarcocornia* was oxidizing.

In terms of the three major successional models discussed above, primary colonization by *Spartina maritima* **facilitates** invasion by *Sarcocornia perennis*, which becomes established only on relatively well-drained sediments. The interior of all the *Spartina* tussocks showed reduced tiller density and vigour with increasing age, a feature consistent with the **inhibition** model of succession, and one which made it much easier for *Sarcocornia* to invade. Finally, the superior competitive ability of *Sarcocornia*, which once established appears to prevent re-colonization by *Spartina*, suggests that a **tolerance** mechanism is operating.

1.6.1 FLUCTUATIONS AND PHENOLOGY

Succession, which is concerned with directional changes over long periods, inevitably intergrades with **fluctuations** which are shorter term and less far reaching, such as those related to rainfall reported from a dune grassland near Oostvoorne, Holland (section 6.3). Here, species such as *Anthoxanthum odoratum* and *Luzula campestris* decrease in dry years and increase in wet ones, while in *Aira praecox* the opposite is true.

Phenology, on the other hand, is concerned with the onset and duration of the activity phases of animals and plants throughout the year. The winter annuals which flourish on many dunes provide a striking example of the efficient utilization of a time period when most plants are in the resting phase (section 7.2). Annual rhythms are also of major importance with regard to the gulls and terns which nest on sand dunes, and migratory water fowl which graze salt marshes.

1.7 STABILITY, SELECTION AND STRATEGY

Superficially, many coastal systems look much the same from one year to the next; to the casual observer the vast expanse of a salt marsh and the tidal flats beyond may seem to change very little in a lifetime. Detailed observations indicate that this is often an illusion; indeed, some coastal ecosystems have been shown to develop with surprising speed. Change in the physical environment is on various time scales; the growth of a shingle spit can take centuries but it may eventually be broken through as a result of a single violent storm. Similarly a series of blowouts may greatly modify a dune system, and isostatic adjustments (section 10.1) may transform saltmarsh communities by altering their heights relative to the tides. Smaller changes, both in the land form and the communities present, can be monitored by such techniques as the repeated levelling of fixed transect lines and the use of permanent quadrats.

Table 1.1 shows how the rapid build-up of substrate resulted in a change of part of the Cefni Marsh, Anglesey, from an area of bare intertidal sand in 1951 to *Juncus maritimus* in lower salt marsh sward by 1968 (Packham and Liddle, 1970). Pioneer *Puccinellia maritima* entered the plot at some time between 1952 and 1954 and was the first higher plant invader. Propagules of all the species listed must have been available – swept to and fro by the tides or carried by birds – for many years before any gained a foothold. Change in sub-

Table 1.1 Changes in the observation plot, a square of c. 16.8 m side, Cefni Marsh, Anglesey, North Wales 1956–1968. (Modified from Packham and Liddle, 1970, by courtesy of the Field Studies Council.)

	Date					
Species	*6.5.1956*	*12.5.1957*	*4.5.1958*	*3.9.1959*	*15.10.1963*	*25.5.1968*
Agrostis stolonifera	•	•	•	•	•	r
Armeria maritima	•	r	r	•	va	va
Aster tripolium	f	a	a	va	va	f
Carex extensa	•	•	•	•	o	o
Cochlearia officinalis	o	o	la	•	•	o
Festuca rubra	•	r	o	o	f	va
Glaux maritima	la	la	la	la	va	lf
Juncus gerardii	•	•	•	r	la	r
Juncus maritimus	•	•	•	vr	o-r	f
Plantago maritima	•	•	f	f	va	va
Puccinellia maritima	d	d	d	ld	f-a	o-lf
Salicornia sp.	•	o	la	o	o	r
Schoenoplectus tabernaemontani	•	•	•	•	•	r
Bolboschoenus maritimus	r	la	la	la	o	lf
Spergularia media	•	•	r	•	•	•
Suaeda maritima	•	•	•	r	•	•
Triglochin maritimum	o	o-f	f	a	f	f
Cladophora sp.	ld	←	l sub d		→ } 5	
Bare ground (%)	10	2	1	5	(5)	} 4
Filamentous algae*	•	•	•	•	•	(4)
No. of vascular species	6	9	11	11	12	15

* Sample taken consisted mainly of *Microspora* and *Oedogonium*, with a very little *Zygnema* and *Oscillatoria*. The first five observations by D.S. Ranwell.

Abbreviations: •, not recorded; d, dominant; va, very abundant; a, abundant; f, frequent; l, locally; o, occasional; r, rare.

strate deposition pattern allowed establishment to begin: increase in species diversity was correlated with rising ground level rather than some other facilitation brought about by pioneer seed plants. Relatively swift change on a much larger scale occurred at Caerlaverock, Dumfries, Scotland, where virtually the whole of a 600-ha salt marsh developed in less than 140 years; the example described below is equally striking.

1.7.1 DEVELOPMENT AND VEGETATIONAL HISTORY OF THE BERROW SALT MARSH, NORTH SOMERSET

The tidal flats seaward of the dunes at Berrow are over 6 km wide at low spring tides: in 1907, C.E. Moss referred to them as possessing no vegetation and attributed this to the constant movements caused by the tides. By 1921, however, a 40-ha area of mud and sand about 1370 m long and 460 m wide had become the pioneer stage of a salt marsh (Thompson, 1922) whose subsequent development is described by Willis (1990, 1998).

A map of 1921 shows that the marsh was sharply delimited from the Gore Sand by a newly developed channel which drained southwards to the River Parrett. The developing vegetation was dominated by *Puccinellia maritima* and there were also very substantial amounts of *Salicornia*, especially at the tapering southern extremity and in parts of the northern end. Since that time the shape of the marsh and the nature of its vegetation have undergone major changes. By 1923 the deep channel forming the seaward limit of the marsh had largely silted up and the Salicornietum of 1921 had almost disappeared. During the 1930s and 1940s dominance passed to *Spartina anglica* in the irregular seaward area of the marsh and to *Bolboschoenus maritimus* (first seen in 1926) on its landward wet muddy border. Even in 1928 a zonation was visible in some places with *B. maritimus* on the landward side, *Aster tripolium* in the centre where it largely replaced *Puccinellia maritima*, and *Spartina anglica* on the seaward side.

Subsequently, the diversity of the marsh greatly increased, and by the 1960s it had a very rich flora; a transect recorded in 1958 (Figure 1.17) gives an indication of this. The clear zonation present then was related to the extent of tidal immersion and salinity over a slope which fell only 0.3 m over a distance of 137 m. The two colonies of *Juncus subulatus* (Somerset rush) discovered there new to Britain in 1957 are believed to have developed in the 1940s. The average salinity of the seaward margin of 'mixed salt marsh' as mapped in 1963 was about 1.0%. The water in the most inland part was at times almost fresh, and

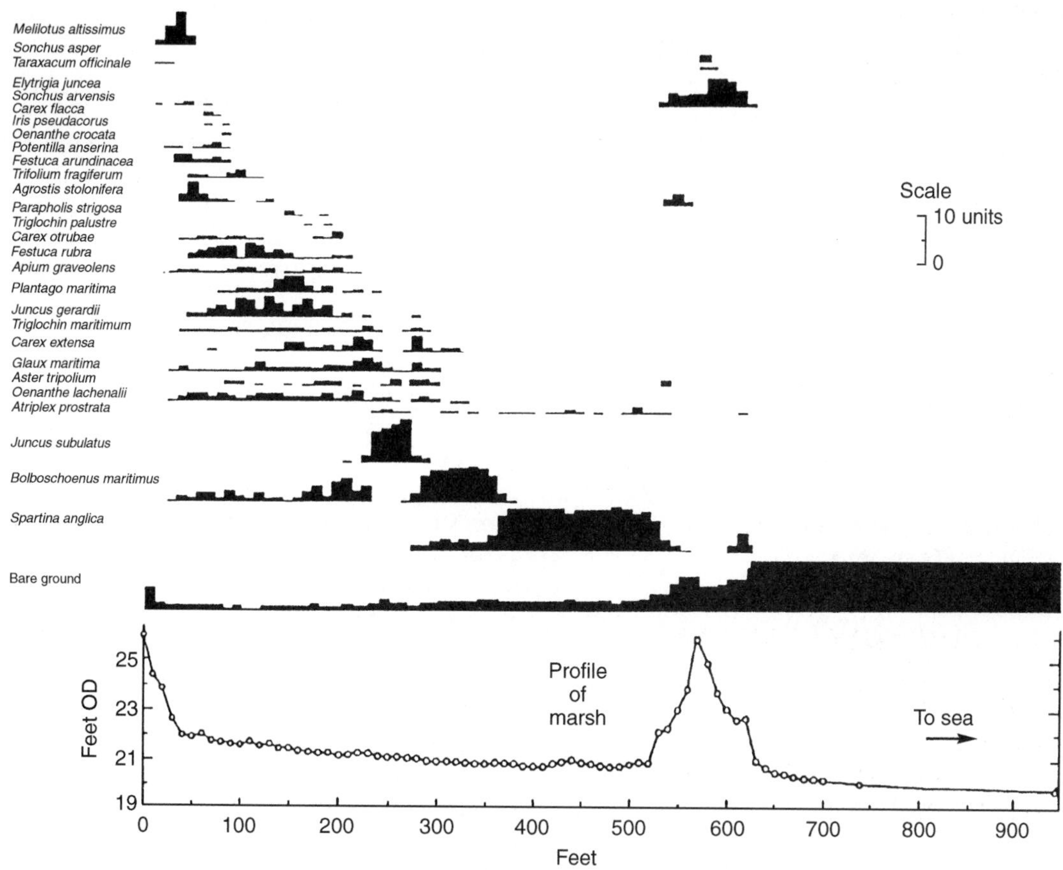

Figure 1.17 Altitudinal profile and vegetation across the Berrow Marsh, North Somerset, in mid-August 1958. Contiguous areas 3.05 m (10 ft) long and 0.91 m (3 ft) wide were recorded, the vegetation being assessed (out of 10 units) by relative bulk '(above-ground biomass)' of the component species, with an allowance for bare ground on an area basis. (Adapted from Willis and Davies, 1960.)

plants common in freshwater conditions – including *Alisma plantago-aquatica*, *Juncus articulatus*, *J. inflexus*, *Triglochin palustre*, *Typha latifolia* and *Schoenoplectus tabernaemontani* – were present by the late 1920s.

Figure 1.18(b) shows the extensive invasion of *Phragmites australis* since 1963. At Berrow – as at Nicholaston Burrows in the Gower Peninsula, South Wales – this vigorous rhizomatous grass has outcompeted many smaller species, and its presence is leading in some parts to higher accretion rates and reduced salinity due to leaching from the raised marsh surface, and everywhere a local loss of diversity. Common reed here is frequently 2.1–2.4 m tall and the number of species co-existing with it is limited, though *Bolboschoenus maritimus* is frequent as an understorey. The two original colonies of *Juncus subulatus* appear to have been overwhelmed by sand but another colony has arisen in a more southerly position and a new colonist, *Juncus acutus*, appeared in the northern region by 1990. *Hippophae rhamnoides*, a markedly aggressive shrub which was originally planted in the area, has greatly extended its distribution on the dunes and has invaded the inland margins of the marsh; it will have raised local soil nitrogen levels. The seaward edge of most of the marsh is now bounded by a well-developed dune line which is much reducing tidal inflow; the enclosed marsh may soon become a freshwater *Phragmites* swamp.

1.7.2 STABILITY AND SPECIES DIVERSITY

Jones, Kay and Jones (1995) reviewed the past and present status of five rare plant species – *Liparis loeselii* (fen orchid), *Limosella australis* (Welsh mudwort), *Rumex rupestris* (shore dock), *Gentianella uliginosa* (dune gentian) and the thalloid liverwort *Petalophyllum ralfsii* – of dune slacks in South Wales, UK, in relation to habitat change and loss. In each case population decline appears to be at least partly related to habitat change through dune system stabilization and the loss of successionally young community types. Other species now in decline are associated with relatively open short-sward vegetation or past disturbance leading to bare soil. The larger dune systems of the South Wales coast around Carmarthen Bay are now highly stabilized with small proportions of bare sand, little blowout activity, and extensive areas of mesotrophic grassland and scrub woodland. Close-mowing with biomass removal, exposure of bare soil, and translocation recovery for extinct taxa have all been suggested as appropriate conservation measures; **soil profile stripping** and **destabilization** of recently quiescent blowout fields might not be politically acceptable.

In terms of population stabilities, an important function of coastal systems is the refuge which they provide for migrating birds; in the *Zostera* and *Enteromorpha* flats and marshes used as wintering grounds by Brent geese, the primary producers are strikingly depleted within a few weeks. Permanent herbivore populations, including those of rabbits which are predated by foxes and stoats on sand dunes, tend to be constrained within limits by complex homoeostatic mechanisms such as prey–predator relationships. These may stabilize herbivore population sizes within the carrying capacity **K**, which is the maximum number of that species per unit area sustainable by the habitat, but can be greatly disrupted by events such as the myxomatosis plague of the 1950s. Myxomatosis lingers on though it is now much less virulent, and may now itself be an important regulatory factor.

1.7.3 SELECTION AND STRATEGY

An organism's habitat can be likened to a mould or templet against which evolution has fashioned for it a specific ecological 'strategy' which maximizes its chances of survival (Southwood, 1977). The components of such strategies can be considered in relation to heterogeneity in the basic habitat dimensions of space and time. For example, adaptations of organisms can be related to an **adversity axis**,

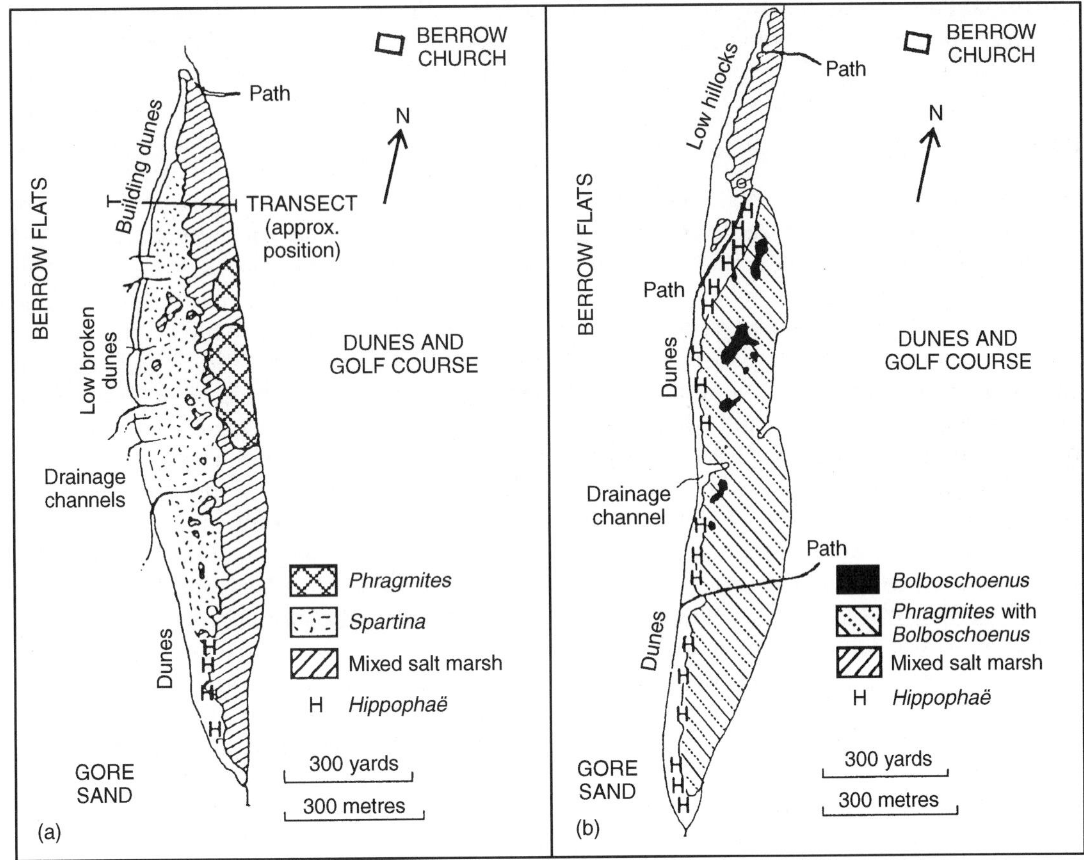

Figure 1.18 The topography and vegetation of the Berrow Marsh in (a) 1963 and (b) 1990, based on aerial photography and ground survey. (Adapted from Willis, 1990.).

as in Raunkiaer's classification of life forms of plants (section 2.1), while the **r–K continuum** of MacArthur and Wilson (1967) is one of many listed by Southwood (1977) as relating species distribution to the durational stability of the habitat. Stability ranges from ephemeral to relatively permanent habitats, and in all cases the important relationship is between the generation time (*t*) of a species and the duration (*H*) for which a habitat remains suitable for breeding.

Extreme **K-species** (i.e. those which display K-strategies, including such characteristics as long life and slow development, evolved through K-selection) are adapted to living in basically stable, 'permanent' habitats, in which *H*/*t* is large, where they tend to maintain their relatively constant populations at or near the carrying capacity (K). They have low mortality and recruitment rates, and have evolved high interspecific competitive ability, making considerable investment in defence. Examples include certain long-lived species of vertebrates and trees, e.g. mangroves. In contrast **r-strategists**, such as the classic pest and weed species, continually colonize unpredictable or ephemeral habitats, for which *H*/*t* is small. Considerable investment in reproduction enables these species to exploit favourable conditions by rapid population growth (hence 'r', from r_m, the symbol for reproductive capacity). They are essentially opportunists, with the ability to migrate to other habitats when carrying capacity is overshot.

1.7.4 PLANT STRATEGIES IN RESPONSE TO COMPETITION, DISTURBANCE AND STRESS

No plants can long survive both high stress and high disturbance, but according to the C–S–R concept of Grime (1974, 1979) three primary ecological strategies – of plants in the established phase – have evolved in response to other combinations of these factors. Plants possessing these extremes of evolutionary specialization are termed '**competitors**' (C), which exploit conditions of low stress and low disturbance as do most shrubs and trees, the '**stress-tolerators**' (S, high stress–low disturbance) of which lichens, very important in grey dunes, show in extreme form the four features (slow growth-rate, longevity, opportunism, and physiological acclimation) associated with this group, and the **ruderals** (R, low stress–high disturbance), e.g. *Poa annua, Juncus bufonius, Funaria hygrometrica.*

While the presence or absence of any one of the following characteristics is not in itself diagnostic of high or low competitiveness, very many plants possessing all four are extremely competitive:

- Tall stature.
- A growth form – such as a densely branched rhizome, as in bracken (*Pteridium aquilinum*) and common reed (*Phragmites australis*) – which allows extensive and intensive exploitation of the environment above and below ground.
- A high maximum potential relative growth rate (RGR).
- A tendency to deposit a dense layer of litter on the ground surface.

Coastal plants are frequently subject to stresses, which may take many forms, such as exposure, salinity, water shortage, or tidal inundation with associated anaerobic conditions and the presence of ferrous ions. The secondary strategy of **competitive-stress tolerator** is exemplified by *Spartina anglica*, frequently dominant at the lower levels of British salt marshes. Under drier conditions it competes much less effectively, while in less saline but still wet conditions it yields to *Phragmites australis*, the tallest native grass in Britain. The much smaller plants of the genus *Salicornia* are extremely tolerant of the stresses associated with high salinity; they may also be common in regions of the shore subject to considerable disturbance from wave slap which tends to prevent establishment from seed. This latter case emphasizes the importance of separating the **established** (adult) **and regenerative** (juvenile) **strategies** of plants, a notable feature of the C–S–R concept (Grime, Hodgson and Hunt, 1988), a potentially useful tool for ecological prediction and the manipulation of vegetation.

Chamerion angustifolium, whose established strategy is that of a competitor, occurs all too often as large clonal patches in grey dunes. It has two regenerative strategies, one vegetative (V) and the other involving numerous wind-borne seeds which are widely dispersed (W), germinate mainly in autumn but may survive ungerminated to the following spring though not becoming part of the persistent seed bank.

Other systems recognize possibilities not fully taken into account by the C–S–R concept. 'Ruderals', for example, can be of different kinds. Though all are monocarpic, *Aira praecox* and *Erophila verna* behave quite differently from *Papaver rhoeas* (usually annual) and *Verbascum thapsus* (biennial, rarely annual). In terms applied by During (1979) to bryophyte life strategies, the first two species – which have restricted but predictable open habitats in which they will reappear year after year, but which can be regarded as being disturbed by summer drought – are annual shuttle species, as are *Salicornia* spp. In contrast, *P. rhoeas* and *V. thapsus* are much larger, have far greater reproductive capacities, and are more opportunistic, tending to move from one open area to another.

In his revised system of bryophyte life strategies, During (1992) emphasizes the size and number of spores, although some bryophytes –including many 'colonists'– also develop asexually from gemmae, tubers and leaf fragments. **Fugitives** (e.g. *Funaria hygrometrica*) have a potential life span of less than a year and move from one ephemeral habitat to another. **Colonists** (e.g. small *Bryum* spp.) also have numerous very light spores, but commonly live for some years. Shuttle species, including **annual shuttle species** characteristic of seasonably suitable microsites within a community (e.g. *Pottia* spp., *Riccia* spp.), have few but large spores. Reproductive effort in all these three groups is higher than in the long-lived groups termed the **perennial stayers** and the **dominants**.

This book considers the continual biotic interplay between the many and diverse organisms of the coast, which exhibit a wide variety of strategies. Of equal importance are the interactions between these organisms and the constantly changing abiotic environment afforded by the dunes, salt marshes and shingle on which they live.

PRIMARY AND SECONDARY PRODUCTION: THE AUTOTROPHS AND THEIR ASSOCIATES 2

2.1 PLANT LIFE FORMS

Various systems have been developed for classifying the life forms of the algae, lichens, bryophytes and vascular plants responsible for primary production in coastal ecosystems. The most widely adopted description and classification of the life forms or types of plant body of vascular plants is that developed by Raunkiaer (1934) who was the first to use life form to construct **biological spectra**. The main feature of his simple, but ecologically valuable, system is the position of the vegetative perennating buds or persistent stem apices in relation to the ground level during the cold winter or dry summer which forms the unfavourable season of the year. The main life forms can be arranged in a sequence, dependent upon the position of the vegetative buds when the plant is dormant, which shows successively greater protection, especially from desiccation.

Raunkiaer assumed that when the flowering plants evolved the climate was more uniformly hot and moist than it is now, and that the most primitive life form is that still dominating the vegetation of the moist tropics. In such a climate large terrestrial plants (**phanerophytes**) can grow continually forming stems, often with naked buds, projecting high into the air. Evergreen or deciduous forms whose buds are protected from cold or desiccation by bud scales can be considered to be more highly evolved. These large plants may be divided into four height classes: **nanophanerophytes**, woody plants with perennating buds between 0.25 and 2 m above the ground; **microphanerophytes** between 2 and 8 m; **mesophanerophytes** between 8 and 30 m; and **megaphanerophytes** of over 30 m. Trees such as the red mangrove (*Rhizophora mangle*) are the dominant plants of mangal associations, whereas phanerophytes are relatively uncommon on salt marshes, though in England alder (*Alnus glutinosa*) and willows (*Salix* spp.) may occur on the landward side of tidal salt marshes.

Parts of mature sand dunes, on the other hand, are often largely covered by phanerophytic scrub, that at Braunton Burrows, North Devon, being quite species-rich (Willis *et al.*, 1959b). *Hippophae rhamnoides* dominates much of the scrub at Gibraltar Point, Lincolnshire, at Cefn Sidan and Tenby in South Wales, and the Murlough dunes in Co. Down, Northern Ireland; once established, its powers of nitrogen fixation greatly influence soil conditions. *Salix repens* commonly occurs in low-lying slacks, reaching a height of 1 m or more in sheltered places but it may be 15 cm or less in unfavourable sites. Sand tends to accumulate around this willow which subsequently often grows progressively upwards so that old plants may form wind-cut 'hedgehogs' several metres above the general level of the terrain. Pines frequently occur on coastal sand; *Pinus sylvestris* and *P. pinaster* (maritime pine) are often planted in European coastal protection schemes. Some of the largest trees growing on dunes are found in Oregon, where

Pinus contorta (shore pine) frequently dominates stabilized dunes along with scattered individuals of *Picea sitchensis* (Sitka spruce), *Pseudotsuga menziesii* (Douglas fir), *Thuja plicata* (western red cedar) and *Tsuga heterophylla* (western hemlock).

Chamaephytes are woody or herbaceous low-growing plants, such as *Sedum acre*, *Calluna vulgaris* and *Atriplex portulacoides*, whose perennating buds are on aerial branches not more than 25 cm above the soil, and frequently much lower, where the wind is not so strong and the air is damper. Being at the surface of the soil the perennating buds of **hemicryptophytes** are even better protected. *Agrostis capillaris*, *A. stolonifera* and *Ammophila arenaria* are protohemicryptophytes (Hp) with uniformly leafy stems but with the basal leaves usually shorter than the rest. *Festuca rubra* and *Puccinellia maritima* are semi-rosette hemicryptophytes (Hs) which have leafy stems with shortened basal internodes, the lower leaves being larger than the upper ones. Rosette hemicryptophytes (Hr), such as *Bellis perennis* and *Plantago maritima*, are often very resistant to grazing and trampling; they have leafless flowering stems and their leaves are very close to the ground. **Geophytes** have their perennating buds buried beneath the soil on rootstocks, corms, bulbs or tubers; those of *Juncus maritimus* (Grh) are on rhizomes. *Elytrigia juncea* (Hp) and *Carex arenaria* (Hs) are classified by Clapham *et al.* (1987) as hemicryptophytes, but can be so buried by sand as to function as rhizome geophytes.

Helophytes, such as *Spartina* spp. and *Phragmites australis*, have their roots in soil saturated with water for most of the time; their perennating buds are immersed in damp soil or beneath the water surface. **Hydrophytes**, including eel grass (*Zostera marina*) and beaked tasselweed (*Ruppia maritima*), have leaves which are either completely submerged or with their blades resting on the surface of the water. Rue-leaved saxifrage (*Saxifraga tridactylites*) and other **therophytes**, which survive the unfavourable season as seeds, are typical of open communities on sand dunes where winter annuals (see section 7.2) are often common. Therophytes are quite frequent in salt marshes also, often as pioneers, e.g. *Salicornia* spp. and *Suaeda maritima*.

Life form is determined primarily by heredity and selection; it may be regarded as an adjustment of the vegetative plant body and life history to the habitat. Under some circumstances, however, the environment directly influences life form, for example *Urtica dioica* and *Trifolium repens* may overwinter as herbaceous chamaephytes under favourable conditions, but are normally hemicryptophytes. Similarly, *Atriplex portulacoides* may develop into a nanophanerophyte rather than a woody chamaephyte. Conversely, severe weather conditions may kill the upper buds so that individual plants fall into the life form class below that normal to the species, as in the dwarfing of trees subject to almost constant wind at the edges of cliffs.

2.1.1 LIFE FORMS OF TERRESTRIAL BRYOPHYTES

Bryophytes often form an important component of the vegetation of older dunes. Five main types have been described by Gimingham and Birse (1957) – **cushions** (e.g. *Grimmia maritima*), **turfs** (e.g. *Tortula ruralis* ssp. *ruraliformis*), **canopy formers** (**dendroid forms** such as *Climacium dendroides*), **mats** (*Brachythecium albicans*, *Hypnum cupressiforme*, *Pellia endiviifolia*), and **wefts** such as *Rhytidiadelphus triquetrus*. Some life forms are better fitted to particular microhabitats than others, as shown by the differential distribution of mosses in sand dune systems (section 6.4).

2.1.2 ALGAE

The algae of salt marshes have not generally been analysed in terms of life form, though systems for doing so are now available (Chapman and Chapman, 1976; Dring, 1992). Nevertheless, these plants often make a substantial contribution to total production, contributing substantially, for example, to the winter diet of the Brent goose (*Branta bernicla*) on the lower marshes of Scolt Head Island (Ranwell and Downing, 1959). All the main groups of algae may be present in salt marsh and mangal associations; the **microalgae** are very important in stabilizing the substrate surface (section 4.2). Large populations of benthic (bottom-living) microalgae are present on salt marshes and mud flats; attached forms, including non-motile diatoms, are rare when fine sediment is accreting. On sand flats, at least in East Anglia, grazing by macroinvertebrates keeps numbers low.

Of larger algae, *Enteromorpha* and *Cladophora* (Chlorophyceae) commonly occur on sandy reaches at a level lower that that colonized by flowering plants, while *Vaucheria* (Xanthophyceae) often gives a green tinge to the firmer areas of low mud flats. Chlorophyceae are the most important algae in salt pans on lower marshes, but they are gradually replaced by Xanthophyceae as the ground level rises. In the lower zones dominated by flowering plants (phanerogams) green algae such as *Enteromorpha* persist; *E. nana* often occurs as a green mat on the vascular plants. Along the English Channel and the East Anglian coast in particular, free-living salt-marsh fucoid seaweeds (Phaeophyceae), mainly variants of *Fucus vesiculosus*, often form a dense covering on the lower marsh with variants of *Fucus spiralis* and *Pelvetia canaliculata* at higher levels. Various red algae (Rhodophyceae) are epiphytic on the fucoids, and also on *Atriplex portulacoides* and *Spartina* growing along creek banks. In the *Spartina* zone a blue-green or blackish soil covering of *Lyngbya*, *Oscillatoria* and *Phormidium* (Cyanobacteria) is common (Chapman, 1976); these organisms can fix atmospheric nitrogen.

On the sand dunes of Braunton Burrows and Harlech the diatom flora is restricted to forms capable of growing epiphytically on the mosses. Round (1958, 1959) showed that these diatoms were all derived from the populations growing on the damp sand of the slacks. At Braunton Burrows strikingly large forms of *Nostoc* also occur in wet dune slacks. **Lichens**, symbiotic associations of algae (or Cyanobacteria) and fungi, commonly occur in the older landward parts of sand dunes where those with Cyanobacteria are of some importance in nitrogen fixation, although most lichens are not nitrogen-fixers.

2.2 BIOLOGICAL SPECTRA

A **biological** (or **life form**) **spectrum** for a particular area is constructed by expressing the numbers of the species in each life form class as percentages of all the species present. Such spectra are also of use in studying succession (section 7.4). As a standard of comparison Raunkiaer used the **normal spectrum** derived from 1000 species taken as a representative sample of the world flora. There is a strong correlation between the climate of an area and the life forms of the plants present: a **phytoclimate** is characterized by the life form which proportionately most greatly exceeds the percentage for its class in the normal spectrum. The phytoclimate of Raunkiaer's native Denmark is hemicryptophytic, like that of most of the cool temperate zone, including Britain. In this system every species carried equal weight regardless of abundance or importance. Consequently, although the relatively few tree species native in Britain since the last glaciation were dominant almost everywhere until the advent of large scale agriculture, the phytoclimate is not classed as phanerophytic. The regions in which the **four major world plant climates** (**phanerophytic**, **therophytic**, **hemicryptophytic** and **chamaephytic**), occur, with their subdivisions, may be delimited by lines along which the biological spectra are similar.

The tropics are **phanerophytic** where rainfall is not deficient. Within this zone a greater proportion of the larger forms is found in the wetter areas. Subtropical desert areas are **thero-**

phytic and spring to life after the very occasional periods of heavy rain. Geophytes are best represented in regions with a Mediterranean climate where the unfavourable season is the hot dry summer. Plants with this life form are at a disadvantage where the soil warms slowly in spring and the growing season is short. The resting buds of **hemicryptophytes** in the cool temperate zone are protected by snow in hard winters, but are warmed by the sun as soon as it melts. **Chamaephytes** characterize the cold zones near the poles where the cushion forms in particular derive protection from snow, but grow as soon as the spring melt commences. Crowberry (*Empetrum nigrum*) is a small evergreen heath-like dwarf shrub which occurs as a woody chamaephyte or nanophanerophyte in some coastal dunes in Scandinavia, and is an important component of the vegetation of sand-covered cliffs in northern Scotland where it is also present on dunes and in slack communities, as at the Sands of Forvie, Aberdeenshire.

Hemicryptophytes are strongly represented in all eight biological spectra shown in Figure 2.1. Three of the sites are on the southern coast of Anglesey, North Wales, where the dune flora of Newborough Warren is richer than that of the slacks, which includes a higher proportion of hemicryptophytes. Myxomatosis spread to Wales in 1954; before then the dunes at Newborough were heavily grazed by rabbits, which have now returned to some extent. Grazing helps to keep the community open and promotes the growth of therophytes, most of which are **winter annuals** whose seed banks provide their only means of surviving the dry summers. Therophyte values of around 20%, as in the two Scottish examples, are more usual for dune areas.

The proportion of therophytes on the Newborough slacks is less than half that of the dunes. Once the vegetation of the slacks forms a complete cover it tends to remain closed unless disturbed, e.g. by rabbits or moles. Moreover, the low winter temperatures on the damp sand of the slacks lowers the growth rate of those winter annuals which do manage to establish. **Helophytes** present in the slacks are favoured by winter flooding and the shallowness of the water table in summer. Mosses are even more prolific in slacks than in fixed dunes.

The life form analyses for the dunes and slacks of the exceptionally species-rich Braunton Burrows system follow the same general pattern as those of Newborough Warren. Of the 339 taxa of flowering plants, 209 were allocated to the dunes and 130 to the slacks; if a species grows in both – as many do – it has been ascribed to its more typical habitat. About half of the species of both dunes and slacks are hemicryptophytes; chamaephytes and geophytes are minor components in both. Therophytes make up more than a quarter of the dune taxa, but as at Newborough Warren, are much less well represented in the slacks. Phanerophytes, which take a long time to colonize, make up an eighth of the taxa found on the dunes though the numbers of individuals are low, but form just under 4% of the slack flora. The strong representation of the helophytes is another indication of the great age of the system; certain of the slacks with essentially permanent water support also a few **hydrophytes**.

Helophytes are also a conspicuous element in the vegetation of the Cefni salt marsh, where a few hydrophytes occur in pools and drainage channels. Salt marshes in widely different parts of the world have hemicryptophytes as the dominant element and therophytes, in terms of species numbers, as the next most important group. This is probably true of the Cefni marsh, where a few ephemerals may have been missed. Table 2.1 shows the markedly different proportions of the life forms in eight sub-habitats of this marsh. The greatest number of annual species occurred on the banks, hummocks and spoil heaps. Although the numbers of species are low, densities of annuals are high in the lower marsh and, to a lesser extent, along the driftline. In Norfolk, *Salicornia* and *Suaeda maritima* occur at quite high densities within the matrix of

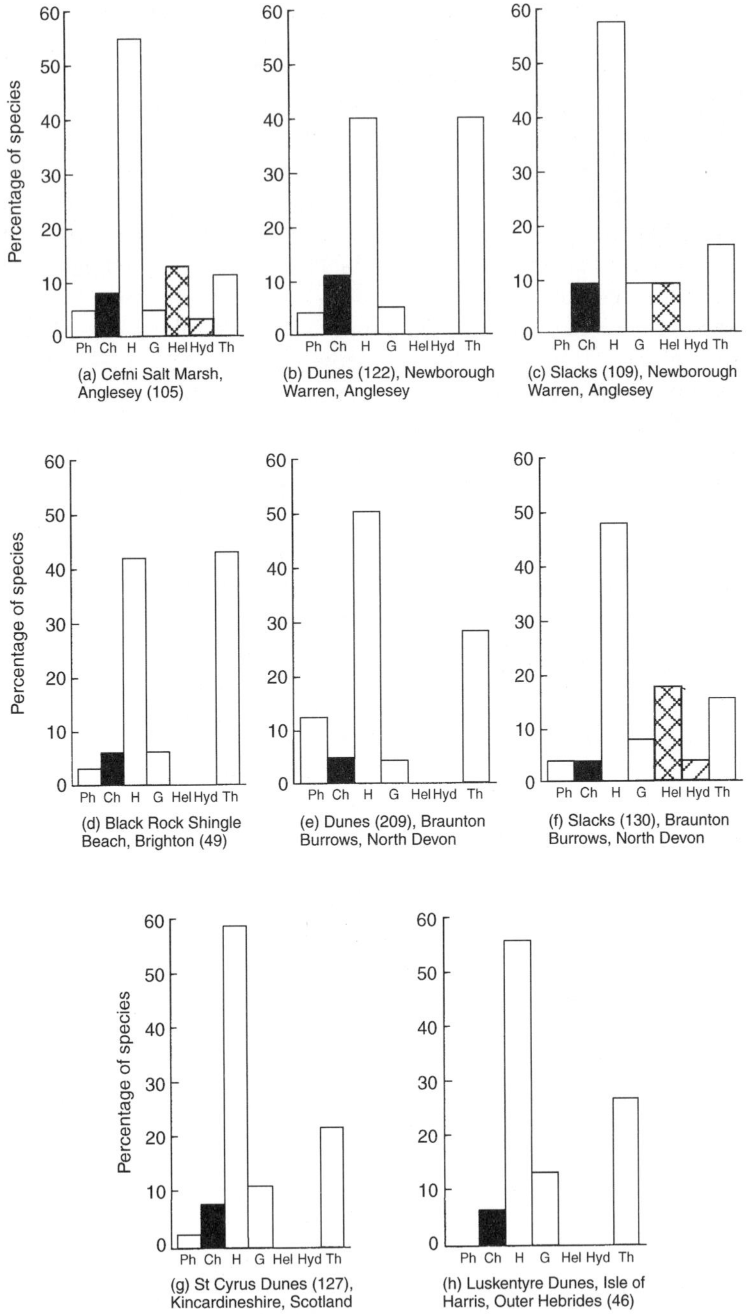
Percentage of species
0
10
20
30
40
50
60
Ph Ch H G Hel Hyd Th
(a) Cefni Salt Marsh, Anglesey (105)
(b) Dunes (122), Newborough Warren, Anglesey
(c) Slacks (109), Newborough Warren, Anglesey
(d) Black Rock Shingle Beach, Brighton (49)
(e) Dunes (209), Braunton Burrows, North Devon
(f) Slacks (130), Braunton Burrows, North Devon
(g) St Cyrus Dunes (127), Kincardineshire, Scotland
(h) Luskentyre Dunes, Isle of Harris, Outer Hebrides (46)

Figure 2.1 Biological spectra in seven life form classes for a salt marsh, a shingle beach, and four areas of sand dune in two of which the dune and slack habitats are analysed separately. Figures in brackets show the numbers of species in each habitat. Ph, phanerophytes; Ch, chamaephytes; H, hemicryptophytes; G, geophytes; Hel, helophytes; Hyd, hydrophytes; Th, therophytes. (Drawn from the data of (a) Packham and Liddle, 1970; (b,c) Ranwell, 1960a; (d) Packham, Harmes and Spiers, 1995; (e,f) Packham and Willis, 1995; and (g,h) Gimingham, 1951a.)

perennials of high marsh 'General Salt Marsh Vegetation'.

The spectrum for the shingle beach at Black Rock, Brighton, is based on records made during the period 1990–1995 and includes species not present in every season. *Lathyrus japonicus* occurred at an earlier stage in the succession and has been included in the spectrum. This beach has built up since the completion of a marina in the early 1970s (section 11.2; see also Figure 8.1) and the vegetation, in which the percentage of therophytes slightly exceeds that of the hemicryptophytes, remains relatively open. Many of these plants establish on patches of humus and sand on the landward side of the beach and their total cover is much less than that of the hemicryptophytes. The imposing and long-lived geophyte *Crambe maritima* (sea kale) which, under favourable conditions, can behave as a semi-rosette hemicryptophyte, contrasts with the fast-growing and deep-rooted halophytic annuals *Cakile maritima* and *Atriplex glabriuscula*. All three species have seeds dispersed by the sea. *A. glabriuscula* is widely distributed on this beach, occurring even further down the shore than the sea kale.

The original concept of the biological spectrum, in which all species are given equal weight, is simple and easy to apply. However, the incorporation of **relative values** for species gives a more realistic result in ecological terms. Such values may be based on cover-abundance ratings or quantitative measures such as density, yield or cover. Point quadrats are especially useful for assessing cover in low growing and very mixed general salt marsh communities. Biological spectra are particularly valuable for comparing different geographical areas. Woody species are much more important in the fixed dunes of the Atlantic coast of the USA than in Britain, for example.

Though one is on the west and the other on the east coast of Scotland, the biological spectra of the Luskentyre and St Cyrus dunes are very similar (Figure 2.1(g,h)). The St Cyrus site, which has far more species, consists of a single ridge behind a sandy beach and its flora contains over 90% of the species found at Luskentyre Banks which has a much higher

Table 2.1 Life forms of plants in relation to habitat in the Cefni salt marsh, Anglesey. (Modified from Packham and Liddle, 1970, by courtesy of the Field Studies Council.)

	Habitat									
Life form	*Whole marsh*	*%**	*s*	*JS*	*FJ*	*F*	*l*	*B*	*M*	*P*
Phanerophytes	6	4.8	1	–	2	1	1	3	–	–
Chamaephytes	10	8.1	3	1	2	–	–	8	3	–
Hemicryptophytes	68	54.8	12	13	29	7	19	40	17	–
Geophytes	6	4.8	2	2	3	1	1	5	3	1
Helophytes	16	12.9	7	4	6	1	3	3	8	2
Hydrophytes	4	3.2	1	1	1	–	1	–	1	3
Therophytes	14	11.3	2	3	3	2	3	9	4	–
Total species/ life forms	124	99.9	28	24	46	12	28	68	36	6
Total species	105	–	22	21	41	12	26	58	29	5

* Percentages shown are those of the **biological spectrum**.
Abbreviations: s, lower salt marsh sward; JS, *Juncus maritimus/Bolboschoenus maritimus* area; FJ, *Festuca rubra* with varying amounts of *Juncus maritimus*; F, grassland (mainly *Festuca rubra*); l, drift line or near it; B, banks, hummocks and spoil heaps; M, moist ditch margins, damp bases of large banks and pool edges; P, pools, ditches and streams.
Where a species has more than one life form listed in Clapham, Tutin and Moore (1987) both are included. The grassland habitat is under-recorded.

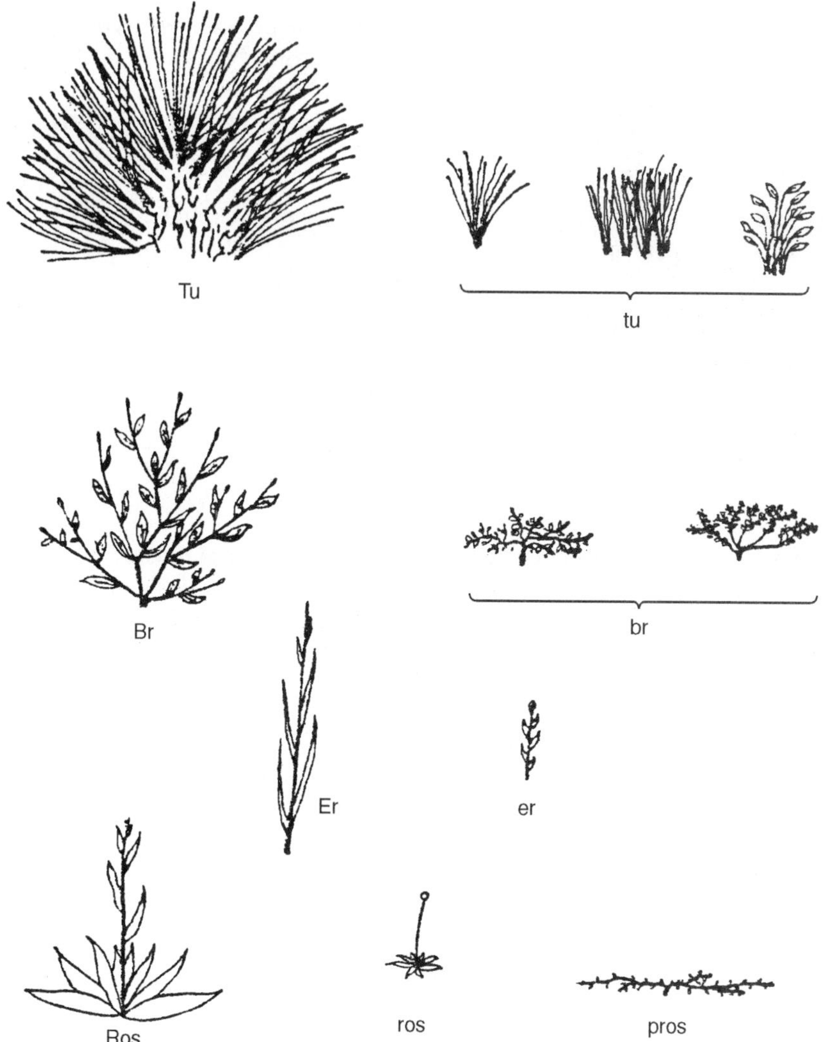

Figure 2.2 Diagrammatic illustration of growth form classes (see text). Tu, large tussocks; tu, tufted growth; Br, large branched forms; br, small branched forms; Er, large erect forms; er, small erect forms; Ros, large rosettes; ros, small rosettes; pros, prostrate forms. (Redrawn from Gimingham, 1951a.)

rainfall. *Ammophila arenaria*, a dominant in both communities and conventionally regarded as a hemicryptophyte (Clapham *et al.*, 1987) was taken to be a tussock-forming **rhizome geophyte** for these spectra by Gimingham (1951a) who considered the Raunkiaer system too insensitive for an adequate comparison of these two communities. His more specialized growth form system (Table 2.2) is illustrated in Figure 2.2; the resulting growth form spectra are shown in Table 2.3. Though the tufted, small erect and small rosette classes are characteristic of both dune systems, Luskentyre has a greater proportion of the smaller and more dense types of growth form (tufted, small-branched, small rosette), together with a higher proportion of species capable of vegetative spread.

2.3 PROCESSES OF PRIMARY PRODUCTION

The life of all components of the food web is maintained by the utilization of organic compounds formed by photosynthesis. This process utilizes only a small proportion of the solar radiation impinging on coastal ecosystems as most of it is immediately transformed into heat, which drives the evaporative processes of the **hydrologic cycle** whereby water from the lakes and seas enters the atmosphere only to be redeposited as rain or snow. Some 40–45% of the energy of this radiation consists of wavelengths in the visible light range, 380–740 nm, which also corresponds approximately to the bands absorbed by the photosynthetic pigments; photosynthetically active radiation, PAR, is generally considered as extending from 400–700 nm

Table 2.2 Growth form classification of sand dune plants. (From Gimingham, 1951a, courtesy *Journal of Ecology*.)

Growth form			*Reference letter*
(a)	(1)	Large tussocks: dome-shaped, frequently over 60 cm diameter (e.g. marram, *Ammophila arenaria*)	Tu
	(2)	Tufted growth: on a smaller scale than (1), though similar in outline. Includes most grasses: also small plants with petioled radical leaves densely aggregated (e.g. common sorrel, *Rumex acetosa*) or with densely aggregated aerial shoots and small leaves	tu
	(3)	Large-branched forms: tall (usually more than 25 cm high), much branched, covering capacity therefore considerable (e.g. prickly sow-thistle, *Sonchus asper*)	Br
	(4)	Small-branched forms: similar to (3) on a smaller scale (usually less than 25 cm) (e.g. prickly saltwort, *Salsola kali*)	br
	(5)	Large erect forms: tall (usually more than 25 cm high), branching restricted, covering capacity therefore slight (e.g. stinging nettle, *Urtica dioica*)	Er
	(6)	Small erect forms: similar to (5), on a smaller scale (usually less than 25 cm high) (e.g. fairy flax, *Linum catharticum*).	er
	(7)	Large rosettes: plants with clusters of large radical leaves, horizontal, erect or ascending, forming a more or less distinct rosette, with a tall, generally leafy and often branched inflorescence axis (e.g. broad-leaved dock, *Rumex obtusifolius*).	Ros
	(8)	Small rosettes: on a smaller scale than (7), forming a typical rosette; inflorescence axis generally leafless or nearly so (e.g. dandelion, *Taraxacum officinale*)	ros
	(9)	Prostrate forms: stems prostrate and spreading from a centre, leaves not long-petioled (e.g. biting stonecrop, *Sedum acre*)	pros
(b)	Presence of a means of vegetative spread		*

Table 2.3 Growth form spectra for two northern dune systems analysed by the method of Gimingham (1951a). (Courtesy *Journal of Ecology*.)

Location	*Growth form*										
	Tu	*tu*	*Br*	*br*	*Er*	*er*	*Ros*	*ros*	*pros*	*	*Total*
Luskentyre	2.2	30.4	–	10.9	2.2	15.2	6.5	23.9	8.7	43.5	46
St Cyrus	1.6	22.0	3.2	8.7	5.6	15.7	11.8	21.2	10.2	22.9	127

See Figure 2.2 and Table 2.2 for symbols.
* Species which have a means of vegetative spread.

(0.4–0.7 μm). PAR is bounded by ultraviolet radiation, UV, on the short wavelength side and by infrared radiation (IR) on the long wavelength side. Almost all the solar radiation received at the Earth's surface is included in the waveband 300–3000 nm.

The proportions of the radiation striking a leaf which are reflected, absorbed or transmitted through it vary according to wavelength, the angle at which light strikes the leaf, and the nature of the leaf itself. Foliage acts as a selective filter; although the thick leaves of many salt-marsh plants are almost spectrally neutral in attenuating incident light, radiation which has passed through a tree or grass canopy of a sand dune association is often rich in green light and in the near IR. It is deficient in the UV and also in the red and blue zones in which the chlorophylls absorb strongly, but the **action spectrum** for photosynthesis is much less sensitive to wavelength in the 400–700 nm range than the **absorption spectrum** of chlorophyll. Leaves absorb the vast majority of the far IR; the heat load often becomes a major cause of stress – especially in dune plants – but is dissipated by outward radiation, evaporative cooling (latent heat of evaporation of transpired water) and convective heat transfer (Jones, 1983).

The **photosynthetic pigments** of the primary producers are frequently well suited to the spectral composition of the incident irradiation. The carotenoids, which are largely responsible for the yellowish colour common in the leaves of sun rather than shade plants, afford protection against **photo-inhibition** (which can damage the photosynthetic apparatus). The **light reactions** of photosynthesis involve the absorption of light by chlorophylls *a* and *b*, leading to the formation of $NADPH_2$ and of the energy-rich compound ATP, both of which are essential in carbon fixation during the so-called **dark reactions** which then ensue. These endothermal reactions are, unlike the light reactions, often limited by temperature under natural conditions. Respiratory rates also increase as temperature rises, thus reducing **net photosynthesis** (gross photosynthesis less respiration) after the photosynthetic temperature optimum has been reached. Most green plants rely principally on the **C-3 pathway**, so called because the fixation of CO_2 (carboxylation), which is catalysed by the enzyme ribulose 1,5-biphosphate carboxylase (rubisco), involves the production of a compound with three carbon atoms (3-phosphoglycerate). **Optimum photosynthetic temperatures** for such plants are in the range 15–25°C. This is considerably lower than those of C-4 plants (between 30–47°C) and crassulacean acid metabolism (CAM; see below) species (centred around 35°C), aspects of whose photosynthetic mechanisms are described below (Salisbury and Ross, 1992).

C-3 species have high rates of **photorespiration** (light-stimulated respiration) in bright light; this phenomenon is restricted to the bundle sheaths of C-4 plants, which use the same CO_2-fixing mechanism as C-3 plants. The surrounding mesophyll cells employ the

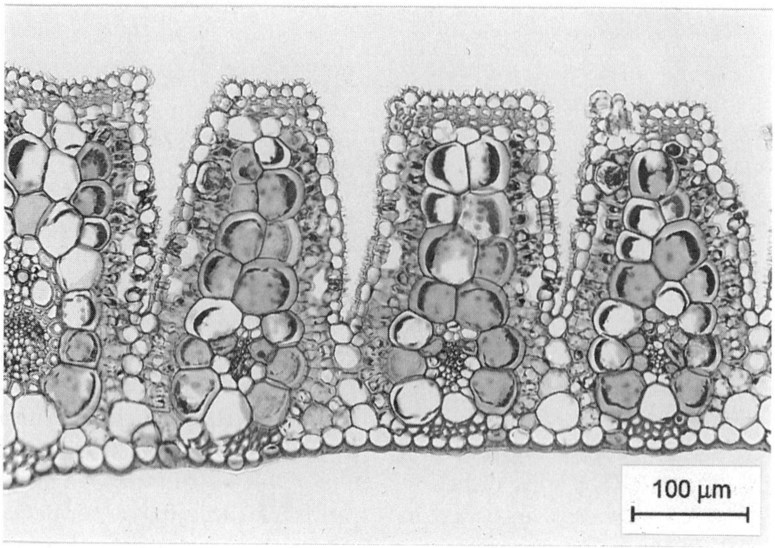

Figure 2.3 Transverse section of part of a leaf of *Spartina anglica* (stained with toluidine blue) showing Kranz ('halo' or 'wreath') anatomy in which the photosynthetic tissues are arranged in concentric cylinders about the vascular bundles. The large bundle sheath cells are fairly thick-walled, contain many chloroplasts and mitochondria, and produce sugars (and starch) by the C-3 route, in contrast to the surrounding radially arranged mesophyll cells which form malate by the C-4 pathway. (Courtesy of Dr L.I. Técsi.)

carboxylation of PEP (phosphoenolpyruvate, a 3-carbon compound), catalysed by PEP carboxylase, in which the initial fixation product is a 4-carbon dicarboxylic acid (oxaloacetic acid). Following reduction or transamination the product is translocated to the bundle sheath cells where it is decarboxylated; the CO_2 released is then re-fixed in the reaction catalysed by rubisco, which in these species is restricted entirely to the bundle sheath chloroplasts. The fixation, decarboxylation, refixation sequence involved in the **C-4 pathway** increases the CO_2 concentration within the inner compartment formed by the bundle sheath cells (Figure 2.3), so that photorespiration is reduced. The steepening of the CO_2 gradient also leads to the CO_2 in the mesophyll spaces being scavenged to very low levels, favouring the maximal operation of the Benson–Calvin cycle producing sugars, from which starches may later be derived.

C-4 species comprise less than 0.5% of the flowering plants investigated, but include *Spartina anglica*, *Atriplex laciniata* and a number of dune grasses (Crawford, 1989). They are important because under high irradiance and warm temperatures they can photosynthesize more rapidly, are more water-use efficient, and produce substantially more biomass than C-3 plants.

The spatial separation of carbon fixation mechanisms of C-4 plants contrasts with the temporal separation found in members of a number of families in which transpiration losses are reduced even more by **crassulacean acid metabolism (CAM)**. Many plants with CAM are succulent, having low surface-to-volume ratios in their photosynthetic organs; a number grow on coastal dunes in warm countries. Most of the photosynthetic cells in their stems or leaves are, or resemble, spongy mesophyll. They have a thin layer of cytoplasm and a large vacuole, which stores malate produced via carboxylation of PEP at night when CO_2 enters through the open stomata. The stomata are closed during the day when CO_2

from malate in the vacuole is used to synthesize sugars via the Benson–Calvin cycle. Many CAM succulents have simple (e.g. columnar) forms intercepting minimal radiation, do not depend on evaporative cooling and tend to have high lethal temperatures.

2.3.1 LIMITING FACTORS

Global variations in **day length** are considerable. Annual variation for mangrove communities at the latitude of Bombay (19°) is less than 3 hours, whereas in Oslo day length changes from less than 6 hours in mid-winter to over 18 hours in mid-summer. Extremes for salt marshes north of the Arctic Circle are even greater; winter temperatures are also adverse. The quality of light received by salt-marsh plants is considerably reduced when they are submerged and this is accentuated when the water is turbid. Many algae obtain their CO_2 from seawater but effective gas exchange in flowering plants, apart from hydrophytes such as *Zostera* and *Ruppia*, virtually ceases until the tide goes down again.

Droughting often limits the growth of dune plants in sites above the water table and its capillary fringe, frequently leading to sparse vegetation (see also section 3.5). In salt marshes water absorption must take place at high concentrations of salt (section 3.2), salt-marsh species having quite high osmotic potentials (> –2 MPa).

Primary production and the competitive abilities of coastal plants are commonly limited by the availability of **mineral nutrients** (sections 3.5 and 6.6). Although mineral elements in salt marshes do not usually show marked deficiencies, the levels of nitrogen, phosphorus and potassium in sand dunes are frequently low. Nitrogen may be the most severe deficiency; symbiotic associations involving nitrogen fixation are discussed in section 2.6. In acidic dune systems calcium and magnesium may be in short supply. Minor (trace) elements – Fe, Mn, B, Cu, Zn, Mo, Co (and Cl) – only rarely limit growth in coastal habitats, but several are deficient in the South Haven Peninsula, Dorset, UK (section 6.6).

2.4 CLIMATE, LIMITING FACTORS AND DISPERSAL

The performance and distribution of particular plant species are limited mainly by their responses to climate, to biotic factors including competition, to soil conditions such as moisture content, pH, aeration and ionic concentration, as well as by their dispersal mechanisms. The ultimate explanations of the distribution and performance patterns of coastal plants – some of which are considered below – are, however, often complex and may involve symbiotic or parasitic relationships with other organisms as well as a consideration of their own life forms, ecophysiology and genetic constitution.

The influence of climate at the community level is clearly illustrated by the vascular flora of the Pacific Northwest coastal sand dunes which extend from Cape Flattery, Washington, in the north to Cape Mendocino, California, in the south (Wiedemann, 1984). The climatic shift from mesothermal to Mediterranean type climate at 46°N is not reflected in the flora – which is in many ways remarkably uniform throughout the dunes – until 43°30′N, near Coos Bay, Oregon. The change is not abrupt but this point marks the southern limit of the Subarctic Beach Flora and the northern limit of the Mediterranean Beach Flora. *Phacelia argentea* (silvery phacelia) and *Artemisia pycnocephala* do not occur north of Coos Bay, whereas *Carex macrocephala* and *Myrica gale* are not found south of it. *Carpobrotus edulis* and *Carpobrotus chilensis* (the former well known to be frost-sensitive in the UK) are confined to the extreme south of the dunes. Some species occur throughout the region but are more important in particular areas. *Thuja plicata* (western red cedar) is prominent in dune forest of the north, while *Baccharis pilularis* (chaparral broom) is far more common in northern

California than in Oregon. There are ten maritime species endemic to the Pacific coast of North America; *Tanacetum douglasii* (dune tansy) occurs entirely within the region between 40°N and 49°N. More strikingly, *Ambrosia chamissonis* ssp. *cuneifolia* is confined to a single degree of latitude, from 46°30′N to 47°30′N.

2.4.1 CLIMATICALLY DETERMINED SPECIES

Figure 2.4 shows Iversen plots (Godwin, 1956) for two species whose presence on Braunton Burrows, North Devon, where mean annual rainfall is about 890 mm and the winters are mild, appears to be climatically determined (Willis, 1985a). *Juncus acutus* (sharp rush) (Figure 2.5), an Oceanic Southern species, occurs in South America and South Africa and is widely distributed in the Mediterranean and western Europe. In Britain it extends northwards to Caernarvon and Norfolk. At Braunton Burrows, as, for example, at Crymlyn Burrows, South Wales, it is plentiful in sites where the water table is close to the surface and flooding occurs in winter. Figure 2.4(a), based on the world distribution of *J. acutus*, shows that in Britain it is near to its thermal limit with respect to both the lowness of summer temperatures and coldness of the winter. During the exceptionally cold winter of 1962–1963, when the mean minimum temperature for February fell to –0.8°C, some 70% of *J. acutus* plants in the slacks on the seaward side of Braunton Burrows were killed by freezing – possibly the highest mortality since the species was first recorded there in 1797.

The only substantial British population of *Scirpoides holoschoenus* (= *Holoschoenus vulgaris*) is present in the southern part of Braunton Burrows, its occurrence probably being governed by a requirement for high summer temperatures (Figures 2.4(b), 2.6(a)) (Willis, 1985a). Elsewhere in Britain it is permanently established only in a low area of the dune system at Berrow, North Somerset, where a single clump has persisted for many years, apparently without any significant spread vegetatively or by seedling establishment. It was first recorded at Braunton in 1695 and has a large distinctive rhizome with tall shoots which may trap sand to build a tiny dune 1–2 m above the general ground level. Seedling establishment is probably a critical phase, typically occurring in sites always within reach of the water table where seedlings and small plants of varying ages are sometimes seen.

The main centres of distribution of *Scirpoides holoschoenus* are the Canaries, the Mediterranean (where it is not particularly maritime), southern Europe and Siberia. The Iversen thermal correlation plot (Figure 2.4(b)) shows that its two British localities have the lowest mean temperatures (c. 17°C) for the warmest month of the year of all sites in the world in which the plant grows. However, the species is also present – perhaps as a different ecotype – in Siberia where winter temperatures are very considerably lower than those in southern Britain (–18°C compared with c. 5°C); the severe winter of 1962–1963 did not damage this cold-tolerant plant on the Burrows. The need for a long hot summer leading ultimately to regeneration by seedlings probably explains the success of this plant on Braunton Burrows – but even here it fails to fruit in many years – and its absence elsewhere in Britain.

Ammophila arenaria is tolerant of very high day temperatures and great diurnal variation (Willis *et al.*, 1959b), but it appears to be limited by low temperatures in the north; its northern boundary coincides with the 0°C January isotherm (Huiskes, 1979). It extends to about 63°N in Norway, giving way to *Leymus arenarius* towards the north. *Spartina anglica*, although possibly damaged by frost (Ranwell, 1967) reaches as far north as northern Denmark (57°N) where it has been planted, and in salt marshes in the British Isles extends to 58°N in the Outer Hebrides and the Scottish mainland (Goodman *et al.*, 1969). *Spartina maritima* reaches nearly 1 m high and

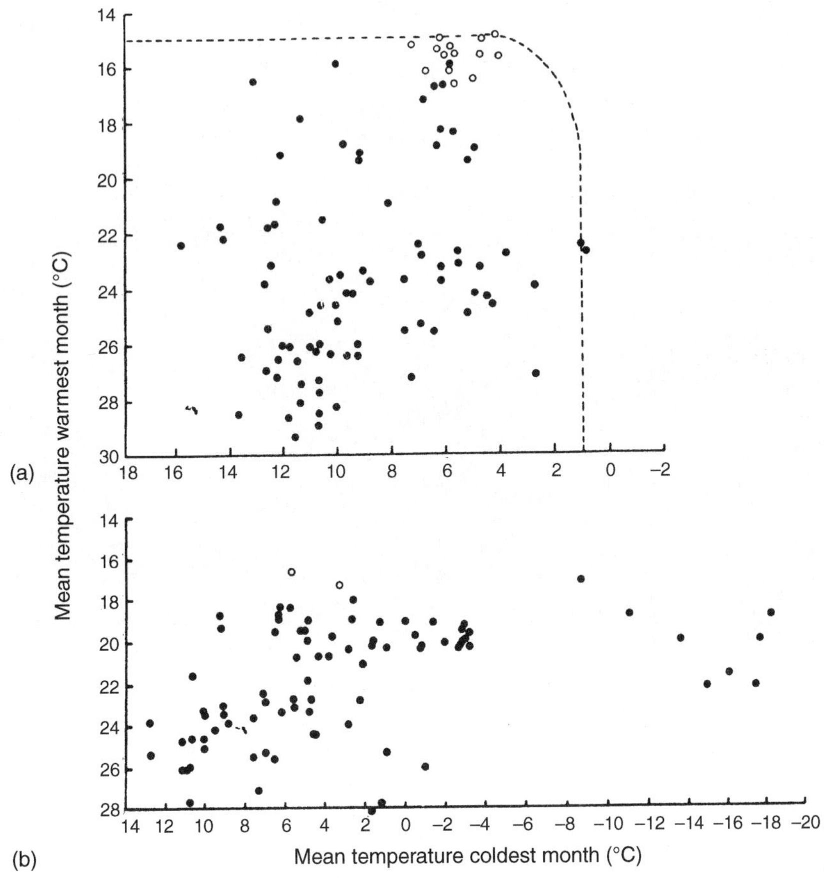

Figure 2.4 Iversen plots showing the occurrence of species relative to the mean temperature of the warmest and of the coldest months. ○, presence in the British Isles; ●, presence elsewhere in the world. (a) *Juncus acutus*. The broken line gives the inferred thermal limits for this rush. Records of occurrence were taken from Floras and herbaria; temperature data were derived from climatological atlases for sites as close as possible to those where the plant occurred. (b) The temperature tolerance of *Scirpoides holoschoenus*. The only two British sites are Braunton Burrows and Berrow. (From Willis, 1985a; courtesy Botanical Society of Edinburgh.)

grows vigorously towards the centre of its range on the Huelva Delta in southern Spain, whereas it is an almost sterile, dwarf form in northern Europe and in Holland was killed extensively by frost in 1963. The exclusion of *Salicornia europaea* from the high Arctic and its distribution in Hudson Bay, where it is restricted microtopographically to the warmer south-facing slopes, appears to be climatically determined (Jefferies, Jensen and Bazely, 1983).

2.4.2 EDAPHIC FACTORS

Edaphic factors are considered in greater detail in section 2.5 but some, such as **toxic ions**, play a major role in controlling species distributions, as in those areas of salt marshes where the growth of the perennial grass *Puccinellia maritima* is inhibited by the presence of sulphide (section 4.5), thus allowing space for the establishment of annual glasswort *Salicornia europaea*. Drainage, salinity and micro-relief are

Figure 2.5 Sharp rush (*Juncus acutus*) in dune slack with eyebright (*Euphrasia* sp.), yellow rattle (*Rhinanthus minor*), and creeping willow (*Salix repens*) at Oxwich dunes, Gower Peninsula, South Wales. Adjacent species included *Betula pubescens, Epipactis palustris* and *Phragmites australis*. (Photograph by John R. Packham.)

particularly important in the distribution of salt-marsh plants (sections 2.5, 3.1 and 4.3). Both **calcicole** and **calcifuge** elements are prominent in particular sand dune floras (sections 6.5 and 7.1); sometimes both are present in old dune systems which have been strongly leached to landward. Organically rich wet slacks of old acid dunes may occasionally have toxic levels of ferrous ions.

2.4.3 DISPERSAL

Seeds and vegetative propagules of both salt marsh and sand dune species may be dispersed over short distances within the local community or may extend the range of the species by long-distance spread, colonizing new sites. Although the fruits of *Aster tripolium* bear pappus hairs which appear well suited for distant transport by wind, they tend to stick together leading to rather restricted movement, largely by water (Ranwell, 1972). *Salicornia* has a **variable dispersal strategy** Ranwell (1972), with a capability for long-distance dispersal that is exemplified by *S. pusilla* whose complete fruiting heads, containing 4–10 seeds, were shown by Dalby (1963) to be capable of floating in sea water for up to 3 months before germination. At least some propagules are retained in the pioneer zone, tides strew others throughout the marsh, while a high proportion are carried to the strandline at the top of the marsh. *Salicornia dolichostachya*, a species of the pioneer zone at the bottom of the salt marsh, has larger seeds and faster rate of radicle elongation than *S. europaea*. The latter is characteristic of more closed habitats and has seeds which mature even in shaded sites, while those of *S. dolichostachya* do not (Ball and Brown, 1970).

Figure 2.6 Plants of damp slacks at Braunton Burrows, North Devon, UK. (a) *Scirpoides holoschoenus* (round-headed club-rush); (b) *Pyrola rotundifolia* (round-leaved wintergreen); (c) *Anagallis tenella* (bog pimpernel). (Photographs by M.C.F. Proctor.)

Spartina anglica spreads locally by seed, but vegetative fragments can be carried to fresh sites and establish successfully, as shown by Ranwell (1964) at Bridgwater Bay, Somerset, where tidal run is fast. Much longer distance transport may occasionally occur. One example is the successful invasion of the salt marsh at Berrow, North Somerset, by *Juncus subulatus*, a Mediterranean rush first found in Britain in 1957 (Willis and Davies, 1960). Propagules must have been carried over a great distance, probably by shipping or by birds, as the nearest source was some 900 km distant.

Huiskes *et al.* (1995) showed that significant interspecific differences occurred when halophyte propagules (seeds, fruits and seedlings) were trapped in nets fixed to the soil surface and in floating nets on a salt marsh in the Oosterschelde, south-west Netherlands. Seeds of *Spartina anglica*, *Triglochin maritimum*, *Limonium vulgare* and *Elymus atherica* were found mainly in the **floating nets** while those of *Salicornia* spp., *Aster tripolium* and *Spergularia* were trapped in the **fixed nets**; floating propagules of any sort were rarely caught in the fixed nets. Transport of propagules was due mainly to tidal currents. In the vegetated marsh there was a net transport of propagules towards the upper marsh with the flood currents. More propagules were transported out of the marsh by ebb currents than were transported into it from elsewhere by flood tides. Substantial exchanges of propagules occurred within the marsh; export of propagules – particularly of species flourishing in the lower parts of the marsh – indicated the potential for exchange between different salt marshes. Koutstaal *et al.* (1987) found that seeds launched from this site were carried a net distance of 60 km in less than a week.

In sand dunes, bare areas may be quickly colonized by seeds carried by wind or other agencies. Distinct rings of seedlings, e.g. of knotted pearlwort (*Sagina nodosa*), may sometimes be seen towards the margins of low-lying bare blow-outs; these arise from seeds stranded at the edge of the water when drying out occurs and germination conditions are favourable. The dune form of round-leaved wintergreen (*Pyrola rotundifolia*) has considerably extended its range in Wales and south-west England in recent decades, seed being carried by birds, or perhaps even by humans. It was present at Braunton Burrows by 1958 and spread rapidly, becoming a major associate of low-growing creeping willow *Salix repens* in sites at levels near to the margin of the water table (Figure 2.6(b)). The 5-year run without cold nights in early spring, and low rainfall in the latter part of the preceding summer in 1957–1961, appears to have favoured the strong establishment of *Pyrola rotundifolia* on the Burrows and its spread elsewhere. Long-distance dispersal must have occurred of the sand dune variety of the fen orchid (*Liparis loeselii* var. *ovata*), first recorded in south-west England when flowering at Braunton Burrows in 1966. Its minute seeds were probably carried by birds or blown by wind from the South Wales dunes (Willis, 1985a).

Found along the driftline throughout the British Isles, *Beta vulgaris* ssp. *maritima* is a common perennial with thick leathery leaves whose seeds can develop successfully following long-distance transport by the tides, and whose case raises interesting questions regarding the initiation of genetically viable populations under such circumstances. The root of sea beet is not conspicuously swollen at its junction with the stem as it is in ssp. *vulgaris*, with which it is interfertile and from which sugar beet, beetroot, spinach beet and mangold have been derived.

The production of **multigerm seed balls** in angiosperms is unusual, but those from sea beet give rise to up to five self-incompatible individuals (Dale and Ford-Lloyd, 1985). It is the capacity of wild beetroot to produce multigerm seedballs which allows its obligately outbreeding populations (some of which are inter-fertile) to run counter to the usual correlation between long distance dispersal, colonization of isolated areas and self-fertilization for which Baker (1955) and Stebbins (1957) have presented data from various species of plants. Though the resulting isolated populations have restricted gene pools these may be enlarged later by contact with other populations.

2.5 SOILS

This section commences with a brief overview of the composition and properties of soil, aspects of which are discussed in other sections. It then considers the deposition of substrate material and the formation of soil, demonstrating by example the information which soil profiles and surfaces can provide concerning the history, longevity and evolution of dune, salt-marsh and shingle ecosystems.

2.5.1 COMPOSITION AND PROPERTIES OF SOIL

Soil is the substrate in which the roots of most terrestrial plants are anchored, and from which they absorb water and mineral salts. Many of their properties, especially aeration and drainage, are related to the nature and size of the particles making up the mineral skeleton of the soil (section 1.3). Colloidal material, whether inorganic clay or humus of organic origin, is important because of its high base exchange and water-holding capacity. Soil organisms, themselves of great interest, play a major role in the development of the soil and influence many of its properties. The ability of certain plants to survive in particular coastal environments is often related to water and mineral nutrient availability, soil aeration and oxygen diffusion rates, redox potential and the presence or absence of **toxic ions** such as HS^-, Fe^{2+} or Al^{3+} (sections 3.1 and 4.5).

Salinity and **soil water** content are two of the most important edaphic factors governing the distribution of salt-marsh plants; both of them are greatly influenced by microrelief (section 4.3) and position on the shore. Salinity is often expressed as NaCl ‰ (parts per thousand) and chlorinity as Cl ‰, but also as simple percentages, as below. Ranwell *et al.* (1964), in studies around Poole Harbour, recorded maximum chlorinity of soil solutions at 10 cm depth for a number of species including *Spartina anglica* (2.44%), *Juncus maritimus* (2.39%), *Bolboschoenus maritimus* (1.71%), *Schoenoplectus tabernaemontani* (1.71%) and *Phragmites australis* (1.32%). The maximum value for the last species, obtained after a period of exceptionally hot and dry weather, was unusually high and the plants were depauperate and dying; the maximum concentration to which *P. australis* was thought to be tolerant was only of 1.2% chlorinity (i.e. 2% salinity) and even less for seedlings, compared with 2.12% chlorinity for normal seawater, and its minimum distance from the mouth of Poole Harbour was 5.3 km. The statement that a tropical form of *Phragmites* had been recorded as tolerating a salinity of 4% on the Red Sea coast (Ranwell *et al.*, 1964, p.639) is misleading; this value was obtained after a period of intense evaporation and was not representative of true tolerance.

In many areas of the Cefni Marsh, Anglesey, North Wales, *Bolboschoenietum maritimi* is quite distinct from *Juncetum maritimi*, although surrounded by it. Sea clubrush often grows on ground a few centimetres lower than that supporting *Juncus maritimus* and it frequently remains flooded with brackish water, even when the ground has not been covered by the tide for several days. This agrees with the observations of Ranwell *et al.* (1964) who placed *Bolboschoenus maritimus*, with *Spartina anglica*, *Schoenoplectus tabernaemontani* and *Phragmites australis*, in a group of plants characteristic of permanently wet or frequently submerged marshland sites. *Juncus maritimus*, on the other hand, was placed with *Agrostis stolonifera*, *Elytrigia atherica* and *Juncus gerardii* in a category in which soil moisture content could fall as low as 20%. Field evidence suggested that the species of this latter group could withstand short periods of submergence by brackish water, as well as surviving the hypersalinity which may develop when an exceptionally high tide is followed by a period of dry weather.

2.5.2 SOIL FORMATION

An introduction to this topic is given by Gerrard (1992, Chapter 8) who points out that the granular deposits of coastal salt marshes, dunes and shingle beaches accumulate as a result of the complex interplay of wind, rivers and sea. In the northern hemisphere many of the materials in coastal systems are **reworked glacial deposits** which form an **offshore store** from which supplies are drawn to feed the depositional coastal system, while in the Tropics huge quantities of silt are carried down to the sea by rivers. Grain size and chemical nature vary from place to place, as do the local energy conditions which help to determine the mix of pebbles, gravel, sand, silt and clay which may be present in any particular deposit.

Many coastal land forms include adjoining areas of salt marsh, sand dune and shingle ridge, whose soils are frequently derived from mineral particles of very different sizes. Even when mineral particle size distributions of salt-marsh and sand-dune soils are initially very similar, subsequent pedogenic processes differ greatly, mainly owing to the much greater water content and frequently poor drainage in salt marshes. Mineral and organic matter accounted for only 4–14% of wetland soils developed in the rapidly subsiding deltaic plain of the Mississippi (Nyman, Delaune and Patrick, 1990); the remainder was pore space occupied by water and entrapped gases. In this area salt-marsh soil had 1.7 times as much mineral matter as brackish marsh soil in which vertical accretion was occurring at similar rates. This may be because *Spartina alterniflora*, the locally dominant salt-marsh plant, has a higher soil bulk density requirement than the species of the brackish marsh.

One of the major processes involved in the '**soil-ripening**' phase is the gradual loss of water, of which a useful index is the **n-value** of Pons and Zonneveld (1965) given by the formula:

$$n = A - 0.2R/L + bH$$

where A is total water content in grams per 100 g dry soil, L and H are respectively the percentages of clay and organic matter in the dry soil, R is the percentage of mineral particles other than clay, i.e. $R = 100 - (H + L)$, and b is the ratio of the water-holding capacity of organic matter to that of clay. Freshly deposited soft muds have n-values of 3–5, mudflats exposed at low tides of about 2.0, the lowest salt marshes 1.2–2.0, whereas those of normal soils drop to 0.3–0.4. Marine soils are often enclosed and used for agriculture, being known as **saltings**.

Variation in accretion patterns also influences the chemical nature and particle size distribution of dune soils, in which low levels of inorganic colloids present permit more rapid leaching and soil development is greatly influenced by climate. As the Ainsdale example later illustrates (see Figure 2.9), freely drained dune soils are in cool humid climates associated with rather rapid leaching, with acidification and finally podzolization; the non-carbonate portion of the sand fraction usually consists largely of quartz grains with very little buffering against hydrogen ions. Where drainage is poor, as it often is in dune slacks, organic matter decomposes slowly, mor humus is common and peat frequently accumulates.

In warm dry climates, clay and soluble soil components accumulate, soil salinity originating from salt spray may be high, and dune soils tend to contain little organic matter; mull-moder humus is the norm (Sevink, 1991). Despite the permeability of dune soils, formation of podzols takes much longer, often several millenia, when conditions are not cool and humid.

Decalcification is an important pedogenic process in salt marshes and also in sand dunes, where soils of the oldest dune ridges are frequently much more acid than those of the young dunes near the sea (see Figure 6.21). There has been vigorous debate about the decalcification of salt-marsh soils, which high levels of CO_2 in the soil atmosphere appear to promote. It has also been proposed that variable conditions during silting were the most important cause of variations now found in the calcium carbonate content and depth of decalcification in Dutch marine clay soils.

2.5.3 SOIL VARIATION IN THE CEFNI MARSH, ANGLESEY, UK

Profile pits and mechanical analyses along the transect shown in Figure 2.7 reveal considerable differences related to changing sources of substrate. The loss-on-ignition figures, obtained by a method designed to avoid loss of structural water from clay minerals and of CO_2 from carbonates (Ball, 1964), are an indication of how much organic matter has built up at the various levels. Even these few pits

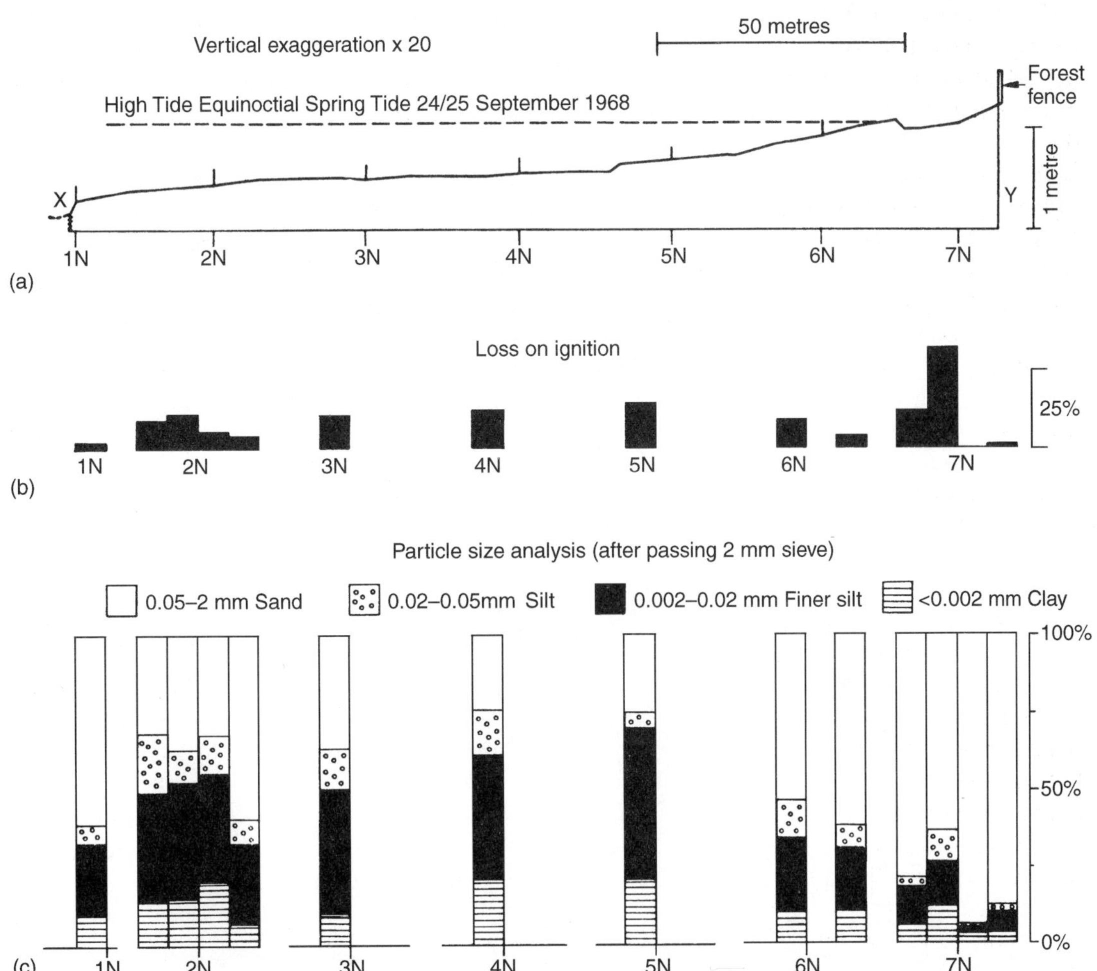

Figure 2.7 (a) Relief of transect X–Y running WNW–ESE across Cefni Marsh, Anglesey. (b,c) Total organic matter (as represented by loss-on-ignition) and particle size analyses of soil samples from depths of 4–6 cm (column on the left of each block), 9–11 cm, 19–21 cm, or 39–41 cm (column on the right of each block). (From Packham and Liddle, 1970; courtesy Field Studies Council.)

showed considerable variation in the soil structure on this sandy Welsh marsh. At Pit 5N there was sloppy mud at 90 cm. Sand commenced at 28–36 cm in Pit 6N, below material containing a good deal of drift, while the amount of silt and clay in the lower levels of Pit 7N was low. Sequences of four samples down to 40 cm were taken for Pits 2N and 7N, whose loss-on-ignition figures contrast sharply. The low values for the bottom of Pit 7N were the first indication of a sudden inundation of sand, only the top of which had since been greatly enriched by organic material. The black sulphide layer was within 2.5 cm of the surface at Pit 1N, close by the erosion cliff caused by constant wave slap, 5–15 cm deep (uneven surface) at 2N and deeper further up the shore. The black zone is within 1 cm of the surface in other parts of the Cefni estuary.

Detailed analysis of this transect of the Cefni marsh (Packham and Liddle, 1970) affords a clear illustration of the variation to be found in coastal sediments. The forestry ditch cut near Pit 7N in 1969 showed that deposition here had been intermittent, there being at least one 'fossil' grassland soil level and two other levels where old *Bolboschoenus* (= *Scirpus*) *maritimus* rhizomes were very abundant, as well as relatively pure sand, within a depth of less than 1 m. Material deposited during successive inundations varies and it is probably this which accounts for the rather random variation in extractable P_2O_5 with depth.

2.5.4 SOIL–LANDFORM RELATIONSHIPS ON THE ROMNEY MARSHES, KENT, UK

Extensive deposits of peat beneath the Romney Marshes contain wood with a radiocarbon date of about 3000 Before Present (BP), indicating that trees grew here during a period of low sea level. Since the formation of the peat, various deposition patterns have produced a number of complex sedimentary structures, of which the largest is the cuspate foreland of Dungeness. Also conspicuous are the complex dendritic creek ridge systems which apparently resulted from a subsequent infilling with sand, silt and clay, of creeks cut into the deep peat. When drainage allowed water tables to fall, the peat between the largely infilled creeks shrank, resulting in an inversion of relief which has left the former creeks standing above the general surface.

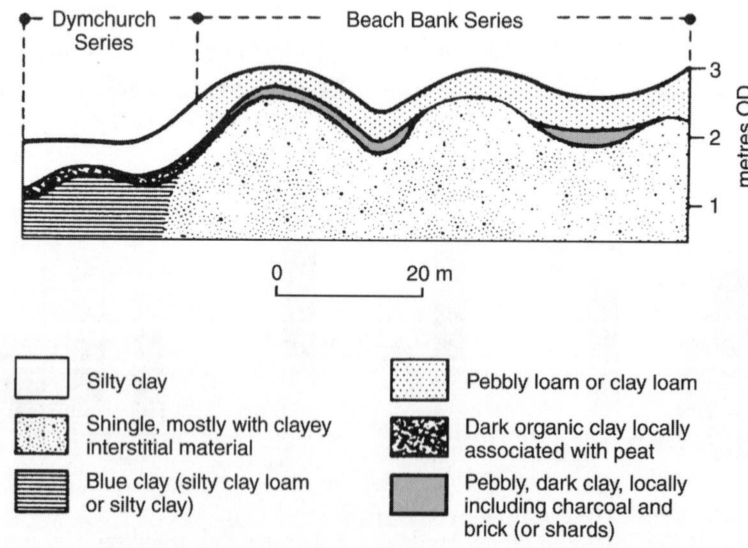

Figure 2.8 Soil–landform relationships from the Romney Marsh, Kent. (From Green, 1968; courtesy Soil Survey and Land Research Centre.)

The soils of the Romney Marshes (Green, 1968) exhibit a wide range of particulate materials; the Beach Bank Series (Figure 2.8) was laid down over a series of shingle ridges very similar to those which still exist on the cuspate forelands of Dungeness and Orfordness (section 8.4). The Appledore Series consists of poorly drained, non-calcareous clay soils over thick peat which is present at a depth of 30–60 cm. The moderately well drained loams or sandy loams of the Snargate Series, on the other hand, occur to a depth of at least 60 cm on the major creek ridges, and beneath them the proportion of sand often increases to a depth of at least 3 m. Detailed study of the Romney Marsh soils indicates repeated changes in the balance of erosion and deposition which can be explained only by a series of alterations in the sea level.

2.5.5 PEDOGENESIS AND THE SOILS OF DUNES, SLACKS AND PINE WOODLANDS AT AINSDALE (MERSEYSIDE), NORTHERN IRELAND AND WALES

The Ainsdale system is calcareous and its oldest stabilized surfaces have been estimated to be 300–400 years old. The prevailing winds are from the south-west, and in the period 1974–1980 the dunes received a mean annual rainfall of 847 mm. They possess features common in other parts of Europe, where soil sequences similar to those shown in Figure 2.9 often occur and pines are frequently planted to stabilize dune soils. **Raw sands** (Profile 1) occur where trampling and rabbit grazing induce erosion, and on naturally unstable dune ridges near the sea. Profile 2 (c. 15 yr) and Profile 3 (200–250 yr) represent different stages in the development of **sand-pararendzinas**. The high humus content and associated high levels of adsorbed cations are restricted to the uppermost 8 cm of Profile 3; immediately below this exchangeable calcium falls to a minimum but at greater depths rises to levels similar to those of unweathered sand. The second of the two slack soil profiles (5b) is drier than the first; such soils frequently have an oxidized and largely mineral A horizon beneath the surface peat.

James and Wharfe (1989), who also present a series of chemical soil profiles, show that while soil pH in the surface layers of the mature sand-pararendzina and the **gley** and **peaty gley** of the slack soils is low, that of the **micropodzol** beneath the pine woodland is even lower. In a soil beneath *Pinus nigra* planted in the mid-1950s the pH of the Ea horizon was 4.1; at 12 cm beneath the mineral soil surface it was 8.3. Iron redistribution represented as a 'micropodzol' was extremely common and in one case had occurred in less than 21 years after planting.

This work demonstrated the limited depth of soil formation and the need to take soil samples from precisely defined depths. It also showed that certain soil properties were correlated with soil age, which was in general considered to increase with distance from the shore. At 0–1 cm depth in a sequence of 20 raw sands and pararendzinas, soil organic matter determined by loss-on-ignition rose with increasing soil age; correlation coefficients were also high and positive for exchangeable K and Mg. They were negative for soil pH, while $CaCO_3$ and exchangeable Ca were not correlated significantly with distance from the shore. At 30 cm depth both Na and K were negatively correlated with relative age.

Wilson (1992) refers to the raw sands (C horizons only) and the sand-pararendzinas with AC profiles of Ainsdale, and notes the presence at Murlough, County Down, in Northern Ireland of all four stages in the sequence raw sand → sand-pararendzina → brown calcareous sand → podzol. Raw sands often have distinct bedding structures unless deposited within a vegetation cover, when they are typically loose and formless. They also have a negligible organic content, although $CaCO_3$ can exceed 30%. In low-carbonate dunes, as at Murlough, the oddly named **brown calcareous sands** (Stage III) have been formed within 75 years of vegetation colonization and surface stabilization. In such soils acidification and decalcification

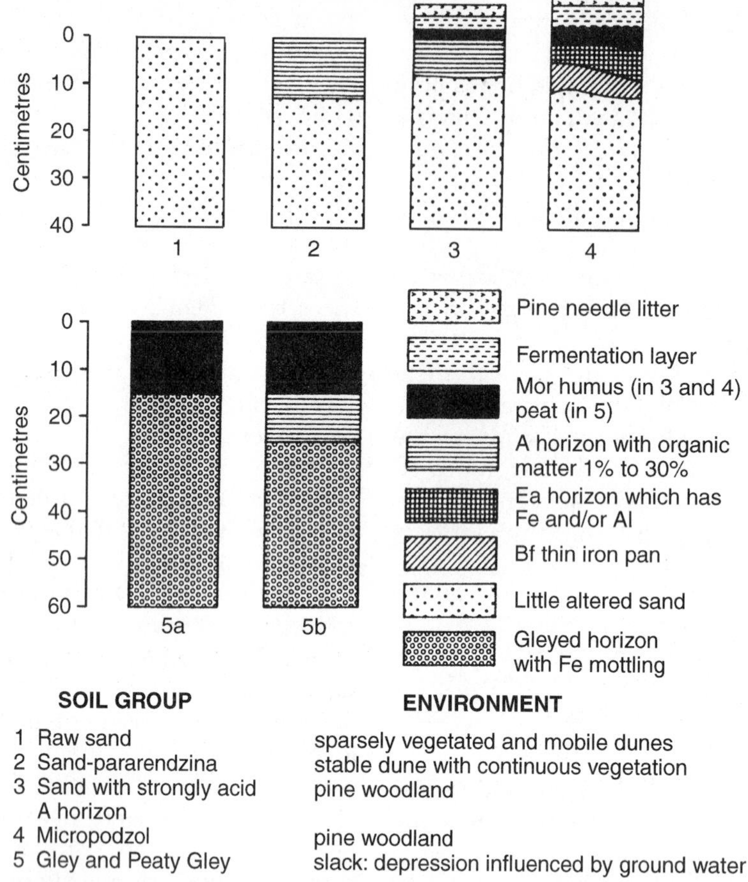

Figure 2.9 Soil groups and profiles of the sand dune system at Ainsdale, Merseyside, UK. (From James and Wharfe, 1989; courtesy SPB Academic Publishing.)

have proceeded to such an extent that a partially or completely decalcified B horizon (Bw) has moved down into the top of the C horizon. The **podzol** at Murlough overlies charcoal dated to 575 ± 60 years BP, which so far represents the most rapid rate of podzolization recorded in Irish sand dunes. Wilson (1992) notes that brown calcareous sands have not been reported from England or Wales, and also that micropodzols have not yet formed under the Corsican pine planted in three areas of coastal dune in Northern Ireland in the 1950s and early 1960s.

In Anglesey, North Wales, the dunes at Newborough Warren, Valley and Aberffraw provide excellent examples of the **Sandwich Association** (Figure 2.10) of deep calcareous and non-calcareous sandy soils, which are everywhere exposed to sea winds and developed on sand dunes, marine shingle and related beach deposits. Young, unstable dunes near the sea are colonized by *Ammophila*, and the most extensive soil is that of the Sandwich Series whose sand-pararendzinas are found on fixed dunes usually over 100 years old. The sand-rankers of the Beckfoot series occur further inland on stabilized non-calcareous dunes, while sandy gley soils of the Formby series occur in wet hollows among the dunes.

2.5.6 BURIED SOILS AND COASTAL AEOLIAN SANDS

Although dune deposits may arise very rapidly as a result of catastrophic gales, a long period of relative stability may then ensue, as in the case of the coastal aeolian sand deposits at the Strand Hotel exposure, Port Stewart, Northern Ireland, described by Wilson (1991). At the base of this profile is a buried stagnogley developed on basalt-rich sandy loam. Above this are two sand units, of which the lower contains a buried humus podzol 1.4 m thick. The stagnogley seems not to have been eroded previous to sand deposition, and the radiocarbon dating of 4780 ± 45 years BP ascribed to the basal sands which contain plant stems appears to represent the maximum age for soil burial and sand accumulation. The land surface at this site then appears to have remained stable for c. 4250 years until inundation of the podzol by the upper sand unit at a time which has been fixed by a further radiocarbon date of 525 ± 45 years.

This exposure was in the low-relief, shallow sand deposits of the plateau dunes on the eastern margin of the extensive dune systems at Portstewart Strand and on either side of the River Bann. Archaeological and palaeoecological investigations have shown that these systems, together with the dunes at Curran Strand, are composite features for which the earliest phase of dune building was before 5300 BP and in which three phases have occurred since.

2.6 MICROBIAL ECOLOGY OF COASTAL ECOSYSTEMS

Microorganisms play a number of important roles in coastal habitats. Some are agents of disease, others are involved in nitrogen fixation or in de-nitrification, yet more are **decomposers** which play a vital role in **nutrient cycling** (sections 1.5, 6.4 and 7.4). Though the fruiting bodies of dune fungi are often conspicuous, many coastal habitats remain under-recorded in respect of this group. Some 117 species have been recorded for the Sands of Forvie and Ythan Estuary NNR, mainly from the dunes, but North (1981) points out that 52 of these were listed in a single visit paid by

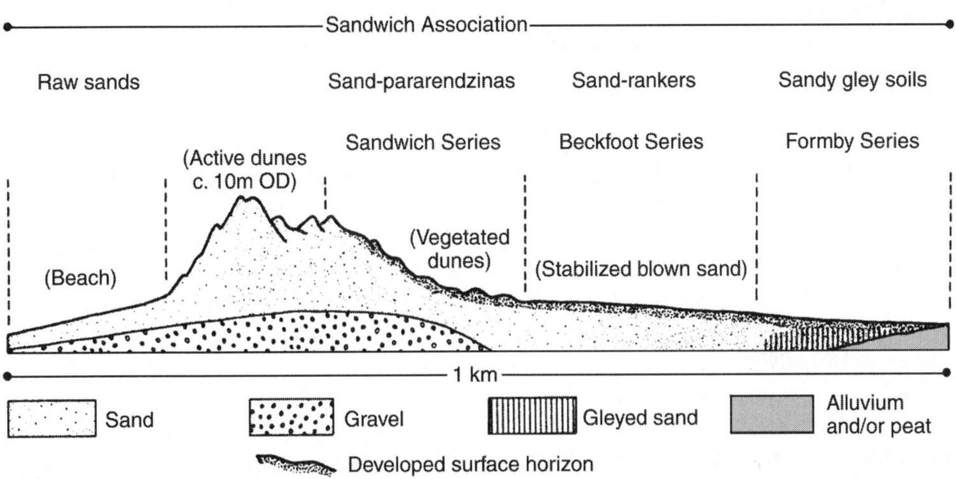

Figure 2.10 Distribution of soil series in the Sandwich association. (From Rudeforth *et al.*, 1984; courtesy Soil Survey and Land Research Centre.)

members of the British Mycological Society in 1975 (section 6.4).

Available nitrogen and phosphorus are frequently in short supply in shingle, dune and salt marsh habitats: the remainder of this section is concerned with interactions involving these two elements.

2.6.1 MYCORRHIZAL RELATIONS OF COASTAL VEGETATION

In sand dune ecosystems, the extent of occurrence, the form and the function of mycorrhizal infection changes with successional development (Read, 1989). The drift line, which may be nutritionally enriched and frequently disturbed, is colonized mainly by non-mycorrhizal ruderal species. In foredunes and mobile dunes, pioneer grasses – including species of *Ammophila*, *Elytrigia* and *Uniola* – appear to be infected to a variable extent and are only facultatively vesicular–arbuscular (VA) mycorrhizal. The major benefit to the grasses is probably in enhanced phosphorus uptake under conditions of phosphorus limitation; another important role of the fungi may be in the aggregation of sand particles, contributing to dune stability. In the fixed dunes, many vascular plants are obligately VA-mycorrhizal; here, seedling establishment may be dependent upon early incorporation into the mycelial network, notably in species with very small seeds and low phosphorus reserves. In the dune slacks, with the accumulation of organic matter and reduced pH, ectomycorrhizal species such as *Salix repens* no doubt benefit from nutrients mobilized from organic macromolecules. In the oldest successional stages, plants with ericoid mycorrhizas gain both nitrogen and phosphorus from organic complexes.

In open sand-dune habitats, however, some species may be adversely affected by infection by mycorrhizal fungi; this antagonistic effect has been shown (Francis and Read, 1995) to reduce growth substantially in, for example, *Arenaria serpyllifolia*, *Echium vulgare*, *Reseda luteola* and *Verbascum thapsus*.

In contrast to plants of sand dunes, many salt marsh species lack mycorrhizal associations or have only low levels of infection. A number of the most frequent species of British salt marshes, including *Atriplex portulacoides*, *Elytrigia atherica*, *Juncus gerardii*, *Salicornia* spp., *Spartina* spp., *Suaeda maritima* and *Triglochin maritimum* are reported as typically non-mycorrhizal (Harley and Harley, 1987). At Stiffkey, Norfolk, UK, however, *Aster tripolium* and *Salicornia europaea* on the higher marsh are infected by both fine and coarse endophyte arbuscular mycorrhizas, associations not found at lower levels in the marsh (Davy *et al.*, 1998). Strong arbuscular mycorrhizal infection of *Aster tripolium* leads to lowering of sodium levels in the shoot (Rozema *et al.*, 1986). Functioning of the sodium exclusion system in the roots may be favoured and the water uptake improved; at high salinity, mycorrhizal *A. tripolium* shows greater leaf extension growth than uninfected plants. In North America, the salt-marsh grass *Distichlis spicata* is usually VA-mycorrhizal, but this association appears to have little effect on the phosphorus concentrations of leaves or on growth rate over a range of salinities (Allen and Cunningham, 1983).

2.6.2 NITROGEN FIXATION IN COASTAL SITES

Nitrogen-fixing cyanobacteria contribute significantly to the nitrogen status of some salt marshes. In pools and on the banks of creeks, as well as to some extent on mudflats and under higher plants, cyanobacteria fix nitrogen when temperature and light conditions are suitable. The highest levels of fixation in a salt marsh at Morecambe Bay, near Lancaster, UK, were found in the pools, attributable mainly to *Nostoc* (Jones, 1974). In the lower salt marsh here 'blue–green algal felts', with abundant *Lyngbya aestuarii*, were shown to be active in nitrogen-fixation, notably during the later part of the summer and early autumn. Consistent levels of fixation by bacteria were also found in many parts of the marsh, but at lower rates than those of cyanobacteria. Nitrogen fixation was found to be stimulated in the rhizosphere of *Puccinellia maritima*; ^{15}N studies showed that the products of fixation are rapidly metabolized by higher plants of the marsh (Jones, 1974). De-nitrification, however, under both aerobic and especially anaerobic conditions in the mud, resulted in loss of nitrogen. Investigations in the Great Sippewissett salt marsh, Massachusetts, showed much greater (up to nine times) nitrogen-fixation by bacteria (especially rhizosphere bacteria) than by cyanobacteria. Here, denitrification caused a major loss of nitrogen, considerably exceeding fixation (Valiela and Teal, 1979a,b).

In the slacks of sand dunes, early colonizing cyanobacteria may fix nitrogen which is assimilated by higher plants, so helping to stabilize the slacks. In studies of a sand dune slack at Blakeney Point, Norfolk, UK, rich in *Nostoc*, it was shown (Steward, 1967) by means of ^{15}N studies that the surface centimetre of soil (the region rich in cyanobacteria) became highly labelled. Within 3 weeks, higher plants (*Agrostis capillaris, Glaux maritima* and *Suaeda maritima*) and the moss *Bryum pendulum* were found to be significantly labelled, indicating transfer of the biologically fixed nitrogen from the soil surface to the vegetation.

The nitrogen-fixing bacterium *Rhizobium*, symbiotic with numerous forbs and shrubs including *Anthyllis vulneraria, Lotus corniculatus, Ononis repens, Trifolium* spp., *Ulex europaeus* and *Lupinus arboreus*, is very important in enhancing the nitrogen status of dune systems. Leguminous trees also benefit from symbiotic associations with *Rhizobium*; South African sand dunes have been planted with a number of *Acacia* species (section 9.4). Also planted on the Mediterranean coast and elsewhere is the tree *Casuarina* (she oak), which like the sea buckthorn *Hippophae rhamnoides* found on many British dunes, has a symbiotic relationship with the nitrogen-fixing actinomycete *Frankia* (also found in the root nodules of *Alnus glutinosa*).

2.7 HETEROTROPHS, SECONDARY PRODUCTION AND ECOSYSTEM PROCESSES

This section is mainly concerned with the functional relationships between broad groups of organisms and their roles in ecosystem processes; more detailed accounts of salt-marsh and dune animals are given in sections 5.2 and 7.3. **Heterotrophs** ultimately depend on the **primary producers**, almost all of which are autotrophs utilizing light energy, for energy and nutrients. Tracing and quantifying the roles of the various links in a terrestrial food web, such as that of a dune system, is invariably difficult. In salt marshes, where the tides both import material from elsewhere and export carbon, other materials and energy largely derived from primary producers growing on the marshes themselves, the situation is even more complicated (sections 1.5, 5.4 and 5.5). **Carbon budget studies** (e.g. Wiegert, 1979; Woodwell *et al.*, 1979) for a number of marshes suggest that net primary production of the marshes on which they live is the major source of material utilized by salt- marsh heterotrophs. Studies of nutrient flux and salt-marsh chemistry are indeed an integral part of salt-marsh ecology and are reviewed in section 4.5.

Vertebrates, particularly birds (section 2.8) and mammals, are conspicuous on dune, shingle and salt-marsh systems, but total numbers of invertebrates – many of which show morphological or behavioural adaptations – are immensely greater. **Secondary production** and the **facilitation of ecosystem processes** by bacteria and fungi are also of major significance.

Ranwell (1972) emphasizes the part played by animals in salt-marsh processes, pointing to the importance of *Hydrobia ulvae*, a small gastropod which occurs in vast numbers on European salt marshes, and which was shown by Newell (1965) to digest the microorganisms but not the organic debris of the sediments which it ingests. Other important salt-marsh molluscs are cockles (*Cerastoderma edule*), mussels (*Mytilus edulis*) and the Baltic tellin

(*Macoma balthica*). All these, together with annelids, e.g. ragworms (*Nereis* spp.), catworms (*Nephtys* spp.) and lugworms (*Arenicola marina*), and crustacea such as the amphipod *Corophium volutator*, are – as described in section 2.8 – prey items for birds. The shore crab (*Carcinus maenas*), which feeds on a wide variety of invertebrates, is itself predated by fish and gulls.

On grazed salt marshes of the Burry estuary, Glamorgan, South Wales, some mites were found to be restricted to particular intertidal zones, while others occurred in all zones including those nearest the sea (Luxton, 1964). These Acarina showed no preference for salt or fresh water conditions in culture; larvae and adults withstood up to 12 weeks' immersion in sea water without apparent harm. They showed distinct preferences for specific salt-marsh fungi as food; the distribution of fungi in relation to salinity was thus a factor which could influence acarine zonation. Luxton noted that many salt-marsh mites are viviparous or ovoviviparous. Direct production of active larvae reduces losses incurred by the ready dislodgement of eggs by tidal action.

The **radio-isotope** ^{32}P, used to label *Spartina alterniflora* and detritus-rich sediment surfaces, proved to be extremely effective in elucidating food chains involving salt-marsh arthropods in the Sapelo Island Marshes, Georgia (Marples, 1966). Four species of insect (one *Orthoptera*, two *Hemiptera* and one *Homoptera*) were dominant grazing organisms, while littorinid snails and species of two families of Diptera (*Dolichopodidae* and *Ephydridae*) were dominantly detritus feeders. Spiders were important carnivores; they obtained energy from both detritus and grazing chains.

In view of the supposed importance of the primary producers of salt marshes to coastal fisheries, considerable interest attaches to the question of whether the existence of major trophic links can be established. In practice, proteins occurring in *Spartina alterniflora* contain only small amounts of the specific amino acids found in marine fish and shell-fish. Burkholder (1956), who followed the fate of *Spartina* litter set out in cages in creeks, concluded that although high-quality protein may not be formed by marsh grasses or phytoplankton there is the possibility that microbial conversion 'may act like a huge transformer to step up the potential value of the pool of protein in the sea'.

2.7.1 CHARACTERISTICS OF ESTUARINE HETEROTROPHS

Salt marshes are frequently situated in estuaries, the activities of whose heterotrophs (section 5.2) are reviewed by McLusky (1989). The bacteria, microfauna, meiofauna and small macrofauna (defined in the caption to Figure 1.13) of the benthos, described by Kuipers *et al.* (1981) as the 'small food web', can be characterized by the small size of the individuals, a high turnover rate, relatively short life spans and a complicated trophic structure. In the Waddensee, this 'small food web' may consume 70–80% of all organic material available.

Although macrobenthic organisms are often classified as suspension or deposit feeders, many can feed in both ways. The most common definition of **deposit feeders** includes those animals that frequently ingest sedimented material of low bulk value (Jumars *et al.*, 1984; Lopez and Levington, 1987). It is generally considered that the main food source is the microbes that are attached to the sediment and detritus particles, and the distribution of deposit feeders often reflects the organic content of the sediments. It has been established that bacteria are digested efficiently in many deposit feeders, that a few intertidal deposit feeders specialize on digesting and assimilating the contents of microalgae, and that at least a few deposit feeders can absorb organic matter from non-living organic detritus (Lopez *et al.*, 1989).

There are long- and well-established relationships between the grain surface area of sediments and organic content, microbial metabolic rate and microbial abundance. The bulk of organic inputs to marine sediments seems to be most valuable to deposit feeders immediately on settlement (Jumars, 1993). There is also an inverse relationship between the feeding rate of deposit feeders and the organic content of the ingested sediments. This suggests that a particular throughput is required to support a unit body size (Jumars, 1993).

Because the ingested sediments are of low bulk value the amounts ingested are considerable. Deposit feeders typically ingest three times their own dry weight in dry weight of sediment per day. This is in excess of full body volume per day. In order to achieve this, such organisms require a particularly large gut. Deposit-feeding polychaetes typically have about one-third of the body volume occupied by the gut, half is not unusual and it may reach 80%. Deposit feeding may therefore represent the natural extreme of chronic limitation by dilute diet and a consequent need for rapid nearly continuous feeding.

2.7.2 MICROARTHROPODS AND OTHER BIOTA OF BEACHES AND DUNES

The larger dune animals are considered in section 7.3, but the microarthropods and associated biota are discussed here because of their role in soil processes. Koehler, Munderloh and Hofmann (1995) investigated the microarthropods of two sandy coastal sites, one in Jutland, Denmark, and the other in the East Frisian Island of Spiekeroog, Germany. Determinations were made along an axis running: tidal line, beach, primary dune, yellow dune, grey dune, **brown dune** (heath dune); in both sites *Empetrum nigrum* and *Polypodium vulgare* were prominent in the vegetation of the brown dune. Microarthropods were mainly found at depths below 4 cm in primary and yellow dunes, but at 0–4 cm in grey and brown dunes. The actual numbers involved were often extremely large; there were some 419 000 mites m^{-2} in the primary dune site at Jutland. Both sites yielded 22 species of Gamasina (predatory mites), of which 10 were in common. The various dune zones were characterized by distinct communities of predatory mites, those of the yellow dune communities of Jutland and Spiekeroog being very similar. Although not so distinctly, Collembola (springtail) species also showed marked zonation in both sites. Springtails were not so common as mites; Protura – whose preferred food is mycorrhiza – were even less abundant although the numbers involved were considerable.

Figure 2.11 illustrates the role played by soil biota in the **biogenic stabilization of sand**, whose individual grains may be glued together by bacterial and algal slimes. Plant root exudates and root hairs hold sand grains together, while further mechanical stabilization is provided by algal, fungal and mycorrhizal filaments. This system is influenced by potworms (Enchytraeidae), mites and springtails which graze the microflora, fungi and algae, and by predators such as Gamasina which act as further controls. Microfloral spores ingested by soil mesofauna are ultimately dispersed rather than digested; nematodes are among the groups which act as both plant root feeders and vectors of pathogens.

2.8 COASTAL BIRDS

The size of coastal bird populations is controlled by many factors including availability of suitable food, roosts and nesting sites, degree of predation and local pollution levels. Dunes, machair and shingle, in all of which the range of species present is influenced by the vegetational succession, are important habitats for breeding birds. Dune systems show a steady rise in the density of breeding birds going inland from the mobile yellow dunes through to grey dunes with scrub. Dunes, especially those that are isolated and inaccessible, may carry huge breeding colonies of gulls and terns. Breeding bird densities are

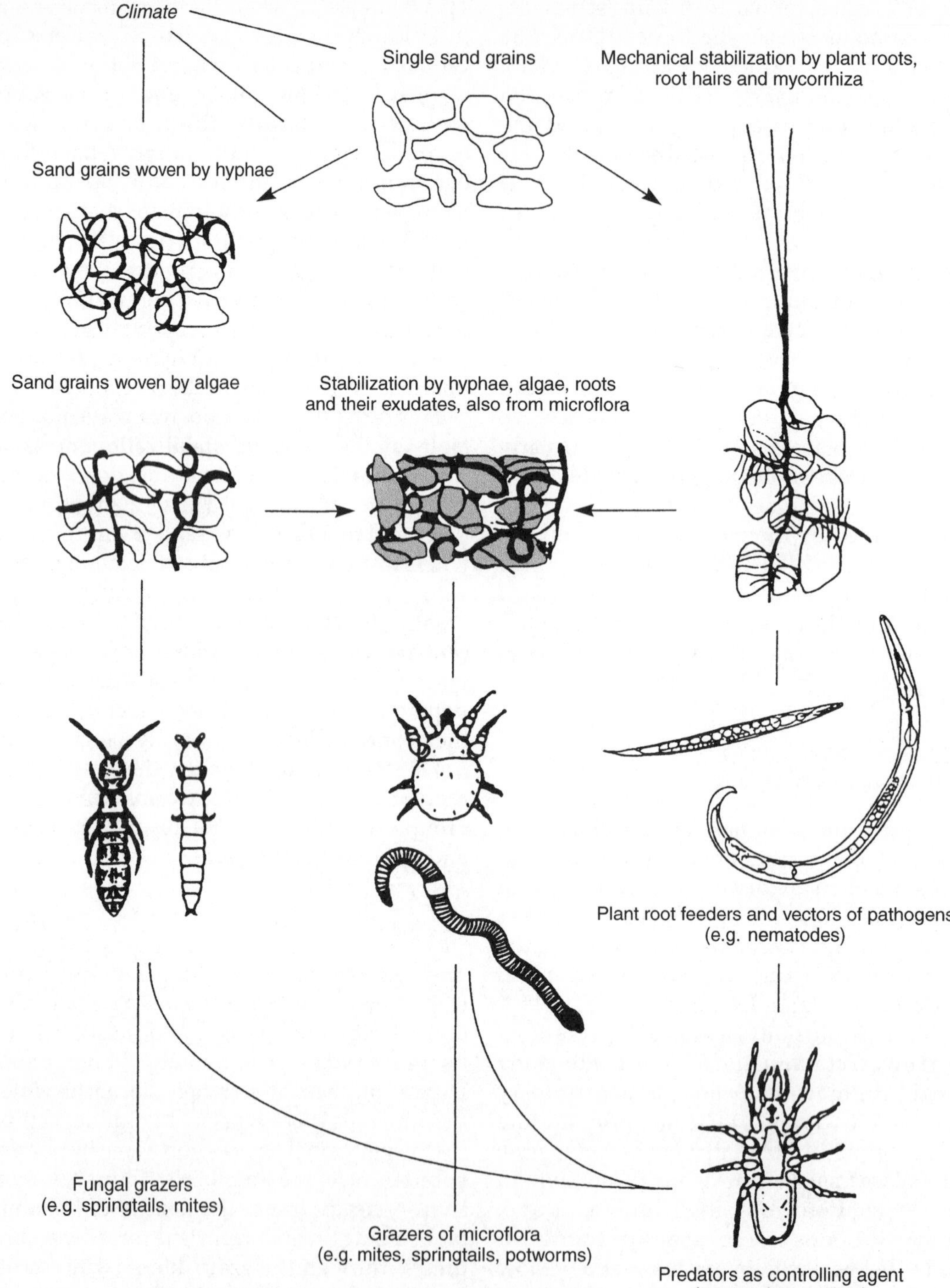

Figure 2.11 Ecological interactions leading to biogenic stabilization of sand in coastal dunes. (From Koehler, Munderloh and Hofmann, 1995; courtesy *Journal of Coastal Conservation*.)

high in dune scrub; this vegetation – particularly that of *Hippophae rhamnoides* – provides food and cover for migrant passerines.

Outer Hebridean machair has more wetland habitats than most dune systems, and scrub seldom features in the succession. Breeding waders and a greater range of wetland species, such as dabbling ducks, moorhen, little grebe and coot, are prominent in bird communities here. Despite the low variety of breeding species on shingle beaches, these habitats hold important populations of, for example, ringed plover and little tern. Many such beaches, however, experience severe human disturbance which reduces their value for breeding birds.

Large numbers of **waders and wildfowl** feed in the intertidal zone. In contrast, salt marshes have few feeding waders, except for the creeks, but many species roost on the marsh. The main feeding areas are the intertidal flats that can carry very large populations of invertebrates. Mudflats formed from sedimentary **particles of intermediate size** are more attractive to invertebrates than are flats of either larger or smaller particles. This is because **invertebrates** are favoured by deposits of high organic and nutrient content, normally formed by fine particles. In beds of very fine mud, however, their feeding and respiratory structures often become clogged and the oxygen supply depleted. Poisoning through hydrogen sulphide released by bacteria and other microorganisms may also occur (O'Connor, 1981).

Wildfowl that graze on salt marshes feed on a variety of plants e.g. Brent geese on *Puccinellia*, *Spergularia* and *Triglochin*, white-fronted geese on *Agrostis stolonifera* and *Puccinellia*, and barnacle geese on clover stolons. Wildfowl such as wigeon and Brent geese also feed on algae and other plants, including *Enteromorpha* and *Zostera*, growing on muddy intertidal substrates. Wildfowl and waders comprise some 40% of the bird species that use salt marshes in winter (Fuller, 1982). The other main groups are gulls, birds of prey, pipits, wagtails and finches. Starlings occur in large numbers; linnets and greenfinches feeding on the seeds of salt-marsh plants are also abundant. On the salt marshes of the Wash, twite winter in large numbers.

The United Kingdom coastline can support between 1.5 and 2 million waders during a winter, probably half the European population. With its relatively mild winter climate, adequate food resources are usually available. The density of waders, which varies within and between estuaries, is related to the density of their invertebrate prey. Well over 90% of the total benthic invertebrate macrofauna is made up of a relatively small range of species which form the main prey types for bird predators of the intertidal zone. Extremes of environmental factors in the intertidal zone tend to go with low biomass and low number of species.

The main **prey species** are **annelids**, especially *Nereis* spp., *Nephtys hombergii* and *Arenicola marina*, **Crustacea**, notably *Corophium volutator* among the many small species taken, and **molluscs**, particularly *Hydrobia ulvae*, *Macoma balthica*, *Cerastoderma edule*, *Mytilus edulis* and *Scrobicularia plana*. Invertebrates are not evenly distributed on the shore; their densities in some areas of estuaries are high, in others low. This is related to the sediment type and inundation. Yates *et al.* (1993) showed that sediment characteristics can be used to predict the densities of shorebirds. This important development facilitates the use of sediment type maps constructed through remote sensing to determine potential bird distribution.

The different species of invertebrate are zoned in relation to shore level. Within a zone that is occupied by a species, the distribution of individuals is usually patchy, sometimes reflecting clear but small differences in the substrate such as nitrogen or bacterial content. In other instances no clear differences in microhabitat have been detected that would explain such patchiness.

Within a zone occupied by a species the distribution of sizes of individuals is uneven. It is common for molluscs such as cockles and tellins to have smaller-sized individuals towards the top of their zone and larger individuals towards the bottom. This may result from differential survival or from movements down shore as the animals get older and larger. Vertical distribution within the sediment may also vary with season, e.g. *Macoma* is nearest the surface in June and deepest in December on the Wash (Reading and McGrorty, 1978). These factors can influence the availability of prey species to their wader predators, e.g. knot have only 4% of *Macoma* available to them in December owing to the depth to which *Macoma* have moved and the length of the bill of the knot (Reading and McGrorty, 1978).

Differences in body **size of prey individuals** reflect rates of growth in relation to environmental conditions as well as increasing age. Changes in prey size and prey abundance with season not only influence the survival of overwintering bird populations, but are important when birds are putting on extra weight preparatory to migration. In spring the abundance of invertebrates tends to be at its lowest for the year, yet the mean biomass of individual prey items tends to be at its highest. In autumn the overall biomass of invertebrate prey is at its highest for the year, but individual body size is smaller than in spring. Density of prey can affect the behaviour of the waders. Many wintering oystercatchers in north-west Europe die of starvation, especially in severe weather. Feeding birds often appear hard-pressed, especially when they should be storing fat for migration and during severe weather, when many starve.

In a winter, redshank are estimated to remove 16–38% of *Corophium volutator* on the Ythan estuary, Scotland, (Goss-Custard, 1969), bar-tailed godwits 25% of *Arenicola marina* (Smith, 1975), and oystercatchers 14% of the mussels, cockles and other molluscs on which they prey (Goss-Custard, 1977). There are predators of the invertebrates other than birds e.g. flatfish grazing *Scrobicularia* siphons and crabs, flatfish, *Crangon* and *Nereis* predating small cockles.

2.8.1 AVAILABILITY AND SELECTION OF PREY ITEMS

The invertebrates, although numerous, are not available at all times. The state of the tide is important, most birds being unable to feed if the substrate is covered by any significant depth of water. A major exception is the shelduck, which varies its technique in relation to the depth of water and the stage in the activity cycle of its principal prey, *Hydrobia*. Temperature can exert a significant effect on the activity of many invertebrate prey so that they are less active at low temperatures, and thus are less evident to searching predators. Probably the best example of this is the reduced activity of *Corophium* at low substrate temperatures and so reduced predation on it by redshank. A further interesting example is the reduced defaecation rate of *Arenicola* at the surface at low temperatures, so that they become less available to foraging curlew and godwits. *Arenicola* also defaecates less often on windy days.

Bird predators do not necessarily take the sizes of invertebrate prey in the proportions at which they exist in the substrate, but show **selection**. For example, Brown and O'Connor (1974) showed that predation by oystercatchers on cockles in Strangford Lough, Northern Ireland falls proportionately more on second year and older animals (above 11–12 mm length), but small first-year cockles are not taken. Usually the smaller sizes are taken at a much lower proportion than they are present, which can be understood in terms of yield in relation to effort. The largest sizes are also frequently predated in reduced proportion, especially with molluscs, probably because of the increased handling time which comes from greater difficulty in extracting the body from a thicker and more resistant shell.

As Hale (1980) suggests, the information required to show differences in feeding ecology is not necessarily the same as that needed for a quantitative study of the predation of invertebrate populations and the impact on the functioning of an estuarine system.

It is important to realize that a particular species of bird predator may change its feeding preferences with time of year as well as the particular part of an estuary where it is feeding. This may result from changes in relative abundance of the different prey species as well as their nutritional value. For example, in spring, knot can feed heavily on spat mussels but by autumn the mussels have grown beyond the size utilized by knot, but are now of a size that oystercatcher will take. Birds may also move off the estuary to feed elsewhere either during certain stages of the tide, e.g. oystercatcher may feed on earthworms on grassy fields inland, or birds may move as the available preferred food is depleted, e.g. Brent geese feeding on *Zostera*. Each species of wader tends to take particular species of prey, whose depth in the sediment often correlates with the length of the bird's bill (Figure 2.12).

2.8.2 BIRD MIGRATION AND ITS IMPACT ON INTERTIDAL INVERTEBRATES

Movement of intertidal birds from their breeding grounds to moulting grounds to wintering areas is an important feature of their interaction with, and impact upon, invertebrate populations. The general pattern seems to be that all species are likely to use a series of estuaries during each year. Birds return to the same estuary, even the same section on a particular estuary at a certain time of year. Many estuaries important in winter support relatively few birds in autumn. In spring, only a few estuaries hold large numbers before many of the birds migrate to Iceland and Greenland.

Only a few estuaries in Europe have large numbers of moulting waders. The Waddensee can have up to 2 million waders in autumn. Perhaps 100 000 can be found on estuaries such as the Ribble, Wash and Morecambe Bay, UK. Waddensee birds disperse in late autumn or early winter. Early spring movements tend to be the reverse; birds move to the estuaries used in autumn, especially the Waddensee, before migrating to their breeding grounds.

Returning birds from Africa, especially ringed plover, dunlin and sanderling, use a small series of sites, particularly the Dee, Ribble, Morecambe Bay and Solway. Goss-Custard and Moser (1990) reported that dunlin

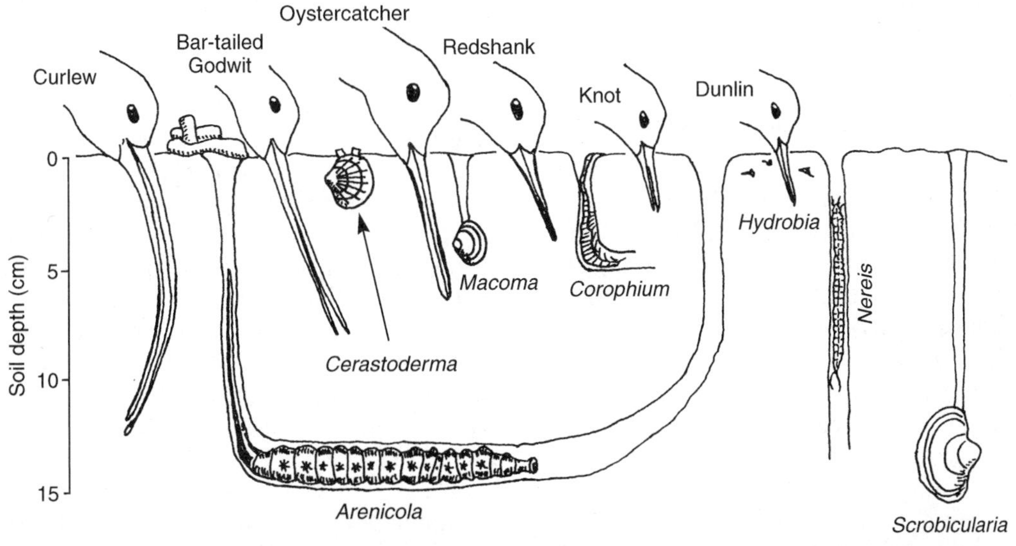

Figure 2.12 Bill lengths of common waders and the depth of common prey species. (After Goss-Custard, 1975.)

numbers had declined in estuaries where *Spartina anglica* had spread, but not in those where the plant had stayed constant or declined (section 5.2).

Durell *et al.*(1993) showed that oystercatchers on the River Exe, Devon, UK, had **differences in bill shape** that reflected the prey taken and how it is attacked. Birds with pointed bills were mudflat feeders on worms, those with chisel-shaped bills were mussel stabbers, individuals with blunt bills were mussel hammerers. The sex ratio for the birds with chisel-shaped bills was 50 : 50, but for those with pointed bills, 70% were females, and birds with blunt bills were 90% males.

The relationship between prey distribution, density and size in relation to the feeding rate of bird predators is complex. Cresswell (1994) showed that redshank feed in different areas with age on the Tyninghame estuary, East Lothian, Scotland. Adults feed on mussel beds, juveniles in creeks on the salt marsh despite the increased risk of predation on them here. Food intake was higher in the creeks than on the mussel beds. This interaction between food availability and factors such as disturbance or risk of predation is also shown by other species of waders.

2.8.3 COASTAL BIRDS: PREDATION AND BREEDING IN A SCOTTISH DUNE/ESTUARY

Many young birds fall prey to predators whose numbers may have to be controlled. So many eider ducklings were killed by foxes in the **Sands of Forvie and Ythan Estuary NNR** in summer 1994 that 42 foxes were culled in the following winter. Casualties among nesting females were also high as these birds stay on the nest to defend their young, while the males go out to sea when moulting begins. Towards the end of July 1995, many of the female eider had dispersed up or down the coast, but a substantial number were nesting at the end of the spit together with some 846 ducklings. At this time the non-nesting females had lost all their primary flight feathers, but mothers with young retained sufficient to ensure mobility. The newly hatched ducklings have different food requirements from the adults and, once deserted by their mothers, form large creches associated with newly arrived females and their large broods.

The position of the Sands of Forvie and Ythan NNR on the migration routes of many species in north-east Scotland makes it an important site. North (1981) lists 225 species, of which 74 are known to have bred and 43 breed regularly. The eider duck population is the largest in Britain, while sandwich, little, common and arctic tern have important breeding sites in the mobile dunes and shingle of South Forvie. Black-headed gulls also breed in the dunes, while oystercatchers breed in the deflation hollows of South Forvie, parts of North Forvie, the estuarine marshes and nearby farmland. Many other species are cliff-nesting or roosting, and the NNR is also an important over-wintering site for many waders and wildfowl, notably the large numbers of greylag and pink-footed geese present between October and April.

2.8.4 CONSERVATION AND MANAGEMENT OF COASTAL HABITATS

Within the UK, the total area of suitable habitat for coastal birds and other organisms has considerably diminished in recent years and the work of government agencies, as well as that of voluntary bodies such as the Royal Society for the Protection of Birds (RSPB), needs efficient coordination, especially as so many of the species concerned are migratory. The European Union is playing an important role in bringing this about. Huggett (1995) discusses the role of the Birds Directive and Habitats and Species Directive in delivering **integrated coastal zone planning and management**. He concludes that the UK coast would benefit from a more holistic and coordinated approach in place of the present piecemeal treatment of its terrestrial, intertidal and marine components.

Figure 2.13 Eider duck (*Somateria mollissima*) and creche. (Drawn by P.R. Hobson.)

WATER AND IONIC RELATIONSHIPS: PLANT ADAPTATIONS TO COASTAL ENVIRONMENTS 3

3.1 LIMITING ENVIRONMENTAL FACTORS

Coastal plants are many and varied but, like normal terrestrial species, require an appropriate water regime, adequate aeration of the root system, and a suitable supply of both major (N, P, K, Ca, Mg, S) and minor (Fe, Cu, Zn, Mo, B, Mn, Co, Cl, and in some C-4 plants, Na) mineral nutrients. These requirements are often not easily satisfied since ecosystems of coastal areas are in some ways among the most changeable and extreme in their environmental conditions. In salt marshes tidal inundation (section 4.2) results in considerable and rapid changes in salinity, as well as waterlogging and the deposition of silt. In dunes the blowing sand leads to instability and its poor water-holding capacity, coupled with high rates of evaporation, may give severely droughting conditions. Shingle banks are characterized by variable salt spray, with occasional inundation, and a usually scanty soil matrix between the pebbles.

Many of the distinctive features of the form, anatomy and physiology of the organisms of all these habitats are related to the prevailing conditions. While many animals can avoid extremes by movement, rooted plants have to withstand the rigours of the environment, numerous species showing adaptive characteristics which enable them to survive. Major traits, both structural and ecophysiological, of plants which suit them to their environment are described below. Morphological, physiological and behavioural adaptations of the salt-marsh fauna are discussed elsewhere (sections 2.7 and 5.2; Daiber, 1977, 1982; Long and Mason, 1983, Chapter 4).

3.1.1 ENVIRONMENTAL IMPACTS ON SALT-MARSH PLANTS

The environmental conditions of saline habitats are very different from those experienced by plants of typical terrestrial habitats (**glycophytes**). These conditions have led to the development of particular strategies by plants for survival on salt marshes. Such plants (and those of other saline environments such as salt pans, salt deserts, and some members of cliff and dune communities), have for many years been described as halophytes, a category now more strictly defined on the basis of their tolerance of high concentrations of salts. There has been considerable argument as to what differentiates halophytes from glycophytes (Adam, 1990); plant response to salt is essentially a continuum from extremely sensitive to extremely tolerant. Nevertheless a useful, if arbitrary, definition of a **halophyte** is based on the ability of a species to grow and complete its life-cycle at salt concentrations in excess of 100–200 mM NaCl (Flowers *et al.*, 1986). As discussed later, halophytes take up salts and use them to maintain low water potentials in vacuoles; the more salt-tolerant glycophytes tend to be able to exclude NaCl at the root level when salinities are low to moderate (Greenway and Munns, 1980).

Figure 3.1 Saline inland marsh, Kiskunsagi National Park, Hungary, with *Bolboschoenus maritimus, Juncus gerardii* and *Puccinellia limosa* in the foreground. Further areas dominated by *Puccinellia limosa* extend beyond the wide zone dominated by the obligate halophyte *Lepidium crassifolium*, here flowering profusely. Many plant communities including halophytes are represented on the seasonally flooded soils (solonchak and solonetz) of Central Europe. The inland saline habitats of Europe, and the physiology of the flowering plants growing in them, are briefly described by Ellenberg, 1988, pp. 364–7). (Photograph by M.C.F. Proctor.)

Figure 3.1 shows the halophytic community growing on an inland saline area which is commonly waterlogged but where tidal influences are absent. The two environmental factors identified in section 2.5 as central in determining the composition of plant communities on maritime salt marshes are **salinity** (salt concentration) and **waterlogging**. These two factors – which in physiological terms largely limit survival of higher plants as well as many other organisms in this demanding habitat – are mainly determined by the influence of the tides. Tidal action not only brings salt (and sediments) onto the marsh as a seawater input, but also tends to flood any air pores present in fine alluvial salt-marsh soils. The main constituents of seawater are given in Table 3.1, with chloride and sodium being the dominant ions. The figures given are from measurements for open ocean. Ion concentrations of samples taken near to land can differ from those of the open ocean; freshwater inputs from estuaries can reduce concentrations. Salinity of the largely landlocked (non-tidal) Baltic Sea is about one-third of that in the neighbouring, but open, North Sea. Salinity is also low in Hudson Bay. In some inland landlocked seas such as the Dead or Aral Seas salinity can approach near-saturation concentrations of about 5 M.

Although salinity of seawater is fairly constant, tides represent an input of seawater to the salt marsh and are part of a complex of factors which bring about substantial spatial

Table 3.1 The major ionic constituents of seawater from the open ocean. (Adapted from Martin, D.F. (1970) *Marine Chemistry*, Vol. 2, Marcel Dekker, New York, p. 27.)

Constituent	*Concentration*	
	$g\ kg^{-1}$	*Molarity (mM)*
Chloride	19.35	548
Sodium	10.76	470
Sulphate	2.71	28
Magnesium	1.29	54
Calcium	0.413	10
Potassium	0.387	10
Bicarbonate	0.142	2
Bromide	0.067	0.8
Strontium	0.008	1.5
Boron	0.0045	0.4
Fluoride	0.001	0.07

and temporal variations in salinity over the area of salt marshes. The height of low and high tides within the tidal cycle varies typically over both a 14-day period and seasonally (see Figure 4.4). This effectively means that the area of salt marsh covered by seawater is constantly changing. In particular, the landward area of salt marshes is less frequently covered than parts closer to the open sea. When salt-marsh soils are not covered by the tide, a range of factors may alter their salt concentration.

Seasonal variations in soil salinity may be considerable and strongly influence the germination patterns of salt-marsh plants; some have seeds which do not germinate under saline conditions (section 5.3). In summer, evapotranspiration is considerable, removing water from the soil system and increasing salt concentration, whereas in winter salt concentration on the open marsh is often diluted by rainfall. Drainage and the addition of freshwater from streams can also be important. Figure 3.2 gives measurements of temporal and spatial changes in soil salt concentration over the course of 2 years on different parts of a salt marsh in North Norfolk, UK. Changes in salt concentration measured as a water potential are also given in relation to soil depth. 'Salinity inversions', i.e. the existence of higher salinity on upper parts of the marsh than near the sea, are common world-wide in Mediterranean-type climates with high summer evapotranspiration and low summer rainfall.

Continual flooding by the tide (Figure 3.3) affects soil aeration and hence the availability of oxygen to the root systems of salt-marsh plants. Anoxic conditions can also substantially change soil chemistry which itself can adversely affect plant growth (sections 4.2, 4.7).

Variation over time in soil **redox potential** – a measure of the capacity to act as a reducing or oxidizing agent, and which also gives an indication of the level of soil aeration – is shown for various depths at three sites on a salt marsh in Yorkshire, UK, in Figures 4.7–9.

The intertidal nature of salt marshes introduces other factors which may affect halophyte physiology. Exposure of the marsh at low tides tends to increase temperatures, particularly in summer months, which then decrease on immersion. In winter, the opposite can be true. Additionally, exposure in the summer can increase the likelihood of desiccation. Although salt marshes are areas of low wave activity, periodic heavy storms can cause mechanical damage, particularly to plants on the lower marsh where seedlings in the establishment phase are often swept away.

3.2 HALOPHYTE WATER RELATIONS

The presence in salt-marsh soils of high concentrations of sodium chloride has substantial implications for the uptake of water and nutrients by the plant. Water in the salt marsh, moving within the soil–plant–atmosphere system, will travel from regions of high water concentration (i.e. dilute solutions) to areas of low water concentration (concentrated solutions). In simple terms, this means that halophytes have to maintain higher osmotic concentrations in their tissues than exist in the already salt-dominated environment outside,

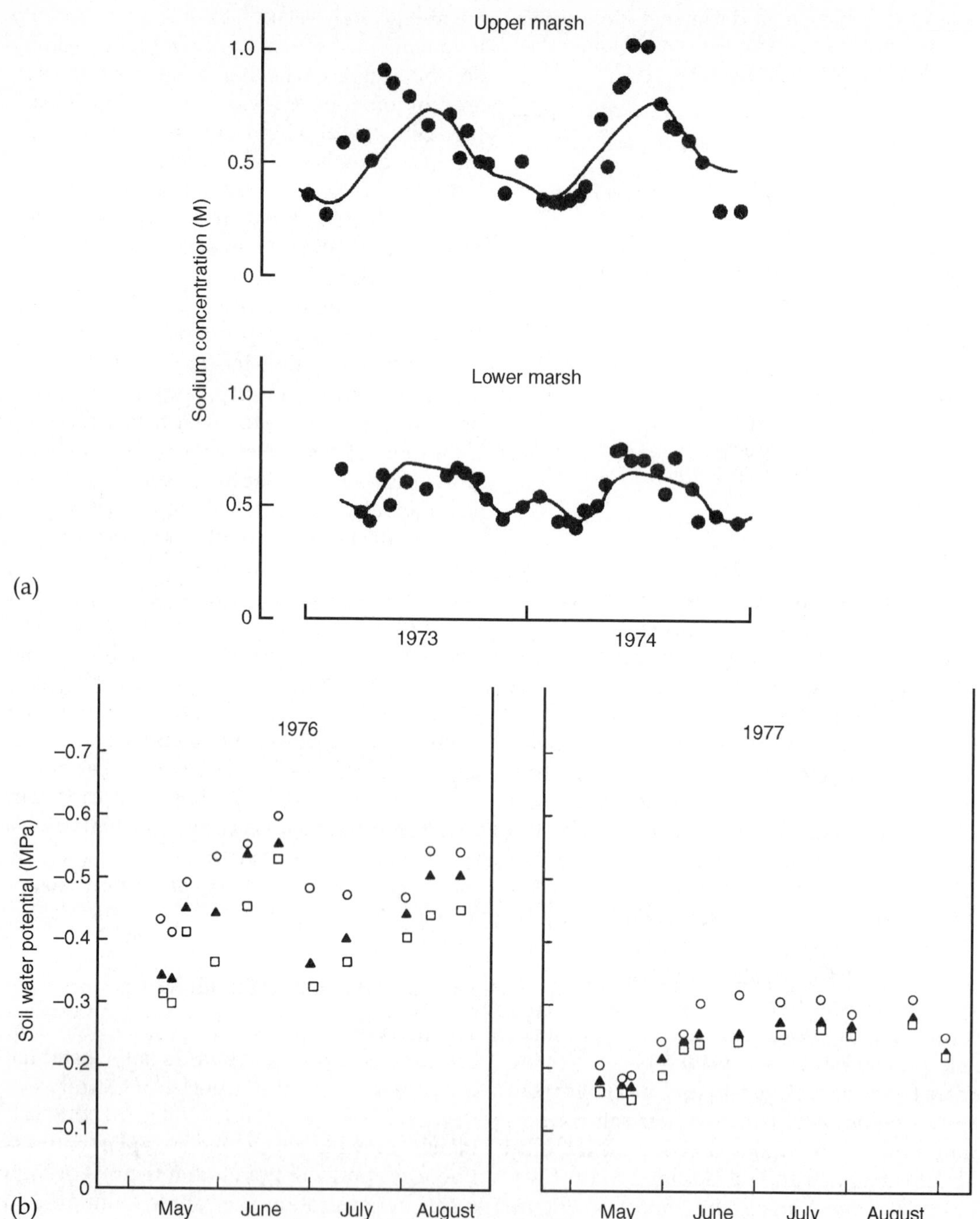

Figure 3.2 (a) Variation in salt-marsh soil salinity on the upper and lower marshes at Stiffkey, North Norfolk, during 1973 and 1974; (b) changes in soil water potential with depth on the upper marsh in 1976 and 1977; ○, ▲ and □ indicate water potentials 0–5 cm, 5–15 cm and 15–30 cm below the surface of the sediment, respectively. (After Jefferies *et al.*, 1979.)

Figure 3.3 Aerial photograph of the backbarrier salt marsh at Stiffkey, Norfolk, UK. The upper marsh is submerged approximately 200 times a year, while the lower marsh (foreground) is submerged some 600 times a year. The main channel on the upper marsh is Cabbage Creek, which carries sediment onto this prograding area and has the mid marsh on its seaward (northern) side. The lower marsh, which has many salt pans and an extensive dendritic and highly sinous creek system, is seaward of the low gravel ridge with conspicuous tussocks of *Ammophila arenaria* and darker lines of gorse thicket which run from left to right across the photograph. *Honckenya peploides* and *Seriphidium maritimum* are notable in more stable areas of sand and shingle on the ridge, where species distribution is strongly influenced by small altitudinal differences. This region of the marsh was unvegetated in 1948, but had been colonized and developed a creek network by 1960. The *Juncus maritimus* zone immediately seaward of the gravel ridge gives way to a poorly drained area of *Limonium/Salicornia/Aster/Spartina*. Nearer the sea the creek density increases, there are more hummocks and the substratum is less consolidated. (See also Jefferies, Davy and Rudmik, 1979, Figure 15.1, p.247.) (Photograph by Simon J. Pittman.)

simply to allow passive uptake of water. The mechanisms that halophytes show which ensure continuous water uptake are central to their adaptive strategy.

Water within the soil–plant–atmosphere system is best considered in thermodynamic terms. The term **water potential** is used to describe the availability of water within any given system and is defined as the difference between the free energy of water within that system as affected by matric or osmotic constraints, and the free energy of pure water. It is conventionally measured in the pressure units of pascals (Pa), although the non-SI units of 'bars' are often still seen (1 MPa = 10 bars). Water potential is normally given the Greek symbol ψ, and is always negative in value (pure water is zero).

In real terms, as the solute concentration of a solution increases (i.e. the water becomes less concentrated), the water potential becomes more negative. When two biological systems are separated (for example the cell wall and cytoplasm by the **plasmamembrane**, or the root from the shoot), the separated water potentials will move towards equilibrium as water diffuses from the less negative to the more negative water potential. This difference in potential gives the **water potential gradient** ($\Delta\psi$), which will be zero at equilibrium when there is no net movement of water. Movements are largely of water and not solutes because water diffuses across membranes much more quickly than solutes. For continued movement of water through the soil–plant–atmosphere system, a permanent water potential gradient is maintained and is the essential driving force for the transpiration of water through any higher plant. Water always moves **down** a water-potential gradient towards a region where water potential is more negative.

Plant cells can be considered as a thin layer of metabolically active protoplasm (adjoining the cell wall), surrounding the vacuole (enclosed by the vacuolar membrane or **tonoplast**). In volume terms, the vacuole may be as much as 80% of the total cell volume. The protoplasm, which contains the metabolic (and genetic) apparatus of the cell including mitochondria, plastids, nucleus and endoplasmic reticulum, is enclosed by another membrane, the **plasmamembrane**, and is surrounded by a semi-rigid cellulosic cell wall. Importantly, the membranes act to control the rate of passage of water and solutes into the compartments which they surround.

Because of the solutes within it, each plant cell has its own water potential, which tends to become more negative upwards through the plant so that a water potential gradient is maintained. The cells of plants, but not of animals, are constrained by semi-rigid cell walls which play an important role in their water relations (Dainty, 1979). If, for example, the cell were to take up solutes which made its water potential more negative than its surroundings, water would tend to flow into the cell to restore equilibrium. The presence of a fairly rigid cell wall serves to prevent a large increase in cell volume, and instead a pressure develops. This is termed **turgor pressure**. It can therefore be seen that within the plant cell, water potential is made up of both an osmotic component (the **osmotic potential**, sometimes termed the solute potential) and a pressure component (the turgor pressure). The tonoplast, separating the cytoplasm from the vacuole, has little structural strength and hence no pressure gradient exists between cytoplasm and vacuole. Cytoplasm and vacuole will tend to come quickly to water potential equilibrium.

Plant cell water potential is expressed by the following equation, routinely used to quantify plant water relations (note that the terms osmotic and solute potential are interchangeable):

$$\psi = \psi_p + \psi_s$$

where ψ = cell water potential; ψ_p = turgor pressure; and ψ_s = osmotic (or solute) potential.

Turgor pressure is not only important for the maintenance of plant rigidity, but is also involved as one of the driving forces for cell expansion and growth (Cosgrove, 1986); it may in addition be involved in the regulation of ion transport processes across membranes (Cram, 1983). It is further necessary to take into account that the walls of living cells are not completely rigid. The **volume elastic modulus** (ϵ) is a measure of how much a cell will expand or contract in response to a change in turgor pressure (Dainty, 1979), according to:

$$\epsilon = dP/(dV/V)$$

where dP is a fractional change in turgor pressure associated with a change in cell volume dV/V.

As the volume elastic modulus of plant cells is generally high, changes in external water potential mean that the cell can alter turgor pressure rather than osmoregulate to change cell solute contents (osmotic potential) to bring about water potential equilibrium. This is particularly important in an environment such as a salt marsh where external water potential continually changes in response to variations in seawater inundation.

Figure 3.4 illustrates some of the water relations values already discussed at a whole plant level in a young seedling of *Suaeda maritima* growing in 400 mM NaCl, a typical salt concentration found on the salt marsh. Figure 3.4(a) shows the root to shoot ion fluxes and the associated water relations values between root and shoot. Note that internal water potentials are more negative than the soil medium and shoot is more negative than root. This maintains the gradient in water potential to drive water through the plant osmotically. At the shoot level (Figure 3.4(b)), water (and ions) are delivered by the xylem stream to the mesophyll cell walls. They are then moved across the plasmamembranes, again driven down the water potential gradient, into the cytoplasm and the vacuoles. The mesophyll cell wall is fairly rigid allowing turgor pressure to develop. Table 3.2 indicates some of the water relations existing through the halophyte shoot. Leaves of different ages maintain different water potentials, osmotic potentials, turgor pressures and water potential gradients. In young tissue, higher turgors are maintained,

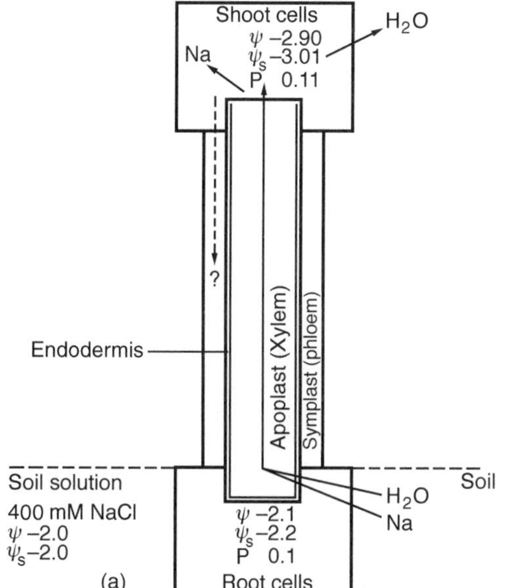

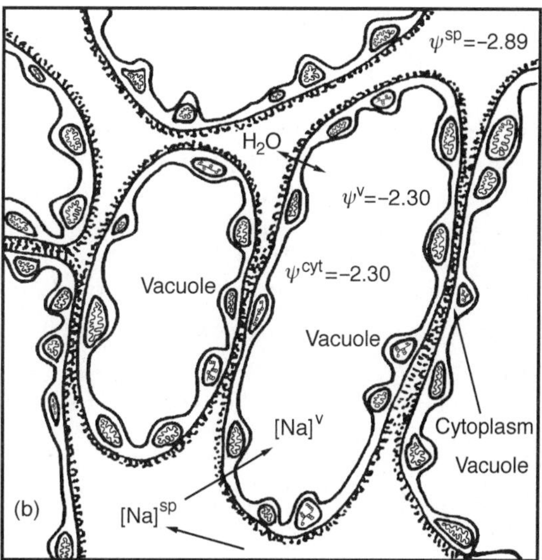

Figure 3.4 Water relations values in young plants of *Suaeda maritima* growing in 400 mM NaC1. (a) Flow diagram with typical water relations values in whole plant; (b) diagram showing typical water relations values in leaf mesophyll cells which often lose water during the day (all values in MPa). (Adapted from Hajibagheri *et al.*, 1984 and Clipson *et al.*, 1985.)

presumably as a component of cell expansion processes, together with larger water potential gradients.

The transpiration stream of a glycophyte with its roots in soil solution of relatively low ionic concentration is driven by the same processes as those described above. If its roots (whose cells have much lower solute concentrations than those of *S. maritima*) were in 400 mM NaCl, however, they would rapidly lose water and the plant would die.

3.3 ADAPTIVE MECHANISMS IN HALOPHYTES

To achieve very negative cellular water potentials halophyte cells have to maintain strongly negative osmotic potentials and hence high concentrations of internal solutes. The most readily available solutes are the sodium and chloride ions from seawater, which are toxic to many components of cellular metabolism. Precisely how halophytes overcome this problem is central to their adaptive strategy and is considered below.

3.3.1 GROWTH RESPONSES IN HALOPHYTES

Adaptation by plants to salt-marsh conditions means that seed germination (section 5.3), seedling development, vegetative growth, and reproduction are all largely tolerant of the salt-marsh environment (though the species concerned commonly perform better when salinity is low or absent). The most studied aspect of halophyte physiology is the vegetative growth response to salt, albeit largely in controlled environments which may not fully reflect the conditions on the salt marsh. The generalized growth responses of plants to salinity in their rooting medium is shown in Figure 3.5.

Figure 3.5 demonstrates that, in terms of their growth, halophytes are much more salt-tolerant than their glycophyte counterparts. 'Growth' under these circumstances is not easy to define, since fresh weight is considerably increased by greater succulence and extra dry weight may to a considerable extent be a reflection of increased salt uptake. In *Salicornia*, for example, 40–45% of dry mass may consist of mineral salts; in many instances it is instructive to examine results on an 'ash-free' basis. The inability of crop plants to grow and yield in the great expanses of saline soils throughout the world represents a major limitation to global food provision. The growth response of halophytes to salt broadly divides into two: species with a positive response to salinity up to about 200 mM NaCl, and those with a growth decline in the presence of salt. Halophytes showing the former response are generally dicotyledonous (often chenopods) whereas species showing growth reductions may be either monocotyledonous or dicotyledonous. Nevertheless, all these species can grow and complete life-cycles over the range 300–600 mM NaCl, typical of the salinity range on the salt marsh.

Table 3.2 Water potential, osmotic potential, turgor pressure and water potential gradient between shoot tissue and external medium in seedlings of *Suaeda maritima* growing in 400 mM NaCl. (See Figure 3.4 for details.)

	Pressure (MPa)			
	Water potential	*Osmotic potential*	*Turgor pressure*	*Water potential gradient*
External	–2.0	–	–	–
Cotyledon	–2.4	–2.46	0.06	0.4
Leaf 3	–2.9	–3.01	0.11	0.9
Apex	–3.7	–3.90	0.20	1.7

The positive growth response of the first group has been termed **succulence** (Jennings, 1976), although no satisfactory physiological basis for the response has been advanced. (Formerly the term 'succulence' implied simply fleshiness, with a high water content in extensive storage parenchyma cells.) This redefinition of succulence is not entirely appropriate, since it fails to include many glycophytic species of succulent habit, especially those in the Crassulaceae and Cactaceae. 'Succulence' may reflect mainly increases in halophyte water and mineral content, although Yeo and Flowers (1980) found that organic matter content increased at low salinities, albeit much less markedly than water or mineral content in percentage terms.

Growth improvement and the increase in succulence with increased salinity suggest that the balance of a range of physiological processes is most favourable at the growth salinity optimum. These processes could include: photosynthetic and transpiration rates; uptake, compartmentalization and regulation of ions; water relations and their regulation; enzyme responses; and general regulation of growth and metabolism. This leads to the conclusion that halophyte salt tolerance is not the product of a single process, but has evolved as an adaptation involving several processes. It is likely therefore that salt tolerance is polygenic in character; the ideas and processes involved are expanded below.

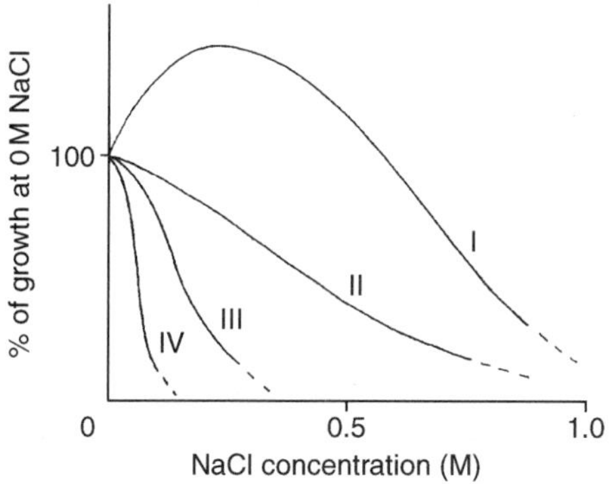

Figure 3.5 Generalized growth responses of plants exposed to salinity. Group I contains extreme halophytes which show a positive growth response to salt, including species of *Suaeda*, *Salicornia* and *Atriplex*. Group II contains halophytes not showing a positive growth response, including dicotyledonous plants such as *Aster tripolium* and *Cochlearia anglica*, and monocotyledonous plants such as *Puccinellia maritima* and *Festuca rubra*. Group III contains salt-tolerant glycophytes such as barley, cotton and sugar beet, while Group IV contains salt-sensitive glycophytes such as rice and orchard crops. (After Greenway and Munns, 1980.)

Halophytes on exposure to salt show substantial morphological changes. At the whole plant level this has simply been considered as an increase in water content of the plant (Jennings, 1976), and is not unique to halophytes, it also being a response of many crop species. Succulence involves widespread changes in morphology; typical examples are shown in Table 3.3.

Attempts have been made to classify halophytes into groups on the basis of their growth responses along with their mineral and organic contents. Glenn and O'Leary (1984) grew 20 different dicotyledonous halophyte species in a range of salinities up to 720 mM NaCl. Species were classified as either **euhalophytes** or **miohalophytes** on their growth responses, water content and cation contents.

Table 3.3 Anatomical and ultrastructural features of halophytes affected by salinity. (Modified from Flowers *et al.*, 1986.)

Response to salinity	
Increase	*Decrease*
Leaf thickness	Leaf number per plant
Leaf area per plant	Stomatal number per unit area
Leaf epicuticular wax	Stelar diameter in stem
Leaf cuticle and epidermal cell thickness	
Leaf cell size	
Stelar diameter in root	

Euhalophytes were able to grow at higher salinities than miohalophytes, and increased both water and mineral contents to a greater extent.

Albert (1975) and Gorham *et al.* (1980) used the term **physiotype** to group halophytes according to their mineral and organic constituents. Physiotypes have a characteristic chemical composition, a primary division occurring between dicotyledonous species which maintain high internal sodium concentrations, and monocotyledonous species which preferentially take up potassium.

At an ecophysiological level, Wyn Jones (1980) introduced the concept of **osmo-conformers** and **osmo-regulators**. These classes depend on the magnitude of the water potential gradient between the shoot and the external salinity. Water potential in osmo-conformers is maintained more negative than that in the external medium but only a small gradient is generated at any particular salinity. In osmo-regulators, on exposure to low concentrations of salt, a large water potential gradient is generated. Ecologically, this may mean that osmo-regulators are better adapted to resist rapid fluctuations in salinities in salt-marsh soils as the substantial gradient ensures that they do not have to take up solutes rapidly to readjust their water potential.

Although generalization of halophytes into classes can be useful, caution is needed as many species do not readily fall into the divisions outlined, but nevertheless are successful halophytes.

3.3.2 INTRACELLULAR COMPARTMENTALIZATION OF SOLUTES

As already stated, water potentials of salt-marsh soils are highly negative. Osmotic concentrations of halophyte cells must therefore be very high if the downward gradient in water potential through the soil–plant–atmosphere system, necessary to maintain the transpiration stream, is to exist. For maintenance of cell solute potentials (often called **osmo-regulation** or **osmotic adjustment**; see Reed, 1984 for a discussion of osmo-terminology), the plant cell can utilize either solutes from the external environment, or can synthesize organic solutes from the products of photosynthesis. A third alternative is a reduction in cell water content to adjust water potential. It is likely that this operates only at higher salinities (Glenn and O'Leary, 1984; Adam, 1990).

High concentrations of sodium and chloride are inhibitory to many aspects of plant metabolism of both halophytes and glycophytes. In numerous studies enzymes have been extracted from plants and their *in vitro* activity assayed at different concentrations of sodium and chloride. In general, it has been shown that for most enzymes (e.g. PEP carboxylase, malate dehydrogenase, nitrate reductase, glucose 6-phosphate dehydrogenase, peroxidase, ATPase) the presence of salt in assay media either increasingly inhibits enzyme activity above 0 mM NaCl, or at best is mildly stimulatory up to 50–100 mM NaCl and then inhibitory (Cavalieri and Huang, 1977; Shomer-Ilan *et al.*, 1985). Some examples are shown in Figure 3.6.

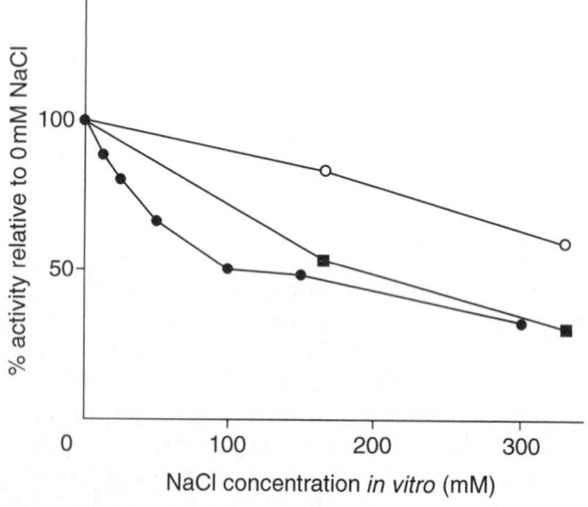

Figure 3.6 The effect of NaC1 *in vitro* on enzyme activity of nitrate reductase (●), glucose 6-phosphate dehydrogenase (○) and malate dehydrogenase (■) isolated from *Suaeda maritima*.

Comparisons between halophytes and glycophytes have suggested little difference in their enzymic responses to salt *in vitro* (Flowers *et al.*, 1977; Shomer-Ilan and Waisel, 1986). These findings suggested that high concentrations of salt similar to those of saltmarsh soils were not compatible with enzyme activity within the **cytosol** (the aqueous part of the cytoplasm with its dissolved solutes) and its associated organelles. Such results must be interpreted with caution as enzyme assays *in vitro* may not reflect metabolism in the living organism. For example, the *in vivo* biochemical environment may substantially affect enzyme activity, perhaps with adapted plants keeping concentrations of enzyme substrate higher, or maintaining higher turnover rates of enzyme synthesis. Both would serve to raise enzyme activity, although neither has been explicitly demonstrated. Nevertheless, metabolism is likely to be particularly affected by the most sensitive enzyme in any metabolic pathway, and it is now considered that cytoplasmic NaCl concentrations much above about 100 mM NaCl have a strong adverse effect on the metabolic functioning of the cell.

Demonstration of metabolic sensitivity led to the hypothesis that halophytes maintain low concentrations of salt in their cytosol. Nevertheless, simple chemical analysis of halophyte tissues does indicate, particularly in dicotyledonous plants, a substantial content of sodium and chloride. Typical values for sodium and chloride content for a range of halophytes are shown in Table 3.4. In general, dicotyledonous halophytes take up larger quantities of sodium and chloride than do monocotyledonous plants, which tend to maintain higher uptake rates of potassium. Comparing sodium with potassium ionic ratios across several plant families, Flowers *et al.* (1986) found that at seawater salinities in 25 members of the Poaceae, Cyperaceae and Juncaceae mean Na : K was 1.2, whereas in 22 species of the dicotyledonous Chenopodiaceae, Asteraceae and Aizoaceae mean Na : K was 12.6. Selectivity for K over Na has been widely seen as a feature of monocotyledonous halophytes, potassium in general being an essential nutrient for stomatal mechanisms, enzyme activation and osmotic balance (Marschner, 1995).

The hypothesis that sodium and chloride were maintained at low concentrations within the cytoplasm, together with high tissue contents, suggested that sodium and chloride were primarily localized within vacuoles, which do not have the metabolic sensitivity of the cytosol. Table 3.5 gives values for compartmental concentrations for the chenopod *Suaeda maritima* and other species growing over a range of salin-

Table 3.4 Sodium, potassium and chloride contents (mM plant water basis) and sodium : potassium ratio of a range of dicotyledonous (D) and monocotyledonous (M) halophytes grown at seawater salt concentration

	Species		*Na*	*Cl*	*K*	*Na : K*
D	*Atriplex spongiosa*	shoot	660	450	100	6.6
		root	350	310	150	2.3
D	*Suaeda maritima*		550	530	75	7.4
D	*Salicornia europaea*		820	965	50	16.4
D	*Spergularia media*		615	480	70	8.8
D	*Glaux maritima*		240	225	100	2.4
M	*Bolboschoenus maritimus*		145	115	183	0.78
M	*Spartina anglica*		345	315	160	2.2
M	*Juncus maritimus*		100	140	234	0.43
M	*Puccinellia maritima*		160	230	159	1.01
M	*Elytrigia atherica*		48	239	152	0.31

ities (data from Flowers *et al.*, 1986). In mature leaf cells, sodium concentrations in the vacuole approached or exceeded those in the external medium, but concentrations within the cytoplasm were typically of the order of 50–150 mM, which is in line with concentrations that cytosolic enzymes can withstand. As already noted, the vacuole is by far the largest of the cellular constituents, often representing more than 70% of the cell volume.

As well as the need for generation of solute potential in the vacuole, water potential equilibrium across the tonoplast has to be maintained. The low levels of ions in the cytosol raise the question as to what solutes generate cytosolic water potential. In halophytes, a variety of organic compounds has been found to accumulate in response to salinity, including the imino acid proline, quaternary ammonium compounds such as glycinebetaine, tertiary sulphonium compounds such as dimethylsulphoniopropionate, sugars, organic acids and polyols such as sorbitol and pinitol (Briens and Larher, 1982; Rhodes and Hanson, 1993).

For these compounds to be important in the osmotic adjustment of the cytosol, it must be shown that they have no effect on cytosolic metabolism and are found there at sufficient concentrations to contribute appreciably to the cytosolic osmotic potential. If a compound fulfils these conditions, it is termed a **compatible solute**. There is now substantial evidence that most of the compounds described have little effect on metabolism, at least *in vitro*. Their localization is more problematic, but at least for glycinebetaine, appreciable quantities have been shown to be present in the cytoplasm (Adam, 1990).

At the cellular level, a final question is how do halophytes regulate their internal concentrations to prevent the build-up of toxic ionic levels in the cytoplasm. Although there is very little information on the regulatory processes themselves, Dupont (1992) has suggested a model of the membrane transport processes which could adjust cytosolic sodium concentrations. The model – which involves the import of sodium into the cell in response to proton gradients generated by a primary pump (as well as efflux at the plasmamembrane to the outside, which has already been shown) – is speculative, but further research in this field can be expected.

3.3.3 ADAPTATION AT THE WHOLE-PLANT LEVEL

Consideration so far has been limited to how halophytes adapt to high salinities at the single cell level. Clearly they also must adapt so that cellular activities integrate with those in the whole plant, internal salt concentrations being regulated both at the cell and the tissue level. Salt is taken up by the roots and if transport to the shoot was not regulated, shoot salt concentrations would rapidly reach toxic proportions. A range of mechanisms has been proposed to restrict or regulate the amount of

Table 3.5 Concentrations (mM) of sodium, potassium and chloride in three species growing in different salinities. (From Flowers *et al.*, 1986.)

Species		*Na*	*K*	*Cl*
Suaeda maritima	external	340	7	340
	cytoplasm	166	27	86
	vacuole	494	20	352
Triglochin maritimum	external	500	–	–
	cytoplasm	148	–	–
Eleocharis uniglumis	external	74	–	–
	cytoplasm	192		

salt reaching the shoot, although a single general mechanism or blend of mechanisms has not been found that describes the situation in all halophytes.

3.3.4 REGULATION OF UPTAKE AT THE ROOT

Clearly the root, or more precisely the plasmamembrane in the root epidermal and cortical cells, forms the first barrier at which uptake of solutes from the external environment could be regulated. Casparian strips developed in the root endodermis, which is situated at the boundary between the cortex and the stele, play an important part in the regulatory system. In *Suaeda maritima* this waterproof thickening is more developed under saline than non-saline conditions (Hajibagheri *et al.*, 1985), while *Puccinellia peisonis* produces a second endodermal layer (Stelzer and Läuchli, 1977). The effect of such thickenings in the endodermis is to tend to force the flow of solutes across the root cortex into the **symplast** (the intracellular compartment consisting of the cytosol of many cells connected by plasmodesmata). Consequently the solutes have to cross membranes in the root which regulate their passage. Solutes are passed across the cortical symplast and are then released to the xylem for transport to the shoot within the transpiration stream. Although the **symplastic route** probably dominates, **apoplastic movement** (via the cell wall continuum) might occur across the cortex and the relatively few passage cells – which lack casparian strips – present in the endodermis which would form an unregulated pathway from the external medium to the xylem. There is no evidence that this occurs, e.g. in the mangrove *Avicennia marina*, Moon *et al.* (1986) demonstrated that apoplastic ion transport in the root was minimal.

Yeo and Flowers (1986) have considered that the benefits of ion uptake at the root are: (i) to provide ions to bring about osmotic adjustment in shoot tissues (particularly young expanding tissue) and to maintain general water potential gradients; (ii) to provide ions of nutritional importance such as potassium, nitrate and phosphate; and (iii) to distribute those ions through the plant, permitting their vacuolar localization away from the salt-sensitive cytoplasm. Calculating the movement of ions on a plasmamembrane area basis in the root epidermis and cortex of *Suaeda maritima*, Yeo and Flowers concluded that ion uptake into the root symplast was more likely to take place across the full cortex than just at the root epidermis. They also concluded that, at high salinities, the availability of ions for transport to the shoot limited the growth rate. The supply of ions determines the volume of tissue which can be osmotically adjusted to maintain water potential gradients.

The root epidermal/cortical plasmamembrane must form a significant barrier to the uptake of ions, being relatively impermeable to them. Membrane permeability is likely to result from carrier and channel proteins present in the membrane which determine both the rate at which ions can pass and membrane selectivity. Little is known about individual carriers/channels operating in halophytes, but quantitative and qualitative differences in their complement between species probably has marked effects on salt tolerance. Many dicotyledonous halophytes take up much larger quantities of sodium than many monocotyledonous species, which – as noted previously – show much stronger selectivity for potassium.

3.3.5 REGULATION OF UPTAKE AT THE SHOOT

Maintenance of relatively constant shoot ion concentrations would suggest that shoots operate as homoeostatic systems, minimizing the effects of transient changes in external salinity. It has been proposed that in dicotyledonous halophytes at least the homoeostat is based on maintenance of turgor through regulation of cell water potential (Clipson *et al.*, 1985; Yeo and Flowers, 1986). Ions are delivered in the transpiration stream to the shoot cell apoplast (cell walls), which is known to be of low volume and hence easily filled with salts, and are then transported across the plasmamembrane and tonoplast into the vacuole. Clearly, this has to be efficiently regulated so that shoot apoplastic ion concentrations do not build up to levels causing dehydration of the cell (Flowers *et al.*, 1991). Interestingly, when transpiration is measured in halophytes there appears to be little correlation between the amount of ion uptake and the amount of water moved through the xylem. Salinity generally reduces transpiration rate when expressed per unit of leaf area (Flowers, 1985).

Many halophytes tolerate high internal salinities by close regulation of ion uptake and internal concentration. Other species have developed mechanisms which remove salt from their tissues, at least complementing internal ion regulation. The two main strategies both serve to remove salts from the plant, either by secretion from specialized structures on the plant's surface, or through the shedding of older, salt-laden parts.

Two forms of specialized salt-exclusion structures are present in halophytes. **Salt glands** have been reported in several families including the Poaceae (Adam, 1990); **salt bladders** occur in the Chenopodiaceae. Examples of salt glands and bladders are shown in Figure 3.7. The simplest salt glands, found in the grasses, may consist of only two cells, whereas multicellular structures occur in dicotyledonous species. Their main function is to secrete salt solution onto the leaf surface; in those species with salt glands, the presence of white crystalline salt on leaf surfaces is a frequent characteristic. The mechanisms by which salt is secreted are not fully understood, but the secretion process is probably energy-dependent. Although salt secretion is widespread among many plant families, particularly those with mangrove members, most halophytes do not have secretory mechanisms (Adam, 1990). Salt bladders are often two-celled structures, a basal stalk and a salt-accumulating bladder cell. When the bladder cell fills with salt it dies or bursts and salt is lost from the plant.

A second potential route for the removal of ions from the plant interior is their sequestration in older parts before their shedding or death. This would imply that ions have to be relocated by retranslocation in the phloem. Although this has been observed in glycophytes and moderately salt-tolerant species (Winter, 1982), retranslocation is not considered a significant mechanism in halophytes. Older leaves of halophytes will obviously be lost as they age, but it is not thought that they are specifically loaded with salt before loss.

3.3.6 MOLECULAR REGULATION OF HALOPHYTE SALT TOLERANCE

Although it is clear that the physiology and biochemistry of salt tolerance in halophytes are controlled by a large number of genes, very few have been identified or their functions determined. Studies have been restricted to a few species including a salt-tolerant relative of wheat *Lophopyrum elongatum*, the chenopod *Atriplex nummularia*, and the CAM plant *Mesembryanthemum crystallinum*. Although these plants are halophytes, they are not strictly salt-marsh species.

In *L. elongatum*, a complementary DNA (cDNA) bank was made from salt-stressed roots, which was then used to identify genes expressed during acclimation to salt. Eleven genes were found to be induced in the first few hours of exposure (Gulick and Dvorak, 1992), although their function was not established. Studies with *A. nummularia* using specific cDNA probes to determine H^+ATPase activity through mRNA abundance showed that induction of H^+ATPase was regulated as a response to salt. This was much more pronounced than in the glycophyte tobacco and it has been postulated that this response to salt by the halophyte may be a salt-tolerance determinant (Niu *et al.*, 1993). *M. crystallinum* has probably been the most studied species at the molecular level, with a range of genes induced by salt stress, including the genes *Imt1* and *Ino1* for polyol synthesis, and a

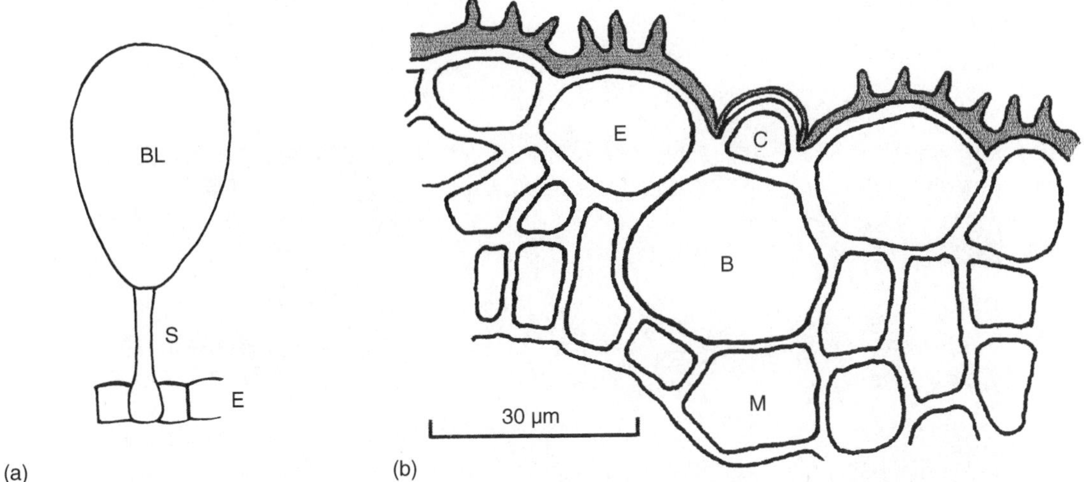

Figure 3.7 Examples of the cell structure of salt-secretory tissues in the leaves of halophytes. (a) Diagram of a young salt bladder (salt hair) from *Atriplex* sp. In others Na^+ and Cl^- are transported via the stalk cell (S) into the bladder cell (BL). The stalk cell is rich in mitochondria, endoplasmic reticulum and vesicles, and has numerous plasmodesmatal connections to both the bladder cell and other adjoining cells. As the balloon-like bladder cell matures it enlarges, then possessing a volume many times that of the stalk cell in such species as the Australian saltbush *A. spongiosa*. The bladder eventually bursts or falls off, removing salt from the plant. In *A. canescens* the unicellular 'vesicular hairs' consist of an expanded bladder with a narrowed stalk-like attachment. (b) Salt gland from *Spartina anglica*. The two-celled gland has a small cap cell (C), and a much larger metabolically-active basal (B) or collecting cell abutting on the mesophyll (M). The epidermis (E) has a thick papillate cuticle. (Drawing by A.J. Willis, from a preparation by Dr L.I. Técsi.) Other halophytes have more elaborate and characteristic salt glands. In dicotyledonous plants the glands are multicellular with up to 16 cells in those of some members of the Plumbaginaceae.

Mip (major intrinsic protein) gene possibly implicated in water transport (Bohnert *et al.*, 1995).

Molecular studies on salt tolerance in plants are still in their infancy. As they develop, they will allow a greater understanding of the role of genes in physiological and ecological aspects of salt tolerance. Interestingly, halophytic genes may be of value in the improvement of salt tolerance in glycophytes. A gene coding for the enzyme mannitol 1-phosphate dehydrogenase, which is involved in the synthesis of the compatible solute mannitol in fungi, has been transformed into tobacco. Transformed tobacco which had the ability to produce mannitol had a greater salt tolerance (Tarczynski *et al.*, 1993).

3.4 WATERLOGGING AND SOIL ANAEROBIOSIS

Inundation by the tide affects halophytes largely through changes in soil aeration and soil chemistry (Figure 3.8). Tidal currents may have mechanical effects on plants, particularly in the lower part of salt marshes, but the presence of thick cuticles on most salt-marsh species is likely to prevent direct entry of salt into the plant. The immediate effect on halophyte physiology is flooding of soil pores affecting the availability of oxygen for aerobic root processes and producing reduced toxic ions.

The physiology of waterlogging tolerance has been quite widely studied in plants of freshwater marshes, but much less is known

Figure 3.8 *Zostera marina* (eelgrass) exposed at low tide, Roscoff, northern Brittany, France. This hydrophytic halophyte is fully adapted to life in water and may be severely damaged by prolonged exposure at low tide. (Photograph by M.C.F. Proctor.)

about salt-marsh plants. Cooper (1982) tested the effect of saline and waterlogging treatments on a range of halophytes of different salt tolerance and found that the more salt-tolerant species such as *Plantago maritima*, *Puccinellia maritima* and *Salicornia europaea* were particularly tolerant of salinity and waterlogging in combination, whereas species generally found higher up the marsh including *Festuca rubra*, *Juncus gerardii* and *Armeria maritima*, were less so.

In many freshwater plants, the presence of air-bearing tissue (**aerenchyma**) has been regarded as an important mechanism avoiding root anoxia. Intercellular lacunae in the aerenchyma of roots allow diffusion of air from the shoot which also has abundant aerenchyma in many wetland plants. Up to 60% of root volume can be intercellular in these species, which itself substantially reduces the amount of oxygen required for root respiration on a volume basis (Adam, 1990). In halophytes, extensive aerenchyma has been reported in some species, e.g. *Spartina alterniflora*, *Distichlis spicata* and *Juncus roemerianus* (Anderson, 1974), but not in others, e.g. *Atriplex portulacoides* and *Elytrigia atherica* (Armstrong *et al.*, 1985). In freshwater plants, synthesis of the plant growth regulator ethylene has been implicated in the production of aerenchyma (Jackson, 1985), but it is not clear whether this is involved in the production of aerenchyma in halophytes.

Many mangroves are characterized by specialized aerial roots (pneumatophores) which aid aeration. It has not been shown whether aerenchyma can provide sufficient oxygen to prevent the onset of anaerobic metabolism fully in roots. Tolerance of anaerobic metabolism itself may be important for survival under waterlogged conditions. Waterlogging brings about changes in soil chemistry, particularly causing reducing conditions. The effects of soil toxicity are considered in section 4.5. Such features and plant responses to waterlogging in both saline and non-saline conditions are fully described by Marschner (1995).

3.5 PLANT ADAPTATION ON SAND DUNES

A major determining factor of the vegetation of sand dunes is the substrate of loose, moving sand, often deficient in essential mineral elements and water. The influence of wind-borne salt is of much less significance than salinity in salt marshes, except on embryo dunes and the seaward face of the foredunes. The general character of the vegetation of coastal dunes is similar to that of inland dunes and to dunes bordering inland freshwater lakes, as in the Great Lakes of North America. The instability of the sand in young dunes leads initially to sparse vegetation; lack of anchorage is frequently exacerbated by limited water supply as well as lack of mineral nutrients in some dune systems. Productivity of old dry dunes is usually substantially lower than that of low areas (slacks) near the water table.

Low levels of macronutrients restrict the growth of a number of species of high relative growth rate (RGR) in many dune systems; this feature, often combined with the influence of grazing, seems largely responsible for the high species diversity seen in such dune systems as Braunton Burrows, North Devon, UK (sections 6.5 and 6.6). Local variations in soil nutrient concentration can lead to marked differences in performance. Low levels of nitrogen, rather than potassium or lack of water, restrict the growth of the two annual halophytes *Cakile maritima* and *Salsola kali* on the foredunes of Morfa Harlech, West Wales, where these plants are small and stunted. This is in marked contrast to the large, rapidly growing plants of these two species found along the strandline, where the concentration of nitrogen in the sand – here largely derived from macro-algal litter – is much higher than in the foredune. Addition of nitrogen led to substantial increase in growth in the foredune populations; there was sufficient phosphorus in the foredune sand to support the enhanced growth of the nitrogen-fertilized plants (Pakeman and Lee, 1991a,b).

The high porosity of sand and its poor water-holding capacity, with its common deficiency of organic matter, coupled often with high evaporation rates, result in water shortage, reflected in distinctive features of morphology and anatomy of dune plants. These **psammophytes** often have thick cuticles, rolled leaves, water storage tissue (succulence), e.g. *Euphorbia paralias*, *Sedum acre*, a spreading growth form with rhizomes or runners and distinctive root systems. Cuticles tend to be thick on the lower (abaxial) leaf surface, e.g. in *Ammophila arenaria* the abaxial epidermis has a cuticle about 8 μm thick, that of the upper (adaxial) epidermis being about 3 μm. In *Elytrigia atherica* cuticular thicknesses are very similar to those of *A. arenaria*, but in the succulent *Euphorbia paralias* are much greater, about 30 μm (abaxial) and 15 μm (adaxial). Thick cuticles may protect from sand-blasting as well as limiting water loss. Leaf rolling in *A. arenaria* appears to restrict transpiration losses (Willis and Jefferies, 1963) even though stomata possibly remain open in the rolled leaf (Rutter, 1981). Substantial reduction (50% or more) in transpiration rate as a result of leaf folding has been reported in the sand-binding grass *Cynodon dactylon* as well as in other species (Walter, 1971).

Salisbury (1952) illustrated the roots of a range of dune species; average rooting depths of annuals are recorded from 5 cm (*Erophila verna*, which like other winter annuals, grows when evapotranspiration tends to be low) to 20 cm (*Anagallis arvensis*, which is occasionally perennial). In contrast, small herbaceous perennials such as *Viola* spp. may have roots extending more than 1 m. Those of larger perennials reach even greater depths, e.g. *Eryngium maritimum*, *Galium verum* and *Ononis repens* (see section 6.3 which discusses the effects of water on plant distribution in dune systems). Root/shoot

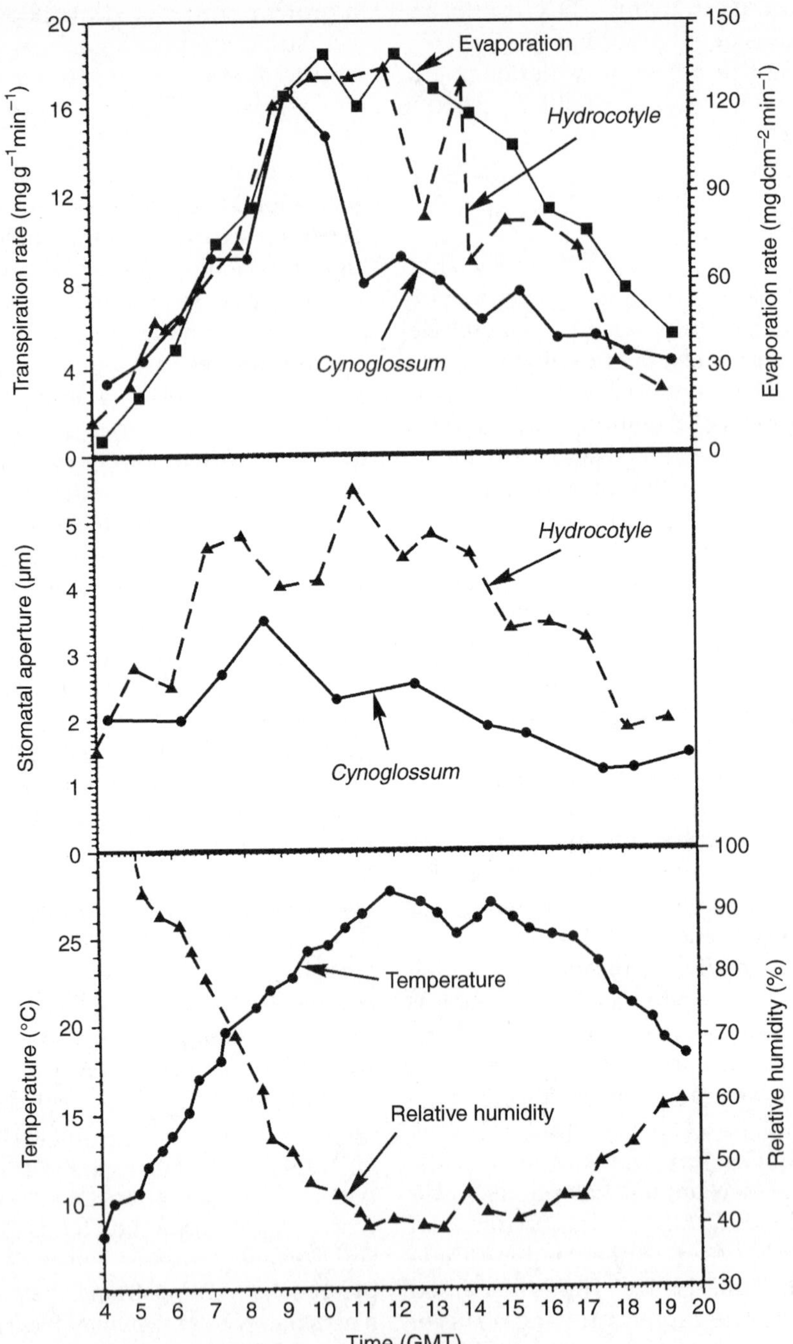

Figure 3.9 Diurnal changes in transpiration rate and stomatal aperture in *Cynoglossum officinale* and *Hydrocotyle vulgaris* on 21 June 1960 at Braunton Burrows, North Devon. The upper part of the figure also shows evaporation rate; the lower part shows temperature and relative humidity. (Data of A.J. Willis.)

ratio is higher in many psammophytes than in mesophytes and osmotic potentials also tend to be more negative (though not usually lower than about –2 MPa as in halophytes and desert plants). Both features favour survival in droughting conditions.

3.5.1 TRANSPIRATION IN DUNE PLANTS

Under conditions of high evaporation in the summer, plants of both the dune slopes and the slacks often show quite high transpiration rates; typical values are about 15–20 mg g fresh wt^{-1} min^{-1} (Willis and Jefferies, 1963). In *Senecio jacobaea*, the water content of the leaf may be completely replaced in 45 minutes. During the day its leaf water deficit may reach some 15–20% until stomatal regulation substantially restricts losses; the water deficits are frequently made good overnight unless droughting is prolonged.

Many dune slope and slack plants show contrasting stomatal behaviour under conditions of high evaporation, as illustrated by the dune slope species *Cynoglossum officinale* and the slack species *Hydrocotyle vulgaris* (Willis and Jefferies, 1963). Measurements of the transpiration rates of these species, as well as of the evaporation rate, were made in the normal 'field' conditions of their habitat at Braunton Burrows, North Devon, UK, on a hot dry day in June (Figure 3.9). Stomatal behaviour was also monitored. At dawn, temperature was about 8°C, relative humidity maximal, evaporation and transpiration rates very low and the stomata of both species largely closed. Until about 10 a.m. the transpiration rates of both species, with progressive stomatal opening, strongly paralleled the rising evaporation rate as relative humidity rapidly decreased and temperature and irradiance increased. After 10 a.m., however, although transpiration rate and stomatal aperture in *Hydrocotyle vulgaris* continued to parallel evaporation rate, stomatal closure in *Cynoglossum officinale*, which set in soon after 9 a.m., led to a sharp reduction in transpiration losses. Cuticular transpiration rates in dune species are fairly small; in plants such as *Senecio jacobaea* and *C. officinale* they are about 0.6–0.9 mg g fresh wt^{-1} min^{-1}, rather less than 10% of the total transpiration (Willis and Jefferies, 1963).

In warm dry weather on clear sunny days, temperature at the ground surface in dunes can become lethally hot (≥50°C); interactions of water vapour loss and heat transfer are very important under these conditions (Gates, 1980). Bryophytes dry out and then survive high temperatures. Leaves of *Tortula ruralis* ssp. *ruraliformis* quickly become tightly spirally infolded in dry weather, photosynthesis virtually ceasing. This xerophytic moss is **poikilohydric**, its water content quickly equilibrating with the environment; on moistening, the leaves untwist and expand, and photosynthesize almost immediately (Willis, 1964). Plants which are fairly tall-growing, projecting above the inhospitable boundary layer of the sand surface, such as *Senecio jacobaea* and *Cynoglossum officinale*, can obtain adequate convective cooling and restrict water loss by stomatal regulation. In contrast, low-growing species need to transpire if overheating is to be prevented. *Hydrocotyle vulgaris*, being normally within reach of the water table, has abundant water supply and keeps stomata open; in exposed sites in full sun the plants would overheat if transpiration was strongly reduced.

In times of drought, plants of the dune slopes may have an appreciable water deficit at dawn but, as in many desert species, the stomata open (facilitating photosynthesis), unlike those of many mesophytes and dune slack species. In this respect *C. officinale* and *S. jacobaea* behave as **euryhydrous xerophytes**, with a large range in osmotic potential, as compared with **stenohydrous** species (with a small range in osmotic potential) whose stomata close immediately with the onset of dry conditions (Walter, 1971). Dune plants with CAM photosynthesis (section 2.3), e.g. *Sedum acre* and *S. anglicum*, close their stomata during the day, with corresponding water economy, and appear to be fairly heat-tolerant.

SALT MARSHES: TIDES, TIME AND FUNCTION 4

4.1 MARSHES AND MANGROVE FORESTS

Maritime salt marshes and mangrove (**mangal**) associations are **wet coastal ecosystems** whose vegetation helps to stabilize the sediments in which it is rooted. Both are largely covered by phanerogamic vegetation and subject to periodic flooding by the sea; on many shores their landward limits are set by the highest point reached by the equinoctial spring tides. These formations provide the best examples of **haloseres** (plant successions developing on saline substrates; section 6.1), though others occur around inland salt lakes, saline springs, and in areas with saline ground water. Unlike salt marshes, mangals are dominated by trees or tall shrubs; both are most extensively developed where a gently sloping coast provides a large surface within the intertidal zone. Where the coast slopes steeply, as in many fjords and Scottish lochs, fringing salt marshes are narrow. In the Mississippi Delta and north-west of the River Amazon very large amounts of sediment are supplied to the coastal zone by rivers. Under these exceptional conditions a wide and shallow nearshore zone absorbs much of the incoming wave energy; muddy sediments accumulate on parts of the open coast and often develop into salt marshes or mangrove forests (Allen and Pye, 1992).

Where tidal ranges are very low, wind direction can be an important influence on inundation, and hence on the extent of saltmarsh and mangal communities. Tidal range on the extensive sandy coast of southern Brazil is only 50 cm and variations in the wind system are strongly influenced by seasonal movements of the dominant ocean currents (Davy and Figueroa, 1993). These lead to inundation of the beach in winter and spring; in summer north-easterly winds tend to expose it for long periods.

4.1.1 MANGROVE ASSOCIATIONS

Mangroves are trees with a canopy up to some 40 m tall, but in marginal habitats are low scrubby plants (Tomlinson, 1994). Their communities may be as much as several kilometres wide. Mangroves flourish on tropical coasts where the mean temperature in the coldest month does not fall below 20°C, though *Avicennia marina* var. *resinifera* tolerates a mean winter temperature of 10°C in New Zealand (Chapman, 1977). Mangrove plants are well suited to the exceptional conditions in which they occur. Some possess pneumatophores ('breathing roots'), while the stilt (prop) roots of *Rhizophora mangle*, for example, not only anchor the tree but aid the deposition of fine alluvium along lagoon and channel margins. Dansereau (1947) described a banded or zonal arrangement of associations inland from the sea or tidal channel margins and, in the New World, general descriptions often place *R. mangle* as the pioneer on the more saline outer margins of lagoons and tidal channels, followed inland by successive belts dominated by *Avicennia germinans* (= *A. nitida*), *Laguncularia racemosa* and, in places above tidal influence, *Conocarpus erectus*. Such zones have frequently been regarded as representing a developmental zonation relating to a complex of factors, notably salinity gradients, tidal exposure, and substrate characteristics. Old World mangals are more species-rich and such zonal distributions less pronounced.

West (1977) doubts the value of applying the concept of developmental zonation to mangal formations, quoting the work of Thom (1967) which showed that in dynamic deltaic–lagoonal environments such as those of Tabasco, south-east Mexico, a simple zonation of mangal communities does not exist. In this deltaic plain a definite pattern of succession is questionable, other than on a short-term basis, and Thom considered long-range trends to be dictated by continually changing physiographic processes which influenced such phenomena as amount of water, soil type and drainage of the surface. Active **distributaries** (river channels reaching the sea – or coastal lagoons – independently) are flanked by low natural levées extending into water bodies as features which are at first submerged, but later build up and become colonized by plants, as do the interdistributary basins.

In this geomorphologically complex situation mangrove species show a definite association with landform and substrate type (Figure 4.1). *Rhizophora mangle* grows best on moist sites with deep, chemically reduced organic litter and peat, and usually lines the stable shores of lagoons and the edges of abandoned distributary channels. It flourishes where salinity is sufficient to reduce competition. *Avicennia* occurs on higher, drier habitats whose soils show marked oxidation and also tend to have very high soil water salinities. Where interdistributary basins are slowly subsiding within the delta mixed mangrove forest with *Rhizophora*, *Avicennia* and *Laguncularia racemosa* is found. *L. racemosa* may also form dense stands on sandy soils near the landward limit of tidal influence, but *Conocarpus erectus* is relatively rare in the entire area. Long-term subsidence in the absence of sedimentation leads to the development and enlargement of water bodies.

4.1.2 SALT MARSH ORIGIN AND FORM

The substrate of most west European salt marshes consists mainly of mineral particles which vary in size from fine sand to clayey silt. Such **minerogenic marshes** contrast with

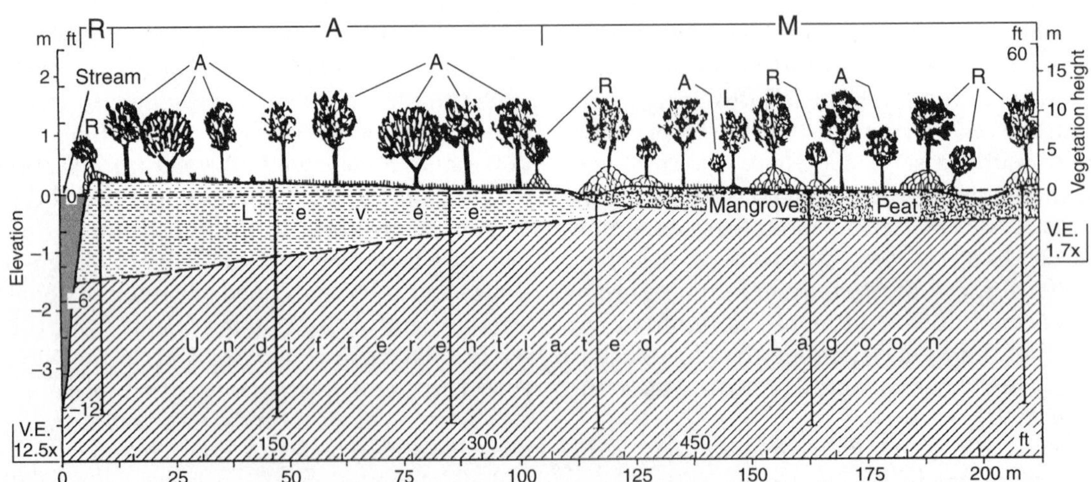

Figure 4.1 Mangal associations in levée and interdistributary basin habitats, west bank of lower Rio Cuxcuchapa, Tabasco, Mexico. Vertical lines represent bore holes. Note the use of differential vertical exaggeration for elevation and vegetation height. R, *Rhizophora*; A, *Avicennia*; L, *Laguncularia*; M, Mixed mangrove forest. (From Thom, 1967; courtesy *Journal of Ecology*.)

organogenic marshes in which a predominance of organic matter derived from plant litter and underground biomass leads to such formations as the peat salt marshes of New England, USA

Active **salt marshes** are mainly found where the direct action of storm waves is greatly reduced – on low-energy coasts, behind spits and barrier islands, in estuaries and shallow bays. They are classified by Allen and Pye (1992) into five types on the basis of their physical settings:

1. **Open coast marshes**. These occur in Essex on the Dengie Peninsula and on Foulness Island, UK, but are usually poorly developed in Britain because of the relatively high wave energy found along most of the coast.
2. **Back-barrier marshes**. Open-coast back-barrier marshes, usually in the shelter of spits or islands, are present in north Norfolk, south Lincolnshire, and at Culbin in Morayshire.
3. **Estuarine-fringing marshes**. Notable British examples include the Severn, Dee, Mersey, Ribble and Solway on the west coast, and the Medway, Thames, Crouch, Blackwater, Humber and Tay on the east coast.
4. **Embayment marshes**. Southern examples include the harbours of Portsmouth, Langstone, Chichester, Pagham and Poole. The Wash and Morecambe Bay Marshes are also regarded as variants of this type.
5. **Loch or fjord-head marshes** occur mainly on predominantly rocky coasts in north-west Scotland.

Salt-marsh ecosystems are particularly sensitive to changes in sea level and isostatic adjustments to the level of the land. Given a sufficiently high **accretion** (sedimentation) **rate**, however, they may form on coastlines which are rising, stable, or even sinking. Wet coastal ecosystems are the products of interactions between the sediments and organisms present, the sea, the winds and factors relating to the land. Thus, an interdisciplinary approach is required if they, and their histories, are to be understood. Allen and Pye (1992) describe the location, character and dynamics of salt marshes as being governed essentially by **sediment supply, tidal regime, wind–wave climate** and the **movement of relative sea level**.

Vegetation, which is unable to colonize sand or mud flats until their level has been raised to a suitable height relative to that of the tides, later plays an important role. Once established, it increases surface roughness, so slowing the water flow and promoting sedimentation (Figure 4.2). Leaves, stems and roots trap and bind sediment particles and vegetation ultimately contributes to the sediment itself, sometimes forming peat. Salt-marsh floras are described in section 4.4, and higher plant strategies in section 5.1. The importance of vegetation in influencing autogenic succession in salt marshes is illustrated by the **Great Marshes of Barnstable, Massachusetts, USA.**

These New England salt marshes have formed over the past 4000 years in an embayment limited to the north by the Sandy Neck sand spit (see Figure 8.15). They show all stages of development from the seeding of bare sand flats, through the establishment of intertidal marsh to the formation of mature high marsh underlain by more than 6 m of peat. Many aspects of the marshes are reviewed by Redfield (1972), who directly observed the first 12 years after colonization by *Spartina alterniflora* (which grows in the upper two-thirds of a tidal range of 2.93 m) and determined the time sequence of later stages by radiocarbon analyses. This work led to the production of a series of maps distinguishing sand dunes, intertidal marsh, high marsh and upland from 1300 BC onwards, as well as observations of variations in tidal height within the marshes and detailed calculations of tidal flows.

Zonation of the marsh at Barnstable is similar to that of other parts of the North Atlantic coast of the USA. The intertidal zone is occupied

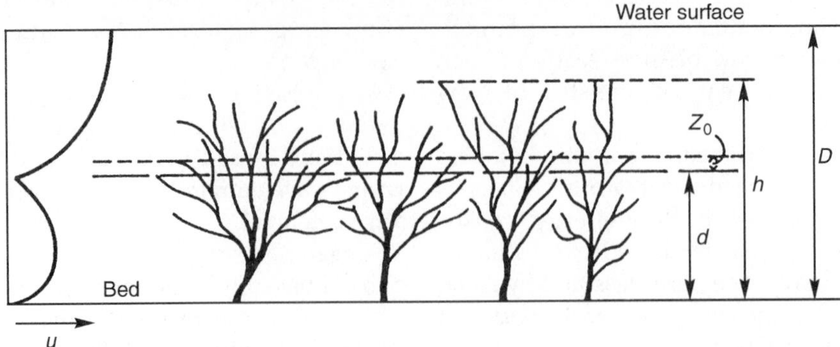

Figure 4.2 Schematic representation of variation in mean flow velocity (u) with height through a salt marsh boundary layer. D, water depth; h, height of vegetation; d, zero plane displacement; Z_0, roughness height. Vegetation increases surface roughness, enlarging the value of the **roughness length**, Z_0, and displacing it upwards by the height d, known as the **zero plane displacement**. Shear stress immediately above the Z_0 layer should increase, owing to a steepening of the velocity gradient, but most of the shear stress is taken up by the vegetation and little is transmitted to the bed. (From Pye and French, 1993b; courtesy of Cambridge Environmental Research Consultants Ltd.)

almost entirely by *S. alterniflora* which here, as elsewhere, exhibits both tall and short forms, the latter growing at or near high water level where the marsh becomes flat. Where the high marsh is sufficiently elevated and well drained, *S. alterniflora* is replaced by vegetation dominated by *Spartina patens* and *Distichlis spicata. Aster subulatus, Atriplex patula, Limonium carolinianum, Plantago maritima, Salicornia* spp., *Solidago sempervirens* and *Suaeda maritima* are scattered among these two species. Pure stands of *Juncus gerardii* occur at high levels along the border of the marsh or where its surface is unusually high. At the margins of the high marsh, where ground water emerges, are species tolerant of limited exposure to brackish water such as *Scirpus robustus, Typha* and, more locally, *Phragmites australis.* Pure stands of *Salicornia europaea* and *Salicornia bigelovii* surround bare salt pans and pathways where grasses have been eliminated by trampling (Redfield, 1972). Reimold (1977) gives a more general description of the salt marshes (and mangals) of the Eastern United States.

During the period of rising sea level in which the Barnstable Marsh and other New England salt marshes were formed, sediment is thought to have accumulated at their margins. When it did so at a rate in excess of the rise in sea level *Spartina alterniflora* established whenever the elevation of the sand flats exceeded the lower limit at which this plant could survive. This critical level increased in elevation as the sea level rose with the consequence that the basement sand, upon which rests a layer of intertidal peat of relatively uniform thickness, rises towards the outer limit of the marsh (Figure 4.3). Above the intertidal peat is the high marsh peat formed at the high marsh surface, which is developed at about mean high water and extends outwards as a wedge of decreasing thickness. The marsh also expanded on the inland side, however, and here high marsh peat frequently overlies sediments with traces of upland vegetation. Investigations of the hydrology of the marsh margin appear to confirm this hypothesis. Peats formed by the high marsh association contain 60–90% water, whereas intertidal peat contains 30–60% water; cores used to examine the structure of a section of marsh conformed to the pattern Redfield expected of the model described above. Subsequent research (section 4.6) using different methods has shown that other east coast

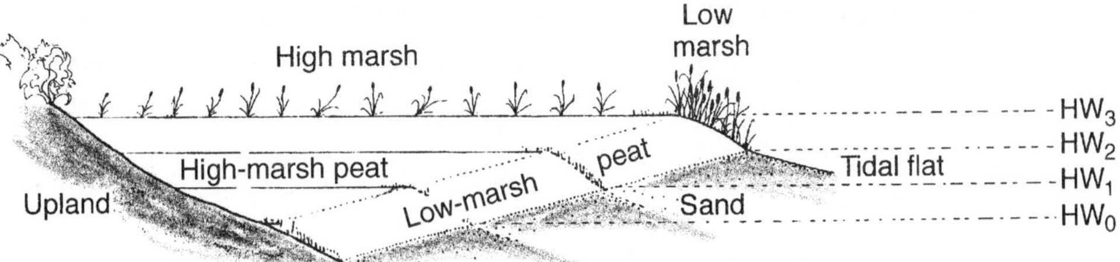

Figure 4.3 Theoretical structure of a 'New England-type' salt marsh. HW_0, HW_1, HW_2 and HW_3 are successive Mean High Water Levels over a prolonged period of rising sea level. The high marsh is extending as a result of (a) transgression over the upland environment as the sea level rises, and (b) high marsh development over areas previously supporting low marsh as deposition causes the level of the marsh to rise at a faster rate than sea level. (After Redfield, 1972, and Clark, 1986a.)

marshes have experienced severe and frequent changes within the last millenium.

4.1.3 TRANSITIONS FROM SALT MARSH TO TIDAL WOODLAND

The upper reaches of the Fal estuary in Cornwall, England, form one of the least disturbed examples of such a transition in Europe. Ranwell (1974) combined the evidence from historical records and maps; tree ageing estimates; levelling survey in relation to plant and animal distributions; current rates of silt accretion at the marsh and soil surfaces; and the stratigraphic record of plant remains in the soil. His conclusion was that salt marsh in this area has been succeeded by deciduous woodland within the last century, and that succession occurred at a rate clearly independent of changes associated with isostatic and sea level adjustments. Chlorinities of surface flood waters at the salt marsh/tidal woodland boundary during equinoctial tides were about one-tenth of those of seawater, while in tidal woodland dominated by *Alnus glutinosa*, *Quercus robur* and *Salix cinerea* ssp. *oleifolia* maximum recorded chlorinity round the tree roots was about one-twentieth that of seawater.

The equinoctial tidal range at the marsh to woodland boundary is reduced to 2.73 m; it is 5.5 m at Falmouth. In extreme landward estuarine situations like this, **reversals of the vertical zonation** normal in salt marshes are common. In this instance *Bolboschoenus maritimus* and *Agrostis stolonifera* occur on the banks of the main channel where they are influenced by diluted tidal water. *Puccinellia maritima*, normally found at a lower level than either, is the pioneer at higher general mudflat levels. Differences in level of not more than 0.7 m and often only 0.2 m suffice to separate salt marsh from tidal woodland. So small is this vertical difference that tree seedlings can often establish at the margin of the marsh on heaps built by ants (*Lasius* and *Myrmica* spp.), or on tussocks of *Deschampsia cespitosa* or the much larger *Carex paniculata*.

4.2 TIDAL CYCLES, INUNDATION AND ACCRETION

Patterns of tidal flows across salt marshes are related to the lunar cycle and to local topography. Tides in most areas are **semi-diurnal,** there being two high and two low tides in each day. As the period between successive low tides is c. 12 hours 25 minutes rather than 12 hours, these tides become about 50 minutes later each day. When successive high and low waters of semi-diurnal tides are of different heights the tides are said to be **mixed**. Where there is one high tide and one low tide per day the tides are **diurnal**. In the Solent, near the Isle of Wight, southern England, and a few other localities, four high tides and four low tides occur every day.

Figure 4.4 shows predicted tide heights at Milford Haven, south-west Wales, for three

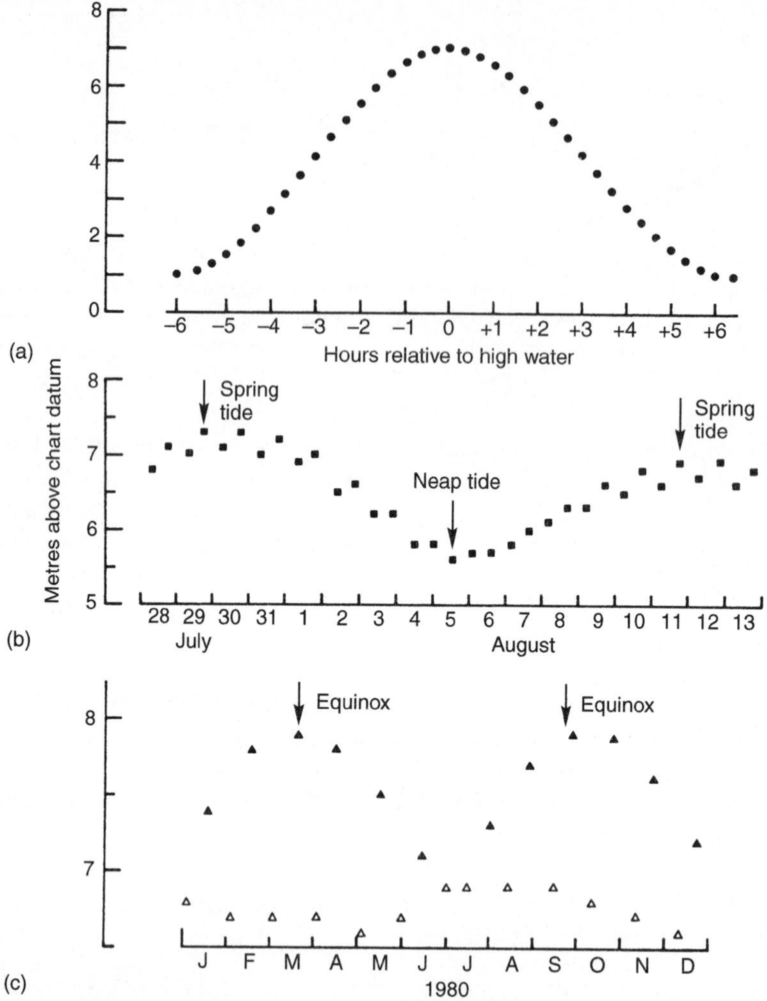

Figure 4.4 Predicted tidal heights in Milford Haven, south-west Wales. (a) Single tidal cycle (1 August 1980); (b) all high tides over 18 days (28 July–14 August 1980); (c) all high spring tides during 1980. Closed symbols show the highest spring tide of each calendar month; open symbols show the lowest of the two spring tides of each month. Chart datum at Milford Haven is 3.71 m below OD (Newlyn). (From Long and Mason, 1983.)

periods of increasing duration beginning with a single tidal cycle of 12 hours 25 minutes. During **neap tides**, which occur twice in a lunar month of 29 days, tidal range in any particular area is at a minimum; these tides occur when the sun and moon are at right-angles relative to the Earth. **Spring tides** occur when the Earth, sun and moon fall along an almost straight line and the combined gravitational attraction of the sun and the moon cause tidal range to reach a maximum. **Equinoctial spring tides** have on average the lowest low and the highest high tide levels of the year. At the equinoxes the sun is overhead at the equator, and the moon, Earth and sun are almost in a straight line, giving the highest spring tides around 21 March and 21 September. At the solstices, when the sun is high, the moon,

Earth and sun are not so well aligned and the lowest predicted spring tides accordingly occur around 21 June and 21 December. Because the Earth is closest to the sun during the northern winter, winter tides tend to be higher than summer tides and the autumn equinoctial tides are likely to be the highest of the whole year.

In coastal areas of unusual geometry, tidal curves may be highly modified and not of the symmetrical type shown in Figure 4.4(a). Curves also steepen going up estuaries, sometimes culminating in periodic **bores** (vigorously advancing floods of powerful waves and breakers, with an almost vertical front, travelling upstream) of which that on the River Severn, UK, is particularly well developed. In consequence, marshes high up estuaries often experience abrupt but relatively brief periods of submergence. The factors which determine the nature of the tides in any particular place are discussed by Pethick (1984). Actual tide levels occasionally differ considerably from those predicted; exceptional combinations of barometric pressure and onshore winds may result in **storm surges** which reach the coast (section 10.2).

4.2.1 TIDAL RANGE AND EXPOSURE

Tidal range varies greatly from place to place, being 0.1 m in the Black Sea, 0.3 m in the Mediterranean, c. 4 m around much of the coast of Britain and Ireland, but 6.8 m in the Wash, UK, 12 m in the Bristol Channel, UK, and up to 15.4 m in the Bay of Fundy, Nova Scotia. The influence of the tides on coastal vegetation is profound. Much depends on the relative duration of submergence and exposure. Exposure occurs daily at the top of the marsh even during spring tides, when a period of exposure also occurs low on the shore in a zone which is continuously flooded during neap tides. During neap tides, the shore above the level of high water neap tides is continuously exposed. The duration of such a period of continuous exposure – and whether it coincides with normal germination time – is critical in the establishment of plants from seed, especially for annual plants such as *Salicornia* spp.

Chapman (1960a, 1976) carried out levelling surveys on the marshes of Scolt Head Island, Norfolk, UK, and of Lynn, Massachusetts, determining the vertical ranges of the major species and communities and relating them to tidal levels. It was then possible to predict maximum periods of continuous exposure at each level and the expected number of submergences during daylight each day. The principal feature used to separate the lower from the upper marshes was the period of continuous exposure. This increases greatly from the top of the lower marshes to the bottom of the upper marshes, the boundary between the two lying near mean high water. At Scolt, the lower marshes commonly undergo more than 360 submergences per year, the maximum period of tidal exposure never exceeds 9 days, and mean daily submergence in daylight is more than 1.2 hours. In this gently sloping marsh the predicted period of continuous exposure can change from a maximum of 8 days at the top of the lower marshes to 16 days at an elevation only 6 cm higher at the bottom of the upper marshes. In practice, recorded tide heights differ from predicted values, such variations becoming highly significant when influenced by extreme weather conditions. Exceptionally high tides, such as that which caused major flooding in eastern England and the Netherlands in 1953, may influence salt-marsh distribution and result in major changes in coastal geomorphology.

Ranwell (1972) points out that data on zonation and numbers of algae and phanerogams at Scolt Head Island given by Chapman (1960a) do not show a really significant change at the proposed demarcation. However, Ranwell considers the distinction between **submergence marsh**, the lower marsh found between about mean high water neaps (MHWN) to mean high water (MHW), and **emergence marsh**, higher marsh situated above mean high water, to be valid. North

American salt-marsh species are known to be more tolerant of submergence in more southern sites; Ranwell (1972) considers this to be the result of higher light intensities and temperatures acting together to allow photosynthesis at deeper levels of submergence further south.

At the Bentlass salt marsh in Milford Haven, south-west Wales, originally described by Dalby (1970), the vegetation is distinctly zoned and has been divided into three regions by Long and Mason (1983; section 4.4). **Low marsh** extends from just above MHWN to roughly MHW, giving way to **middle marsh**, which yields to **high marsh** at MHWS (Figure 4.5). While the lowest part of the marsh is inundated by most of the 700 or so tides a year, high marsh at 7.5 m or more above chart datum will be reached only by the equinoctial spring tides.

4.2.2 MEASUREMENTS OF TIDAL FLOWS

Accurate measurement of tidal flows across salt marshes is extremely difficult, involving as it does both creek flow and mass movement across the general surface of the marsh. Carpenter (1993) discusses methods of measuring these flows, demonstrating the considerable potential for error when estimating instantaneous discharge from creeks using gauges – often placed at only one or two positions along the cross-section of the creek – to provide velocity readings when the water level is rising (flood tide) or falling (ebb tide).

The **integrating float technique**, as modified by Sargent (1981) and by Carpenter (1993), can conveniently be used to gauge discharge through a channel using the horizontal displacement of a buoyant float which is released from the creek bed. The float consists of air

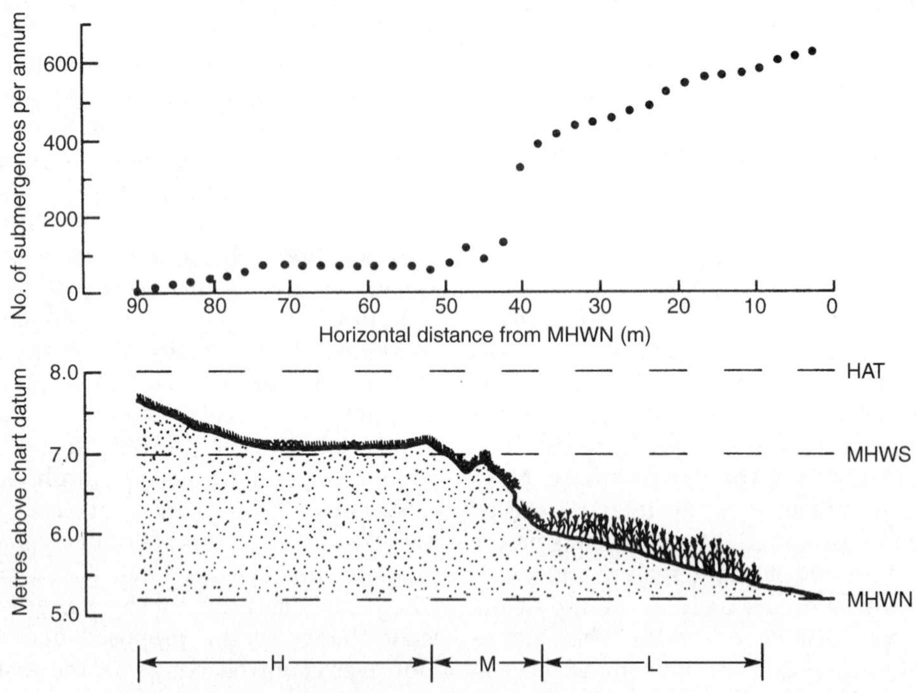

Figure 4.5 Profile of the salt marsh at Bentlass in Milford Haven, south-west Wales. Low marsh (L) dominated by *Spartina anglica*, middle marsh (M) dominated by *Puccinellia maritima* or mixed communities including *Armeria maritima* and *Plantago maritima*, and high marsh (H) dominated by *Festuca rubra*. (After Dalby, 1970). The number of submergences per year for points along the transect are indicated by the graph above the profile. HAT, highest astronomical tide; MHWS, mean high water spring; MHWN, mean high water neap. (From Long and Mason, 1983.)

bubbles of specific size released from nozzles in a plastic pipe laid across the bottom of the creek. Total displacement of the float – which is usually greatest in midstream where the water is normally deepest and the flow fastest – is related to water depth, water velocity profile, and the vertical velocity of the air bubbles as they rise through the water column. The positions of the surfacing bubbles are recorded by a camera on the channel bank; discharge across the stream can then be calculated using computer programs.

Tidal inequality involving substantial asymmetry between flood and ebb discharge has frequently been reported. Such asymmetries may result from real volume changes as when an ebb flow is either augmented by substantial land-derived inputs, as at the Sapelo Island marshes, Georgia, or reduced, as occurs when the first 'over bank' tide of a neap to spring tide cycle involves replenishment of water in salt pans and infiltration into unsaturated surface sediment. They may, however, be artefacts resulting, for example, from using a velocity meter in a fixed position. Ebb flows in a creek are concentrated near the surface, while flood flows are either vertically uniform or bottom intensified. Thus, regression between the velocity recorded at the position of the meter and the velocity of the whole creek section will not be constant for both halves of the tidal cycle. Error between actual and calculated discharge rates would differ for the flood and the ebb and might well indicate a tidal inequality even if the actual volumes were the same. Carpenter (1993) analyses the causes of tidal asymmetry – whether real or merely apparent – and concludes that in many marshes, unless major precipitation or evapotranspiration has taken place, it is better to equalize the flood and ebb flows in calculating fluxes of materials, such as sediments, heavy metals or nutrients, over a tidal cycle.

4.2.3 SAMPLING PATTERNS FOR NUTRIENT FLUX DETERMINATION

In **tidal material flux calculations** the amount of a substance leaving a salt marsh during the ebb tide is subtracted from that which enters during the flood tide. In order, for example, to calculate the amounts of phosphate, ammonium or nitrate being imported or exported it is necessary to determine both the volumes of water involved in the two flows and the concentrations of the nutrients within the water. Water in large creeks may not be well mixed so sampling should take account of spatial variation within the channel and temporal variation over an **ebb/flow cycle**. A wide spread of sampling stations is required when attempting to determine the pattern of fluxes within a marsh. The tendency is for younger **lower marshes to be net importers of nutrients** and for older upper marshes to be net exporters. Variation in nutrient cycling within such areas is related to biogeochemical factors such as redox potential and zonation of the plant communities.

4.2.4 DRAINAGE AND PORE WATER

Concentrations of soluble inorganic nutrients in pore water are much higher than in creek water. For this reason studies of the role in nutrient cycling of **groundwater seepage** from salt-marsh sediments into tidal creeks, which have largely been conducted on the east coast of North America (Harvey *et al.*, 1987; Nuttle and Hemond, 1988), are of particular interest. Carpenter (1993) considers seepage to be of even greater importance in British marshes whose morphology and climate are different. **Water-table measurements** can be made by installing perforated plastic tubes, usually arranged in a grid or along a transect, into augered boreholes. The spiral of holes in each tube is kept below the surface of the marsh so that water cannot flow directly into the capped tube during flood tides. The average hydraulic conductivity around each well can be determined by means of a pump test. Total water fluxes through the creek bank during the tidal cycle can be calculated using the hydraulic conductivity and the slope of the water table (hydraulic gradient). Such **wells** can also be used as sources of pore water for chemical analysis. (A **piezometer** differs from a well in that it records the hydraulic pressure of pore water at the specific depth at which its opening is set.)

Figure 4.6 shows changes in the position of the water table, determined at low water, between an over-marsh tide and a neap tide 5 days later. The over-marsh tide substantially recharged the water table, leaving it almost at the surface in the middle of the transect but curving downwards towards the creek channel on the left, the sides of which drained rapidly at low tide, and a smaller channel slightly to the right of the region shown.

4.2.5 INUNDATION, SOIL AERATION AND REDOX POTENTIAL

Soil aeration, one of the most important of the environmental features affecting vascular plants, is greatly influenced by the tidal cycle and there have been many attempts to measure the extent and rate at which it alters. Chapman (1976) summarizes his investigations of **movements of the soil water table**, to which soil aeration is inversely related, by measuring the water levels in blind-ending tubes sunk into marsh soils. The tubes had side openings through which water entered as the water table rose; the water level was measured by removing the capping cork at the top of the tube and using a graduated stick. Chapman's results led him to believe that even during a flooding tide the water table never rises completely to the surface. Other workers have since shown that such an aerated layer is not a general feature of salt-marsh soils. Chapman showed by analysis that oxygen concentrations in bubbles of gas released by agitation from the soil of subsurface zones in Norfolk salt marshes (UK) are much lower than in normal air, while CO_2 content is greatly increased. The soil water table in emergence marsh soils gradually rises during a sequence of spring tides, only to fall again during the neaps. Upon this fortnightly cycle are superimposed the effects of single tides, whose influence is mainly upon areas of marsh immediately adjacent to the creeks, into which water begins to drain once the ebbing tide has lowered the surface of the creek sufficiently.

Polarographic techniques pioneered by Lemon and Erickson (1955), were used by Packham, Willis and Poel (1966) in grassland soils for periodic measurements of soil oxygen

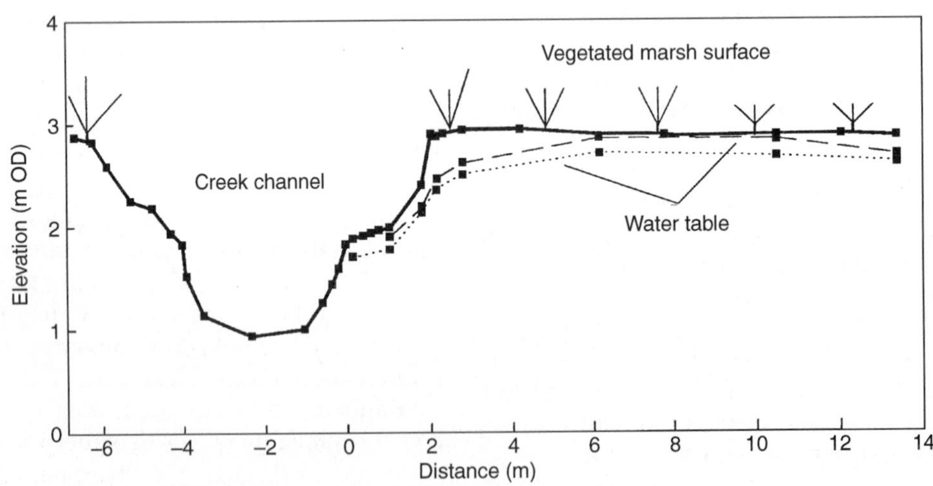

Figure 4.6 Changes in the position of the water table at Stiffkey Upper Marsh, Norfolk, UK, determined at low water in August 1992, between an over-marsh tide and a neap tide 5 days later. There was a minor channel to the right of the region shown. ···■··· , neap water table; ––■–– , spring water table. Elevation is in m (OD). (From Carpenter, 1993.)

diffusion rates, which were inversely related to soil moisture content and of considerable importance in influencing species distribution. Armstrong *et al.* (1985) used permanently installed platinum microelectrodes to measure redox potential as an index of soil aeration at various depths in the well-zoned ungrazed salt marsh at Welwick in the Humber estuary, Yorkshire. (**Redox potential** is a quantitative measure of ability to gain or lose electrons, and hence, respectively, to be reduced or oxidized. For further discussion of the use of Pt microelectrodes to measure redox potential and soil oxygen diffusion rates, see Armstrong, 1982.)

The results obtained from Welwick Marsh indicated that very substantial fluctuations in redox potential can occur over short periods. Potentials of 600 mV are characteristic of very well-aerated soils. There is little free oxygen in soils at potentials lower than 200 mV, and strongly reducing anaerobic salt-marsh muds often have potentials more negative than –100 mV (at which stage sulphate is being reduced). The rate at which redox potential drops on flooding and rises when flood cycles have passed indicates that oxygen *per se* has the greatest single influence on these fluctuations.

The three main soil aeration patterns which emerged are illustrated in Figures 4.7–4.9. At *Spartina anglica* sites reducing conditions were present throughout much of the profile; only near the surface (<5 cm) and at neap tides did phases of oxidation occur. At *Puccinellia maritima* sites, creek bank (*Atriplex portulacoides*) and *Elytrigia atherica* sites, high spring tides caused a monthly lowering of redox potentials. In creek bank and *E. atherica* sites potentials rarely fell below 200 mV, and then only in response to high or very high spring tides. In the **general salt marsh** and its sub-sites (including *Festuca rubra* hummocks) there was a third pattern in which long periods of oxidation were interrupted by very high spring tides only.

Oxygen availability in many salt-marsh soils is so depleted that there is a progressive development of bacterial populations which utilize **electron acceptors** other than oxygen for respiratory oxidation, thus reducing various inorganic and organic materials. Redox potential becomes ever lower in a sequence in which nitrate is reduced to ammonia, manganic ions are altered to the manganous state, and ferric ions change to the ferrous state. At still lower redox potentials sulphate-reducing

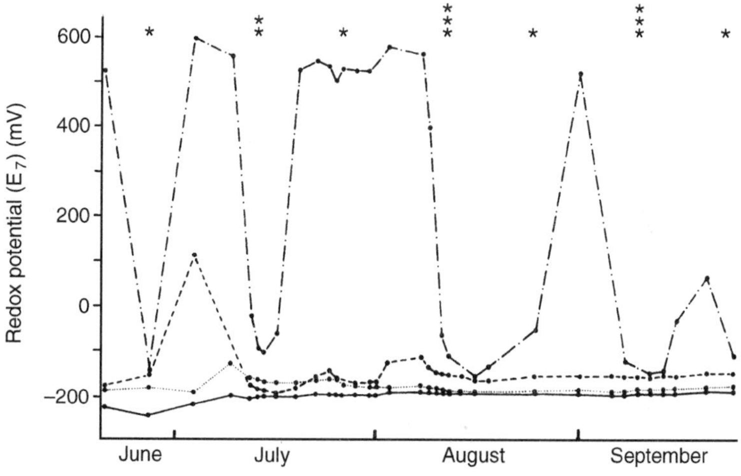

Figure 4.7 Fluctuations in soil redox potential between June and September 1979 in a *Spartina anglica* sward at Welwick Marsh, Humber Estuary, UK. Measurements taken at •-·-• at 1.5 cm; •---• at 5 cm; •····• at 10 cm; •——• at 30 cm depth. *, low spring tides (≤6.9 m above chart datum); **, high spring tides (≥7.1 m above chart datum); ***, very high spring tides (≥7.3 m above chart datum). (Redrawn from Armstrong *et al.*, 1985.)

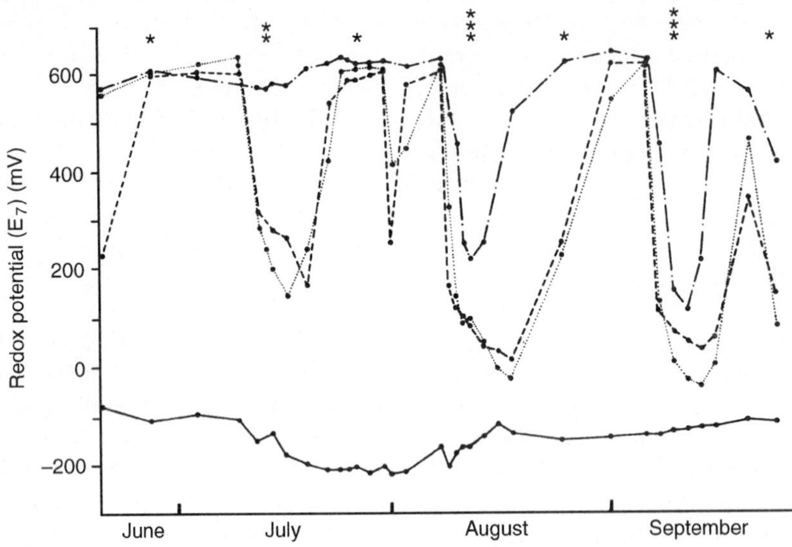

Figure 4.8 Fluctuations in soil redox potentials between June and September 1979 in a sward at Welwick Marsh dominated by *Puccinellia maritima*. Symbols as for Figure 4.7. (Redrawn from Armstrong *et al.*, 1985.)

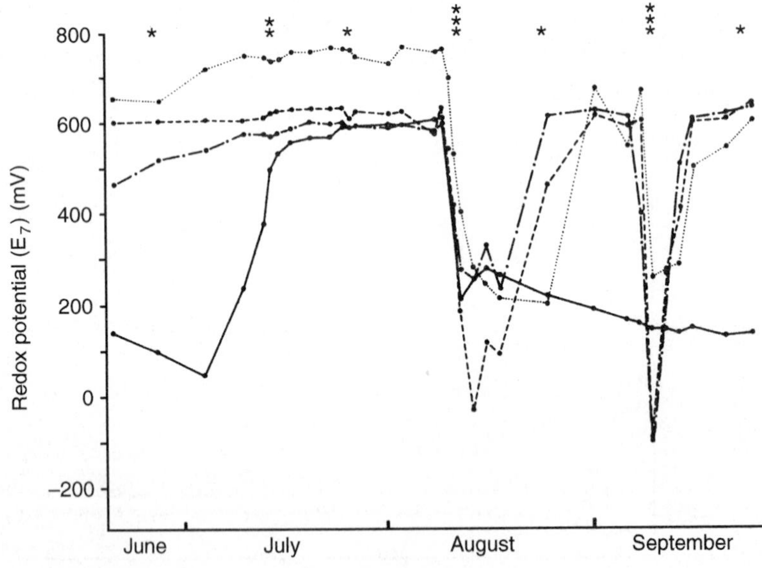

Figure 4.9 Fluctuations in soil redox potential in the general salt marsh community at Welwick Marsh for June to September 1979. Symbols as in Figure 4.7. (Redrawn from Armstrong *et al.*, 1985.)

bacteria produce toxic **sulphide**, only some of which is converted to pyrite (FeS_2) in salt-marsh soils and whose influence on plant distribution is discussed in section 4.5.2.

4.2.6 PATTERNS OF ACCRETION AND EROSION

Under suitable circumstances accreting salt marshes can both gain in height and **prograde** (increase in seaward extent). Suspended soil colloids often undergo physicochemical deposition caused by the high concentration of sodium in seawater. Mass flows of turbid tidal waters over salt marshes are slowed by vegetation causing sediment to settle out (Figure 4.2). Rates of sedimentation can thus be expected to be inversely proportional to elevation, because the upper marsh receives fewer tides and much of the sediment has been trapped in the lower marshes before reaching the higher levels [in several marshes, e.g. Scolt Head (Chapman, 1976, p.96) this has proved to be so]. Similarly, overbank flow from creeks initiates the formation of creek levées. Moreover, coarser sediment is likely to settle out first so the lower marsh and creek levées can be expected to have a higher proportion of sand than in the upper marsh where active sedimentation involves predominantly silt and clay fractions. In practice the highest rates of sedimentation are often achieved at the lowest level of continuous vegetation.

Stumpf (1983) suggests that sediment initially impacted onto vegetation during the short period of slackwater at high tide may then reach the marsh surface by being washed off by rain, deposited in the faecal pellets of gastropods grazing on the vegetation, or by the collapse of dead leaves and stems. Up to 50% of the material lost from suspension in the small Delaware marsh, USA, that Stumpf investigated could be accounted for by retention on *Spartina alterniflora*. Salt-marsh accretion is fully discussed by Adam (1990), who shows that while long-term averages suggest a relatively simple inverse relationship between accretion and elevation, repeated measurements at short intervals provide a much more detailed indication of the factors causing seasonal variation in local patterns of accretion and erosion. In the long term, climatic change will lead to further interest in vertical accretion in marshes with varying rates of sea level rise, a subject discussed by Stevenson *et al.* (1986) and by French (1993).

French and Spencer (1993) investigated the sedimentation dynamics of Hut Marsh, a tide-dominated back-barrier salt marsh on Scolt Head Island, using **buried marker horizons** and **surface sediment traps**. On this marsh, retention of sediment on plant surfaces is minimal, 95% of total deposition being by direct settling. Over a 5-year period annual accretion varied from 8 mm yr^{-1} adjacent to larger channels to <1 mm yr^{-1} on the highest surfaces remote from the creek network. Suspended sediment 'exhaustion' occurred along extended pathways of water movement, while the underlying link between predicted tidal height and sediment deposition was occasionally disrupted by unusual combinations of winds and barometric pressure and by resuspension of muddy sediments within the creek system. Aperiodic storm events accounted for a significant fraction of long-term sedimentation on the highest surfaces; this should be allowed for in **sediment budget calculations**.

4.2.7 BIOTIC INFLUENCE ON THE STABILITY OF SALT-MARSH SEDIMENTS

Fixation of salt-marsh sediments, which is not effective until they have undergone compaction and been colonized by flowering plants, often commences with the activities of microalgae, of which filamentous forms are commonly present in the upper 2 cm of apparently bare sand. Experiments by Ginsberg *et al.* (1954) demonstrated that a surface mat of sediment 4 mm deep could be established within 1 day by the cyanobacterium *Phormidium*. Coles (1979) showed that in sheltered areas organisms can strongly influence the type of sediment on the upper intertidal zones. On the salt marshes and upper mud flats of the Wash, eastern England, which have large populations of benthic microalgae, there is more or less continual net accretion of fine sediments. There are, however, no appreciable net gains of sediment on the sand flats, where grazing by macroinvertebrates keeps numbers of benthic microalgae low. Motile **epipelic** (mud-dwelling) **diatoms** – the most abundant of the microalgae in this situation – produce copious mucus which forms a surface layer that 'traps' fine sediments; stabilization is also assisted by release of mucus as the algae migrate through freshly deposited sediment. In experiments where the diatoms were killed with bleach or formalin, the mucus disintegrated and accretion of fine sediment – previously observed relative to a marker of white silica flour – ceased on the salt marsh and mud flat plots concerned.

Coles (1979) also investigated the feeding behaviour of *Hydrobia ulvae* and the amphipod *Corophium volutator*, among the most common macroinvertebrates on the inner sand flats of the Wash, UK. Both these deposit-feeders consumed large numbers of microalgae, and *Hydrobia* had a distinct preference for sediments with large numbers of epipelic diatoms. Reduction in population levels of epipelic diatoms caused by this continual grazing greatly reduces sedimentation, particularly of fine particles. Conversely, mud was found to accrete on sand flats when large microalgal populations built up following removal of macroinvertebrates. Mud-dwelling **deposit-feeders** (sections 2.7, 2.8 and 5.2) also have a direct effect in rendering the surface substrate less compact and liable to scour. In contrast, **suspension** (= **filter**) **feeders** – such as mussels (*Mya*) – enhance accretion by forming suspended clay particles into faecal pellets.

Hughes (1998) considers that recent increases in the rates of erosion experienced by the estuarine salt marshes of south-east England may, at least in part, be related to increased population levels of macroinvertebrates, particularly of the polychaete *Nereis diversicolor*. The **bioturbation** caused by *N. diversicolor* and *Corophium volutator*, both burrowing species, which bury seeds and spores – on which they may also feed – adds to the instability of the sediment, promoting erosion. In addition, both of these animals feed on epipelic diatoms and other microflora which secrete mucopolysaccharides serving to increase the shear strength of the sediment (Gerdol and Hughes, 1994). Experiments in both the field and the laboratory have also shown that the activities of *C. volutator* are important in preventing the establishment of *Salicornia europaea* at its lower limit (Gerdol and Hughes, 1993). Also adversely affected by *N. diversicolor* and *C. volutator* are *Spartina anglica*, *Zostera* spp. and sporelings of *Enteromorpha*. Plant colonization and succession on mud flat pioneer zones dominated by these macroinvertebrates appear to be retarded but, on the other hand, dense beds of *Zostera* and *Spartina* may deter burrowing so that two stable states, one plant- and one animal-dominated, may be present in the pioneer zone (Hughes, 1998).

Underwood and Patterson (1993) found epipelic diatoms to be the dominant microphytobenthos on the intertidal mudflats of the Severn Estuary, UK. During locomotion these organisms produced exopolymers (mucopolysaccharides) which enhanced the cohesion of sediment particles. These compounds were also produced by bacteria. Algal biomass (measured as chlorophyll *a* concentration), which varied seasonally, was strongly related with sediment shear strength, critical shear stress and position on the shore. Numerous further reports also emphasize the role of complex couplings between sedimentology, benthic biology and climate in modifying the behaviour of estuarine sediments.

4.2.8 SUBSIDENCE AND CHANGES IN MARSH ELEVATION

Surface elevation, an important factor in determining the future development of salt marshes, is increased by vertical accretion and diminished by subsidence resulting from compaction of shallow sediments. Accretion rates are usually determined by measuring the thickness of sediment laid down over a **marker horizon** of feldspar, coloured sand or some other material previously placed on the marsh surface. Variation in the elevation of the marsh can be measured by a **sedimentation–erosion table (SET)**; changes with time of such values at particular positions represent increases owing to accretion minus decreases caused by autocompaction of the sediment or to erosion (Reed and Cahoon, 1993). The base of an SET is an aluminium tube which is driven deep into the marsh and underlying sediments to give it a firm foundation and thus provide a reliable reference point. In the microtidal salt marshes of Louisiana, Florida and North Carolina it has been shown that, 2 years after installing an SET, shallow subsidence ranging from 0.45 to 4.9 cm had occurred, the greatest rate of subsidence being in the Mississippi delta (Cahoon *et al.*, 1995).

4.3 TOPOGRAPHY: CREEKS AND SALT PANS

The topography of a salt marsh at a particular time results from the interaction of the processes of deposition, erosion and sediment consolidation, of which the first two are most strongly influenced by the organisms living on the marsh (section 1.2). Two highly characteristic features of an unmodified salt marsh are the creeks and salt pans (Figure 4.10). The creeks are frequently supplemented by artificial ditches, such as those used to drain North American salt marshes in the interests of mosquito control.

Yapp, John and Jones (1917) listed four types of pan found in the Dovey Estuary, Wales, and elsewhere in Europe. **Primary salt pans** were described as developing while marshes are being formed. This process can still be observed in the young marshes of the Dovey Estuary where sand and silt accumulate around pioneer plants, such as *Puccinellia maritima*. Once a bare hollow has developed it tends to become gradually lower in relation to surrounding vegetated marsh. There is little to consolidate the substratum which swirls around these hollows as the tide ebbs, and increased summer salinity discourages plant colonization.

Secondary pans form in secondary marsh which commonly develops below and in front of erosion cliffs. **Creek** or **channel pans** are formed when vegetation dams off a minor creek or tributary, and residual pans develop when vegetation grows across and breaks up the bare area of any other form of pan. Sequences of elongated creek pans are not uncommonly connected by irregular drainage tubes, overgrown former creek bases through which water may seep quite rapidly. Pans form early in the development of salt marshes and it is often easy to stumble into the deeper portion of an old pan largely overgrown by *Bolboschoenus maritimus* or some other tall species.

'**Rotten spots**' form a further type of pan, notable in the Atlantic marshes of the USA from Virginia northwards. Here, they appear to result from vegetation die-off in areas where snow persists regularly for longer than average (as in some New Hampshire marshes), where the soil is continually waterlogged because the ground is slowly depressed without outlet, or where tidal litter is left lying for many months (Chapman, 1960b,1976). The final situation results in a **trash pan**; forms with this origin have been described from Bridgwater Bay, Somerset (Ranwell, 1964) and the Norfolk Marshes (Pethick, 1974).

Pethick (1974) analysed the distribution and densities of salt pans in 75 random quadrats (each of 1000 m^2) in the area between Scolt Head island and Blakeney Point. As Figure 4.11 demonstrates, pan density was positively related to marsh height and negatively to distance from the seaward margin of the marsh (2.05 m OD is the mean surface height of the 75 quadrats). Yapp, John and Jones (1917) considered that primary pans were formed as the

Figure 4.10 Salt pans on the Cefni Marsh, Anglesey, North Wales, UK, with the Cefni estuary in the background. (Photograph by M.C.F. Proctor.)

marsh itself developed. If so, their density would either stay constant or, more probably, decrease as the marsh aged, became higher and had some of its pans overgrown by vegetation. Pethick's results suggest that some factor operating subsequent to marsh formation increases pan density. This factor is most probably deposition of tidal litter by high spring tides, causing the development of trash pans. Such litter commonly stays in position long enough to create bare patches in the underlying vegetation; as long ago as 1904 Warming believed that turf could be broken through by the presence of 'putrefying masses of algae or *Zostera* or by the treading of cattle'.

The ground falls only 10 cm over a distance of 300 m or more in parts of these Norfolk marshes, so litter is most likely to be deposited on any high ground near the sea, or else carried right across the marsh by very high tides. There is still much to learn regarding salt pan formation and flux; long-term observations of fixed quadrats and artificial retention of tidal litter for varying periods of time would both yield valuable insights. In Norfolk, bait diggers searching for lugworms (*Arenicola*) often turn pans over and help to keep them open. The most bizarre agent of pan enlargement observed so far is a small terrier which moved steadily round the margin of a pan on Llanrhidian Marsh, South Wales, chewing away the edge of the turf.

The classification of salt pans given above is morphogenetic, but other features may also be used. **Soft-bottom pans**, for example, appear to have far more worm and crab burrows than **hard-bottom pans**. Nichol (1935) showed how extreme are daily and seasonal variations in pH and oxygen content (both of which show a rise around mid-day), temperature and carbonate levels to which organisms resident in salt marsh pools are subject. The salinity of the surface mud is often much higher than that of the subsurface when pools dry out in summer;

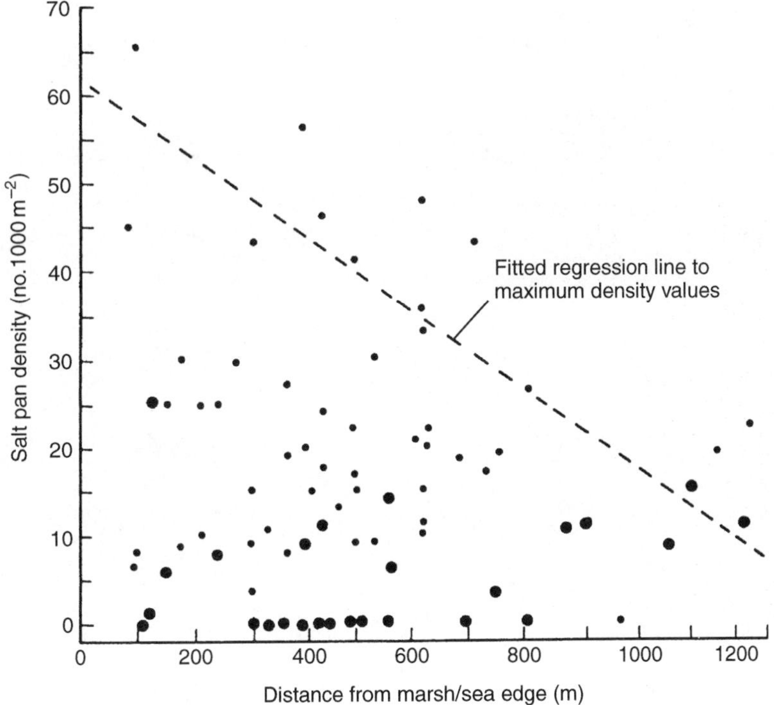

Figure 4.11 Scatter diagram showing limiting factors of salt pan density, North Norfolk. • , observations on marsh surfaces higher than 2.05 m OD, the mean height of the 75 quadrats involved. ●, observations on surfaces lower than 2.05 m OD. (From Pethick, 1974; courtesy *Journal of Biogeography*.)

such differentiation does not occur in sandy-bottomed pools. Water in salt marsh pools is frequently layered with fresh water over salt; the marked environmental variations in these microcosms are in stark contrast to the buffering found in the open sea.

Creek formation occurs when receding tidal waters scour the slight grooves into which some of them are deflected, thus preventing colonization by the plants which grow on their margins. As these grooves are eroded more deeply and form narrow channels, their banks gradually rise as soil particles are deposited among the vegetation. Plants growing on the well-drained creek banks are frequently larger than those of the same species elsewhere, so they are particularly effective in trapping silt carried in water flooding from the creeks as tides rise, thus accelerating levée formation. 'Tall' *Spartina alterniflora*, for example, occupies some 6–7% of the total marsh surface of the barrier island salt marshes of Georgia, USA (Wiegert, 1979), being found on creek banks, levées and the marsh surface behind the levées.

Creek banks are often undercut or are eroded as water pours off the marsh after a flooding tide. If they collapse the creek has, in section, an almost terraced appearance for a time and secondary marsh may develop on the fallen sediments. Heads of developing creeks tend to erode towards the land and branch creeks frequently develop. In periods with very high tides headward erosion of minor creeks along the Dovey Estuary may be as much as 3–5 m within a few days; huge volumes of water drain into these creeks when tides begin to ebb.

Figure 4.12 Creek patterns shown in the sandy Dovey marshes, near Ynyslas, North Wales, which are far simpler than those of many Norfolk marshes (Figure 3.3). The marsh surface has scattered *Armeria maritima, Aster tripolium, Puccinellia maritima, Salicornia* sp. and *Spergularia marginata*, together with some short *Spartina anglica*, a species which grows well in the shallow creeks whose banks it lines. Banks of the deeper creeks in the distance are relatively bare. (Photograph by John R. Packham.)

The form of creek systems varies according to tidal range, type of vegetation and the nature of the substrate. Creeks of sandy marshes such as those on the Welsh (Figure 4.12), Lancashire and Solway coasts are simple and not very numerous. Some creek systems on peat in New England, USA have a similar form in plan view. In contrast, the muddy marshes of the Humber, the Wash and North Norfolk (Figure 3.3) have more numerous minor branches and the whole system has a tree-like (dendritic) pattern. The growth habit of *Spartina* appears to favour development of tortuous dendritic creeks in south coast marshes such as Poole Harbour, whose tidal range is only 2 m. However, the plant is also dominant in marshes bordering the Bristol Channel, where the tidal range is more than six times as great and small, branched parallel creeks run at right angles to the shore line (Chapman, 1976).

In salt marshes, the ratio of total creek length to ground area is, according to Pethick (1984), often ten times as high as that of an upland river system. The annual total of sea-water draining off a salt marsh exceeds the rainfall of an upland river system by a much higher factor. With spring tides, however, there is a **mass flow** of the water that floods and ebbs over the general surface of the marsh, largely independent of the creeks, which carry tidal flows below the level of the marsh surface and act as small tidal estuaries. Additionally, high creek densities increase the chance that salt pans will be drained and subsequently grown over.

4.3.1 MICRORELIEF

The general slope of a marsh and the presence of creeks and salt pans strongly influence the distribution of animals and plants. The use of a topograph (Boorman and Woodell, 1966) simplifies the simultaneous recording of plant distribution and topography, enabling less obvious effects to be recognized. Figure 4.13 shows results from two short transects in an area of the Cefni Marsh, North Wales, where *Festuca rubra* occurs intermittently. The distribution of the plants is influenced by two scales of relief, the relatively minor variations within each transect and broader-scale variation which accounts for the difference between transect means, leading to major differences in the numbers of the various species present. The order in which the mean species altitudes occur, with the lowest on the left, is the same in both transects for *Salicornia* (lowest), *Puccinellia maritima*, *Armeria maritima* and *Festuca rubra* (highest). The distribution of *Puccinellia* is markedly different from that of *Festuca* in that some individuals occurred well below the mean transect height, whereas *Festuca*, which responds very adversely to impeded drainage and poor soil oxygenation under experimental conditions (Pigott, 1969), was here entirely restricted to the higher levels, though *Armeria* was not.

Gray and Scott (1977b) made use of the topograph in investigations of the salt marsh grasses of Morecambe Bay, which confirmed the competitive advantage of *Puccinellia* over *F. rubra* at high water tables. Their results suggested that both the large- and small-scale distribution patterns shown by *Agrostis stolonifera*, *F. rubra* and *Puccinellia maritima* are determined by their competitive abilities under various waterlogging and salinity regimes. *Puccinellia* was found largely on immature soils at low elevations and on more organic soils with high sodium contents; the other two species occurred on more organic and less saline soils at higher elevations.

In the high level saltings, *Festuca* occurred on the humps, *Agrostis* on hump edges, and *Puccinellia* in the hollows. Relationships between the species pairs were determined by growing them in replacement series (de Wit, 1960). The five pots of each species pair contained 4 : 0, 3 : 1, 2 : 2, 1 : 3 and 0 : 4 tillers of each species respectively arranged in a line across the pot (e.g. FFFF, FPFF, FPFP, PPFP, PPPP). *Puccinellia* was the most competitive when seawater was applied, whereas *Agrostis* was favoured by freshwater conditions.

4.4 DEVELOPMENT, ZONATION AND AGE OF SALT-MARSH ECOSYSTEMS

4.4.1 SALT-MARSH FLORAS

The total numbers of species in many salt-marsh floras are very considerable; Adam (1990) noted 325 in British salt marshes (for distributions of the marshes see Figure 4.14). Although some 250 of these are widespread, only 45 constitute the halophytic element. Nevertheless, salt-marsh floras the world over show a remarkable consistency in that most of the important species present belong to a few cosmopolitan genera such as *Salicornia* and *Sarcocornia* (glassworts), *Spartina* (cord-grasses), *Juncus* (rushes), *Plantago* (plantains) and *Limonium* (sea-lavenders). Genera such as *Salicornia*, *Sarcocornia* and *Limonium* are restricted to saline habitats, whereas many species of *Juncus* and *Plantago* are common elsewhere. Long and Mason (1983) describe the remarkable geographical range and ecological amplitude of some of these genera. *Salicornia europaea* (Western Europe), *Salicornia australis* (New Zealand) and *Salicornia virginica* (California) all occur at the lowest levels of salt marshes, while *Sarcocornia perennis* (= *Arthrocnemon perenne*, Western Europe), *Salicornia ambigua* (New Zealand) and *Salicornia subterminalis* (California) grow in the higher zones.

Grasses are the most abundant herbaceous plants in many salt marshes and in western Europe *Spartina anglica*, *Puccinellia maritima*,

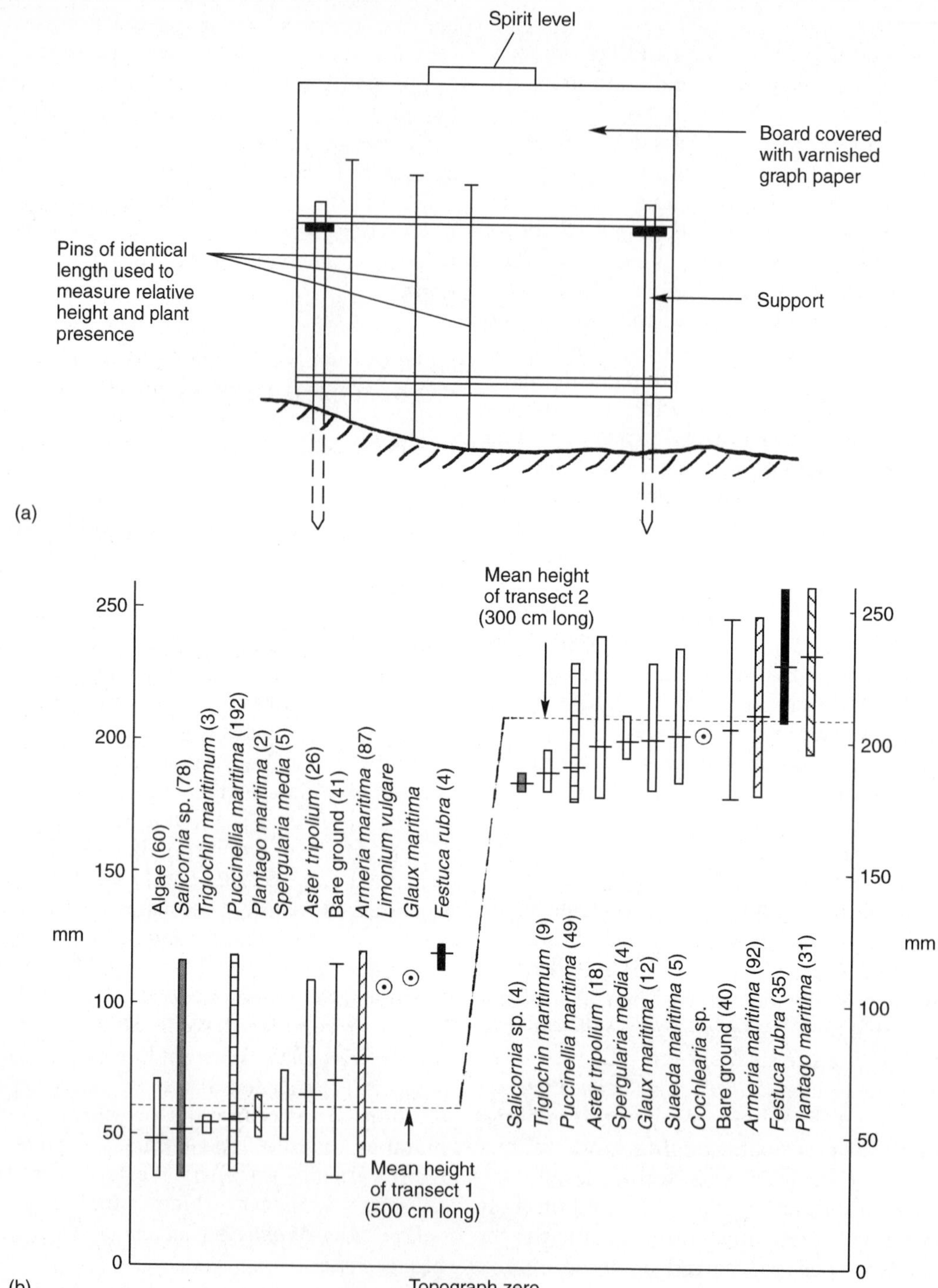
Spirit level
Board covered with varnished graph paper
Pins of identical length used to measure relative height and plant presence
Support
(a)
Mean height of transect 2 (300 cm long)
Algae (60)
Salicornia sp. (78)
Triglochin maritimum (3)
Puccinellia maritima (192)
Plantago maritima (2)
Spergularia media (5)
Aster tripolium (26)
Bare ground (41)
Armeria maritima (87)
Limonium vulgare
Glaux maritima
Festuca rubra (4)
Salicornia sp. (4)
Triglochin maritimum (9)
Puccinellia maritima (49)
Aster tripolium (18)
Spergularia media (4)
Glaux maritima (12)
Suaeda maritima (5)
Cochlearia sp.
Bare ground (40)
Armeria maritima (92)
Festuca rubra (35)
Plantago maritima (31)
Mean height of transect 1 (500 cm long)
mm
Topograph zero
(b)

Festuca rubra, *Agrostis stolonifera* and *Phragmites australis* are very widespread. *Spartina alterniflora*, which commonly dominates the lower levels of salt marshes on the east coast of North America from the Bay of Fundy to the tropics, shows a remarkable geographical range. This is paralleled by plants in other families such as *Triglochin maritimum* (Juncaginaeae), a familiar species in the middle zones of west European salt marshes, which is also found in similar situations on the Atlantic coast of Canada, Hudson Bay, the coasts of Japan and China, and the Pacific coast from Alaska to California.

4.4.2 ZONATION

Salt marshes have often (section 4.2) been divided into **low marsh** (the 'pioneer zone' commonly between MHWN and MHW), **middle marsh** (between MHW and MHWS), and **high marsh**, the zone above MHWS. Long and Mason (1983), aware that the situation varies with tidal amplitude and wave action, divide the vertical range of the marsh – from the seaward limit to the highest point influenced by the tides – into three zones of equal altitudinal width. Figure 4.15 demonstrates the similar zonation patterns found in three salt marshes: in the Canadian Arctic, the east coast of England and California. The low marshes are dominated by very few species (two, five and four, respectively) and have significant amounts of bare ground. The low marsh species have reduced cover or are absent from the middle marsh, which supports many more species than the low marsh, while the high marsh typically contains both halophytic and non-halophytic species. At Devon Island and other arctic marshes the large freshwater input resulting from high precipitation and low evapotranspiration pushes the saline influence seawards so that almost no halophytes occur on the high marsh, and species not generally considered to be halophytic (such as *Carex aquatilis*) are present in the middle marsh. In contrast, low precipitation and high evapotranspiration cause high salinity even on the high marshes of California, whose vegetation is predominantly halophytic.

One of the most striking features of the emergent marshes of north-west Europe is the occurrence of large areas of a spatially homogeneous community of long-lived perennial species typically including *Armeria maritima*, *Atriplex portulacoides*, *Limonium vulgare*, *Puccinellia maritima*, *Spergularia media* and *Triglochin maritimum*, of which usually none is obviously dominant. An example where *P. maritima* is plentiful is afforded by the middle marsh of Kirby-le-Soken, Essex (Figure 4.15(b)). Chapman (1960a), who provided diagrams of excavated root systems, considered that this **codominance**, in what he termed the **General Salt Marsh** community (**GSM**) and described for Scolt Head Island in 1934, resulted from niche separation arising from the different rooting depths and phenologies of these species. Davy and Costa (1992) point out, however, that there appear to have been no experimental investigations supporting this hypothesis, and that the longevity of these clonal species – together with the probable antiquity of the marshes – are likely to obscure present-day relationships.

Stands dominated by *Armeria maritima* and *Plantago maritima* occur spatially between areas of Puccinellietum and Festucetum in the salt marshes of Milford Haven, Pembrokeshire, generally occupying the lower range of the latter. Such stands resemble the Plantaginetum

Figure 4.13 (a) Structure of a topograph. (b) Distribution of species on the Cefni marsh in relation to microtopography. The two transects were 12.3 m apart and a levelling telescope was used to maintain constant topograph zero. The mean transect altitude of each species is indicated by a horizontal line. Where a species predominated in more than one square, the number of 1-cm squares it occupied is given. (From Packham and Liddle, 1970; courtesy *Field Studies*).

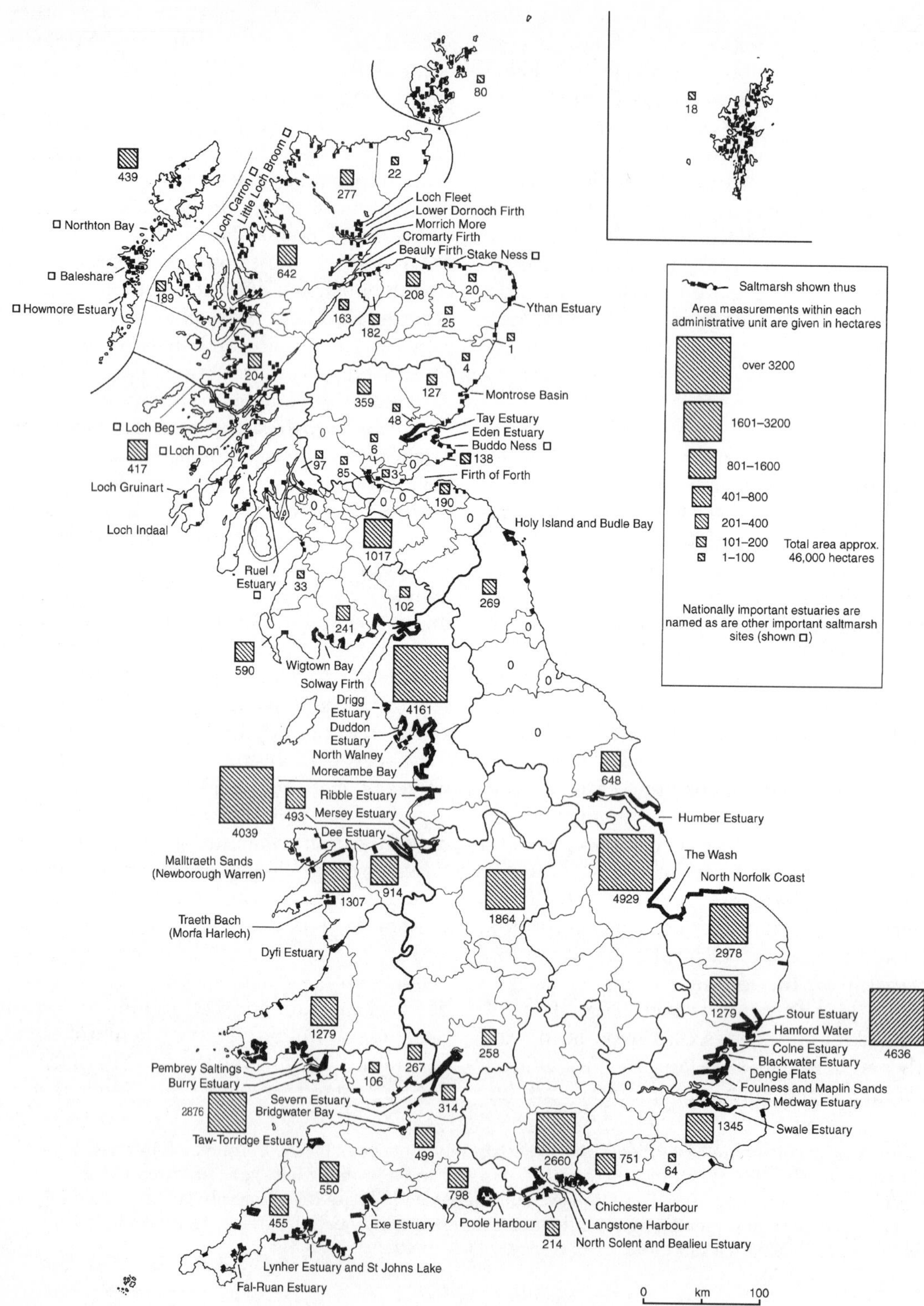

Saltmarsh shown thus
Area measurements within each administrative unit are given in hectares
over 3200
1601–3200
801–1600
401–800
201–400
101–200
1–100
Total area approx. 46,000 hectares
Nationally important estuaries are named as are other important saltmarsh sites (shown □)
Northton Bay
Baleshare
Howmore Estuary
Loch Carron
Little Loch Broom
Loch Fleet
Lower Dornoch Firth
Morrich More
Cromarty Firth
Beauly Firth
Stake Ness
Ythan Estuary
Montrose Basin
Tay Estuary
Eden Estuary
Buddo Ness
Firth of Forth
Loch Beg
Loch Don
Loch Gruinart
Loch Indaal
Holy Island and Budle Bay
Ruel Estuary
Wigtown Bay
Solway Firth
Drigg Estuary
Duddon Estuary
North Walney
Morecambe Bay
Ribble Estuary
Mersey Estuary
Dee Estuary
Humber Estuary
The Wash
North Norfolk Coast
Malltraeth Sands (Newborough Warren)
Traeth Bach (Morfa Harlech)
Dyfi Estuary
Stour Estuary
Hamford Water
Colne Estuary
Blackwater Estuary
Dengie Flats
Foulness and Maplin Sands
Medway Estuary
Swale Estuary
Pembrey Saltings
Burry Estuary
Severn Estuary
Bridgwater Bay
Taw-Torridge Estuary
Exe Estuary
Poole Harbour
Chichester Harbour
Langstone Harbour
North Solent and Bealieu Estuary
Lynher Estuary and St Johns Lake
Fal-Ruan Estuary
0 km 100

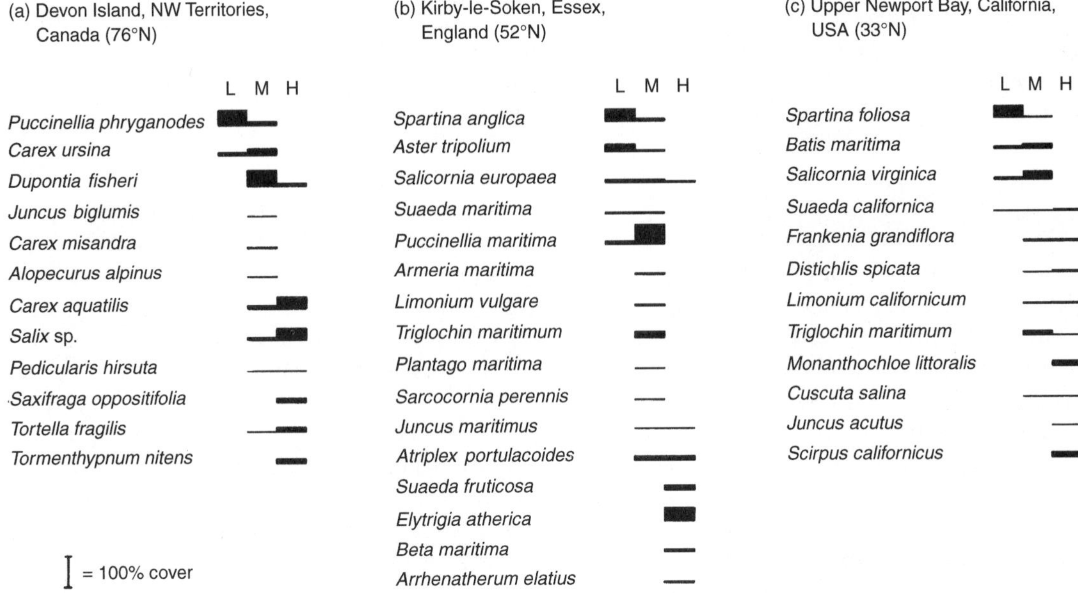

Figure 4.15 Ground area covered by different plant species in low (L), middle (M) and high (H) levels of three salt marshes in contrasting latitudinal locations. (From Long and Mason, 1983; data for (a) from Jefferies, 1977; (b) from Long (unpublished); and (c) from Vogl, 1966.)

described by Chapman (1960a) from Scolt Head Island in lacking both *Puccinellia maritima* and *Festuca rubra*. In the Milford Haven marshes the dominants of these somewhat uncommon stands are *Armeria, Plantago, Juncus gerardii* or even *Limonium*. Dalby (1970) terms this the **Forb Salt Marsh (FSM)** community and points out that dominants in the FSM are also important members of the GSM, the distribution of grasses varying independently of these. (**Forbs** are non-graminoid herbs). Salt-marsh species of *Armeria, Aster, Glaux, Plantago,*

Figure 4.14 Distribution and areas of British salt marshes. (From Doody, 1992.) There is an inventory of British salt marshes (Burd, 1989), which are further described in NCC site reports and by Rodwell (1998). **Literature sources listed below are given in the References.** Sites are listed in clockwise order starting at the top right. SM, salt marsh. Scottish Coast: Gimingham (1964); North Sea coastal margin: Doody, Johnston and Smith (1993); Dornoch Firth SM: Mudge and Allen (1980); Cromarty Firth SM: Kessel and Smith (1978); Inner Moray Firth: Tilbrook (1986); Ythan Estuary SM: North (1981), Gorman and Rafaelli (1983); Firth of Tay: Ingram *et al.* (1980); North Norfolk SM: Pye (1992), Pethick (1992); Scolt Head Island and Blakeney Point SM: Allison and Morley (1989); Maplin Sands, Essex SM: Boorman and Ranwell (1977); Essex and north Kent SM; Burd (1992); Solent SM: Tubbs (1995a,b), Brewis, Bowman and Rose (1996); Exe Estuary SM: Proctor (1980); Llanrhidian Marsh: Gillham (1977); Bridgwater Bay: Ranwell (1964), Cadwalladr and Morley (1974); Severn Estuary SM: Allen (1992); Berrow SM: Willis (1990); Milford Haven SM: Bassindale and Clark (1960), Dalby (1970); Cefni Marsh SM, Anglesey: Tunnicliffe (1952), Packham and Liddle (1970), Jones (1990); Dee Estuary: Doody (1992); Conwy Estuary SM: Howells (1988); Mersey Estuary SM: Rankin (1986); Ribble Estuary SM: Greenhalgh (1975), Doody (1992); Morecambe Bay SM: Gray (1972), Gray and Adam (1974), Gray and Scott (1977b); Inner Solway Firth: Bridson (1980).

Spergularia and *Triglochin* are all much more resistant to repeated oil pollution than *P. maritima* or *F. rubra*; Dalby postulates that in certain natural situations these two grasses are at a comparative disadvantage.

4.4.3 DEVELOPMENTAL ZONATION AND SUCCESSION

The formation of salt marshes is typically viewed as resulting in a series of zones parallel to the coast with the youngest at the lowest altitude nearest the sea, and the oldest at the top of the shore. This concept, which seeks to make a connection between the zonation of plant and animal communities in space, and succession involving directional change in the vegetation of a particular area with the passing of time, is termed **developmental zonation**. An alternative hypothesis is that the species composition of the various zones has been maintained since establishment early in the life of the marsh. The question of whether zonations of species and their communities represent true chronosequences is considered by Davy and Costa (1992, p.164) who demonstrate many pitfalls in an apparently plausible concept.

Since *Spartina anglica* was discovered in Southampton Water just over a century ago, it has established on mudflats to seaward of many north European salt marshes in positions where it appears to be the primary colonist. The communities higher up the shore, however, have never passed through a *Spartina* stage. The generalized successional diagram for Norfolk, England (Figure 4.16(a)) is based on an interpretation of plots found at a single time and is not supported by long-term evidence of change. In postulated successions for two Welsh marshes (Figure 4.17) Packham and Liddle (1970) assumed that *Spartina* would yield to other species as the level of the Cefni Marsh rose. Dalby (1970) left the future of areas of the Milford Haven marshes occupied by *Spartina* as an open question, and considered that *Suaeda maritima* and the annual *Salicornia* spp., which show enormous variations in numbers from year to year, to be opportunists which, like *Zostera* – whose rhizomes stabilize some of the more fluid substrates in the Milford Haven marshes – play little part in the main flowering plant succession.

It is also apparent that many salt marshes are much older than was previously appreciated. The oldest features of the *Spartina alterniflora*-dominated Great Marshes of Barnstable, New England (section 4.1), were in existence 4000 years ago. The protective sand spit of Sandy Neck grew to half its present length in the first 1000 years, and has elongated at ever-decreasing rates of extension since then. On the North Norfolk coast freshwater peats accumulated some 8500 years ago after the glaciers retreated. Rising sea levels led to inundation about 6000 years ago, when silty sands and muds appear in the stratigraphic sequence. Funnell and Pearson (1984, 1989) have demonstrated that considerable variations in sea level, both positive (rising) and negative (falling), have occurred since then. Subsequently, much of the sedimentary environment has been remarkably persistent, though some erosion and roll-over of the coastal barrier system has occurred. Remarkably, the positions of major channels, tidal flats and marshes appear to have been stabilized in their present positions for 4000 years or more. Strongly zoned salt-marsh vegetation may have remained essentially similar for thousands of years, and measured rates of accretion may be largely the results of isostatic or eustatic changes in sea level.

4.4.4 ALLOGENIC VERSUS AUTOGENIC FACTORS

Classical concepts of succession emphasize the role of organisms in modifying the environment; the operation of such **autogenic factors** leads to the gradual development of plant and animal communities (Clements, 1916), which are considered to exhibit increased species diversity, yield and amounts of organic matter with time (Odum, 1969). Such features are indeed likely to predominate in the high marsh

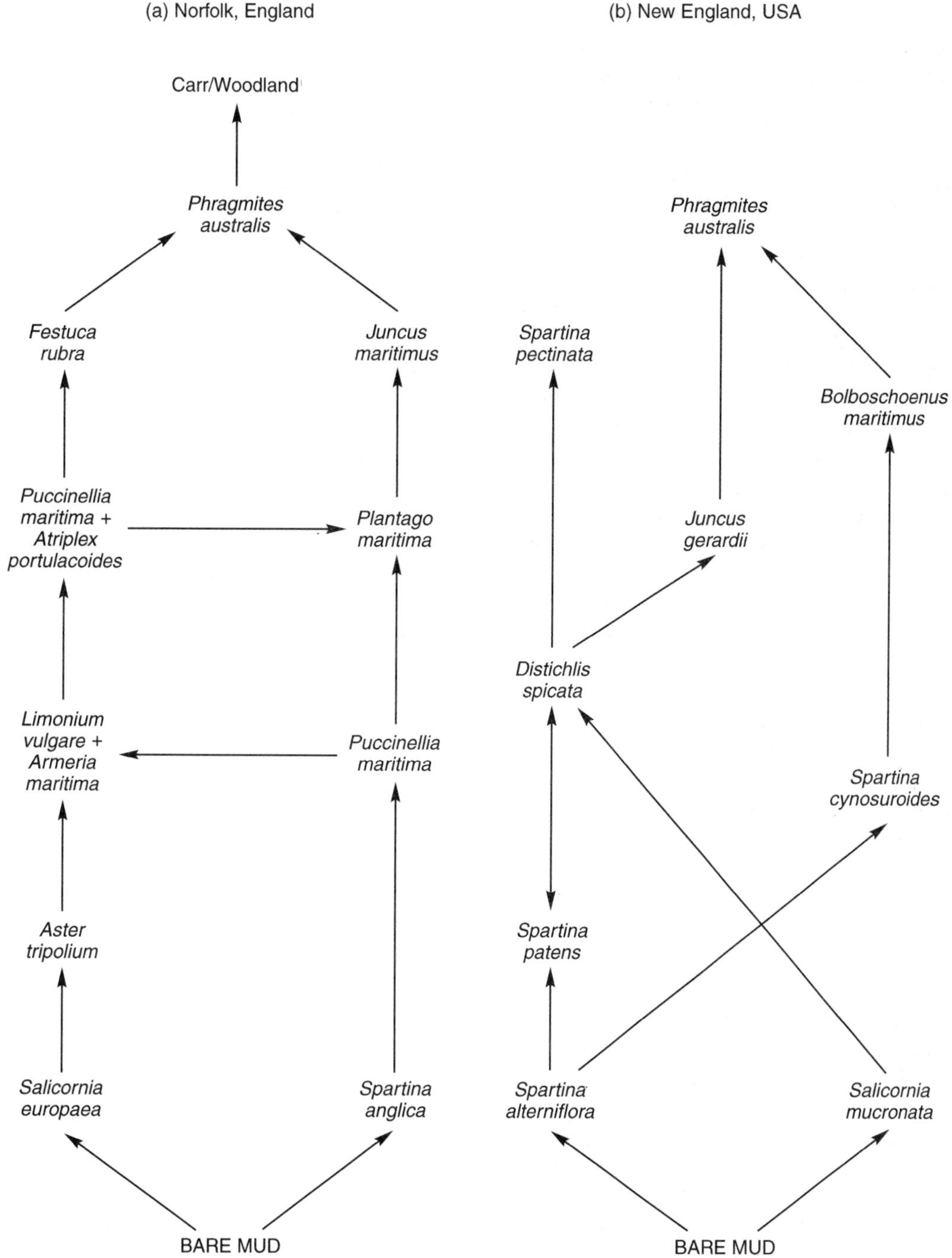

Figure 4.16 Generalized successional changes in dominant species of vascular plants, following the colonization of bare tidal muds, projected from observed zonations at (a) Norfolk, eastern England, UK and (b) New England, USA. (After Long and Mason, 1983.)

where tidal influence is reduced. Low marshes, however, experience frequent tidal inundation, associated with more complete litter removal and greater likelihood of inorganic sedimentation. This increase of the **allogenic** (environmental) influence leads to pulse-stabilization by the tides, as Eilers (1979) found in the lower Nehalem Marshes, north-west Oregon.

4.4.5 CYCLIC CHANGE IN A TROPICAL SALT MARSH

Watt (1947) investigated the relationships between the distribution patterns found in various plant communities and the processes by which they were produced. In many such instances there is a mosaic of patches or phas-

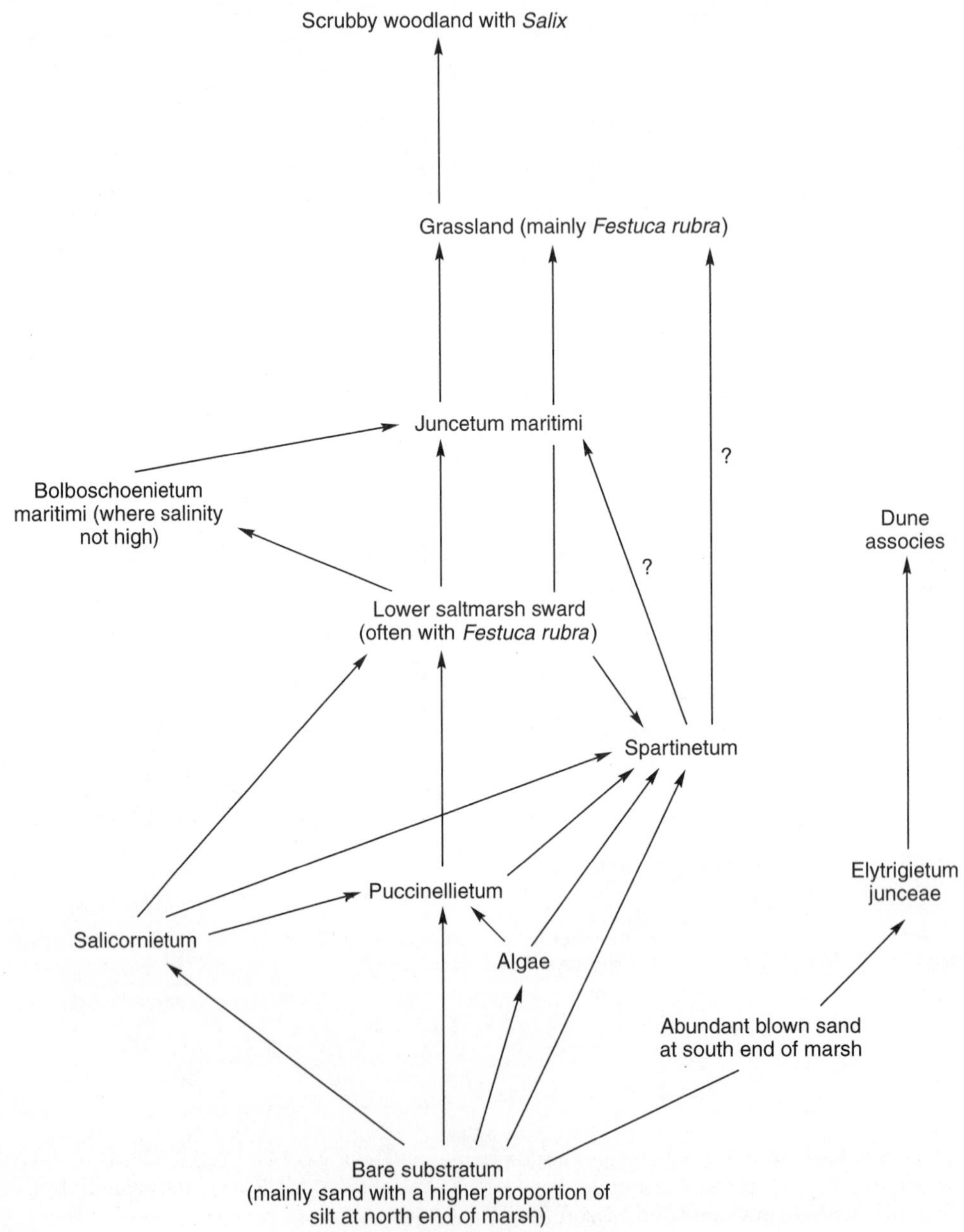

es which are dynamically related to each other and the vegetation is in a state of **cyclic change**, with the dominant species exhibiting the phasic series pioneer, building, mature and degenerate during which its competitive ability changes. Such a pattern is exhibited by the low-growing shrub *Sarcocornia indicum*, which often contributes more than 80% of the plant cover (about 50% of the ground is bare) to the frequently dry saline vegetation it characterizes on the coast of Sri Lanka (Pemadasa, 1981). The two therophytes *Cressa cretica* and *Heliotropium supinum* occur during the rainy season; of the perennials *Cynodon dactylon* and *Salicornia brachiata* are more common than *Suaeda monoica* and *Suaeda nudiflora*. *Sarcocornia indicum* spreads by means of radiating prostrate shoots which bear upright aerial shoots with clusters of branches, and often forms clumps on hummocks about 20 cm high in the centre. The amount of fine sand and organic matter in the hummocks is far greater than in the bare areas between the hummocks, where the water-holding capacity of the soil is low. These edaphic factors influence the establishment of *Cynodon dactylon*, a grass more frequent on *S. indicum* hummocks than on the bare ground between them.

Pemadasa (1981) concludes the vegetation pattern of the dominant *S .indicum* to be auto-

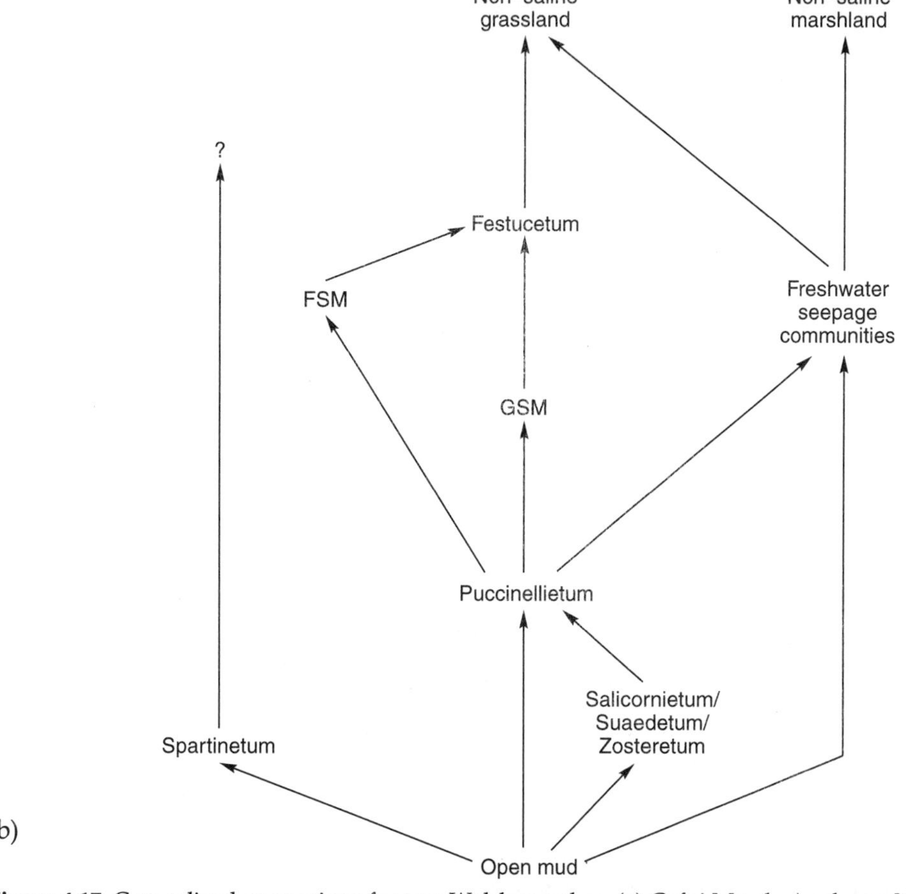

Figure 4.17 Generalized successions for two Welsh marshes. (a) Cefni Marsh, Anglesey, North Wales. In a few parts of the shore where conditions are often brackish *Bolboschoenus maritimus* acts as a primary colonizer. (After Packham and Liddle, 1970.) (b) Putative seres in salt marsh communities at Milford Haven, south-west Wales. Local facies and unusual modifications are omitted. (After Dalby, 1970.)

genic; that of the subordinate *C. dactylon*, however, is allogenic as it results from a differential response to microenvironmental variations caused by the phasic growth of *Sarcocornia*. *Cynodon* is usually more abundant when either the density or the performance of the dominant species is low, and its growth is better during the downgrade of the *Sarcocornia* cycle than the upgrade. It also begins to disappear as the dead *Sarcocornia* plants decay, as *Cynodon* is unable to prevent wind-induced disintegration of the hummocks and the associated soil.

4.5 NUTRIENT FLUX AND SALT-MARSH CHEMISTRY

Nutrient, fluvial and groundwater fluxes between salt marshes and the estuaries or coastal waters adjacent to them vary from marsh to marsh and from month to month. Marshes around basins with little freshwater throughput and which are connected to the open sea by a narrow channel are frequently **net importers**. In contrast, those surrounding estuaries which have a large throughput of freshwater and which widen and deepen towards the sea are likely to be **net exporters**. Storms, which are difficult to monitor, may remove considerable amounts of matter; export is promoted by a combination of high winds, rainfall and elevated tides (Odum, Fisher and Pickral, 1979). Major differences in estimates often result from the use of different techniques to determine nutrient flux; it is also clear that some marshes receive a net import of materials that are exported by others (Nixon, 1980; Carpenter, 1993).

Many studies of nutrient fluxes (Carpenter, 1993; Table 1.2) have concentrated on the annual **import/export balance** of particular nutrients or other materials, but the determination of fluxes on a seasonal basis often provides information of greater value to an understanding of the ecology of the organisms involved. Figure 4.18 illustrates an example where 15–20% of the above-ground net production of a *Spartina anglica* marsh was exported as particulate matter, with 70% of the carbon involved passing to the estuary and 30% to the driftline in 1980. In summer, however, there is a period of net import. Marshes with an approximately neutral annual balance for particular nutrients are net exporters at certain seasons, affecting marine organisms in adjacent shallow waters. In view of many differences between marshes, it may be more rewarding to study the operation of the processes involved in particular examples than to attempt to draw conclusions about salt marshes in general.

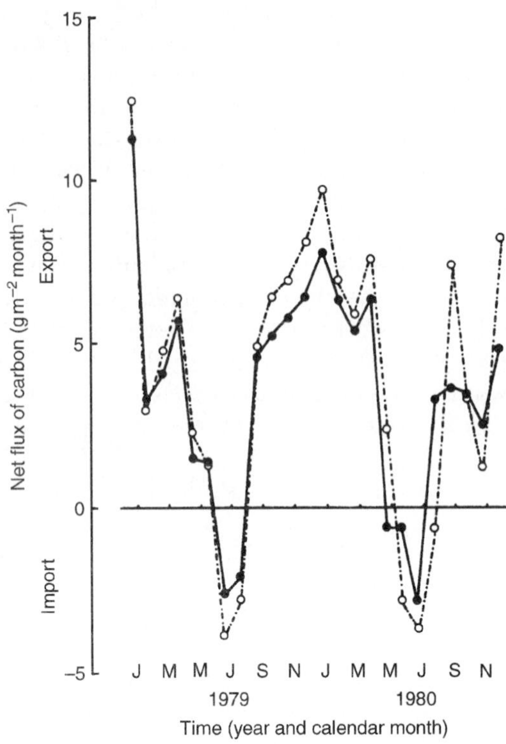

Figure 4.18 Net tidal flux of carbon derived from the vegetation of the salt marsh at Seafield Bay, Suffolk, UK. •——•, *Spartina anglica* only; ○–·–○, all vegetation (including mainly *Atriplex portulacoides* and *Puccinellia maritima*). (From Jackson, Long and Mason, 1986; courtesy *Journal of Ecology*.)

A recent example of such a study is provided by Carpenter (1993) who investigated nutrient, fluvial and groundwater fluxes between Stiffkey Upper Marsh, Norfolk, and the North Sea, using an integrated rising float technique to gauge water flow and obtaining more accurate estimates of flow rate than those given by velocity meters. Pore water draining from the sediment into the creek was a source of silicate and total dissolved inorganic nitrogen (TDIN) throughout the year. Measured net flux through the tidal creek indicated a net import of silicate when diatom blooms were occurring in the North Sea. TDIN import to the Upper Marsh also coincided with the algal blooms which occur in the sea during the spring and early summer. The sediment was a source of phosphate in summer, when concentrations in the coastal water were low, and a sink in winter, when coastal phosphate concentrations were relatively high. Creek water column processes were important in controlling fluxes between the salt marsh and the North Sea. Stiffkey Marsh appeared to be a long-term sink for phosphorus, and also for nitrogen, the import of particulate nitrogenous material exceeding the export of TDIN.

Flux studies in general have mainly been concerned with flows of carbon, nitrogen, phosphorus, silicon and sediment into or out of marshes. Flux values are conventionally given a negative value if the marsh is a source of the material, and a positive one if it is receiving it and thus behaving as a sink. The four main techniques used to quantify net exchanges between salt marshes and coastal waters are community budgeting, direct tidal creek measurements, flume studies and diffusion chamber studies. A critical appraisal of these methods, made by Carpenter (1997), is outlined below.

Community budgeting is a long-term, indirect approach which derives nutrient flux as the difference between annual production and aerobic decomposition. This method was used in studies of salt-marsh ecosystems in Georgia, USA by Teal (1962), who concluded that salt marshes were exporters of organic carbon. The method ignored anaerobic decomposition, nitrification and denitrification, and was superseded by **direct tidal creek measurements** which have been employed by numerous authors (Valiela and Teal, 1979b; Woodwell *et al.*, 1979) who have measured the amounts of nutrient flowing into and out of marshes, ideally through closed tidal creek systems (section 4.2). Such measurements should (in theory) provide net flux values for the marsh as whole, but throw little light on the nutrient-transforming processes and whether they occur in the water column, sediment, vegetated surface or creek bank. Direct tidal creek measurements are of great value in that they enable fluxes to be determined in marshes of appropriate topography, but many salt marshes are not of an easily monitored form and in them **mass flows** over the surface are a great complication. There is also the question of **storm tides**, which result in considerable fluxes that are difficult to assess.

Flume studies have been used to investigate surface processes in the marshes of North Inlet, South Carolina (Wolaver *et al.*, 1983). This technique, which can be used only for tides high enough to spill over onto the marsh surface, involves pushing the bases of two walls of plexiglass into the sediment parallel to the flow of overtopping water, thus creating a flume leading from a creek or mudflat towards the interior of the marsh. As in the previous method, flux estimation involves combining discharge and nutrient concentration data, but in this instance flow is calculated from changes in the volume of the water within the flume, topography of the marsh surface within the flume having been carefully determined, rather than by using velocity meters. Flume experiments are useful in assessing surface wash-off, uptake and diffusion from the vegetation, and can well be used in conjunction with direct tidal creek studies. However, they do not take account of water column, creek bank or sediment processes and so cannot legitimately be extrapolated to indicate whether a whole marsh is a source or sink of a particular material. Wolaver and Spurrier (1988) ignored this when reporting, on the basis of flume studies, that North Inlet marsh, South Carolina, imported both phosphate and particulate phosphorus in all seasons. Whiting *et al.* (1985), who investigated eight tidal cycles in a single month (May) within the same marsh using the integrated direct tidal creek approach, recorded an export of these two nutrient species.

Diffusion chamber studies can be used to measure changes in nutrient concentration during tides which flood the vegetated marsh surface. A perspex cylinder is partially embedded in the surface of the marsh and filled with water of known initial composition. The choice of this water is important as its ionic concentration will strongly influence the gradients initially established, consequently affecting the passage of materials into or out of the water in the chamber. This technique will perturb the system, particularly during installation, but can yield useful information regarding small areas. It can thus be used to compare the influence of particular plant species, while an ingenious modification introduced by Chambers (1992) enables a natural flooding cycle to be employed, whereas in other experiments (for example, see Scudlark and Church, 1989) the depth of water remained constant and flooding was continuous.

4.5.1 NITROGEN FLUX IN A *SPARTINA ALTERNIFLORA* MARSH IN NEW ENGLAND

Valiela and Teal (1979a,b) used direct tidal creek measurements by mechanical flow meters in studies of nitrogen flux in the Great Sippewissett Marsh, Falmouth, Massachusetts, which has a single tidal entrance. Sampling was at frequent intervals during tidal cycles and repeated monthly. Tidal waters and samples collected from rain gauges were analysed for ammonium, nitrite, nitrate, and dissolved organic nitrogen (DON); for tidal waters estimates were also made of particulate N. Nutrients, particularly nitrate, were provided by ground and rain water; an additional input of nitrogen was provided by bacterial nitrogen fixation. Loss of nitrogen by **denitrifaction** just exceeded gains by fixation. Losses of nitrogen by volatilization of NH_3 from marsh water were thought to be small. The pattern of seasonal uptake and release of tidally transported nutrients was driven by the activities of the vegetation (mainly *Spartina alterniflora*) and the decomposers; total particulate losses were equivalent to 40% of the net above-ground production of *S. alterniflora*.

Major inputs and outputs of nitrogen are shown in Figure 4.19. Standing stocks and fluxes of nitrogen among some major components of the ecosystem on a day in early August are given in Figure 4.20. Most of the nitrogen was in the sediments where turnover rates were slow. Interconversions and exchanges within the marsh were active and complex. The nitrogen content of the vegetation is determined by the balance between uptake, which is high in early summer, and leaching rates which increase during the flowering period of mid-August. Fragments of dead grass became particulate nitrogen, some of which was consumed by bivalves or laid down in the sediments. The shellfish removed particulates which would otherwise have been flushed out to sea, but they excreted ammonia and produced faecal pellets which were not readily resuspended. The over-riding conclusion, however, was that nitrate in the ground water received by the marsh is converted to ammonium and particulate nitrogen which are then exported to coastal waters by the semi-diurnal tides. These exports are of considerable significance; particulates act as the nitrogen source in detrital food webs and production of coastal phytoplankton is nitrogen-limited.

Nitrogen fluxes are very important elsewhere. There is, for example, a tendency for dissolved nitrate in waters entering mature British marshes to undergo denitrification in the sediments, whereas in marshes where

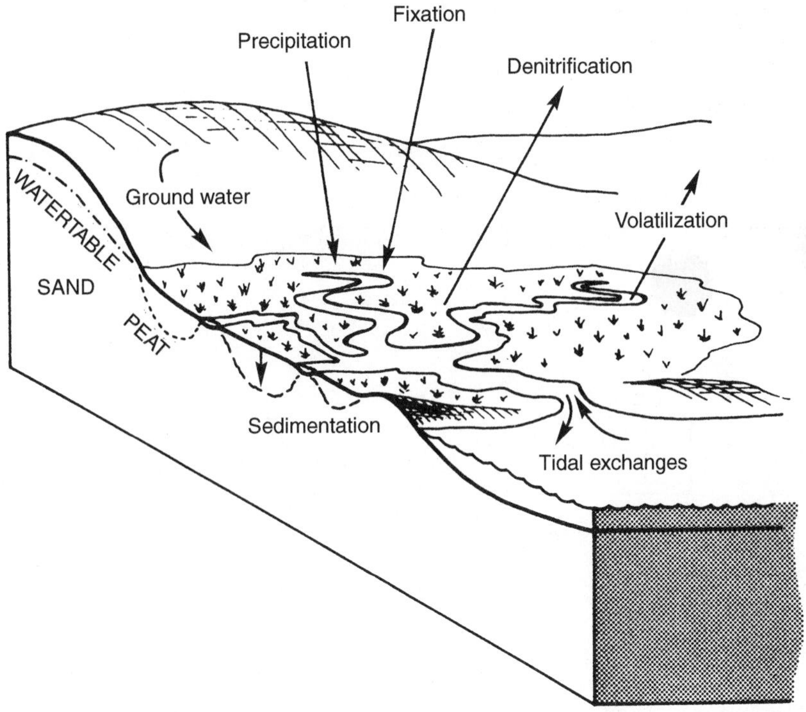

Figure 4.19 Diagram of major inputs and outputs of nitrogen into a salt-marsh ecosystem. (From Valiela and Teal, 1979b; courtesy Blackwell Scientific Publications.)

nitrate is at a lower concentration it usually undergoes ammonification. The switch between the two processes, which is largely dependent on the nitrate concentration of the water, helps to regulate the nitrogen status of the marsh and is an important feedback mechanism in marshes contaminated by sewage.

4.5.2 INFLUENCE OF SOIL SULPHIDE ON THE DISTRIBUTION OF HIGHER PLANTS

Apart from its importance in salt marshes dissolved sulphide is known, under waterlogged conditions, to cause the death of *Acacia* trees and wilting disease in rice. Sulphide is formed in anaerobic environments, such as the soils of lower salt marshes (section 4.2), where there is a supply of sulphate ions and of organic matter adequate for sulphate-reducing bacteria of the genus *Desulphovibrio*. Sulphide is released into the soil where it may combine with ions of the transition metals such as iron, copper and manganese, forming insoluble inorganic sulphides. Under such conditions soil sulphide exists in dynamic equilibrium partly determined by the activity of hydrogen ions:

inorganic sulphide	acid sulphide	hydrogen sulphide

$$\underset{\text{(solids)}}{S^{2-}} \rightleftharpoons \underset{\text{(soluble)}}{HS^{-}} \rightleftharpoons \underset{\text{(soluble gas)}}{H_2S}$$

Under field conditions some sulphide remains in solution either as the acid sulphide ion (HS^-) or the dissolved gas which, although highly soluble, is gradually lost to the atmosphere in areas where sulphate-reducing activity is high. Although high concentrations of dissolved sulphide in salt-marsh sediments have been correlated with reduced growth of *Spartina townsendii* agg. (Goodman and Williams, 1961) and of *S. alterniflora* (King *et al.*, 1982), these studies did not provide direct

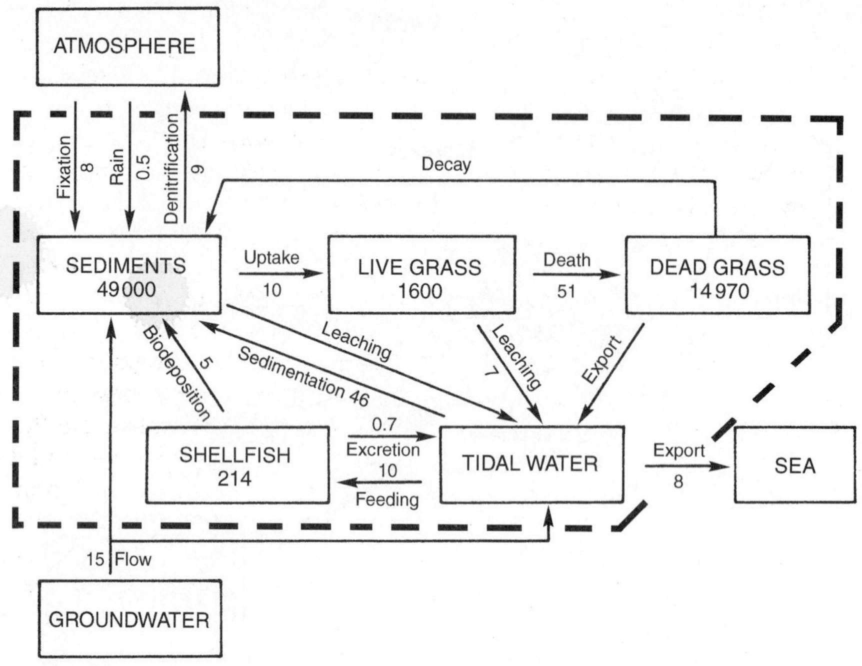

Figure 4.20 Standing stock and fluxes of nitrogen among some major components of the Great Sippewissett Marsh, Massachusetts. Values are in kg N for the entire marsh (22.6 ha), while exchanges are kg day^{-1}. The values are for one day in early August. Dotted line represents margin of salt-marsh ecosystem. (Modified from Valiela and Teal, 1979b, as adapted by Long and Mason, 1983.) The annual budget for this marsh is shown in Table 1 (Valiela and Teal, 1979a).

experimental evidence that sulphide is toxic to plants at the concentrations found in the field. They also fail to demonstrate whether reduced growth of some plants in salt-marsh sediments is caused by sulphide *per se* or to other effects of low soil redox potential.

Ingold and Havill (1984), whose main field site was at Canvey Island Point, on the Essex coast of the River Thames, investigated these points. Three different electrodes were used to measure sulphide concentration, soil pH and redox potential *in situ* in the top 5 cm of soil. Soluble sulphide at this depth was found only in the lower marsh, salt pans and creek beds, and its presence showed no significant correlation with redox potential. On the lower marsh only *Salicornia europaea* among the vascular plants present was rooted in sulphide-containing sediments, although it is likely that other *Salicornia* species can grow in these conditions. In a survey of eight salt marshes the association between soil sulphide and *S. europaea* was positive, while that with *Puccinellia maritima* was negative. Moreover, free sulphide was not detected in surface soil associated with *Aster tripolium*, *Atriplex portulacoides*, *P. maritima*, *Spartina anglica* or *Suaeda maritima*.

In liquid media the growth of *Atriplex patula*, *Festuca rubra* and *P. maritima* was significantly inhibited by sulphide, while that of *S. europaea* was not.

4.6 SUCCESSION, STABILITY AND PALAEOECOLOGY

Many studies of coastal vegetation seek to describe present vegetation patterns and to understand the mechanisms which have led to their development (topics introduced in sections 1.6, 1.7 and 4.4). Chapman (1976), who discusses succession in the salt marshes of the Arctic, Europe and North America, considered the communities of any one marsh to be related to each other in space and time. His successional diagrams illustrate the broad outlines of the sequences occurring in particular areas, often indicating the influence of type of substrate or of variations in salinity. Beeftink (1979, p.84) states that in the salt marshes of south-west Holland, 'each main community, arranged according to its present zonational position, may be restored via a series of developmental stages after disturbance' with an eventual return to the original vegetation if the soil and tidal conditions remain the same. The model proposed by Redfield (1972) for the development of 'New England-type' marsh (Figure 4.3) stresses biotic control over floristic composition and a predictable sequence of vegetation zones with time. Odum (1969) recognized the role of tides in limiting the extent to which vegetation at particular sites within salt marshes may change, describing them as '**pulse-stabilized systems**'. Jefferies, Davy and Rudmik (1979) point out that, while varying in different seasons and places in the marsh, the 'salinity, water potential and osmolarity of the sediments all fluctuate around the corresponding values for sea water, which change little during a year'. To this extent salt marshes are highly predictable, stable environments displaying **cyclical stability** in the sense of Orians (1975) – 'the property of a system to cycle or oscillate around some central point or zone' – though physical instability occurs near drainage channels and at the seaward margin of the marsh where rates of accretion or erosion may change rapidly.

The views of these authors have been interpreted by Clark (1986b) as emphasizing the 'inherent stability of salt marshes and the importance of autogenic succession as a driving mechanism of vegetation change', and contrasted with those of other coastal ecologists who, studying different salt marsh systems, give less weight to the value of traditional successional concepts. In doing so he quotes the work of Martin (1959, p. 43) who considered that succession on barrier beaches 'appears to be initiated by physiographic causes and directed perhaps by environmental changes .... over which the vegetation has little if any direct control'.

On the basis of their pollen, macrofossil and ^{210}Pb analyses of a salt marsh on the north side of Long Island, USA, Clark and Patterson (1985) questioned the concept of stability and the importance of autogenic succession in this ecosystem, where coastal vegetation responds to frequent and severe fluctuations in the physical environment and historical factors partly determine vegetation pattern. Similar conclusions are drawn from further work on Long Island (Clark 1986a,b), but in other parts of the world coastal ecosystems may be disturbed less frequently and more moderately.

In pollen analyses of the Great South Beach, Long Island (Clark, 1986b), samples were taken from the surface and from buried peat and silt; pollen in sandy sediments was too degraded to provide useful information. Data from 66 cores arranged in 16 transects from the 60-km barrier beach, which is fringed by salt marsh on the side away from the ocean, enabled a reconstruction of the vegetation on both regional and local scales. Samples were dated both by ^{210}Pb and pollen analysis of sediments laid down in the last three-and-a-half centuries, thus allowing comparisons with documented changes in the structure of the inlets and barrier islands. Three separate lagoons fringed by marshes and with distinct inlet histories now open to the sea via Fire Island inlet, Moriches inlet and Shinnecock inlet (Figure 4.21). From the assembled results it was possible to build a stratigraphic history of the mosaic vegetation of the coast, and to evaluate current concepts of coastal vegetation dynamics. The latter are clearly relevant to informed policies for the management of barrier beaches in eastern North America (Dolan *et al.*, 1973; Kerr, 1981; and section 11.3).

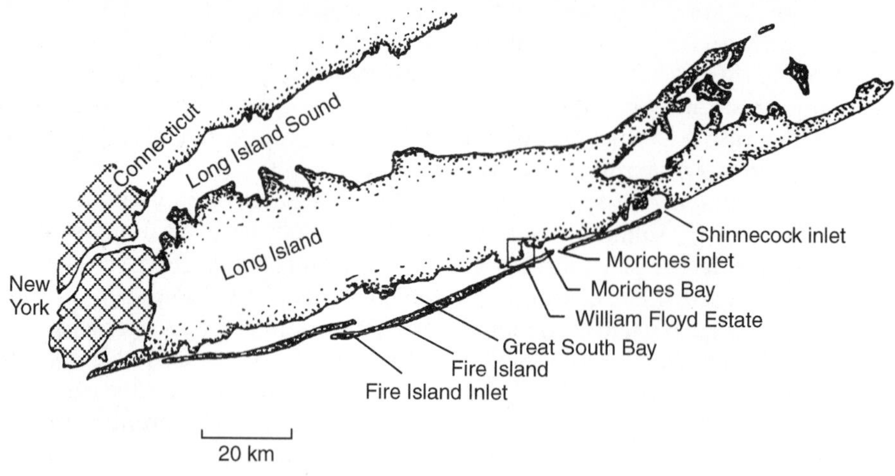

Figure 4.21 Map of Long Island showing the barrier beaches and the position of the William Floyd estate, the site of the two salt marsh cores (FE1, FE2) discussed in section 4.5. (Modified from J.S. Clark, 1986a.)

Moriches inlet, which was open to the sea from 1755 to 1836, 1931 to 1951, and from 1953 to the present, opens into Moriches Bay on the north-west side of which are salt marshes owned by the Floyd family since 1724. Clark (1986a) compares two high salt marsh cores from this estate, showing that the vegetation changed frequently in response to such factors as inlet opening and closing, change in the average rate of sea-level rise, the 18.6-year tidal cycle and such cultural influences as the clearing of the first fields by the Floyds in 1730.

When the inlets were open pollen percentages of *Salicornia* were high; *Limonium* pollen, Foraminifera, and unknown types sphere A, B1 and B2 were also present. Type A closely resembles *Cymatiosphaera* found in samples from Dutch barrier islands, where its presence indicates marine conditions. Figure 4.22 compares documented events and the principal pollen curves used for dating the sediments of one of the cores (FE2), and Figure 4.23 gives a reconstruction of vegetation development and environmental change at this point, as well at the position from which core FE1 was taken. Many species are present in the full pollen profile; pollen of the upland taxa are grouped as forest trees and shrubs, or agricultural weeds. Then come groups indicating moisture thresholds, salt thresholds, or moisture and salt. The reciprocal dominance of Gramineae and Cyperaceae at FE2 between 1835 and 1930 may be evidence of a pulsating sea-level rise associated with the 18.6-year tidal cycle in which Cyperaceae flourished when the sea-level dropped for a sufficiently long period. Since 1930 the increased rate of sea-level rise has prevented the alternating dominance of Gramineae and Cyperaceae and the seaward migration of salt-intolerant fringe populations on the high marsh: periodic sea-level declines no longer occur.

Clark (1986a) considers that analyses of the two cores from this 'New England-type' salt marsh are compatible with the view of Redfield (1972) that high marsh transgresses uplands with a rising sea. However, constant sediment supplies, constant sea-level rise and absence of major disturbances which the Redfield model assumed on the basis of sedi-

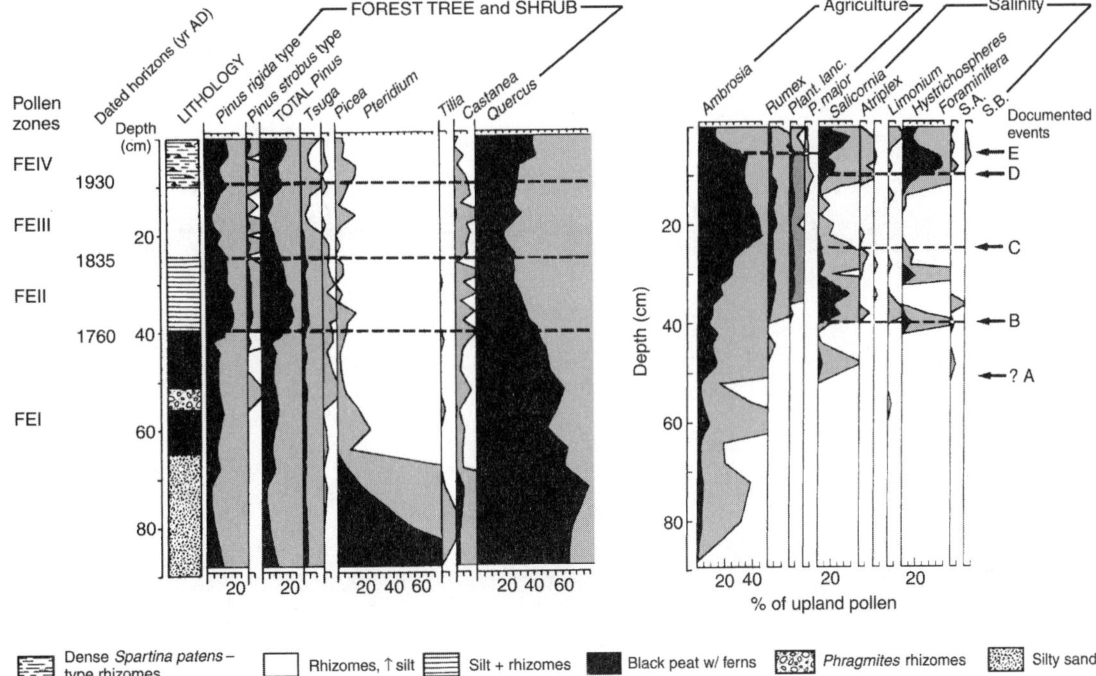

Figure 4.22 The lithology and pollen curves for some of the forest species together with comparison of documented events and principal pollen curves used for dating of core FE2 from the high salt marsh of the William Floyd Estate, Long Island, New York. A, Initial settlement of the estate in 1724 may be the factor responsible for the *Ambrosia* rise at 48 cm; B, Moriches Inlet opened in 1755, identified by increasing percentages of salinity indicators; C, Moriches Inlet closed in 1836, identified by decreasing percentages of salinity indicators; D, Moriches Inlet opened in 1931, identified by a second rise of salinity indicators; E, invasion of abandoned fields by shrubs and arboreal taxa in 1950, identified by an abrupt *Ambrosia* decline and a *Quercus* rise. Pollen values are given as percentages of total plant pollen counted (curve infilled with black) and as ×10 exaggerations of these values (curve infilled with stippling). (Adapted from J.S. Clark, 1986a; courtesy *Journal of Ecology*.)

ment water content have not been evident on the short time scales investigated on the William Floyd marshes. Water content measurements of sediments may not be sensitive to frequent and severe changes clearly detected by pollen stratigraphy.

Pollen studies such as those just discussed may to some extent supply the type of information which would be given by long-term studies of coastal vegetation. There is a dearth even of successional records for particular plots over periods as short as the 12-year record given for the Cefni Marsh by Packham and Liddle (1970, Table 3); integrated studies of larger areas such as those of the Barnstable Marsh, New England (Redfield, 1972) are rarer still. Repeated observations such as those concerning the salt marshes of Cold Spring Harbor on the north shore of Long Island (Johnson and York, 1915; Conard, 1924; Conard and Galligar, 1929) are of particular value. The two resurveys made by Conard revealed a seaward migration of the salt-intolerant fringe populations and the fact that drift lines no longer influenced upper reaches of the marsh. On this basis, Conard postulated an emergent coastline, whereas the sea level is rising but was, as Clark (1986a) points out, subject to a decline of

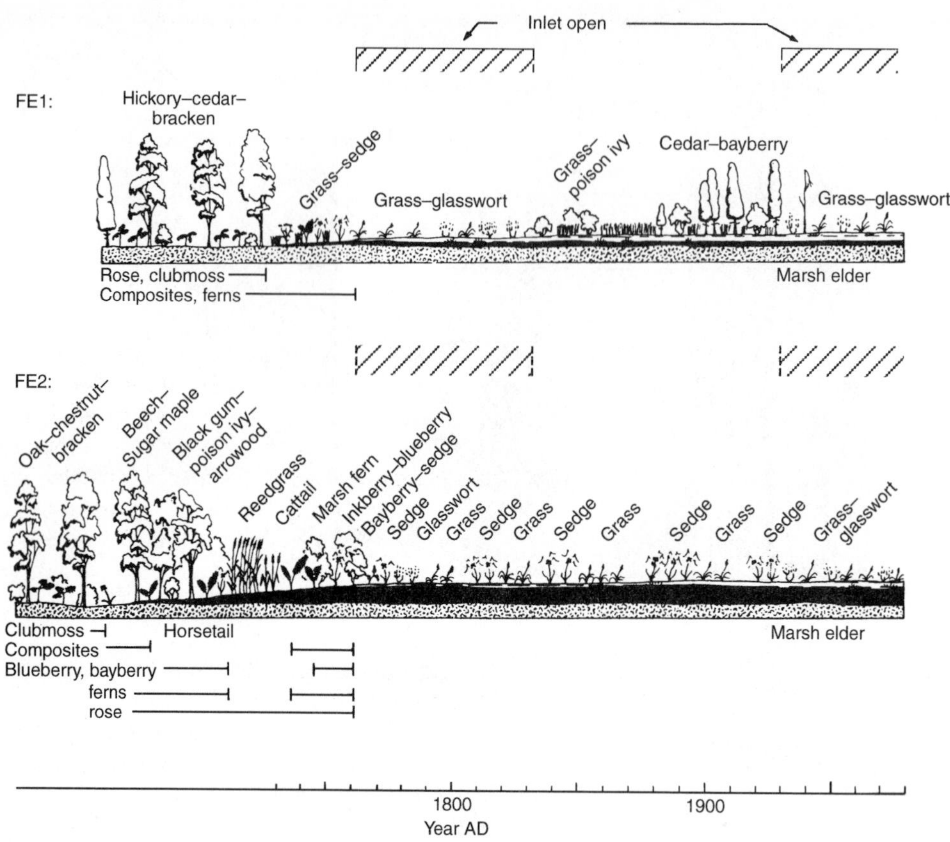

Figure 4.23 Vegetation development and environmental changes at FE1 and FE2, Long Island, reconstructed from pollen profiles and sediment stratigraphy. The time-scale was determined by pollen-dated horizons. (From J.S. Clark, 1986a; courtesy *Journal of Ecology*.)

some 10 cm in the annual mean high water recorded by tide gauges from 1920 to 1930. This short-term decline represents the influence of the 18.6-year tidal cycle that can be observed in tide-gauge curves globally (Hicks, Debaugh and Hickman, 1983).

SALT-MARSH DYNAMICS AND COMMUNITIES 5

5.1 AUTECOLOGY OF MAJOR PLANT SPECIES

Duration of life (ephemeral, annual, biennial or perennial), life form, phenology, reproductive methods and capacity, and competitive ability under a range of environmental conditions are all important in relation to the **strategies** of plants of a particular habitat (section 1.7; Harper, 1977; Grime, 1979). Maritime salt marshes are unique in that all parts of them are to some degree subject to the action of tides which influence sedimentation rates, ionic relationships, water regime, photosynthesis and the ability of seedlings to establish on bare surfaces. This section concerns the major plant species of European salt marshes and the characteristics which influence both the strategies they adopt and their success or failure in the environmental mosaic.

There are normally two relatively narrow zones where **salt-marsh annuals** are abundant (section 7.2). The pioneer zone in which *Salicornia* is commonly dominant is located around mean high water neap tides, and is subject to a high incidence of wave slap. There is also a high incidence of wave break at mean high water spring tides in the area where *Atriplex* spp. are associated with the strandline, but here intermittent smothering by tidal litter also helps to keep the vegetation open. Between these two zones, areas away from creeks and salt pans usually have relatively few gaps in plant cover unless recently trampled by horses or driven over by motor vehicles. Moderate grazing by sheep greatly modifies plant form but seems seldom to expose bare soil. Ranwell (1972) points out that if the bulk of salt marshes were open habitats the proportion of annuals in the flora would be much higher than it is, and that in such relatively closed vegetation domination by perennials is to be expected. Chapman (1960b, Table 30) gives quite high values for therophytes, but such values may well be inflated by casual species not specific to salt marsh which occur along the strandline. In fact, relatively few species of annuals appear to be adapted to the true salt marsh habitat, which is dominated by hemicryptophytes.

The majority of salt-marsh annuals are in the Chenopodiaceae. *Atriplex*, *Suaeda* and *Salicornia* all include annual species able to survive the saline and tidally disturbed conditions of maritime salt marshes: all show succulence and have cell sap with high osmotic potentials. *Salicornia* is frequently much branched and has a form which offers low frictional resistance to water currents; its succulent fused opposite leaves ensheathe the stem, giving these annual glassworts a jointed appearance. In most species the flowers are embedded in threes in the tissue of the shoot. Successive generations do not overlap; the size of *Salicornia* seed banks persistent for more than the year following their production varies in different parts of the world (section 7.2). Wiehe (1935) found that plants of this annual genus needed 2–3 days without flooding after germination so that they could root effectively and avoid being swept away by the tides.

Taxonomically this is a very difficult genus which shows great phenotypic plasticity and a strong tendency to inbreed (cleistogamy), forming distinctive local populations occupying contrasting environments created by a range of tidal regimes. Noble, Davy and Oliver (1992) provide evidence for genetic differentiation between such populations that is related to life history and physiological characteristics that affect fitness. In experiments using nuclear ribosomal DNA variants as genetic markers, analysis of rDNA in 38 maternal plants from Stiffkey, Norfolk, UK, and 2112 of their progeny provided no evidence for outcrossing.

The population biology of the *Salicornia europaea* complex, which shows marked phenotypic plasticity and in which isoenzyme studies have revealed a high degree of genetic uniformity, is discussed by Davy and Smith (1988). In reciprocal **transplant experiments** between two diploid populations found respectively on the lower and upper marshes at Stiffkey, whole-life selection coefficients were decisively against the alien transplants. Probability of survival to establishment, growth phase survival, fecundity and finite rate of population increase all showed negative density-dependence.

The perennial glasswort *Sarcocornia* (*Salicornia*) *perennis* is a somewhat woody perennial with usually little-branched aerial stems up to c. 30 cm high. Its thin extensive rhizomes can give rise to patches up to 1 m in diameter, but its growth form is very variable; some Spanish plants resemble small trees.

Atriplex (= *Halimione*) *portulacoides*, a woody chamaephyte (Chw) or nanophanerophyte (N), is a very mealy small shrub with a short creeping rhizome. It reaches a height of some 80 cm, or rarely 150 cm, and is commonly found on levées fringing channels and salt marsh pools. Its recent spread in the eastern salt marshes of the UK is attributed to a run of mild winters.

Aster tripolium (sea aster, Asteraceae) is a very variable species. Its flower heads typically have purple ray florets and yellow disc florets, but forms with either purple ray or yellow disc florets only also occur. Tall plants in ungrazed Norfolk marshes do not flower until late summer while dwarf plants in closely grazed swards such as those of the Llanrhidian Marshes, Gower Peninsula, South Wales, often fail to flower at all. In the upper estuarine marshes of the River Humber, Yorkshire, and other places with lime-rich silty soil where salinity does not exceed 2%, *A. tripolium* is robust, flowers freely and reaches 1.8 m. Wind dispersal is usually over relatively short distances; most fruits detach from the parent plant, fall, and are dispersed by water. This 'compromise' enables *Aster* to colonize areas of open ground within a salt marsh and without undue wastage of seeds. This, together with the fact that the species may behave as an annual or a short-lived perennial, enables it to behave as a particularly well-adapted weed or opportunist species (Gray, 1971, 1985). The plant is often a temporary colonist of gaps in emergence marsh created by drought or tidal litter. Intolerant of shade and unable to persist as a weed in agricultural land flooded by the sea, it dies out as the salt is leached away.

Spartina (cord-grass) is arguably the most important genus of herbaceous perennial present in the salt marshes of the world. Species of this genus dominate large areas of salt marsh in the temperate zone (Goodman, 1969). Such marshes are remarkable for their high production rates: an exceptionally high estimate of net primary production exceeding 60 tonnes ha^{-1} yr^{-1} for a *S. alterniflora* low marsh in Georgia, USA is mentioned by Long and Mason (1983). Moreover, the *Spartina* species present in salt marshes commonly occur in monotypic stands, providing relatively simple systems for the field investigation of primary production and for ecosystem modelling. The high productivity of *Spartina* is largely owing to the occurrence of C-4 photosynthesis in this genus, in which 'Kranz' leaf anatomy, high $^{13}C/^{12}C$ ratios and a CO_2 compensation point close to zero have all been reported for many of the 16 species (Long and Woolhouse, 1979). *Spartina* differs from most C-4 grasses in having a distribution extending into cool temperate regions, rather than being confined to tropical and sub-tropical areas. At temperatures of 10°C and below, the leaf photosynthetic rates of the European *Spartina townsendii s.l.* are equivalent to those of temperate C-3 grasses; at higher temperatures they greatly exceed the values found in many such grasses.

Five taxa of *Spartina* occur in Britain (Stace, 1991). Three of these are of restricted distribution; *S. maritima* is native whereas *S. alterniflora* and *S. pectinata* are exotics from North America. The other two taxa are of extremely wide distribution, having been planted in numerous reclamation schemes, despite their relatively recent hybrid origin. One, *S. townsendii* (Townsend's cord-grass 2n = 62), was first collected as a male-sterile hybrid on the shores of Southampton Water in 1870. It arose as a cross between *S. maritima* (2n = 60) and the American *S. alterniflora* (2n = 62) which was first noticed near Southampton in 1829. This hybrid spread widely and gave rise to *S. anglica* (2n = 120, 122, 124; Figure 5.1), which is a fertile amphidiploid, by chromosome doubling (Marchant, 1975). This is larger and more vigorous than *S. townsendii* and has been planted in attempts to stabilize mudflats and salt marshes in many parts of the world. It has a relatively large seed with considerable food reserve, rapid shoot and root growth, possesses strong anchor roots, and its shoots have abundant aerenchyma – all features favourable to survival at the seaward limit of the submergence marsh.

The late D.S. Ranwell, who had a particular interest in the recently evolved *Spartina anglica*, pointed out its remarkable achievement in 'spanning survival from the absolute seaward limit of salt marsh growth on open high level mudflats to the landward limit of salt marsh growth at high water equinoctial tides, and from fully saline to brackish water conditions' (Ranwell, 1972, p.106). This contrasts with the situation in east coast North American marshes where different and longer established species of *Spartina* dominate specific zones

Figure 5.1 Common cord-grass (*Spartina anglica*) growing in margin of the river estuary at Sandside, Milnthorpe, Cumbria, UK. (Photograph by John R. Packham.)

(see Figure 4.15(b)). *S. anglica* can spread over long distances by means of fragments carried to new sites along the tideline. Gaps in the ensuing colony may rapidly fill with seedlings once the pioneer clones have established. A.J. Davy (personal communication) notes that *S. anglica* produces seed very erratically and that the seeds tend to lose viability if they dry out. *Spartina maritima*, on the other hand, sets little or no fertile seed and even in favoured sites only widely separated clones develop from the occasional established fragment.

Spartina has been variously classified as a rhizomatous helophyte, a hemicryptophyte, or a geophyte. Its seedlings each have a basal rosette of leaves whose origins are separated by short internodes. The bud in the axil of each leaf may develop either into a green leafy tiller or into a white scaly rhizome which grows horizontally. After a few internodes, first-generation rhizomes turn upwards close to the parent stock, outward growth being continued by second-generation rhizomes arising from buds on the lower and outwardly-directed side of the first-generation rhizomes. Fresh generations of rhizomes develop in uncolonized substrate at the periphery of the colony, which if growing without competition takes the form of a circular **auxoclone** (an enlarging or increasing group of individuals of the same genetic origin) that develops alternating 'rings' of high and low shoot density as it enlarges. Rings of low shoot density are mainly occupied by degenerating tussocks of plants that have seeded, broken off at the base and degenerated. The situation in large *Spartina* colonies, which may exceed 4 m in diameter, becomes complex, but the inner rings of high shoot density appear mainly to result from an inward migration of later generation rhizomes.

The number of rings of high density is proportional to the diameter of the colony; Caldwell (1957) found that auxoclones reached the same complexity at a similar diameter on both bare mud and muddy sand. *Spartina* appears to be unusual in the development and duration of complex concentric zonation. Nevertheless, early stages of development of concentric pattern occur, when competition is absent, in a number of other rhizomatous or semi-rhizomatous species including sea arrow-grass (*Triglochin maritimum*). As in many other salt-marsh plants, the below-ground biomass (rhizomes, fibrous absorptive roots and the larger penetrating and anchoring roots) of *Spartina* have a far higher dry weight than the aerial shoots (section 5.4).

Several British populations of *Spartina* have in recent years been heavily infected by *Claviceps purpurea* (ergot), whose sclerotia are themselves sometimes colonized by the hyperparasite *Fusarium heterosporum*. *S. anglica*, which shows little genetic variation and may even comprise a single genet throughout most of its range, appears to be prevented by its genetic programming from developing effective resistance to the parasite (Gray, Drury and Raybould, 1990). As the grass spreads mainly by vegetative growth and loss of vigour caused by the fungus is not substantial, the ergot epidemic is unlikely greatly to influence the distribution of *S. anglica*.

The recession of *Spartina anglica* as a result of '**dieback**', notably in parts of the south coast of Britain, e.g. Lymington Estuary, The Solent (Goodman, Braybrooks and Lambert, 1959); Poole Harbour, Dorset (Gray, Benham and Raybould, 1990), has been known since 1928. However, the cause of this condition, with failure of rhizomes, little tiller production and yellowing, leading to death and decay, and baring of the marsh area, is incompletely understood. Pathogens do not appear to be the primary cause of dieback, which may arise from the effect of the plant itself in creating unfavourable, waterlogged, anaerobic, reducing conditions in the substratum from the accretion of fine-grained sediment. There remains the possibility that colonization, development and recession may be cyclical on a very long time scale, of the type of upgrade and downgrade phenomena described by Watt (1947). Dieback can be seen also in *S. maritima* at Odiel, south-west Spain, and has been reported in *S. alterniflora* in the USA

Puccinellia maritima extends over a very wide range of salt marsh habitats, but is often more abundant on the lower marsh. It is much smaller that *S. anglica* and even more reliant on vegetative, rather than sexual reproduction. In a survey of over 40 British populations of common salt-marsh grass all cells were octoploid 2n = 56, apart from a few in the range 53–55 (Gray and Scott, 1977a). *P. maritima* is tolerant of waterlogging, high salinity and accretion rates of up to 5 cm a year. It is well adapted to the salt-marsh environment, having a mixed and variable breeding system with a combination of clonal spread and sexual reproduction. As a colonist it frequently expands radially, forming hummocks on which *Salicornia*, *Suaeda* and other species often develop. During expansion of the Cefni Marsh, Anglesey, roughly circular patches 120–150 cm in diameter formed within 3–4 years in the newly developed Puccinellietum. Upward growth was rapid and the lower rooting levels often became anaerobic and blackened by sulphide as accretion proceeded. *P. maritima* has considerable phenotypic plasticity; pioneer forms have stolons up to 50 cm long, while under intense sheep grazing a mat-like growth little more than 1 cm high develops (Ranwell, 1972). In clone trials and a crossing programme Gray (1985) found that *P. maritima*, with its fairly wide ecological amplitude, shows adaptation by both genetic differentiation and phenotypic flexibility. Plants in the pioneer zone are more variable than those of the mature marsh, where there is directional selection for vigour and 'competitive ability', leading to taller, wider, longer-leaved plants with increased yield and seed production.

Festuca rubra grows over wide areas of high marsh, but is absent from badly drained areas of salt marshes where soil oxygenation is poor (section 4.3). Davies and Singh (1983) found its growth to be depressed by soil waterlogging, while that of *Agrostis stolonifera* was stimulated. They concluded that salt-marsh populations had evolved a degree of tolerance, however, as yields of these populations of *F. rubra* were less depressed by soil waterlogging than were those from free-draining soils. In the continuously waterlogged treatment, shoot concentrations of Fe and Mn were lower in salt marsh populations than in those from inland.

The different tolerances of *A. stolonifera* and *F. rubra* to waterlogging are clearly reflected by their frequently very close association in the upper regions of salt marshes, where *A. stolonifera* occurs on the lower edges of slightly raised humps bearing *F. rubra*. Plants of *F. rubra* from salt-marsh populations have also been shown to be more salt-tolerant than those from more inland areas, tolerance being heritable and dominant to non-tolerance (Humphreys, 1982). Populations of *A. stolonifera* on salt marshes also have been found to have greater tolerance to salt, partial anaerobosis and osmotic stress than those of the spray zone, which are themselves more tolerant than plants from inland habitats (Ahmad and Wainwright, 1977).

Phragmites australis is the tallest non-woody species in the British flora, reaching well over 3 m under the most favourable conditions and commonly 6 m high in the much warmer Danube delta (Haslam, 1972). Its shoots do not grow in cold weather and are killed by frost. In Britain, *P. australis* grows well with water tables ranging from c. 1 m above ground level to c. 1 m below it – an important characteristic since many of its potential competitors cannot survive prolonged flooding. It is very long-lived, has extensive rhizomes, and can form virtual monocultures on more fertile sites.

The species shows considerable genetic variation but is in the UK sometimes excluded from the more saline areas. Indeed, its remains in peat or clay horizons often indicate a marine–freshwater contact and it is commonly important in the succession from salt marsh. The seeds frequently have low germination rates in this country, but vegetative reproduction by rhizome extension is extremely effective and *P. australis* also forms **legehalme**, long stolons or runners which may reach a length of 10 m and are more common in brackish than fresh water. *Phragmites australis* is a very formidable competitor under damp conditions of low salinity when it can suppress *Spartina anglica*, but is outcompeted by many much smaller species when the soil is dry.

Of the 12 species of eelgrass (*Zostera*) in the temperate seas of the world, three occur in Britain, all now classed as scarce (Stewart *et al.*, 1994), but *Z. marina* may well be under-recorded. These truly aquatic flowering plants are **hydrophilous** (water-pollinated; in at least *Z. angustifolia* pollen is water-repellent, pollination probably occurring in the surface film). The top of the zone which they occupy is immediately below that of true salt marsh. *Zostera marina* occurs down to a depth of 4 m in Britain, but to 10 m in the Mediterranean. It remains abundant on the west coast of Scotland and around the Outer Hebrides. *Zostera* is intolerant of drying out; the upper limit of *Z. marina* was found by Tutin (1942) to be correlated with the time of low water spring tides, being lower down the shore where these occurred at mid-day. *Zostera angustifolia* and *Zostera noltii* are both intertidal species of sand and muddy flats, the former tending to be favoured by depressions and the latter by slightly raised areas. Invasion of intertidal mudflats by *S. anglica* has considerably diminished the area available to *Z. marina* and *Z. noltii* in Britain; Adam (1990) provides evidence of several examples where their communities have been invaded and replaced by this cord-grass. The shoot demography of *Z. marina*, a clonal plant like many seagrasses, was studied by Olesen and Sand-Jensen (1994). Many leaf shoots, which form most rapidly on side-branches in May/June, are short-lived; a high proportion of over-wintering shoots survive until flowering. Seedlings contribute to patch maintenance in the event of local declines; unless there are gaps in the canopy few seedlings establish. Extensive populations of *Z. noltii* occur at Maplin Sands, Essex, and it sometimes grows among *S. anglica*, indeed, a number of mudflats created by the decline of *Spartina* marsh have subsequently been recolonized by *Zostera* and algal mats.

'Wasting-disease' severely reduced eelgrass populations on both sides of the Atlantic in the 1930s, and to a lesser extent in the 1980s; the second epidemic was in America conclusively demonstrated to be caused by an infectious species of the protist *Labyrinthula*. In the Solent, UK, the epidemic was over and recovery had begun by 1994. In the early 20th century European eelgrass meadows were prolific and economically important. Sublittoral communities supported several exploitable species of fish, molluscs and crustacea, while intertidal meadows sustained wintering populations of Brent geese (*Branta bernicla*) and wigeon (*Anas penelope*). Tubbs (1995b) suggests that the pathogen of wasting-disease in eelgrasses persists as a low-level parasite subject to periodic population explosions, of which that in the 1930s was particularly large.

5.2 PLANT AND ANIMAL COMMUNITIES

The modern concept of the **ecological niche**, developed from the work of Hutchinson (1957), provides an effective way of viewing the roles of individual species within the communities to which they belong. Gray (1992) examined the difference in salt-marsh plants between the **fundamental niche** (the total biological and environmental space or *n*-dimensional hyper-volume which a species could potentially occupy) and the **realized niche**, which is the space which the species actually occupies. Many aspects of the niche in salt marshes are tide-related factors, and tolerance of these seems largely to control the ability of a plant to grow at the lowest levels at which it occurs (Gray, 1992). Occupancy of much of the fundamental niche of a particular species may be prevented by interspecific competition or lack of opportunity to spread there.

Niche overlap is calculated from use by individual species of particular niche dimensions, while **niche breadths** are measurements of evenness of resource utilization along given niche dimensions. Both concepts were employed by Russell, Flowers and Hutchings (1985) when comparing halophytes growing within the same tidal ranges (4.56–5.06 m above tide datum) on two emergent salt marshes in Chichester Harbour, Sussex. Of the 16 species which occurred at Hayling Island, those not found at East Head – *Aster tripolium*, *Cochlearia officinalis*, *Elymus atherica*, *Inula crithmoides*, *Juncus gerardii*, *Triglochin maritimum* and possibly *Festuca rubra* – all occurred in the upper part of the studied area. *Ammophila arenaria* was present at East Head but not at Hayling Island, in whose more diverse vegetation mean niche overlaps were significantly smaller and most species had lower niche breadths. These results indicate that the presence of a larger number of species increases the role of competition in determining **resource partitioning between species**, whose realized niches become smaller. If it were not so the addition of species to a community would produce coexistence, increased niche overlap and no change in niche breadths.

5.2.1 INTERACTIONS BETWEEN PLANT AND ANIMAL COMMUNITIES

Facultative mutualisms between *Spartina alterniflora* and the marsh mussel *Geukensia demissa*, and between this salt-marsh cord-grass and fiddler crabs *Uca pugnax* are important determinants of *S. alterniflora* production in the Rumstick Cove salt marsh, New England (Ellison, Bertness and Miller, 1986). Decrease in production up the shore is directly correlated with decrease in nutrient and oxygen availability and the declining abundance of marsh mussels and fiddler crabs with increasing tidal height. Burrowing activities of the fiddler crabs increase soil aeration and the mussels deposit nitrogenous wastes; both activities increase production in *S. alterniflora*.

5.2.2 INFLUENCE OF *SPARTINA ANGLICA* ON SALT-MARSH COMMUNITIES

The recently evolved cord-grass *Spartina anglica*, whether actively sward-building or dying back, now occurs through most of its range (whose northern limit appears to result from susceptibility to frost) as a belt of vegetation immediately seaward of other salt-marsh communities (Gray, Marshall and Raybould, 1991), in a zone which is crucial to the salt-marsh ecosystem as a whole. The positive and negative features resulting from a century of spread of this species are evaluated by Doody (1990), who points out that its monocultures replace potentially more diverse pioneer communities, that it invades intertidal flats rich in invertebrates, and also promotes the reclamation of land for agriculture, thus destroying species-rich high-level salt marsh. Moreover, some ungrazed *Spartina* marshes eventually change into tall communities of *Phragmites australis* or *Bolboschoenus maritimus* which are equally poor in species.

The spread of *Spartina* over intertidal flats has for a number of years been seen as a threat to wading birds that winter on British estuaries. Decline in numbers of dunlin (*Calidris alpina*) since the early 1970s is highest in those estuaries where *Spartina* has spread most (Goss-Custard and Moser, 1990). Dunlin numbers have failed to increase in a number of estuaries where *Spartina* has receded since 1971; the condition of mudflats resulting from the recession of well-established swards of this plant may be unsuitable for effective colonization by many invertebrates.

5.2.3 LOSS OF *SPARTINA ANGLICA* MARSH

Despite concern about the continued spread of *S. anglica*, it is losing ground in a number of British sites (Figure 5.2). In the Solent, southern England, where it originated, it has suffered dieback owing both to the accumulation of very fine-grained sediment and, much more locally, to chronic oil pollution (sections 5.1 and 9.1). Large areas have died back, eroded and reverted to mudflat along the English Channel; other losses here result from successional processes on the landward side of the *Spartina* zone or to lessened vigour, erosion and fragmentation on the seaward side. In many other parts of Britain, particularly on the eastern coast, only erosion and fragmentation are of significance (Adam, 1990). In the Dovey estuary, North Wales, creeks in *S. anglica* marsh sometimes erode rapidly headwards; this loss in area is accentuated when *Festuca rubra* replaces *S. anglica* on the sandy levées of the creeks. When growing with *Phragmites australis* in damp soils that are not unduly saline *Spartina* is shaded out by the taller grass; if the *Phragmites* stems are cut, *Spartina* resumes tillering, flowering and active growth.

5.2.4 SALT-MARSH ANIMALS

Salt marshes are well known for the number of birds which visit them; 130 species are listed for the marsh and tidal sand flats of the relatively small Cefni Marsh, Anglesey, UK. They also provide frequently extreme and highly variable habitats for animals of many different groups; the large spider population of the Cefni Marsh includes at least 19 species (Packham and Liddle, 1970). Frid and James (1989) recorded 32 species of marine invertebrates in the submergent *Spartina–Salicornia* salt marsh at Stiffkey, Norfolk, UK. The animal community was dominated by deposit-feeders, of which *Hydrobia ulvae* and *Littorina littorea* fed at the surface. Many of the others were burrowing worms (*Capitella capitata, Tubifex costatus*), as were the predators *Nereis diversicolor* and *Nephtys hombergi*. The numbers of all these annelids increased in spring, declined in June–July and then increased again in late summer. The sand goby, *Pomatoschistus minutus*, came onto the marsh early in summer, moved away in August–September and was again abundant in autumn. The spatial heterogeneity of the marsh, which is a mosaic of habitats, may help

Figure 5.2 *Spartina* dieback exposing areas of bare mud near Hurst Castle Spit, Hampshire, UK. (Photograph by M.C.F. Proctor.)

to prevent local extinctions resulting from competition or predation.

The following brief account is based on animals of the **Blakeney Point** and **Scolt Head Island marshes**, for which terrestrial species are described by Foster (1989) and those of marine origin by Barnes (1989). Details of shore organisms are given by Fish and Fish (1989). The remains of many marine animals are cast up along the strandline, others actually live there. Sand-hoppers (*Talitrus*) shelter in burrows excavated in sand at or just above this level, moving down the shore at night to scavenge for food when the tide is low.

Numerous **deposit-feeding invertebrates** subsist on bottom-living communities of microalgae and blue–green algae, and on organic particles which settle out of suspension during tidal cover. Most common of the gastropod molluscs which move over the surface is *Hydrobia*, whose density here can exceed 130 000 m^{-2}. This snail crawls up salt-marsh plants and wooden posts while browsing on filamentous algae which cover them. If marooned when the tide recedes, they attach themselves by a cord of mucus, withdraw into their shells and await the next flood tide. Below the surface of frequently flooded areas live small oligochaete and polychaete worms, crustaceans and the larvae of crane-flies (Tipulidae). These are preyed upon by carnivorous worms, including the nemertines *Tetrastemma* and *Lineus*, and also by dolichopodid flies. Ragworms (*Nereis*), which live deeper in the sediment, are opportunistic, feeding on smaller animals, browsing on larger green algae, and acting as both filter and deposit feeders.

Bivalve molluscs including *Cerastoderma*, *Macoma*, *Mya* and *Scrobicularia* are pre-eminent among the **suspension feeders**. These utilize material suspended in the water column; several can also deposit feed by means of their inhalant siphons. Deposit feeders such as *Nereis* and *Corophium* can also filter feed; the polychaete worm *Lanice* – known as the 'sand mason' because the tubes in which it lives are made of cemented sand grains – also feeds in both modes. Males of the isopod *Paragnathia*, which have large heads and huge jaws, guard harems of females with tiny heads and abdomens swollen with developing larvae at the top of creek banks. These larvae live for a period as ectoparasites of fish, mainly gobies, when released into creek water on high spring tides. Flounders and grey mullet are among the other fish which swim up the creeks and onto the intertidal flats at high water.

The shore crab (*Carcinus maenas*) and gastropods including *Assiminea*, *Ovatella* and the rough periwinkle *Littorina saxatilis* live on the marsh surface beneath the vegetation. The marine fauna is, however, more abundant in the **salt pans** – hostile environments which can fill with rainwater during heavy storms or dry out in summer neaps. Lugworms (*Arenicola*) occur in deep salt pans where huge mounds of their castings accumulate in the relatively still water. The small isopod *Sphaeroma*, which swims upside down, is a specialist pan species.

Foster (1989) describes the chironomid midge *Clunio marinus* as the most fully marine insect on the British coast. It lives below the salt-marsh zone at extreme low water mark and is common in the exposed peat beds at Brancaster, Norfolk; its adult life coincides with the few days in the tidal cycle when this habitat is uncovered by the retreating tide. Emergence from the pupal cases, mating, egg laying and death occur within a few hours. On the mudflats above this zone are found a group of flies including *Erioptera stictica*, a crane-fly whose larvae are eaten by dolichopodid flies of which *Hydrophorus* can sometimes be seen in huge numbers riding the incoming tide.

Small horizontal tunnels and little heaps of soil near pioneer *Salicornia* at the bottom of the salt marsh are associated with a group of beetles restricted to salt marshes. *Bledius spectabilis* (Staphylinidae) digs a vertical burrow with a long narrow neck surmounting a wider living chamber below. This burrow, which is plugged when the tide come in, acts as an air-store. The female uses it to rear her eggs and young, fighting off attacks from a parasitic ichneumonid wasp and predator beetles. *Anurida maritima*, a small dark violet springtail which also retreats to underground refuges when the tide comes in, searches for edible fragments when the marsh is exposed. Aphids are well adapted: some surround their colonies with water-repellent wax, their young are dispersed floating on the surface of the sea.

A range of butterflies, moths and bees collect nectar and pollen from the extensive stands of *Limonium vulgare* and *L. humile*, which are specifically associated with a number of insects. In some years sea-lavender suffers from invasion by such enormous numbers of the aphid *Staticobium staticis* that it fails to flower. These aphids tolerate tidal submergence, but whether this is true of their predators and parasites – including parasitic wasps and the larvae of hoverflies and ladybirds – is unknown. The weevil *Apion limonii* is confined to *Limonium*, whose leaves are eaten by larvae of the plume moth *Agdistis bennetii*. *Suaeda vera*, which dominates the highest and driest salt-marsh zone, is colonized by a variety of insects; in death its woody stems are bored by common woodworm (*Anobium punctatum*). On the Norfolk coast seeds of *Salicornia*, *Suaeda maritima* and *Atriplex portulacoides* are extensively predated by caterpillars of the microlepidopteran *Coleophora atriplicis* (Proudfoot, 1993).

Rabbit populations at Scolt Head Island, which once grazed very extensively on the marshes, were estimated at between 8000 and 10 000 before 1954. Very few survived the myxomatosis epidemic; although their numbers subsequently built up they have fluctuated considerably since then, reaching a high point in summer 1987. A more virulent strain of the disease appeared in November 1987, and by the summer of 1988 the population was reduced by 80%. Hares also graze the marshes, as they do the Cefni Marsh.

Ants, particularly *Lasius flavus*, occur quite frequently on upper salt marshes where their mounds provide variation in the physical habitat. Woodell (1974) showed that such anthills on the north Norfolk coast had more *Puccinellia maritima* and *Festuca rubra* and less *Juncus gerardii*, *Limonium vulgare* and *Plantago maritima* than surrounding salt-marsh vegetation. The south-facing sides of these hills were also colonized by the ericoid chamaephyte *Frankenia laevis*, here at its northern limit. Many salt-marsh ants feed on honeydew from the root-feeding aphids which they maintain in galleries. Air is often trapped in nest mounds when anthills are flooded, but ant species found on marshes can endure considerable periods of submergence.

Ranwell (1972, p.122) emphasized that many salt-marsh animals of terrestrial origin have only slight adaptations to marsh conditions, describing the 'mass escape of terrestrial animals by flying, crawling up stems, swimming, or walking over the water surface supported by surface tension, one quiet evening on the equinoctial flood of the Fal marshes in Cornwall'. Daiber (1977) gives a very wide reaching review of the distributions of animals in relation to tidal flooding, salinity and vegetation in wet coastal ecosystems. A similar review by Adam (1990) also emphasizes the various respiratory adaptations, many of them behavioural, seen in salt-marsh animals.

5.3 SEASONAL CHANGES

Seasonal changes in temperature, tidal levels and salinity – which vary greatly from time to time and place to place within a marsh – exert a major influence upon the production and germination of seeds (Adam, 1990) and on the activity, growth and distribution of salt-marsh organisms in general.

5.3.1 GERMINATION AND SEEDLING ESTABLISHMENT

Germination of salt-marsh plants tends to occur in the spring and is encouraged by low salinity levels. Many of these species are **halo-**

phytes which normally live permanently in what are predominantly saline environments (section 3.1). Plants characteristic of non-saline environments are known as **glycophytes**, but though their germination and often seed viability are usually more strongly affected by saline conditions than those of halophytes, many grow in salt marshes, especially at the top of the upper marsh. Figure 5.3 shows the seedlings of some important British salt-marsh dicotyledonous plants, many in the family Chenopodiaceae.

Germination of *Salicornia* spp. is high even at high salinities. This is a feature of strong adaptive value in many of the habitats of these annuals; only after exceptional rain storms is soil surface salinity appreciably lowered in the low marsh and even then frequent tidal flooding soon restores it to normal levels.

Woodell (1985) conducted germination experiments employing freshwater, half-strength seawater, full-strength seawater, and one-and-a-half strength seawater. After 18–25 days in these treatments the seeds were transferred to fresh filter papers in Petri dishes and moistened with distilled water. In *Rumex crispus* there was almost no germination under saline conditions but the final high germination in distilled water was not influenced by previous saline treatment. *Juncus maritimus* showed optimal germination in distilled water; after transfer to distilled water germination of the treated seeds was positively correlated with treatment salinity. Germination patterns shown by the strandline species *Salsola kali*, in which germination after transfer to distilled water is inversely correlated with the salinity of the original treatment, and by *Limonium vulgare*, are very different (Figure 5.4). Seeds of the latter can survive at least 21 weeks immersion in sea water and show '**salt stimulation**', a higher percentage germination after transfer to distilled water following pretreatment in salt water than if they had been in fresh water throughout.

Few **seed bank** studies have been as thorough as that of Hutchings and Russell (1989) who investigated the seed regeneration dynamics of an emergent salt marsh at Hayling Island, Sussex, UK. The marsh receives fewer than 360 submergences per year; periods of exposure never exceed 9 days. Flow charts given for eight major species studied on the high, middle and low sites showed the number of seeds produced, the number viable, numbers of seeds germinating and seedlings surviving, as well as numbers of seeds present in the seed bank at various periods in the investigation. No seeds were recovered from *Atriplex portulacoides* at any site, highest seed output being by *Juncus gerardii*, *Plantago maritima* and *Puccinellia maritima* on the high, middle and low sites respectively. A higher proportion of seed appeared in the seed bank for species with seed ripening in the summer than those with seed ripening in the autumn. The strong correspondence between the composition of the growing vegetation and the seed banks on all sites reflected the transient nature of the seed banks, which derived from current vegetation. This contrasts with most plant communities dominated by perennials, whose seed banks include at least some long-lived propagules produced by previous populations.

5.3.2 SEASONAL GROWTH PATTERNS IN PLANTS

Seasonal biomass accumulation patterns within the same species may vary according to locality. *Spartina alterniflora* shows two seasonal peaks in below-ground biomass in southern marshes, such as those of North Carolina, on the Atlantic coast of North America. The first is in early summer at the time of maximal above-ground growth, and the second in late autumn following winter dieback of above-ground parts (Stroud, 1976; Gallagher, 1983). Further north in New England, where the growing season is much shorter, the pattern is different. Figure 5.5 shows that *S. alterniflora* populations investigated by Ellison, Bertness and Miller (1986) in a salt marsh at Rumstick

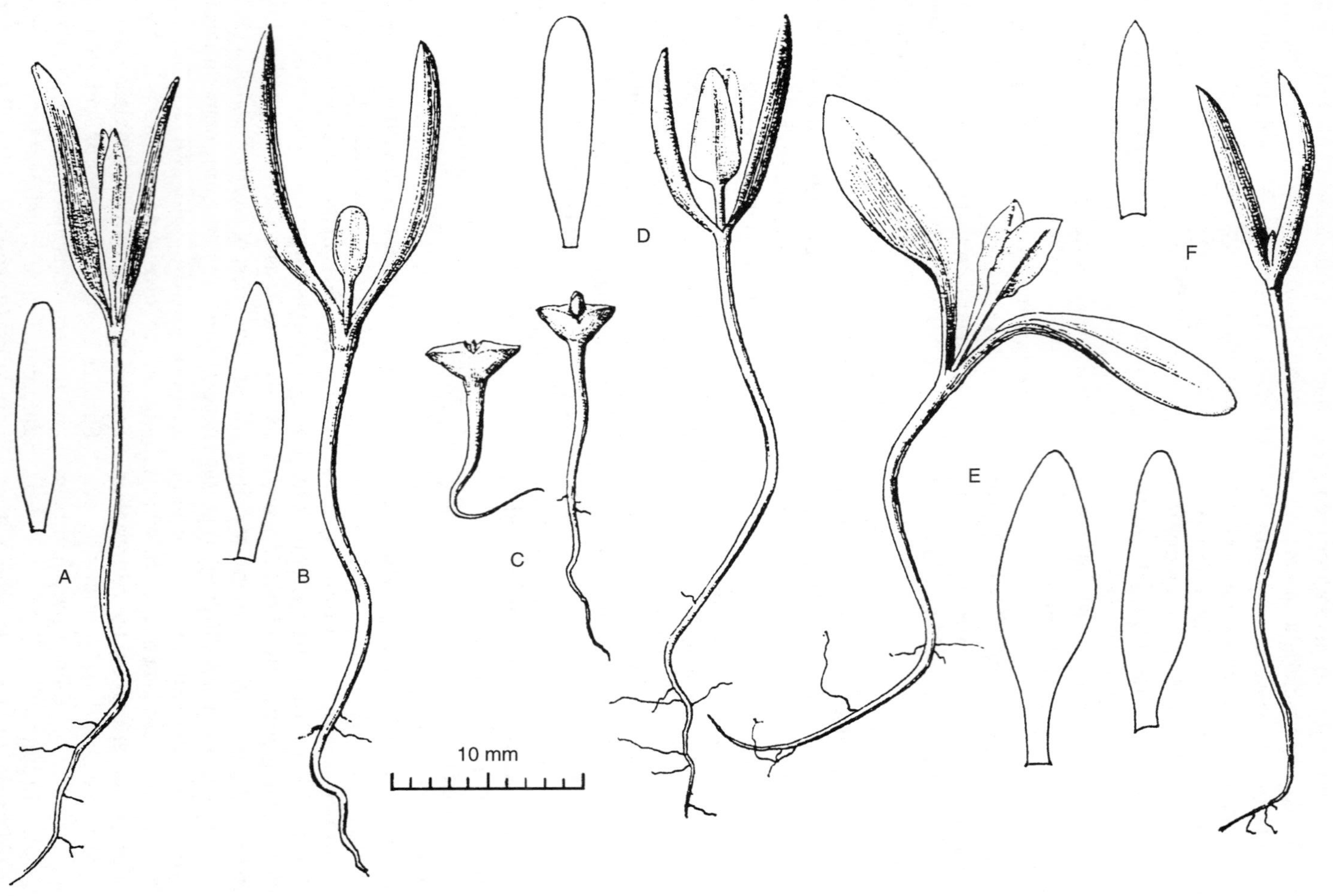
A
B
C
D
E
F
10 mm

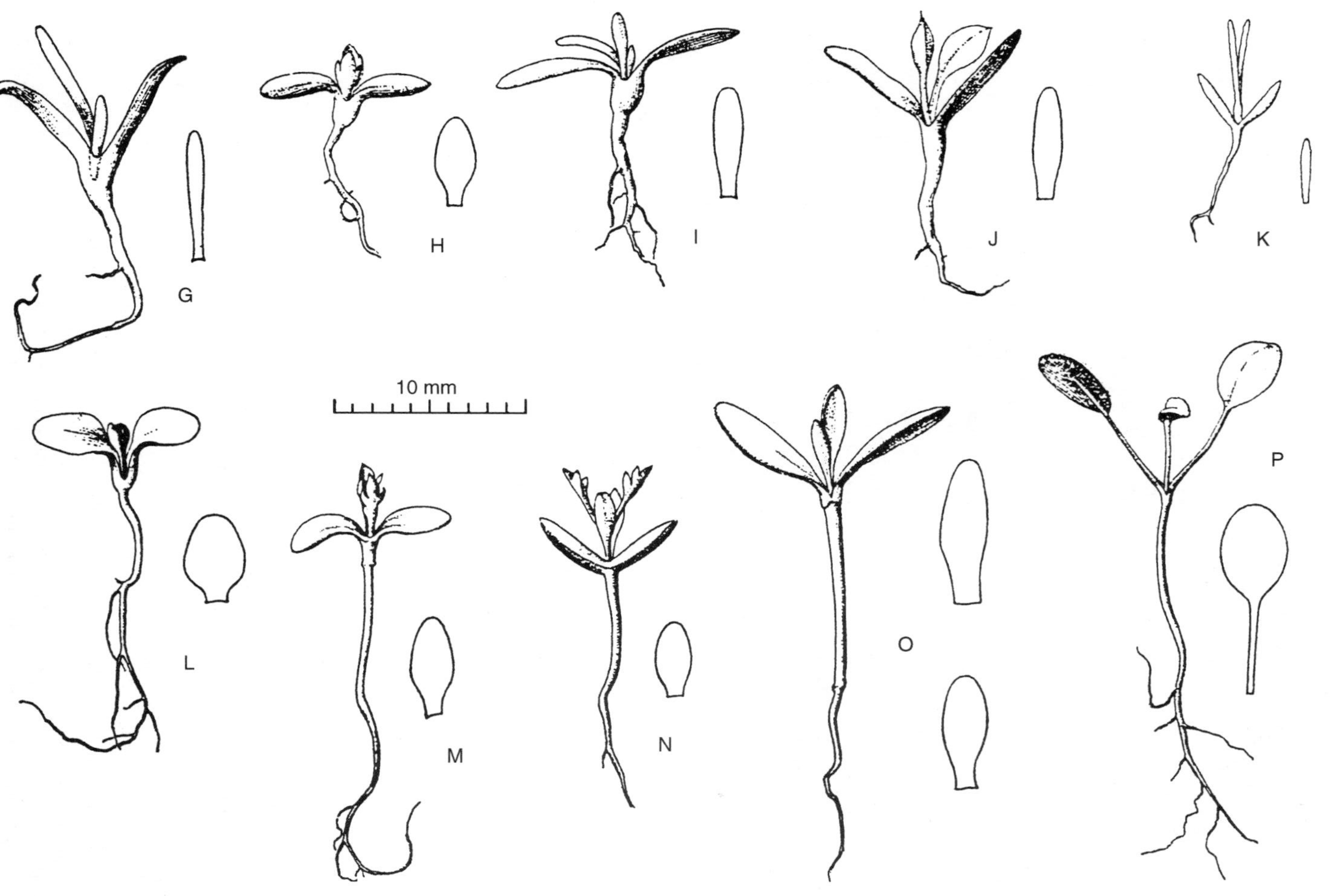

Figure 5.3 Seedlings of salt-marsh dicotyledonous plants. I. Chenopodiaceae. A: *Atriplex littoralis*; B: *A. portulacoides*; C: *Salicornia* spp.; D: *A. prostrata*; E: *Beta vulgaris* ssp. *maritima*; F: *Suaeda maritima*. II. Other families. G: *Plantago maritima*; H: *P. coronopus*; I: *Armeria maritima*; J: *Limonium humile*; K: *Spergularia media*; L: *Sonchus* sp.; M: *Tripleurospermum maritimum*; N: *Seriphidium maritimum*; O: *Aster tripolium*; P: *Cochlearia anglica*. Outlines are of cotyledons seen from above. (From Dalby, 1970; courtesy *Field Studies*.)

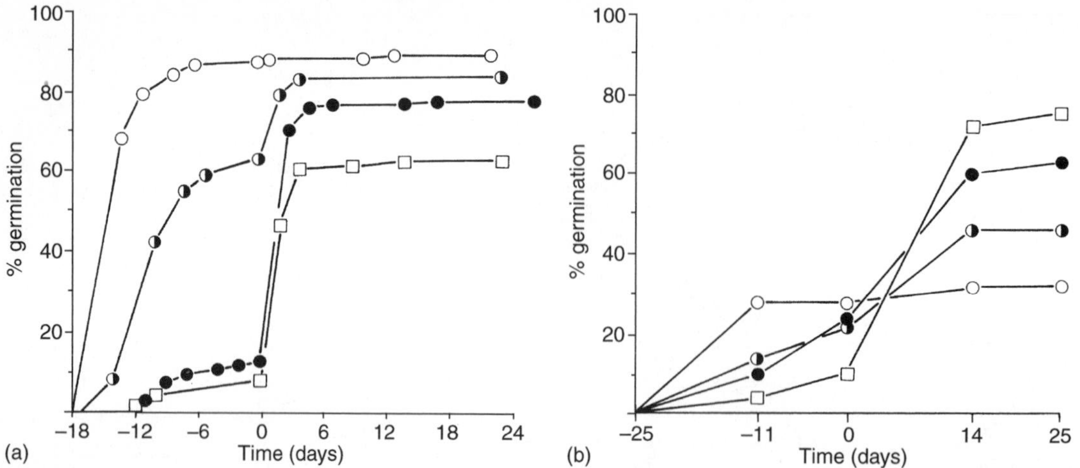

Figure 5.4 Germination in relation to salinity. (a) *Salsola kali.* (b) *Limonium vulgare.* ○—○, fresh water; ◑—◑, half-strength sea water; ●—●, full-strength sea water; □—□, one-and-a-half-strength sea water. Day 0 is the day of transfer to distilled water after the seeds had been in the appropriate treatment for 18–25 days. Experimental temperatures within 20±3°C. (Redrawn from Woodell, 1985.)

Cove, Rhode Island, exhibited a seasonal cycle in which below-ground biomass was generally highest in midsummer (as was above-ground biomass), but without a secondary peak corresponding with the autumn dieback of the above-ground parts. This effect is most clearly seen in the tall height form population (mean height 150 cm) at the lower marsh edge. The short height form had a mean height of 80 cm, occurring up to +1.2 m mean tidal height above which *Spartina patens* dominated the marsh vegetation. Ellison, Bertness and Miller (1986) postulated that the shortened New England growing season compressed the time available for root growth and eliminated the summer 'depression' in below-ground biomass seen in southern marshes.

Seasonal and spatial **root foraging patterns,** although difficult to study, are of great importance. In *S. alterniflora* populations at

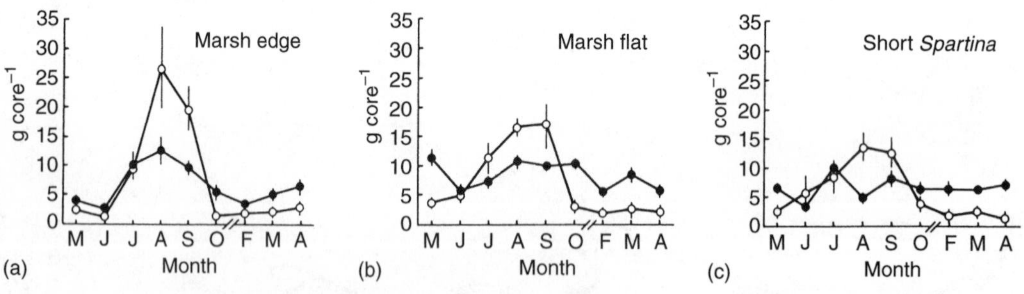

Figure 5.5 Rumstick Cove, Rhode Island, USA. Total above-ground (○) and below-ground (●) dry biomass of *Spartina alterniflora* (g core^{-1}, multiply by 301.4 to obtain g m^{-2}) in each of the three marsh zones over the course of the year. (a) Tall-height form on the marsh edge (+0.1 m to +0.6 m mean tidal height); (b) tall-height form on the marsh flat (+0.6 m to +1.0 m mean tidal height); (c) short-height form (+1.0 m to +1.2 m mean tidal height). Values are means, bars indicate 1 S.E. bar. (Redrawn from Ellison, Bertness and Miller, 1986.)

Rumstick Cove, seeker root elongation proceeded most rapidly in early spring before onset of rapid above-ground growth; fine root elongation followed in summer and early autumn. When seeker roots encounter suitable areas for growth, fine roots develop for nutrient absorption.

5.3.3 RAINFALL, SALINITY AND SURVIVAL

The increasing openness of salt-marsh vegetation going south from northern Europe to the sub-tropics appears to be dependent on the degree to which evapotranspiration exceeds precipitation for substantial periods. Apart from the pioneer zone, vegetation in north European marshes is normally interrupted only by discontinuities in the marsh surface. In the Camargue, southern France, where evapotranspiration is greater than precipitation for much of the summer, cover is often less than 80% and may be as low as 50%. In exceptionally warm dry summers, exposed surface soils of British marshes may become hypersaline, dry out and crack, as in the Dovey Marshes, North Wales, in July 1995 (Figure 5.6). Indeed, as Ranwell (1972) points out, it is the timing and amount of rainfall, rather than the tidal influence, which dominates salinity levels in the upper salt marsh.

In the dry summer of 1967, **drought** on the northern Irish coast caused extensive dieback of species such as *Glaux maritima*, *Limonium vulgare* and *Spartina anglica*, all of which are tolerant of salinities in excess of those of seawater. Ranwell (1972, p.17) quotes this example in support of the possibility that successive replacement of species in progressively higher marsh zones may be linked with tolerance of drought rather than salinity. **Transplant experiments** have demonstrated that drought is a limiting factor of growth for *L. vulgare* on the sandy soil of an upper marsh (Boorman, 1967).

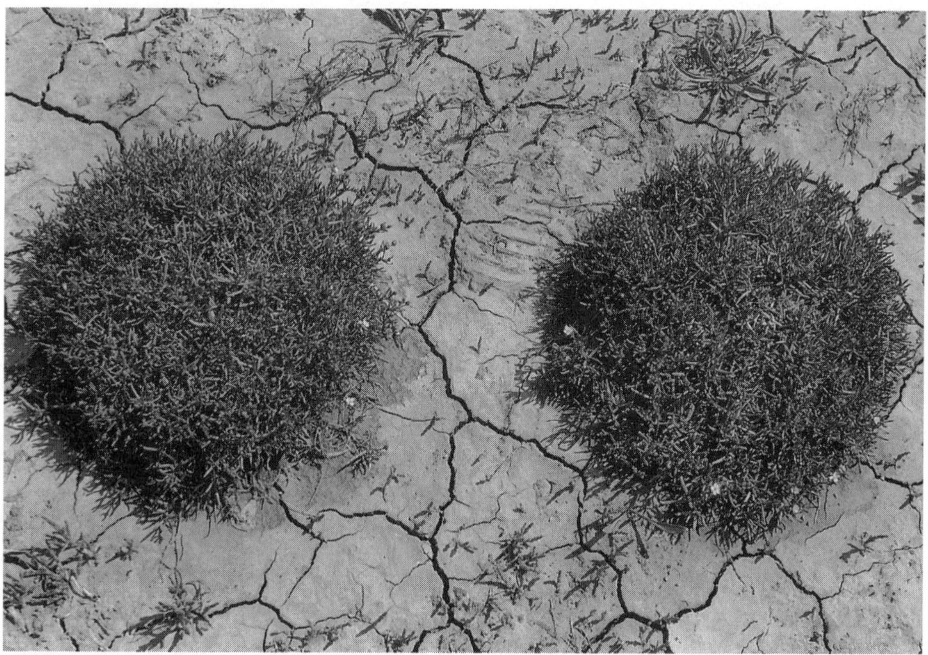

Figure 5.6 Two clumps, each about 30 cm in diameter, of greater sea-spurrey (*Spergularia media*) with small plants of sea aster (*Aster tripolium*) and glasswort (*Salicornia* sp.). Note dry cracked surface of the soil. Dovey marshes, near Ynyslas, North Wales. July 1995. (Photograph by John R. Packham.)

5.3.4 SEASONAL BEHAVIOUR IN ANIMALS

Seasonal changes influence the annual **migratory and reproductive patterns of birds**, many of which feed on the invertebrates and plants of salt marshes, sandflats and mudflats (section 2.8). The reproductive and growth patterns of the invertebrates – and in many cases the depth at which they occur in the sediment – are also strongly influenced by the weather.

Annual rhythms in the bird populations of Scolt Head Island and Blakeney Point, Norfolk, UK

These two coastal features and their immediate environs afford a remarkable combination of inshore waters, sandy cliffs, dunes, shingle strands, mud flats, salt marshes, reedbeds, freshwater marshes and freshwater and brackish pools. They are especially well known for the large and complex bird populations present, many of them migrating great distances every year (Seago, 1989). There is a continual change in these bird populations throughout the year with January seeing wildfowl returning from their annual moult migration in the German Waddensee, the moulting place for almost all the shelduck in north-west Europe. Wigeon, pintail, mallard and teal are common at this time; goldeneye, velvet and common scoters, long-tailed duck, eider, and red-breasted merganser are found inshore. Weary fieldfare and redwing which have flown from southern Norway are often preyed upon by great black-backed gulls. Wintering birds of prey include merlin, peregrine and hen harrier.

Waders are particularly conspicuous in May with godwits, knot, grey plover and spotted redshank; colourful and attractive vagrants may include the hoopoe, tawny pipit, alpine swift, subalpine warbler, and red-footed falcon. June is the month for breeding birds including terns, ringed plover, oystercatcher and shelduck. A complete lack of disturbance is essential for successful breeding of Sandwich terns; their nests are usually very close together on the embryo dunes or shingle and the parent birds will hunt over a radius of 65 km for sand eels or sprats if food is scarce locally. Young Sandwich terns often form groups guarded by a single adult. In contrast, straying young common terns are brutally treated by other adults. Terns defend their nesting sites vigorously; people are driven off by repeated high speed passes and young rabbits may even be killed. Food for the majority of the large bird populations comes largely from the sea; eiders, for example, are known to dive to 9 m or more.

By the end of July there is a good showing of waders – greenshank, spotted redshank, whimbrel, bar-tailed godwit, turnstone and sanderling – but a succession of other arrivals extends into October with gannet, razorbill and guillemot all represented. The predatory skuas disrupt the feeding of many other birds, especially terns, which are forced to disgorge freshly caught fish.

5.4 PLANT–HERBIVORE RELATIONSHIPS IN AN EVOLVING SUB-ARCTIC SALT MARSH

A series of studies made on the sub-arctic salt marshes of La Pérouse Bay, Manitoba (58°4′N, 94°4′W), are illustrative of the conditions prevailing in the southern region of Hudson Bay, where extensive salt marshes continue to form in a very flat area subject to a post-glacial rate of isostatic uplift of 1.2 m 100 yr^{-1}. These marshes frequently develop on the landward side of offshore bars and ridges on an emergent coast with a gradient often no more than 0.5 to 1 m km^{-1} (Jefferies, Jensen and Abraham, 1979). The unconsolidated water-saturated sediments are colonized by *Carex subspathacea*, *Hippuris tetraphylla* and *Puccinellia phryganodes*. *Salix brachycarpa*, the dominant species of the low willow tundra characteristic of the coastal strip, and *Leymus arenarius* var. *mollis* establish on elevated mounds resulting from frost heaving close to the limits of the highest tides. *Salicornia europaea* agg. and *Triglochin maritimum* are here close to their northern limits in Canadian coastal sites; the former occurs on those parts of the upper marsh where

drainage is impeded and relatively high salin-
...arsh is an
...inated by
...*ia fisheri,*
...l species of

...E ON
...ERSITY

...esser snow
...s) – which
...5000 pairs
...ds on the
...from June
...he vegeta-
...en migrate
...*Puccinellia*
...en the *Salix*
...ed surface
...d ice accu-
...lonized by
...*um.*
...xclude the
lesser snow goose from 5 m × 5 m salt marsh plots at La Pérouse Bay the composition of the vegetation changed rapidly (Bazely and Jefferies, 1986). In the oldest fenced area, 16 species of higher plant were present after 5 years of exclusion, compared with six species in the adjacent control area where grazing continued, and in which frequencies of both graminoids and dicotyledons showed little change from 1982 to 1984. In the absence of grazing, *Carex subspathacea* replaced *Puccinellia phryganodes* as the dominant graminoid, while dicotyledons such as *Potentilla egedii* and *Plantago maritima* increased substantially in frequency (Figure 5.7). When an exclosure was erected immediately after snowmelt in May 1982, the uninterrupted stolon growth of *Potentilla egedii* and *Ranunculus cymbalaria* caused these species to have a higher frequency than in the adjacent grazed sward when both were assessed in August of the same year. Because *Anser caerulescens* is a generalist in its foraging, only those species able to tolerate heavy grazing and trampling grow well in grazed areas. Salt marsh and sedge-meadow communities can be degraded and destroyed if geese grub out the roots, rhizomes and swollen shoot bases of graminoids. If these plants are replaced by mosses and *Potentilla palustris*, which are not eaten by geese, there is a permanent loss of grazing.

In a particular season the standing crop was highest in exclosures erected 1 or 2 years before. At this stage, dicotyledonous plants comprised 30% of the total biomass; several of these species flowered and seeded profusely in the same way as in many sand dune species released from rabbit grazing. Peak standing crop in older exclosures is less and so is the proportion contributed by dicotyledonous plants. Plant litter accumulated in the ungrazed plots where the ratio of maximum standing crop to litter fell from more than 4 : 1 to less than 2 : 1.

Net above-ground primary production (NAPP) varied from year to year, that of an ungrazed sward dominated by *Puccinellia phryganodes* and *Carex subspathacea* being c. 100 g m^{-2} yr^{-1} in 1979 and 50–60 g m^{-2} yr^{-1} in 1980; NAPP for grazed marsh is considerably higher (Cargill and Jefferies, 1984a,b). The changes which occur in the vegetation when the grazer is excluded make it less attractive to the geese. When the fence of a 2-year-old exclosure at La Pérouse Bay was removed in spring the sward was grazed very little in the following two summers.

5.4.2 GROWTH RESPONSES OF ARCTIC GRAMINOIDS FOLLOWING GRAZING

An important prediction of the **herbivore-optimization model** (Dyer, 1975) is that up to a certain limit moderate levels of grazing lead to enhancement of net above-ground primary production (NAPP) over that of ungrazed plants. In plots of *Puccinellia phryganodes* subject to grazing by goslings of the lesser snow goose for a limited period early in the season, it has been shown that the nitrogen content of the regrowth in grazed plots is higher than

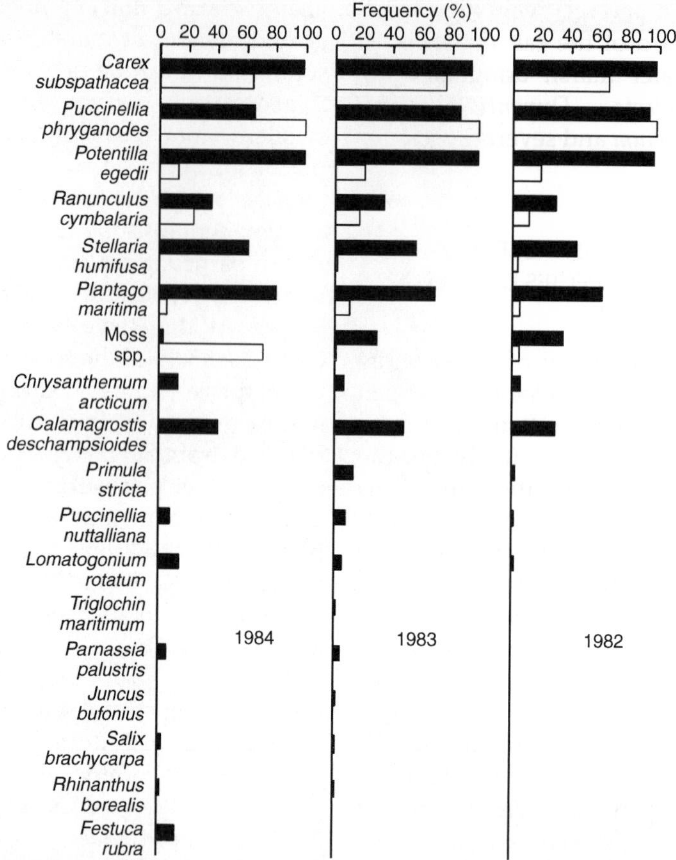

Figure 5.7 Percentage frequencies of species in a salt marsh at La Pérouse Bay, Manitoba in 1982–1984; solid bars indicate data recorded in an exclosure (5 m × 5 m) erected in 1979; open bars indicate data recorded in an adjacent plot where the vegetation is grazed by lesser snow geese. Results based on presence or absence of species in 200 quadrats (5 cm × 5 cm). (From Bazely and Jefferies, 1986; courtesy *Journal of Ecology*.)

that of shoots in the ungrazed plots, and that grazed plots accumulate biomass in excess of ungrazed plots provided faecal droppings remain within the plot. These positive effects are not observed when droppings do not remain within the plot, depend on rapid recycling of nutrients via the faeces, and are augmented by higher rates of nitrogen fixation by cyanobacteria. These are better able to colonize the surface of the sediments in the relative absence of plant litter at grazed sites, and largely replace the nitrogen loss experienced when the geese migrate south. The ability of swards to recover from grazing decreases during the course of the summer, as does nitrogen content of the forage (Hik and Jefferies, 1990).

The effects of altering the frequency and timing of multiple grazings (including faecal input) were also investigated. A factorial pot experiment in which leaves of individual tillers of half the plants were clipped to mimic the demography of shoots grazed by wild birds, and ammonium nitrate solution was received by half the plants, threw further light on the natural situation.

Addition of nutrients ameliorated the adverse effect on shoot growth and NAPP of clipping *per se*; experimental plants which received nutrients, but whose leaves were not clipped, produced the highest amount of above-ground biomass (Hik *et al.*, 1991). In similar experiments on the effects of grazing by captive goslings on swards of *Carex subspathacea*, *Festuca rubra* and *Calamagrostis deschampsioides*, there were no significant differences in standing crop or NAPP between grazed and ungrazed plots at the end of the summer. Nitrogen contents in leaf tissue from grazed and ungrazed plots were similar for all three species, suggesting that faecal nitrogen was not rapidly incorporated into plant biomass during the growing season (Zellmer *et al.*, 1993).

Lesser snow geese play an important role in the **maintenance of *Puccinellia* sward** at La Pérouse Bay. When grazed, this stoloniferous grass, and *Carex subspathacea*, a rhizomatous sedge particularly abundant in brackish sites, regrow from basal meristems; axillary shoots and new leaves are also produced. As in other natural grasslands subject to grazing, tillers grow to form a prostrate dense canopy. Intensive grazing early in the summer removes the apical meristems of dicotyledonous species, thus tending to maintain the community in an early successional state, while the 6 weeks which remain at the end of summer after the geese migrate enables the sward – unless it has been grossly over-grazed – to recover before the onset of winter.

5.4.3 INFLUENCE OF GRAZING ON SUCCESSION IN EMERGENT VEGETATION IN THE HUDSON BAY LOWLANDS: AN UNCERTAIN FUTURE

Negative effects include the destruction of large areas of salt marsh and sedge-meadow communities by the grubbing, especially in spring, of underground organs of graminoids together with very heavy grazing of species such as *Carex aquatilis*. In extreme cases the vegetation is stripped to expose the underlying peat. Near Eskimo Point, Northwest Territory, goose activities have degraded or totally destroyed plant communities in an area of some 5000 km^2. The combination of isostatic uplift and erosion of organic sediments makes it unlikely that the original plant communities will reappear at these degraded sites, even if the present very high numbers of geese diminish.

The colonial foraging of the geese, when not too severe, helps maintain the salt marsh communities and delays – but ultimately cannot prevent – the successional processes resulting from isostatic adjustment which have been studied elsewhere, notably on the Bothnian coast (Ericson, 1980; Cramer and Hytteborn, 1986). As the land rises the salt marsh communities of the tidal flats will be replaced by 'low willow' communities of *Salix brachycarpa* and fen associations of various graminoids (Kershaw, 1976; Jefferies, Jensen and Abraham, 1979). The future of the exclosure vegetation, however, is worth separate consideration.

Because the shore slopes so gently, wave action is minimal and large quantities of litter are not deposited along the shore line. Despite their tidal inundation in September and August, litter continues to accumulate in exclosures. The combined leaf area index (LAI) for live and dead shoots in exclosures exceeds 3.0 late in season, while that for the standing crop in grazed sites is less than 1.0. When LAI approaches or exceeds 3.0 it is probable that attenuation of photosynthetically active radiation (PAR) through the canopy will limit the rate of photosynthesis within the sward (Johnson and Tieszen, 1976).

The low willow community beyond the strand line is dominated by *S. brachycarpa*, *Calamagrostis deschampsioides* and *Festuca rubra* agg. These and other low willow associates can all be found at very low frequencies in the older exclosures, whose vegetation is unlike any found elsewhere in La Pérouse Bay and may represent a cul-de-sac in community development. Bakker (1978) described abrupt changes in salt-marsh vegetation on a Dutch Frisian island in response to mowing and grazing, with little further change over 5 years. Jensen (1985) reports that in Denmark a *Puccinellion maritimae* showed no qualitative changes in species composition following cessation of sheep and cattle grazing, though other communities gained and lost species when exclosures were erected. Such evidence indicates that the structure of plant communities often changes abruptly and may then remain in a steady state for long periods. Vegetation in the older salt marsh exclosures at La Pérouse Bay certainly remains stable for a period, but may well later change in ways similar to other ungrazed communities elsewhere in the area.

Between 1985 and 1991 extensive grubbing and grazing by geese, followed by erosion, led to the loss of 50% of the vegetation on the La Pérouse Bay salt marsh. Because the organic layers and some of the underlying sediments have eroded, Hik *et al.* (1992) consider it unlikely that the assemblages of plants described above will re-establish within 50 years. When such areas are re-colonized the changes caused by excessive grubbing and erosion, combined with the progressive effects of isostatic uplift, are likely to encourage the establishment of vegetation different from that which has been lost. Thus, on a time scale of 100–150 years, **non-equilibrium conditions** will prevail.

5.4.4 ENVIRONMENTAL DEGRADATION, FOOD LIMITATION AND THE COST OF PHILOPATRY

The breeding population of lesser snow geese at La Pérouse Bay increased from 2000 pairs in 1968 to 8000–9000 pairs in 1993; this was associated with a real increase in the density of birds using the brood-rearing areas. Over the same period there has been concurrent marked habitat degradation resulting in a significant decrease in food availability (Williams *et al.*, 1993). Female snow geese usually exhibit strong **philopatry**, an adaptive conservative behavioural strategy in spatially heterogeneous habitats in which individuals tend to exhibit long-term use of particular areas, employing the nesting and feeding areas in which they were reared as goslings. Colonial breeding and foraging, and philopatry to specific feeding areas is held to reflect a strategy which both minimizes mortality due to predation by predator saturation, and reflects a strongly synergistic relationship between lesser snow geese and their principal forage plants.

When local environments change over time, philopatry may become maladaptive and behaviour may change. Between 1979 and 1991 above-ground biomass of graminoid salt-marsh sward at La Pérouse Bay declined from c. 50 g m^{-2} to 25 g m^{-2}. During this same period the mean proportion of goslings in a brood that survived from hatch to ringing, at 5–6 weeks of age, decreased from some 65% to 40–55%. Cooch *et al.* (1993) found that, in recent years, increasing numbers of family groups have dispersed from traditional feeding areas at La Pérouse Bay. Goslings at the dispersed sites, where available suitable forage per capita was greater, were found, in each of 5 years, to be larger, heavier (7.3%) and to have significantly greater first-year survival than those at La Pérouse Bay.

5.5 BIOMASS, PRODUCTIVITY AND ENERGY FLOW

Plant biomass greatly exceeds animal biomass in the standing crop of salt marshes, in which **decomposers** are responsible for a far higher proportion of total energy flow than are members of the herbivore subsystem (section 1.5). In the example shown in Figure 5.8 the dry weight of animals was shown by Paviour-Smith (1956) to be about 2% of that of the plants, while dead organic matter exceeded the combined weights of both by a factor of 22.1. Nevertheless, the role of animals and other heterotrophs (section 2.7) in facilitating energy flow and nutrient cycling is of crucial importance to the proper functioning of the ecosystem.

The inorganic contents of salt-marsh plants can exceed 50% of the dry weight, so **plant biomass** for this habitat is best expressed on an **ash-free basis**; ash weight being determined after heating to 500°C. Above-ground biomass (W_s) is usually determined directly by clipping vegetation from randomly selected quadrats at ground level and then drying the live material, which has to be carefully separated from dead tissue (Long and Mason, 1983). This time-consuming process may be partially superseded by remote sensing techniques based on the infra-red reflectance of live shoots.

5.5.1 PRIMARY PRODUCTION

Though many authors comment on the high productivity of salt marshes, numerous investigations have failed to consider important aspects of this complex process. After making this point, Long and Mason (1983, pp. 90–103) discuss the results of the many studies which have been made – largely in North America and concerning above-ground organs – and evaluate the methods used in measuring the

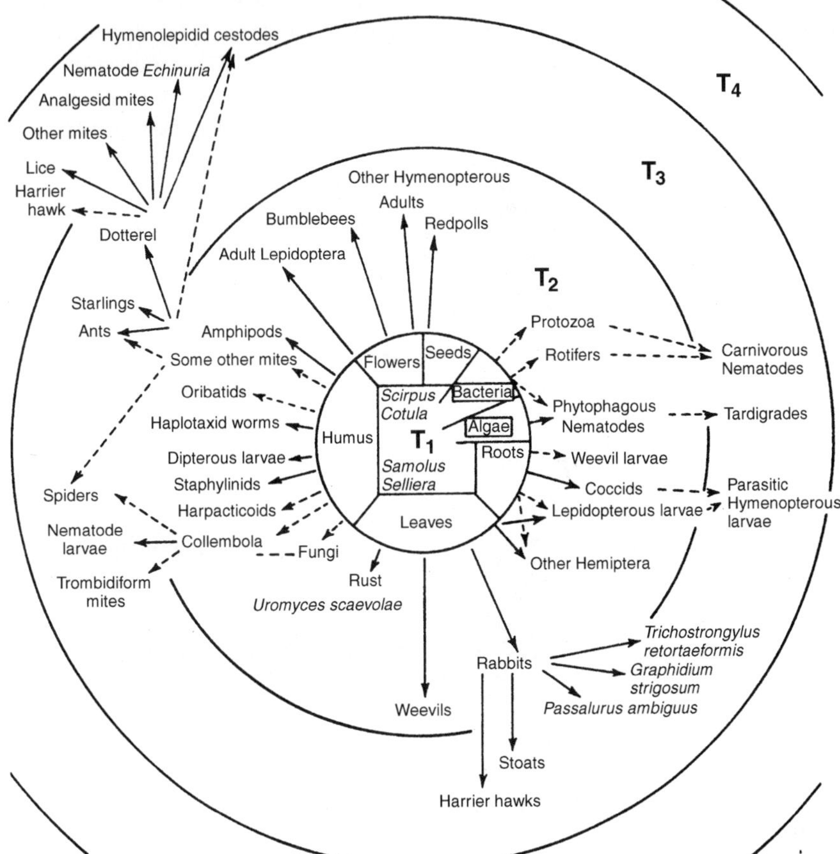

Figure 5.8 Food web of a salt meadow community at Hooper's Inlet, Otago Peninsula, South Island, New Zealand. Producer organisms (T_1) are enclosed in boxes. Succeeding trophic levels (T_2, herbivores (and detritivores); T_3, carnivores; T_4, top carnivores and parasites) are enclosed in succeeding concentric circles. Solid lines represent known relationships and broken lines assumed relationships. (From Ranwell, 1972, after Paviour-Smith, 1956.)

carbon exchanges involved in primary production by vascular plants in salt marshes (Figure 5.9).

Net primary production (P_n) is the mass (or energy) incorporated by photosynthesis (**gross primary production, P_g**) less that lost in **respiration** (*R*) during a given time period, and is usually expressed per unit area of the marsh. As P_n represents a gain of material by the plant community, it can be estimated from the resultant of the changes in plant biomass (ΔW) and all losses over a specified time interval, being redefined (Long and Mason, 1983):

$$P_n = \Delta W + L + G + E$$

where ΔW = change in biomass; L = losses by death or shedding; G = loss to grazers, i.e. predation; and E = loss through root exudation.

Losses through organic material exuded into the rhizosphere – which in arable crops may account for up to 50% of the total carbon assimilated – are particularly difficult to estimate and have normally been ignored in salt marsh studies. Direct observations of the shoots of plants within permanent quadrats make it possible, though laborious, to measure the loss of individual stems or leaves. Loss of mass can then be derived by regressing weight on a non-destructive measure such as leaf length or area, using parallel destructively-tested samples. An alternative is to measure change in the amount of dead material. Significant losses in most ecosystems are limited to those caused by decomposition, including consumption by macro-invertebrates, but in salt marshes tidal import and export of plant litter must be measured:

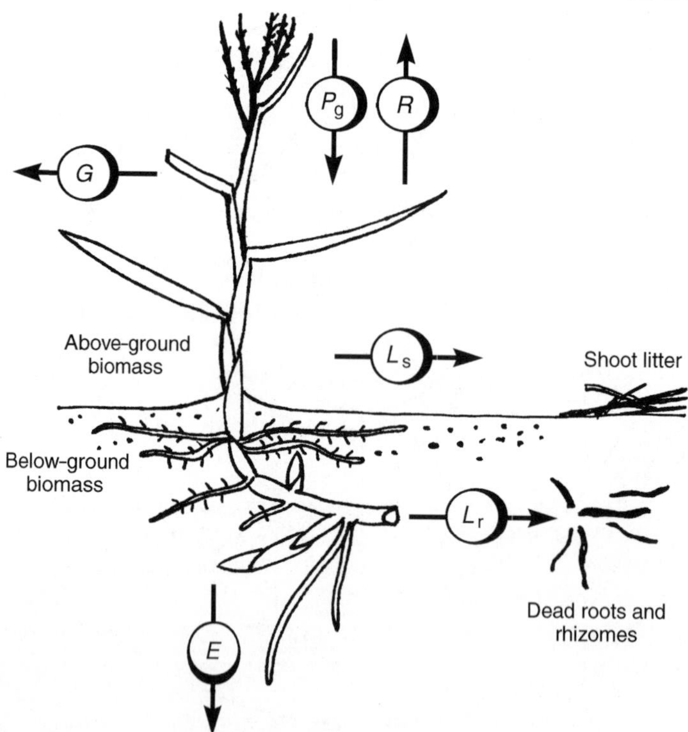

Figure 5.9 Diagram illustrating carbon exchanges between salt marsh vascular plants and their environment. Stand photosynthesis or gross primary production (P_g), respiration (*R*), grazing (*G*), shoot death (L_s), root and rhizome death (L_r) and root exudation (*E*). *Spartina anglica* is shown, with shallow fibrous roots, deep penetrating anchoring roots and young rhizomes with protected apices (*E*). (Redrawn from Long and Mason, 1983.)

$$L = W_d + D + T_e - T_i$$

where W_d = change in dead vegetation; D = loss due to decomposition; T_e = tidal export of material; and T_i = tidal import of material.

Repeated and regular measurements over the whole year are necessary if losses are to be estimated accurately by either direct or indirect methods. In consequence, attempts are frequently made to estimate net primary production by extrapolating a value from measured biomass, making specific assumptions which are outlined by Long and Mason (1983, p.98).

Despite the apparently high net production of *Spartina*, the fact that in this plant up to 70% of gross production may be accounted for by respiration implies that environmental changes causing quite moderate alterations in the balance between assimilation and respiration could cause a large drop in net primary production. The high respiratory demands of *Spartina* appear to be linked to the maintenance of a high below-ground biomass, and the costs of soil re-oxidation, symbiotic nitrogen fixation, and of operating mechanisms for maintaining salt balance (Long and Mason, 1983).

Algal productivity makes an important contribution to total net production on many salt marshes (Adam, 1990, p.338). Algae occur as **phytoplankton** in creeks and salt pans, grow on the sediment surface beneath vascular plants, occur epiphytically on plant stems and also as **epibenthos** on creek sides and the bottoms of salt pans. Although the standing crop of epibenthic algae is low compared with that of vascular plants, productivity tends to be high because virtually the whole algal biomass is photosynthetic and turnover is also rapid. Seasonal variation in production in algae is less varied than in vascular plants. Consequently, algae may in temperate regions be responsible for the major part of total marsh production in autumn and winter. Many microalgae on mud can migrate several millimetres to the surface, which assists survival during active sedimentation and tidal flooding.

Productivity is closely linked to the supply of nutrients, particularly of nitrogen, as the work of Valiela and Teal (1979a,b; section 4.5) on the Great Sippewissett Marsh demonstrates.

5.5.2 PRODUCTIVITY AND TIDAL GRADIENTS

The above-ground (shoot) dynamics of *Spartina alterniflora* vary markedly across tidal gradients. At Rumstick Cove, New England, shoot biomass and shoot height decrease from the lower intertidal region, where a tall-height form is recognized, to the upper intertidal zone occupied by the short form (Figure 5.5; sections 5.2 and 5.3). This reduction in production is associated with decreases in nutrient levels (except for P), percolation rate, decomposition rate, and oxygen levels (as measured by redox). In contrast, substrate hardness, soil acidity, and peat accumulation all increase further up the shore. It is postulated that the more rapid rate of decomposition at the lower marsh edge and on the marsh flat, together with the consequent decrease in below-ground debris, may both increase available soil nutrients and increase available space for root growth and water movement in these zones (Ellison, Bertness and Miller, 1986).

A very different situation was revealed when the production ecology of intertidal salt marsh at Nehalem Bay, north-west Oregon, USA, was investigated by Eilers (1979), using the harvest method with allowances for production losses resulting from mortality and disappearance during the growing season. Vegetation was sampled at intervals along an elevation gradient which corresponded to maximum submergence periods per day of 17 hours at the lower border of the marsh to near zero hours at the upper margin. Calculated above-ground net production, which ranged from 230 g m^{-2} yr^{-1} (*Triglochin maritimum* adjacent to mudflat) to 2800 g m^{-2} yr^{-1} (mixed high marsh dominated by *Aster subspicatus*, *Potentilla pacifica* and *Oenanthe sarmentosa*), increased with higher elevation and decreasing submergence period. Production in low, middle and high marsh averaged 1200, 1400 and 1700 g m^{-2} yr^{-1}, respectively.

5.5.3 SECONDARY PRODUCTION BY AN AVIAN HERBIVORE

The grazing chain of the herbivore subsystem is in salt marshes generally responsible for a smaller proportion of total energy flow than is the detritus chain of the decomposition subsystem. Nevertheless, salt marshes often provide food at critical stages in the lives of particular herbivores, as the following example illustrates.

Observed growth rates of vertebrates beginning life in boreal and arctic regions are often rapid in the short snow-free season. This requires provision of adequate supplies of nutrients and is in many species accomplished by parental subsidies, including lactation in mammals and food provision in birds. A yolk-sac, which can support a gosling for a few days only, is the only nutrient subsidy received by the young of the arctic-nesting lesser snow goose (*Anser caerulescens caerulescens*). In consequence, these largely herbivorous birds have rapidly to convert a great deal of plant material – generally low in protein and high in fibre – to animal tissue high in protein and low in

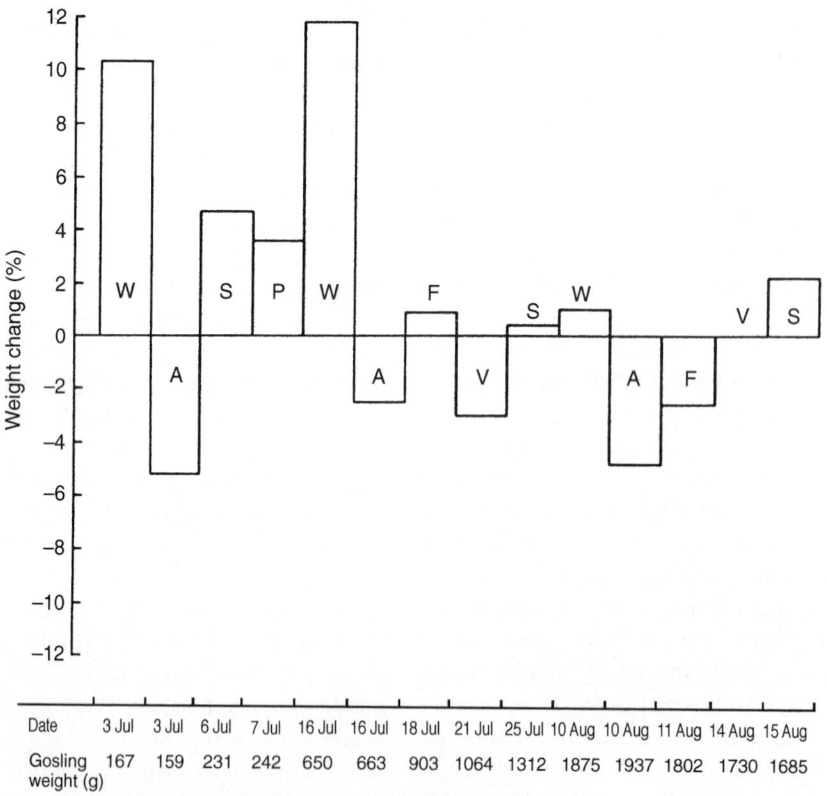

Figure 5.10 Weight changes of captive goslings of the lesser snow goose in feeding trials at La Pérouse Bay after 6 hours of feeding followed by 2.5 hours of food deprivation. Bars represent direction and magnitude of weight change. Symbols within bars represent forage type: W, duck chow; A, *Carex aquatilis*; S, *C. subspathacea*; P, *Puccinellia phryganodes*; F, *Festuca rubra*; V, *C.* × *flavicans*. (From Gadallah and Jefferies, 1995b; courtesy *Journal of Applied Ecology*.)

fibre. Efficiency of digestion of forage in the small, immature guts of the goslings is low. When the goslings hatch, usually around the end of the third week in June, each weighs about 85 g; this increases to between 1500–1800 g at the start of the autumn migration 7 weeks later (Figure 5.11). **Selection of high-quality forage** between hatch and migration is essential if the goslings, and indeed adult females (which may lose over 32% of their body weight during incubation and replace all their flight feathers before migration), are to survive.

This example of **secondary production** is of considerable importance, particularly with regard to the nutrient contents of the principal forage plants utilized and the growth of captive goslings (Gadallah and Jefferies, 1995a,b). Captive goslings grazed for 6 hours on salt-marsh swards of *Puccinellia phryganodes* or *Carex subspathacea*, or fed in pen trials on standard duck food, gained or maintained weight (Figure 5.10). These salt-marsh forages had higher contents of the essential nutrients nitrogen, calcium and phosphorus, as well as lower fibre and silica contents, than inland swards of *Festuca rubra* and *Carex* × *flavicans*, on which – along with pen trials in which leaf tips of the tall salt-marsh species *Carex aquatilis* were supplied on trays – they either maintained or lost weight. These results supported field evidence (section 5.4) that decline in gosling weight at La Pérouse Bay since 1980 results from a change in diet, the birds having been driven to the use of alternative forage plants in the absence of sufficient biomass of the preferred species.

5.6 ECOLOGY OF THE SAPELO ISLAND SALT MARSHES, SOUTH GEORGIA, USA

These salt marshes have been extensively investigated, particularly in terms of the dynamics of the *Spartina alterniflora* communities which dominate the salt marshes of the barrier islands along the coast of Georgia

Figure 5.11 Goslings of blue phase lesser snow goose (*Anser* (= *Chen*) *caerulescens caerulescens*) feeding on graminoid vegetation, Hudson Bay, Canada. (Photograph by R.L. Jefferies.)

(Pomeroy and Wiegert, 1981). Following the approach of Wiegert (1979) this discussion will view the ecological processes involved in terms of the overall carbon balance. These processes may be classified into three groups:

1. those that operate in the water that inundates the marsh twice a day;
2. those that occur in or on the sediments, including interstitial water retained at low tide; and
3. those which are the least important in determining marsh productivity and the fate of fixed carbon, and involve terrestrial species (salt-marsh grasshoppers, ephydrid flies, long-billed marsh wrens, wading birds and others).

Processes in category 1 largely determine the relative proportions of fixed carbon degraded in the water, the amount leaving the marsh through tidal flushing, and the growth and migration of motile species. Processes in category 2 are also important in regulating the dynamics of the marsh ecosystem: the productivity of *Spartina* is limited substantially by processes operating in the sediment, which is also the site of temporary storage of much surplus carbon and of its anaerobic degradation.

Productivity of the 'tall' *Spartina* that occupies creek banks, levées and marsh areas immediately behind the levées, which together form 6–7% of the total marsh surface, exceeds that of 'short' *Spartina* by over 70% as determined by the harvest method used by Gallagher *et al.* (1980). Net annual productivity (g C m^{-2} yr^{-1}) of tall *Spartina* on Sapelo Island was 1658 above ground and 938 below ground. Corresponding figures for 'short' *Spartina* were 593 and 895. An estimate of 1491 for total net annual productivity of the short *Spartina* based on CO_2 exchange (Wiegert, 1979) was in close agreement. Tall *Spartina* grows in sediments with relatively low salinity and high nutrient concentration; such conditions result from the greater tidal flushing associated with levée formation. The 'short' *Spartina* varies in height from medium in the main marsh to very short plants near the upper, landward boundary of the marsh where *Juncus roemerianus* (black needle rush) is often present in areas of low salinity. Where the sediment is highly saline *Salicornia* is common; neither of these latter plants is present in large amounts in the Sapelo Island marshes.

Net input of fixed organic carbon derived from *Spartina alterniflora* for the marsh as a whole, allowing for open water and mud banks, is calculated as 1573 g C m^{-2} yr^{-1}, of which some 40% is aerial shoot production and 60% is by roots and rhizomes in the sediments. Rate of input is limited by lack of nutrients or by high interstitial salinities; the marsh sediment water contains abundant phosphate but nitrogen enrichment stimulates production. Grazing by terrestrial organisms including grasshoppers removes only a very small proportion of net primary production (NPP), probably less than the 31 g C m^{-2} yr^{-1} shown in Table 5.1.

5.6.1 ECOLOGICAL PROCESSES IN THE MARSH WATER

The water which covers the entire marsh at high tide and that in the larger creeks, which never entirely drain, contains many aquatic organisms as well as CO_2, CH_4, DOC (dissolved organic carbon) and POC (particulate organic carbon). Organic carbon enters the water by photosynthetic fixation of gaseous inorganic carbon (CO_2), by the release of metabolites from living *Spartina* shoots and from microorganisms living on dead shoots (DOC) and, most importantly, as particulate material (POC) resulting from the death and degradation of the *Spartina* shoots. The entire *Spartina* shoot biomass eventually passes into the water; storms help break up large stems but heterotrophs appear to be important in particle formation. (In more northerly marshes ice shears and packs dead *Spartina*, thus initiating peat formation.)

Phytoplankton contribute very little to the total photosynthetic input of the marsh, being

Table 5.1 Estimated net production and degradation of carbon* by ecological processes in the air, water and sediments of the Sapelo Island Marsh

Category	*g C m^{-2} yr^{-1}*
Net primary production of *Spartina*	1573
Net primary production of algae	180
Total net primary production	1753
Degradation in air	31
Degradation in water	209
Degradation in sediments	533
Total degraded	773
% of *Spartina* net production	49
% of total net production	44
Surplus carbon	980

* All values are g C m^{-2} yr^{-1}. Conversions from Kcal, where necessary, were made assuming 5 Kcal g^{-1} for detritus and a carbon content of 50%. Thus, g C = Kcal × 0.1. (Modified from Wiegert, 1979.)

limited by the constant turbidity of the water. Diatoms and other benthic algae contribute up to 10% as much as the *Spartina*, maximum illumination and predation on the algae occurring at low tide. All fixed carbon that is initially part of the standing crop of the algae and the *Spartina* shoots is eventually transferred to the abiotic component as a result of mortality and secretion. What happens to this abiotic carbon? Is it removed by the physical processes of sedimentation and tidal flushing or do detritivores transform it to biotic carbon again?

Abiotic carbon from the water enters various heterotrophs as a result of filter feeding, particle feeding or **microbial assimilation**. The **filter feeders** of the marsh consist of the zooplankton in the water (which grazes on the phytoplankton, whose biomass is small, but whose reproduction is rapid), and polychaete worms, mussels (*Modeolus demissus*) and oysters (*Crassostrea virginica*) on and in the sediments. All the filter feeding organisms together are estimated to account for only 23 g C m^{-2} yr^{-1} of the material derived from *Spartina*.

Small mud and fiddler crabs (*Uca, Sesarma, Eurytium* and *Panopeus*) are the major **particle feeders**, though the most abundant of them (*Uca*) derives at least part of its energy from benthic algae rather than *Spartina* detritus, as does the mud snail *Nassarius*. The respiratory energy loss of the '**detritus eaters**' in this group converts to a mean of 17 g C m^{-2} yr^{-1}. It is calculated that 11% of the net production of shoots becomes available for rapid conversion to bacterial biomass when the *Spartina* dies and is submerged, which represents a further 69 g C m^{-2} yr^{-1}. Even when a figure representing the amount of material degraded by slower-acting organisms which utilize cellulose is added to these three sub-totals, it is apparent that only a small proportion of net primary production by *Spartina* shoots is degraded within the water itself.

5.6.2 ECOLOGICAL PROCESSES IN THE MARSH SEDIMENTS

When the roots and rhizomes which form 60% of total NPP of *Spartina alterniflora* die, they gradually degrade into POC and DOC within the sediment. As the standing crop of living roots and rhizomes shows little change from year to year it can be assumed that the annual rate of abiotic carbon input to the sediment, where it becomes available to saprophages, is relatively constant. Most of the marsh sediments are anaerobic; methane from the deeper sediments is not oxidized, even in the thin but active aerobic zone at the surface. There are also aerobic oxidized zones surrounding the roots of *Spartina*; these result from the diffusion of oxygen down the stems and into the sediments. Much of the oxygen demand of the sediment results from the microbial degradation of secreted DOC to CO_2 in the rhizosphere of *Spartina* roots.

The figure for degradation of carbon within the sediments (533 g C m^{-2} yr^{-1}) is based upon the CO_2 that they evolve; an independent estimate obtained from the amount of oxygen which they absorb gave results of a similar order. Thus, the figure of well over 50% for degradation of root and rhizome material within the sediment is much higher than that for degradation of shoot material in water and air.

5.6.3 TIDAL TRANSPORT OF ORGANIC MATTER AND A SIMULATION MODEL OF CARBON FLUX

A computer simulation model using the 14 compartments shown in Figure 5.12 was used in the cooperative study of the Sapelo Island salt-marsh system to identify 'sensitive' processes such as **tidal exchange**, which was thought to cause transfers of carbon between the waters of the marsh and the estuary. These exports included algae, DOC and POC. The first model, which postulated a daily loss of 25% of the DOC and POC in the marsh water, together with 12.5% of all suspended algae, gave reasonably good agreement with field data for standing crop and flux rates (Wiegert and Wetzel, 1979). However, various lines of evidence ran counter to the view that there

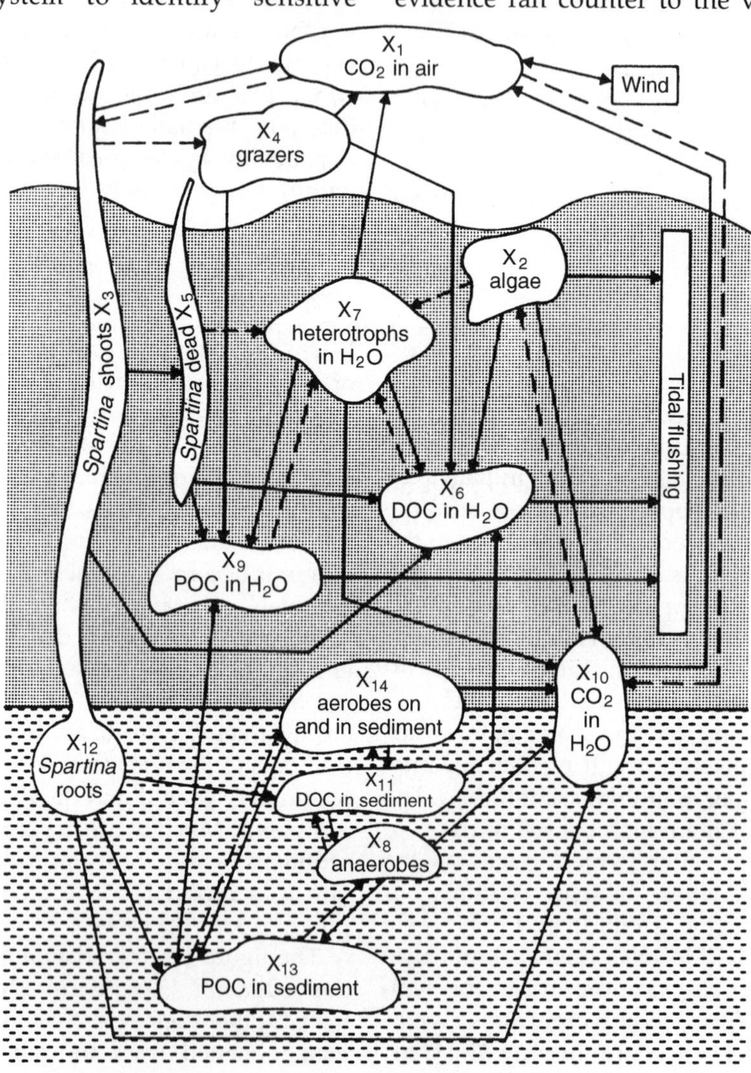

Figure 5.12 Food web of the *Spartina alterniflora* marsh, Sapelo Island, Georgia, salt marsh model showing the 14 components and their association with the air, water and sediment flows of carbon. (From Wiegert, 1979; courtesy Blackwell Scientific Publications.)

was a large daily transport of DOC and POC mainly derived from *Spartina* off the marsh, out of the tidal creeks and into the estuaries.

Fresh research into the hydrology of the Duplin River showed that daily convective exchange between marsh water and other Duplin River water was much less than 25%. Only when 100 mm or more rain fell at low tide would a quick (1–2 tide cycles) overturn or displacement of marsh water downstream occur. New computer simulations were now made as an approach to the question of whether surplus carbon – produced but not degraded within the marsh – is exported as a result of a small daily tidal exchange; the simulations incorporated a series of **catastrophic events** such as the coincidence of intense rainfall, high winds and spring tides. A low value of 1% was used for daily export of standing stocks of algae (X_2), heterotrophs in H_2O (X_7), DOC in H_2O (X_6) and POC in H_2O (X_9); four catastrophic events a year were assumed in each of which 80% of the four standing stocks mentioned was exported from the marsh. The 'catastrophe' was also assumed to remove, by scouring the marsh and creek bottoms, all particulate carbon of the sediment (X_{13}) above the measured average standing stock of 17 637 g C m^{-2}.

Simulated fluxes, particularly exports of the revised (catastrophic) model, are in reasonable balance with imports, though the seasonal behaviour of algae and POC in the water is not adequately represented. A significant point to note is that mathematical modelling is of most use to the practical ecologist when it highlights those areas of knowledge most crucial to the problem. Seasonal behaviour of the algae, rates of sedimentation and POC production, and the behaviour of water masses under a variety of weather conditions are among such considerations in this ecosystem.

Total missing ('surplus') carbon shown in Table 5.1 is 100 g greater than that originally given by Wiegert (1979). However, in respect of this, Wiegert (personal communication) and Chalmers later concluded that 'particulate carbon is moved back and forth from the marsh to the upper creek by tidal deposition on the marsh, and erosion by rain at low tide. Aerobic decomposition of this material and incorporation into mobile consumers can account for most or all of the missing material'.

SAND DUNES: INITIATION, DEVELOPMENT AND FUNCTION

6

6.1 BLOWN SAND AND ITS FIXATION

Coastal dunes form where there are adequate supplies of well-sorted beach sand and where onshore winds capable of moving it blow for at least part of the year. While not essential for coastal dune formation, vegetation has an important controlling influence on dune morphology (Pye, 1983). Though these wind-dominated systems have traditionally been regarded as distinct from the adjacent wave–current-dominated beaches from which their sand is derived, there is increasing recognition that dunes and beaches are often strongly coupled and mutually adjusted. Sediment exchanges between them may have important repercussions for the evolution of the **integrated beach–dune system**, which from a coastal engineering standpoint is often best considered as a whole (Sherman and Bauer, 1993)

Wind erodes by **deflation**, in which loose particles are either forced from the ground and carried through the air or simply rolled along, and **abrasion** whereby airborne particles wear away exposed rock and soil surfaces. This latter process operates close to the ground (usually up to 0.3–0.6 m), so metal sheathing 1 m high is sufficient to prevent the abrasive felling of telephone poles in sandy areas.

In humid regions, evidence of deflation is seldom seen away from coasts; the ground is held in place by vegetation and ample soil moisture. Coastal sand dunes are found in many parts of the world in arid, semi-arid and temperate climates, though on semi-tropical and tropical coasts luxuriant vegetation, low wind velocities and damp sand make them less frequent (Pethick, 1984). They exist in zones bordering high water mark and may extend inland for up to 10 km, but may have a width of only 1 km or less. In some arid areas coastal dunes are devoid of vegetation and consequently possess features more typical of desert dunes. The coastal dunes of Baja California with their extensive barchan fields are a case in point (Inman *et al.*, 1966). This, however, is unusual; most coastal dunes are essentially **phytogenic**, evolving with a partial cover of vegetation which both helps to fix the ground and modifies its surface properties with respect to air flow. At a late stage in their development – and seldom permanently – dunes often become inactive or **fixed**, held by the roots of the plants covering them. The great arid deserts of the world are generally devoid of permanent vegetation and their dunes are active or **live**, constantly changing under wind currents and fed from blowouts (**deflation hollows**).

Almost all wind-blown sand travels quite close to the ground as a result of **saltation**, in which individual grains move in a series of leaps. Once airborne, a grain describes a curved path, hitting the ground at a low angle but with sufficient force to rebound into the air again. At the same time the grains of the surface layer also move downwind, driven by repeated impacts from the higher energy saltating grains (Figure 6.1). The impact of a jumping grain can roll forward a grain up to six times its diameter. Of the total amount of sand in motion, roughly three-

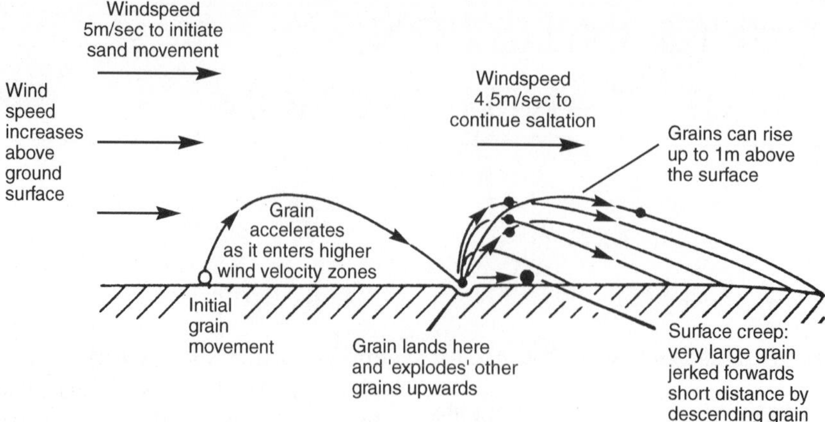

Figure 6.1 The process of saltation. (Redrawn from Pethick, 1984.)

quarters – belonging to the finer fractions – moves quickly downwind by saltation, while the remaining quarter moves more slowly by **surface creep**. The resultant sorting of the sand grains may be of at least local significance.

6.1.1 CLASSIFICATION OF DUNES

Blown sand gives rise to a considerable variety of landforms; many are found in coastal systems and complex intermediates occur. A simple treatment of these landforms is given by Strahler (1975); that of Olson and van der Maarel (1989) is more elaborate and accompanied by a map of the dune coasts and major dune systems of Baltic and Atlantic Europe. Pye and Tsoar (1990, p.162) provide a fundamental review; in their dune classification sand is shown as accumulating in relation to **topographic obstacles**, to **bed roughness changes** or **aerodynamic fluctuations**, and to **vegetation**. The third group, of major importance in the present context, consists of **vegetated linear dunes**, **parabolic dunes** and **hummock dunes**, such as the **dune hedgehogs** developed around bushes of *Salix repens* which have grown gradually upwards, sometimes from positions at or near to the level of the ground water table where *S. repens* colonizes. The causes of sand accumulation are often complex; at Pennard Burrows, Gower Peninsula, South Wales, UK, the sand accumulated in relation to the underlying Carboniferous Limestone, but was also subject to the influence of vegetation and marked changes in bed roughness as well as aerodynamic fluctuation.

Barchans are desert dunes, shaped like blunted crescents in plan form, which occur as isolated units or in groups (see Figure 6.4(a)). The rounded ends of the crescent point in the direction in which the dune and the prevailing wind are travelling. Sand moves up the gentle slope on the windward side of the crest, and once it has been blown over the top, falls or slides down the steep **slip face**, which has angle of about 35° from the horizontal.

Transverse dunes are normally linear and formed where sand completely covers the ground; they are separated by trough-like furrows and have their crests at right-angles to the prevailing wind. Individual transverse ridges of free sand, as opposed to those in which vegetation plays a part, have sharp crests with gentle windward slopes and steep slip faces on the lee side. Coastal transverse dunes are almost invariably vegetated and, as shown in Figure 1.9, have a rather steep windward face and a gentler lee slope.

In **parab** [illegible] mmonly develop fro [illegible] ies and play a majo [illegible] Warren, Anglesey, the [illegible] down-wind. This is [illegible] n in a barchan: cas [illegible] with low ridges ru [illegible] rgins in a direction [illegible] entic horns to whic [illegible] unes are switching t [illegible] bolic dunes often lac [illegible] side and may be c [illegible] but sometimes migr [illegible] nd assume the long [illegible] ide form of a **hairpin** [illegible] ra-bolic family.

Longitudinal du [illegible] th the prevailing wind [illegible] nt on desert plains and [illegible] p-ply is meagre and the [illegible] n develop from extraorc [illegible] pin dunes in which [illegible] s have become dominan [illegible]

6.1.2 SAND SUPPLY FOR [illegible] FORMATION

The four zones on sandy s [illegible] Krumbein and Slack (1956) [illegible] 6.1. Their definition of the backshore is critical in terms of sand movement: many authors appear not to distinguish it from the foreshore. Each of the four zones feeds the next highest, so in the long term all require an adequate sand supply if dunes are to continue to grow. However, it is from the backshore that most of the sand for dune building is driven by the wind. Seaweed dumped with other litter along the highest limit reached by an individual tide is often covered by sand dropped from onshore winds slowed by the drag which it causes. Deposited seaweed may be reworked by later tides or remain as a buried source of moisture for strandline plants.

6.1.3 PLANT HABIT AND DUNE FORMATION

Two growth habits are particularly effective in dune-forming plants (Cowles 1899). Although it does not normally occur in high dunes, *Elytrigia juncea*, like *Ammophila arenaria*, produces horizontal rhizomes of potentially unlimited growth. Though lacking this capacity, species such as lyme-grass (*Leymus arenarius*), *Salix repens* and *Populus* spp. readjust to sand burial by oblique or vertical rhizome or stem growth, which in the case of woody species may be almost unlimited as long as

Table 6.1 Sand shore zones. (F [illegible] ck, 1956)

Shore zone	*Limits*	*Tidal relations*	*Agents of sand movement*
1. Nearshore bottom	Mean low water to minus 9 m	Nearly always submerged	Currents and breaking waves
2. Foreshore	Mean low water to high tide line	Alternately submerged and exposed	Currents, breaking waves, occasional wind action
3. Backshore	High tide line to dunes	Nearly always exposed but occasionally submerged during storms or exceptionally high water	Breaking waves, wind action
4. Dunes	Above highest tide limit	Always exposed	Wind action

accretion is not too rapid. These plants have a tight growth form resulting in steep-sided hummock dunes, rather than the broad-based dunes resulting from the spreading growth of *Ammophila*.

The potentially unlimited horizontal and vertical rhizome growth of *Ammophila* has enabled *A. arenaria* and *A. breviligulata* to create many of the really high vegetated dunes known. Another plant capable of such growth is the sterile hybrid purple marram × *Calammophila baltica*, a naturally occurring intergeneric hybrid between *A. arenaria* and *Calamagrostis epigejos* (wood small-reed or bush grass), which has been used for dune stabilization on the coasts of Suffolk, Norfolk and Hampshire, UK. *Leymus arenarius* produces large dunes in Iceland, where it is the main stabilizing grass (Greipssòn and Davy, 1994a,b). Likewise *Uniola paniculata* forms substantial dunes in southern USA, as does the C-4 grass *Panicum racemosum* in southern Brazil (Cordazzo, 1994).

6.1.4 INITIATION OF SALTATION

The following account of sand dune formation draws on descriptions by Ranwell (1972) and Pethick (1984), both of whom build on the pioneer work of Bagnold (1941). Pye and Tsoar (1990) provide a review of the complex mechanisms involved in dune formation, erosion and subsequent reworking of aeolian deposits. Air flow over a stable sand surface – which Bagnold obtained experimentally by dampening the sand – is slowed by frictional drag whose influence becomes progressively less away from the surface. Wind velocity profiles follow logarithmic curves. When plotted on a log scale and thus transformed to straight lines (Figure 6.2(a)), the profiles converge at a point where frictional drag is producing **zero wind velocity** at a small, but very significant, height above the sand surface. This height (Z_0), which is related to the roughness of the surface, can be called **effective surface roughness** (Olson, 1958) or **roughness length** (Monteith and Unsworth, 1990). On a flat beach with no sand movement, Z_0 is approximately 1/30 of the average surface grain diameter – some 0.03 mm for a sand composed of particles 1 mm in diameter (Pethick, 1984).

The existence of an air-flow velocity gradient above the beach surface causes a force to be applied to grains lying on that surface, in a way analogous to dragging a pliable stick

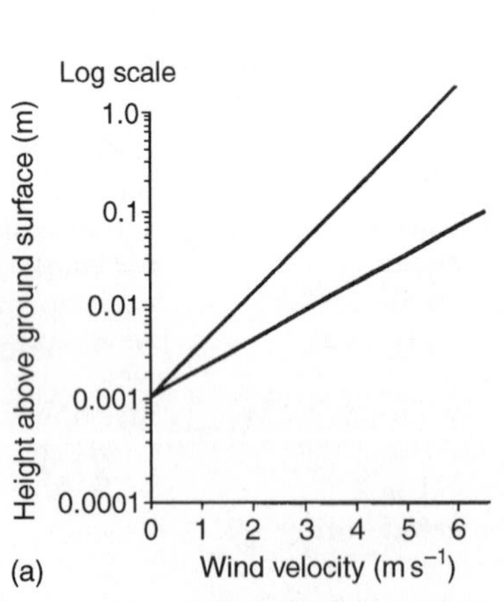

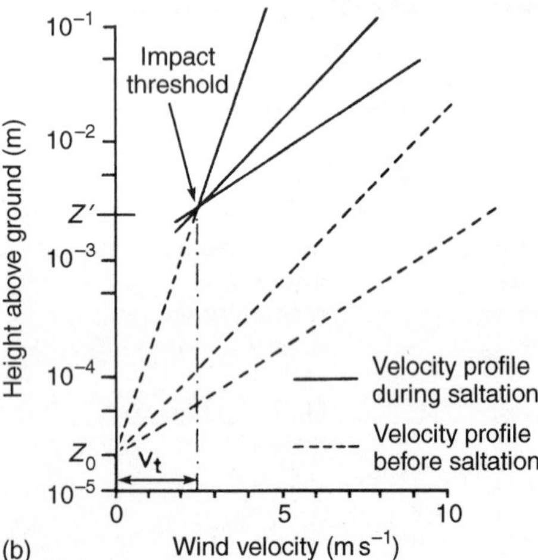

Figure 6.2 Wind velocity profiles over sand with height plotted on logarithmic scales. (a) Over a stable bare sand surface where saltation is not occurring; (b) before and after onset of saltation (based on measurements by Bagnold over a bed of 0.25-mm uniform sand). (After Pethick, 1984; Pye and Tsoar, 1990.)

across a rough surface. The curvature of the stick increases the rougher the surface, or the harder the stick is pressed down. The amount of curvature – of the stick or wind profile – is measured as its gradient, which is called the **shear velocity** ($u_.$) or **friction velocity** (U_*) in the case of wind. Greater force (or shear stress), giving a greater value of $u_.$, is exerted on the sand surface when the wind velocity (the pressure at the top) or the surface roughness, or both, are increased.

Surface sand grains begin to move when shear velocity increases to a critical value ($u_{.crit}$), the **fluid threshold velocity**, which depends on the square root of the grain diameter. These grains, which vary in size, project through the Z_0 layer of zero velocity air into the velocity profile. Once a given shear velocity is applied, projecting grains with diameters at or below the critical size will start to roll or slide forward. Because sand grains are 2000 times heavier than a similar volume of air they behave in an extremely bouncy manner, so the impact caused when moving grains meet larger immobile grains flicks the smaller ones into the air – sometimes almost vertically – thus commencing the process of **saltation**.

As such grains rise they are accelerated by the increasingly fast wind velocities they encounter, reaching the same speed as the wind at each grain's apogee. The grains then fall in a characteristic gradual trajectory so that they plough into the beach surface and force other groups of sand grains into the air. These too shoot up and forward before landing with an impact that shoots up yet more grains, so that soon the whole beach surface downwind of the original rolling grains is on the move. The landing impact of the finer, saltating grains causes any larger grains they hit to roll or slide forward, initiating the process of **surface creep**.

Where there is a chance projection on the original beach more grains will be ejected upwards than elsewhere. Such saltating grains tend to land in a cluster around the mean horizontal length of their pathway and to continue to do so as they saltate downwind. Their landing impact will jerk larger grains forwards so that each landing zone forms a shallow depression with a ridge of larger grains downwind. As a result the original plane surface of the beach gradually becomes covered by regularly spaced asymmetrical **sand ripples** 1–2 cm high, and with wavelengths of 2–12 cm. Ripple wavelength increases with wind velocity, though ripples tend to flatten and disappear when wind speeds are very high.

6.1.5 SALTATION BOUNDARY LAYER

Saltating grains absorb energy from the air flow just above the beach, act as a brake upon it, and consequently slow it to the point where shear velocity, $u_.$, drops below the critical level at which grain movement is initiated. Saltation continues, however, but now it results from the impact of the grains descending on the beach surface rather than the direct action of the wind, which at this stage merely accelerates the grains forwards once they are airborne. The energy supplied by the wind to a saltating grain at the top of its trajectory is in fact balanced by that lost when the grain falls back to the beach.

Whereas wind velocity gradients over a smooth level sand surface converge at a point of zero velocity (Z_0) just above the ground surface, convergence occurs at a much higher point, Z', at least 0.2–0.4 cm above the surface, once the sand begins to move and saltation commences (Pye and Tsoar, 1990, p.110). Moreover, this convergence (Figure 6.2(b)) is at a height where wind speed corresponds to the **impact threshold velocity** for maintenance of saltation in moving sand (V_t). Because of the kinetic energy of the grains in motion this is less than the fluid threshold velocity required to initiate movement. V_t is roughly 4.5 m s^{-1} (c. 10 m.p.h.) for average dune sands (Ranwell, 1972). Wind speeds below this cannot maintain saltation. Effective surface roughness is increased as a result of sand ripples developed as a consequence of saltation.

6.1.6 RATE OF SAND TRANSPORT

The rate at which sand transport occurs depends mainly on wind speed, though particle size and shape are also important. Although wind velocity at Z′ is constant during saltation, the faster the air flow encountered by a saltating grain after it has risen to this height, the greater is its speed when it falls. A high energy grain will 'explode' more grains upwards, and for a greater distance, than one travelling less rapidly. Rate of sand movement is normally calculated from the gradient of the wind profile. Several authors, from Bagnold (1941) onwards, have shown the amount of sand transported per unit beach width per unit time to be related to the **cube** of the shear velocity:

$$q = (C\sqrt{D}) \times (u'_{.})^3$$

or, to include actual velocities:

$$q = C(V_{100} - V_t)^3$$

where q = weight of sand moved per unit width per unit time; C = a constant; D = grain diameter; u′. = shear velocity during saltation; V_{100} = velocity measured at 1 m above the surface; and V_t = critical threshold velocity for a given grain size.

This cubic relationship is responsible for the extreme sensitivity of sand transport to wind speed. For example, a wind velocity (V_{100}) of 50 k.p.h. (13.9 m s^{-1}) will move 0.5 tonne of sand per metre beach width per hour. If wind velocity is increased by only 16%, to 58 k.p.h. (16.1 m s^{-1}), the sand transported is doubled to 1.0 tonne per metre per hour (Pethick, 1984). Thus, once above impact threshold velocity, a relatively small increase in wind speed causes a considerable extra volume of sand to be eroded from the beach, while a slight decrease results in a mass of sand being deposited from the saltating cloud. In practice, substantial sand movement occurs as a result of occasional periods of high-velocity wind, rather than as a continuous low activity process caused by light to moderate winds. For this reason detailed reports of transport during specific storms are particularly valuable, for example, that given by Ranwell (1958) for the 3-day south-west gale during 15–17 May, 1953 at Newborough Warren.

Ground surface energy losses also influence transportation rates; softer surfaces such as loose sand absorb a great deal of energy and result in less transport than occurs over a hard surface, such as that provided by pebbles off which sand grains easily bounce. There is still much to learn about the aeolian transport of sand. Anderson (1989) has developed a mathematical model involving the four sub-processes of: (i) aerodynamic entrainment; (ii) grain trajectories; (iii) grain-bed impacts; and (iv) momentum extraction from the wind. This helps to elucidate the physics of the saltation process, but has yet to deal with problems such as the influence of moisture, saline coatings which tend to render the particles cohesive, and the behaviour of various size particle mixtures.

6.1.7 EMBRYO DUNE FORMATION

Saltating clouds of sand grains blow from the foreshore past high tide mark and its associated drift. Any inanimate rubbish – plastic bottles, baulks of timber, holed and deserted boats – on the backshore between high tide mark and the permanent dunes tends to be buried by sand deposited as a result of the slowing of the air flow that it causes. Dunes of any permanence are formed only by vegetation, whose growth can keep pace with the accreting sand and thus continue to provide a barrier that will evoke further deposition. Far-creeping rhizomes of *Elytrigia juncea* and *Leymus arenarius* – two of the few perennials which can survive the not infrequent saline inundations or heavy spray – frequently initiate the growth of embryo dunes which are initially unconnected and gradually accumulate piles of sand 1 or even 2 m high. The fleshy stolons of *Honckenya peploides* also survive the winter here, growing up through the sand in late spring when seedlings of such annuals as *Cakile maritima* (sea rocket) and *Salsola kali* (prickly saltwort) are beginning to appear. Other perennials behaving similarly include *Blutaparon portulacoides* and *Spartina ciliata* in southern Brazil. *Ammophila arenaria* is unable to survive repeated inundation by sea water and is often absent from the very lowest dunes near the sea, though it may enter the succession very soon afterwards and form an intimate mixture with *Elytrigia juncea* in low dunes not far from the shore.

6.1.8 INFLUENCE OF VEGETATION ON SAND MOVEMENT

When the saltating cloud leaves the embryo dune and meets the foredune the vegetation there again increases the surface roughness and also intercepts the descending saltating grains, acting as a very soft springy surface which absorbs a large proportion of their energy. Z_0, effective surface roughness, for bare sand is increased to about 1 cm by the presence of sand ripples. Olson (1958) found Z_0 = 1 cm under *Ammophila breviligulata*; Bressolier and Thomas (1977) found Z_0 = 2.5 cm under *Euphorbia* spp. and Z_0 = 16.7 cm under *Elytrigia* spp. Willis *et al.* (1959b, p.260) measured wind velocities at known heights above the sand surface in a number of dune vegetation types and calculated the heights at which the wind was effectively zero. The values obtained are similar but not identical to Z_0, and varied from 0.02 cm (dune pasture, 1-cm high *Tortula–Homalothecium*) to 15 cm (*Pteridium aquilinum*, 49 cm high).

Loss of energy from sand grains as they hit the leaves alters the balance between wind-energy input and impact-energy export to the saltating cloud so that a new lower transport rate results, with consequent rapid deposition. Moreover, the increase in Z_0 means that the saltating grains must rise to much greater heights before reaching the wind current which will accelerate them forward (Pethick, 1984). Some grains do bounce this high but others fall back to the ground, most wind-borne sand being trapped by vegetation within a few metres of its edge. Deposition rates can reach 0.3 to 1.0 m per year on a foredune, so a new dune may fairly rapidly build up as a distinct ridge up to some 10 m high.

6.1.9 MAINTENANCE OF DUNES AND SLACKS

Figure 6.3 shows windflow over the first and second dune ridges. There is often a complete sequence of sub-parallel dune ridges with the youngest near the shore and the oldest on the inland side of the system. The absolute time interval between the formation of any pair of ridges varies between systems, but is commonly between 70 and 200 years.

As the wind approaches the steep windward dune face its speed near the ground increases. Once over the crest a flow separation occurs with the higher-velocity air pulling away from the surface. As a dune ridge gets higher so the winds at its crest become more severe, causing more vigorous saltation and erosion of sand grains from the crest itself. The maximum height reached by an individual ridge is determined by this self-limiting effect. Saltation rates drop suddenly on the lee side of the dune where rapid deposition occurs. The effect of these erosive processes is to cause a 'gradual "rolling over" of the whole dune which subsequently advances landwards' to use the evocative phrase of Pethick (1984).

Cooper (1967) used experiments in wind tunnels and '**smoke flow visualization**' in the field to investigate air flow in the lee of the crest. It was formerly thought that hollows between dune ridges were maintained by reverse eddies in air flow at this point. What actually occurs is that a 'wind-wave' is developed with its crest over a position halfway down the lee slope. Its trough, where the streamlines come close together and high-velocity air approaches the ground, is in the hollow itself where sand is liable to be eroded rather than deposited as it is on the lee slope. Any loose sand is swept towards the next ridge and the hollow may eventually be eroded down to the damp sand of the water table.

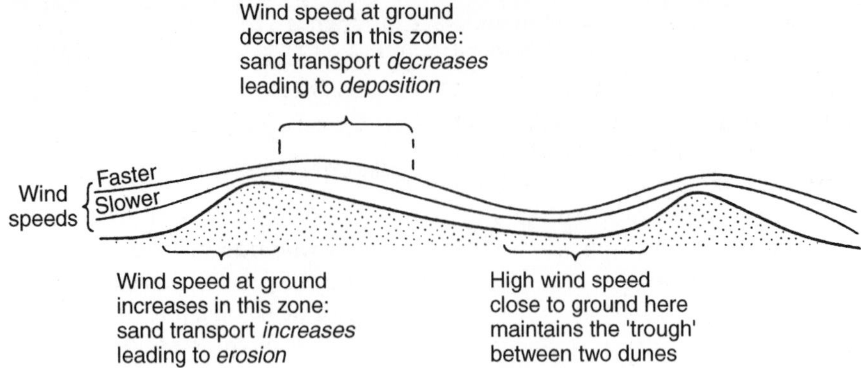

Figure 6.3 Wind stream lines over a dune system. These 'wind-waves' are responsible for maintenance of the characteristic dune ridges. (Redrawn from Pethick, 1984.)

6.1.10 OLDER DUNE RIDGES AND SLACKS

The first dune ridge (that next inland from the foredune) is normally the highest and the most complete. The older second and third ridges inland from it are frequently lower, because of the reduction in sand supply and gradual loss of sand. Saltation cannot be initiated beneath vegetation, only perpetuated, so once a cloud of grains loses all its energy owing to the effects of vegetation saltation ceases altogether (Pethick, 1984). Older dune ridges become fragmented when blowouts and parabolic dunes develop. Conversely, established slacks of great vegetational interest may be overwhelmed by fresh sand when blowouts occur upwind of them, a tendency seen at Ynyslas, Wales, in the 1970s.

6.1.11 TOPOGRAPHIC CHANGE AND PARABOLIC COASTAL BLOWOUT DUNES

Complete dune ridges seldom erode uniformly and even at Newborough Warren, Anglesey, where the system is perfectly orientated for maximum uniform erosion, parabolic dune units develop (Ranwell, 1959). Prevailing south-westerly winds blow at right-angles to the dune ridges, so that poorly stabilized regions are eroded more rapidly than the more completely vegetated areas on either side. As the bare sand of the central region moves inland, the two 'horns' at the outer margins of the parabola remain, at first attached to the relatively stable sand ridge. Eventually the parabolic unit breaks free, moving inland independently and leaving a developing slack in its wake (Figure 6.4(b)). The horns continue to remain upwind of the centre of the dune, in contrast to the crescentic barchan dunes of unvegetated desert sand in which they are downwind (Figure 6.4(a)).

The proliferation of parabolic dune units from partially stabilized dune ridges causes a general break-up of the ridge structure at Newborough Warren, resulting in a series of skeletal dune ridges and parabolic dunes separated by erosion hollows or slacks (Ranwell, 1958, Figure 2). According to Greenly (1919) vertical sand thickness reaches a maximum of 15 m (50 ft) in most parts of Newborough. Around the 'Rock Ridge', however, dunes reach 30 m (100 ft) or more above mean sea level, about the same maximum as at Braunton Burrows which is a large free granular system resting on an impermeable subsurface and which has no rocky spine (see Figure 6.18).

In systems with powerful onshore winds, dunes move slowly inland, becoming gradually lower and sufficiently remote from strong winds and saline spray to stabilize at a height intermediate between dune and slack. Once

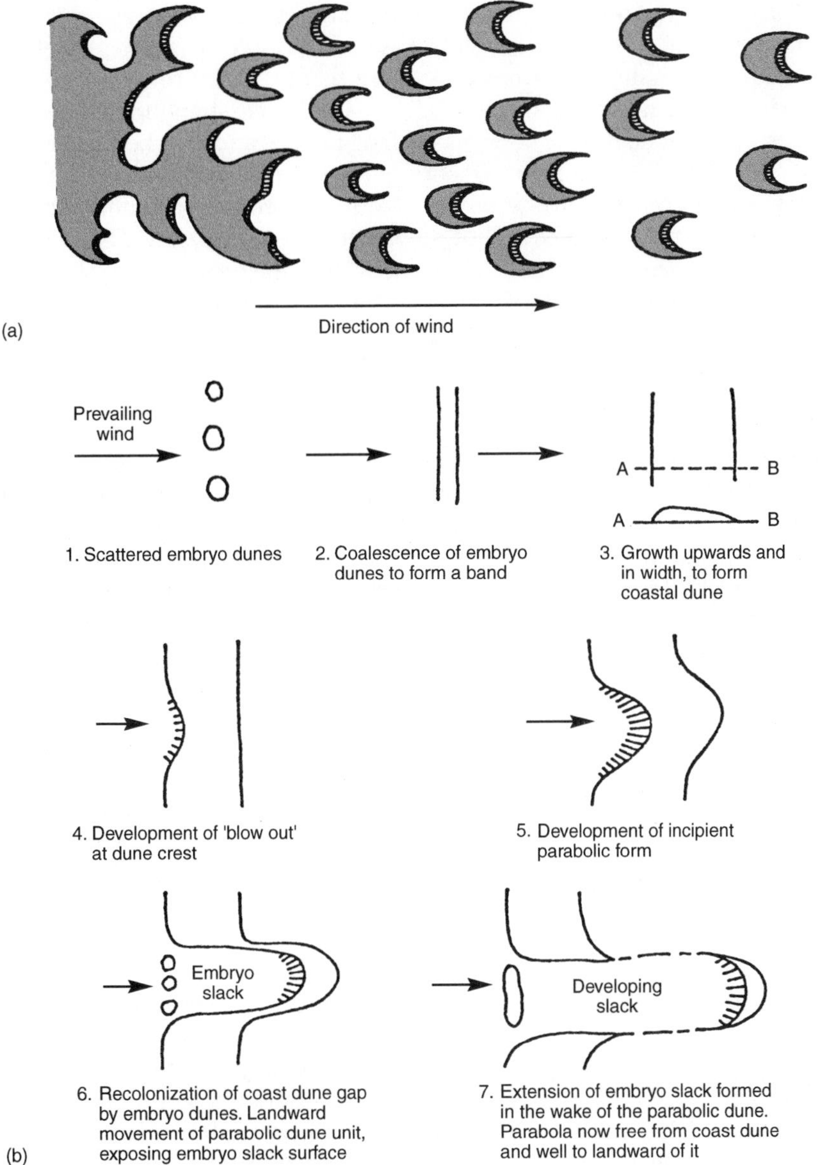

Figure 6.4 Crescentic dunes. (a) Plan of a typical procession of barchans in the Libyan desert. (From Holmes, 1965.) (b) The formation and development of a parabolic dune unit. (From Ranwell, 1972.)

this has occurred, long-term plant succession ultimately leading to woodland, and of the same general type as that found in dunes that stabilize *in situ* on a sheltered prograding shore, can be expected. Nearer the shore, however, there is a **cyclic alternation of dune and slack** in many places (Figure 1.15). Using direct measurements of dune movement and studies of growth and age patterns of *Salix repens* on either side of mobile dunes, Ranwell (1960a) estimated the duration of the cycle as eighty years at Newborough, the same as the period of the dune-building cycle at the coastline. The situation becomes very complex once the dune ridges have broken up and Ranwell points out that there may not be a regular alternation of dune and slack at any one point thereafter. It is noteworthy, however, that since the occurrence of myxomatosis, with much reduced rabbit activity, the dune system at Newborough is now considerably less mobile.

Evidence of temporal alternations between dune and slack comes from many sources. Oosting (1956), for example, provides a photograph of a 'ghost' or 'graveyard' forest near Florence, Oregon. A closed stand of dead trees, mainly lodgepole pine (*Pinus contorta*), is shown complete with a characteristic litter and soil layer developed on relatively stable sand. The mobile dunes which killed and completely inundated the trees have now moved on, uncovering them and leaving the forest soil vulnerable to blowout. At Coto Doñana, south-west Spain, substantial groves of *Pinus pinea* grow in slacks. Most are killed by dunes advancing at 5–6 m yr^{-1}, eventually re-emerging as dead tree skeletons when the ridges pass on (Garcia Novo, 1979).

The **wandering dunes** of the Sands of Forvie, Aberdeenshire, (Landsberg, 1955; North, 1981) and the Culbin Dunes, Nairn and Morayshire (Patton and Stewart, 1917) are the best Scottish examples of dunes which have moved extensively within historic times. Both are situated close to the mouth of a river where blown sand has accumulated for long periods of time over rather extensive raised beaches (Gimingham, 1964). The Culbin Dunes have now been largely stabilized by plantings of Corsican pine (*Pinus nigra* ssp. *laricio*). Such afforestations have been effective in many other places, notably the use of *Pinus radiata* to stabilize the dunes behind 90 Mile Beach which faces the Tasman Sea in North Island, New Zealand.

6.2 THE STRANDLINE AND GENERAL ZONATION: EMBRYO DUNES

Colonization of the **marine strandline** by plants, which is discussed by Davy and Figueroa (1993), may be transient or mark the start of a true chronosequence. This physically demanding habitat is susceptible to disturbance by wind and waves; its porous sediments retain little water and there is little protection from the sun. The surface of the strandline is essentially dry for long periods and may be subject to large diurnal temperature fluctuations; nutrient capital – apart from that released by mineralization of organic detritus – is usually low. The availability of nitrogen from this irregularly distributed source is a major determinant of success here. On a world basis, strandline colonists show a diversity of life-histories, with many ephemeral or annual species completing their life-cycles in the 5 months between successive periods of equinoctial spring tides. Clonal perennials such as *Elytrigia juncea* often recolonize from vegetative fragments after catastrophic disturbance, which is less common in the Tropics where vines and small shrubs may colonize the strandline.

Disseminules of strandline species are frequently buoyant and long-lived in seawater; seeds tend to be large, providing reserves sufficient to allow emergence from considerable depths in the sand. Extensive root systems may exploit deep groundwater, which may move upwards as 'internal dew' (section 6.3), thus helping to ameliorate this harsh and unpredictable environment.

Passing from the sea, a transect of a typical European dune system, such as that at Ynyslas, North Wales (Savidge, 1976), runs through the **foreshore**, the **strandline** at high tide level, the **backshore** and **embryo dunes** to the high **mobile** (**young** or **yellow**) **dunes.** Beyond these are the rather lower **unconsolidated dunes**, with the **consolidated** (**mature**) **grey dunes** between them and the **maritime sward** (**dune heath** in areas with more acid soil) on the landward side. Wind causes blowouts from time to time (sand may be blown away until water-stabilized) and may lead to the formation of the interspersed **dune slacks** which later acquire a rich flora. Along such a transect ground wind speed, influence of salt spray, soil pH and levels of soil calcium and sodium diminish in an inland direction, while the extent of vegetation cover, amount of soil organic matter, number of plant and animal species and overall stability increase. Although dune systems of this general form exist in many parts of the world, the species involved in coastal foredune zonation and succession vary considerably (Doing, 1985). The distributions of British dunes are shown in Figure 6.5; the form of many of them differs considerably from the 'typical' dune described above.

Strand zones, and even the embryo dunes nearest the sea, afford a very hostile environment in which the plants may be splashed by seawater and take the full force of onshore gales. The westerly gales at Ynyslas are often severe and the area occupied by strand vegetation varies considerably from year to year according to the weather, particularly that of the spring and early summer. In this respect, 1971 was a good year and there was a dense vegetation on the beach with the following six species predominating:

Ammophila arenaria P (marram)
Elytrigia juncea P (sand couch-grass)
Honckenya peploides P (sea sandwort)
Cakile maritima A (sea rocket)
Euphorbia paralias P (sea spurge)
Salsola kali A (prickly saltwort)
(A = annual; B = biennial; P = perennial.)

Other species, such as *Atriplex glabriuscula* A (Babington's orache) and *A. laciniata* A (frosted orache), are also found here. Taschereau (1985) gives a guide to the taxonomy and distribution of British species of *Atriplex* and notes that *A. patula* , although fairly salt-tolerant and a colonizer of disturbed soil, is not a member of saltmarsh communities.

The sand covering the shingle at Ynyslas is shallow and often mixed with rotting seaweeds. The plants present have stiff or fleshy leaves and even the annuals have extensive root systems going down among the underlying pebbles and finer shingle. *Cakile maritima* (Figure 6.6) is frequently the most abundant strandline species here, but often a narrow back-shore results in a very condensed sequence between the drift line and the steep west-facing slope of the tall mobile dunes, that in some years have very little vegetation seaward of them. In the north of the system embryo dunes up to 2 m high may be scattered in front of the larger dunes after several months of calm summer weather. Sometimes they are consolidated into the outer mobile dunes by marram; more frequently they are suddenly obliterated by gales.

Salisbury (1952, p.210) lists 'dune face and drift line plants' in which *Euphorbia peplis* (purple spurge) and *Polygonum maritimum* (sea knotgrass) are very rare. The list does not contain marram, but includes species mentioned above and also:

Atriplex littoralis A (grass-leaved orache)
Beta vulgaris ssp. *maritima* P (sea beet)
Crambe maritima P (sea kale)
Polygonum oxyspermum A/P (Ray's knotgrass)
Raphanus raphanistrum ssp. *maritimus* B/P (sea radish)
Silene uniflora (= *S. maritima*) P (sea campion)
Solanum dulcamara P (bittersweet, woody nightshade)
Tripleurospermum maritimum B/P (sea mayweed)

The most typical strandline species are, however, annuals, notably *Cakile maritima, Salsola kali, Atriplex glabriuscula* and *Atriplex laciniata*. The longer-lived *Crambe maritima, Raphanus raphanistrum* ssp. *maritimus* and *Solanum dulcamara* are frequent on shingle and fairly stable sandy habitats and are major species in foredune succession. Salisbury (1952, Plate 22) shows extensive embryo dunes at Harlech, North Wales, formed around plants of *Cakile maritima* and *Salsola kali* growing over a former driftline, besides two further driftlines to seaward. A few marram plants have colonized the rear of these embryo dunes, themselves only a few metres in front of a foredune dominated by *Ammophila* (Figure 6.7). Ranwell (1960a) did not find such a continuous band of strand flora species at Newborough during 1950–1953, but subsequently something approaching it developed at the toe of a dune

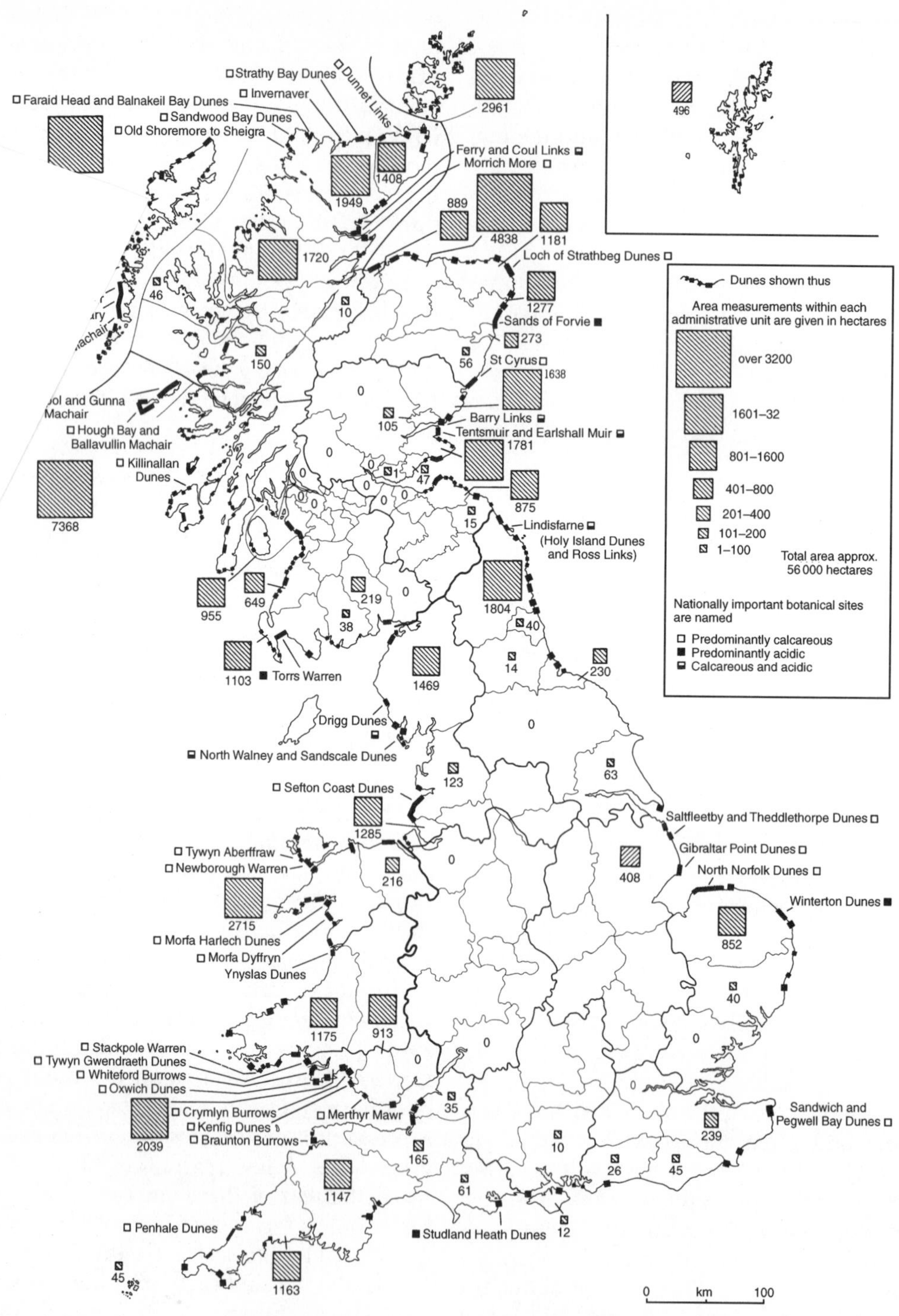
Strathy Bay Dunes
Dunnet Links
Invernaver
Faraid Head and Balnakeil Bay Dunes
Sandwood Bay Dunes
Old Shoremore to Sheigra
2961
496
Ferry and Coul Links
Morrich More
1949
1408
889
4838
1181
1720
Loch of Strathbeg Dunes
Dunes shown thus
Area measurements within each administrative unit are given in hectares
over 3200
1601–32
801–1600
401–800
201–400
101–200
1–100
Total area approx. 56 000 hectares
Nationally important botanical sites are named
Predominantly calcareous
Predominantly acidic
Calcareous and acidic
46
10
1277
Sands of Forvie
273
150
56
St Cyrus
1638
Hough Bay and Ballavullin Machair
Killinallan Dunes
105
Barry Links
Tentsmuir and Earlshall Muir
1781
7368
47
875
15
Lindisfarne
(Holy Island Dunes and Ross Links)
955
649
219
38
1804
40
14
230
1103
Torrs Warren
1469
Drigg Dunes
North Walney and Sandscale Dunes
123
63
Sefton Coast Dunes
1285
Saltfleetby and Theddlethorpe Dunes
Tywyn Aberffraw
Newborough Warren
216
408
Gibraltar Point Dunes
North Norfolk Dunes
2715
Winterton Dunes
852
Morfa Harlech Dunes
Morfa Dyffryn
Ynyslas Dunes
40
1175
913
Stackpole Warren
Tywyn Gwendraeth Dunes
Whiteford Burrows
Oxwich Dunes
Crymlyn Burrows
Kenfig Dunes
Braunton Burrows
Merthyr Mawr
35
2039
Sandwich and Pegwell Bay Dunes
239
10
26
45
165
1147
61
Penhale Dunes
Studland Heath Dunes
12
45
1163
0 km 100

formed against fences erected by the Forestry Commission on the western margin of the Warren. Locally abundant here was *Elytrigia juncea*, with frequent *Salsola kali*, occasional *Honckenya peploides*, and rare *Atriplex laciniata*, as well as other typical strandline species.

Sand-dune systems never attain full stability but change is usually most evident on their seaward margins. At Braunton Burrows, North Devon, the equilibrium between sand stabilization and erosion in the foredunes was destroyed when a minefield was cleared in 1946–1947. Subsequently, strand vegetation was represented by only a few scattered plants until 1956, when it developed into a well-marked zone some 25 m wide. *Cakile maritima* and *Salsola kali* were the chief species and *Elytrigia juncea* was sporadic, but the strand vegetation was much sparser in 1957: population sizes of annual species frequently fluctuate strongly. A line of foredunes dominated by marram – some naturally established and some planted – and interrupted by washouts and blowouts has gradually built up again since 1947. This line separates the shore from the eroding parabolic dunes inland and the general situation is similar to that described by Watson (1918). In many dune systems in which the foredune ridge is eroding there is no significant strandline community (or embryo dunes), but once there is some degree of stability *Cakile maritima* and *Salsola kali* can quickly colonize.

6.2.1 SCOTTISH STRANDLINE VEGETATION

Gimingham (1964) treats the sandy 'foreshore' of the Scottish maritime zone separately since it can occur independently of dune development, and is subject to a special set of environmental conditions. The upper portions of sandy beaches are frequently colonized by one, two or more parallel belts of mainly annual plants. A similar limited assemblage of species occurs wherever suitable sandy beaches are present around the Scottish coastline. This includes all the species listed for Ynyslas except *Euphorbia paralias*, a plant of more southern distribution, together with *Cirsium arvense, Festuca rubra, Leymus arenarius* and *Rumex crispus*. Amongst the more constant species only *Honckenya peploides* is perennial; indeed, on some highly exposed beaches including Bettyhill, Sutherland, it is the only one of this group of species present, presumably because high winds make the surface sand so unstable that colonization by annuals is prevented. Even where shores are less exposed, as at St Cyrus, Kincardineshire (Gimingham,

Figure 6.5 Distribution and area of sand dune systems in Great Britain. (From Doody, 1989c.) The Sand Dune Survey of Great Britain resulted in inventories for the dunes of England (Radley, 1994), Scotland and Wales (Dargie 1993, 1995). These summaries, and the numerous site reports on which they were based should be consulted for descriptions in terms of the National Vegetation Classification (Rodwell, 1998). **Literature sources for important dune systems listed below are given in the References.** Sites are listed in clockwise order starting at the top left: Luskentyre, Harris, Outer Hebrides: Gimingham, Gemmell and Greig-Smith (1949); Scottish Coast: Gimingham (1964); North Sea coastal margin: Doody, Johnston and Smith (1993); Culbin Sands: Patton and Stewart (1917), Stewart and Patton (1927); Sands of Forvie: Landsberg (1955), North (1981); St Cyrus: Gimingham (1951b); North Norfolk Dunes: Moore (1971), Jefferies (1976); Winterton, Norfolk: Boorman and Fuller (1977); Hampshire: Brewis, Bowman and Rose (1996); South Haven Peninsula, Studland Heath Dunes: Diver (1933), Alvin (1960), Willis (1985b); Braunton Burrows: Willis *et al.* (1959a,b), Hope-Simpson and Yemm (1979); Kenfig Pool and Dunes: Gillham (1982), Jones and Etherington (1989); Margam Burrows, Crymlyn Burrows: Gillham (1982); Oxwich Dunes: Gillham (1977); Whiteford Burrows: Gillham (1977); Newborough Warren, Anglesey: Gibbons (1994); Ranwell (1958, 1959, 1960a,b); Aberffraw Dunes, Anglesey: Liddle and Greig-Smith (1975a,b); Sefton Coast Dunes: James and Wharfe (1989); Atkinson and Houston (1993).

Figure 6.6 Sea rocket (*Cakile maritima*) at foot of foredune at Ynyslas, North Wales. (Photograph by John R. Packham.)

Figure 6.7 Vigorous marram grass (*Ammophila arenaria*) on partially eroded foredunes at Ynyslas, North Wales, with embryo dunes dominated by the annual *Cakile maritima* nearer the sea. (Photograph by John R. Packham.)

1951b) and Longniddry, East Lothian (Hulme, 1957), large fluctuations in the numerical proportions of the various species result from instability of the surface, variations in tidal submergence and wave action, together with differences in seed supply.

6.2.2 EXTENT OF THE GROWTH ZONE

In the northern hemisphere, annuals colonizing the strandline germinate in April and May, well after disturbance by the equinoctial spring tides in March. Hulme (1957) investigated colonization by one of the coastal forms of the *Atriplex prostrata* group (*A. hastata s.l.*) of the beach at Longniddry, charting its downward extension until 16 June, 1954, when the high-water mark was at its lowest for the summer and seedlings were present just above it, and then its gradual recession until 28 September as tidal scour took its toll. These observations show that, in the main growing season, conditions for growth of annuals are suitable from maximum high water spring tides to near mean high water spring tides, a zone up to 11.5 m wide at Longniddry beach. It is, however, the plants at the higher levels which live the longest and are most likely to set ripe seed.

Tidal litter frequently contains or traps viable seed. Moreover, daily temperature variations at the sand surface of 25°C in open sand were found to be reduced to 7°C beneath tidal litter on a summer day at Piver's Island, North Carolina, USA (Barnes and Barnes, 1954). Plants of particular species often occur in rather dense groups in strandline vegetation. These may develop from seeds attached to complete inflorescences buried in sand at the end of the season or from aggregations resulting from redistribution of surface sand and other particles, including seeds brought in as litter, during the winter.

6.2.3 PIONEER DUNE GRASSES AND DIFFERENTIAL GRAZING

Elytrigia juncea and *Ammophila arenaria* are rhizomatous sand-fixing grasses whose leaves roll up, probably reducing transpiration and reducing damage by sandblasting, when the relative humidity of the air is low. The ligule of *E. juncea* is truncate and less than 2 mm long, whereas that of *A. arenaria* is pointed, about 10–30 mm long and sometimes forked. The closely related North American species *Ammophila breviligulata*, which has a short truncate ligule, has been planted on some British sand dunes. Lyme grass (*Leymus arenarius*) is a robust bluish-grey perennial up to 200 cm high. Its leaves have pointed auricles and its long stout rhizomes enable it to flourish in loose sand on the seaward side of dunes. This species hybridizes with *E. juncea* on the shores of the Baltic. When used to prevent erosion, small leafy pieces of lyme grass rootstock are planted at 1-m intervals; in Iceland its seeds are sown on a massive scale in dune restoration programmes. It may be infected by stem smut fungus *Ustilago hypodytes* and produce culms covered by black fungal spores rather than bearing seeds, as in plantings observed at Lytham St Annes, Lancashire, in the very wet summer of 1985.

At Ynyslas, *Elytrigia juncea*, the main colonizer of the embryo dunes, did exceptionally well in the virtual absence of rabbits. By the early 1970s, however, these grazers had largely recovered from the effects of the myxomatosis epidemic; now sand couch-grass is again being heavily grazed and producing very little seed. Rabbits feed on the young tender shoots of *Ammophila* but do not graze it to the ground – older spiny shoots are left. At Crymlyn Burrows, South Wales, there is little evidence of rabbit grazing south of the slack, and here *E. juncea* does so well that at the end of the summer most of its individuals can be easily picked out by the characteristic inflorescence axis with the laterally inserted spikelets on both sides.

Invasion of strandline communities by *Elytrigia juncea* is often the first sign of permanent establishment of dune-forming plants, though the ability of this species and of marram to establish on the open sand of the backshore means that both may be strandline species. *Ammophila arenaria* frequently succeeds *E. juncea* in the embryo dunes. At Braunton Burrows, on some very exposed Scottish beaches, and at Coto Doñana, southwest Spain, dune formation is often initiated directly by *Ammophila* even though *Elytrigia* is present. The capacity of these and other pioneer grasses, such as *Uniola paniculata* in North Carolina, USA, to bind sand depends on their ability to perennate and to develop long-ranging horizontal, and to some extent vertical, rhizome systems.

6.2.4 ACCRETION RATES AND SURVIVAL OF DUNE GRASSES

Except for January, *Leymus arenarius* makes vegetative growth at Aberdeen throughout the year (Bond, 1952), and although *Elytrigia juncea* was found by Nicholson (1952) to be dormant during the winter at St Cyrus, its dead shoots are persistent and help to retain sand. Viable buds have been found at a depth of 60 cm in both these species, which also have similar tolerance of sand accretion – about 60 cm a year. *Ammophila breviligulata*, on the inland dunes round Lake Michigan (Laing, 1954), and *A. arenaria* at Newborough Warren (Ranwell, 1958), have both been shown to survive an absolute accretion limit of 1 m sand per year, though density soon diminishes if these conditions persist.

Germination of *Leymus arenarius* has previously been described as slow and erratic. Clarke (1965) considered that the presence of a water-soluble germination inhibitor in seeds and glumes of this species tended to delay production of seedlings until the spring when they are more likely to survive, the maximum depth of sand burial the seedlings can tolerate being 7–8 cm. Greipsson and Davy (1994a, 1995), however, found that almost 100% germination could be obtained when caryopses were soaked in water for 24 hours, stratified for 2 weeks (5°C), and then kept in continuous darkness for 2 weeks under alternating temperatures with an amplitude of 10–20°C on a 12-hour cycle. Such diurnal fluctuations in temperature occur under average weather conditions in black volcanic sands in Iceland during the growing season. The requirement for darkness may be a selective response to adverse conditions for establishment at the surface, while the alternating temperature response tends to ensure dormancy under deep burial with accreting sand.

6.2.5 EMBRYO DUNE FORMATION BY *ELYTRIGIA JUNCEA*

Dune initiation and development by *E. juncea* is described by Gimingham (1964) on the basis of investigations of its autecology at St Cyrus by Nicholson (1952). Germination may commence within 2–4 days of seed deposition if the sand is moist. It often follows heavy rain which imbibes the seed and leaches salt: germination is completely inhibited by sea water and reduced in rate and amount when sodium chloride concentration exceeds 0.5%. The roots reach a depth of 7 cm and also often almost permanently humid sand within 10 days of germination. Where moisture conditions are favourable the primary root quickly reaches about 15 cm deep, but the first formed lateral roots grow horizontally closely below the sand surface. After a rosette of tillers is formed, short rhizomes extend for 5–30 cm, new groups of tillers arising (Figure 6.8(a–c)). Such growth may continue for two seasons, but ultimately long horizontal rhizomes much increase the spread of the plant (Figure 6.8(d)). In autumn their tips often turn upwards, breaking the surface and forming a new group of shoots in spring. If sand accretion is slight, such development sometimes continues indefinitely; elongation of shoots can bring them to the surface in sand up to about 23 cm deep. If, however, burial is very rapid, shoots are killed and rhizomes extend vertically rather than laterally until the surface is reached, when tillering occurs. This sympodial development can keep pace with sand accretion up to depths of 1.8 m or more. Consequently, by lateral exten-

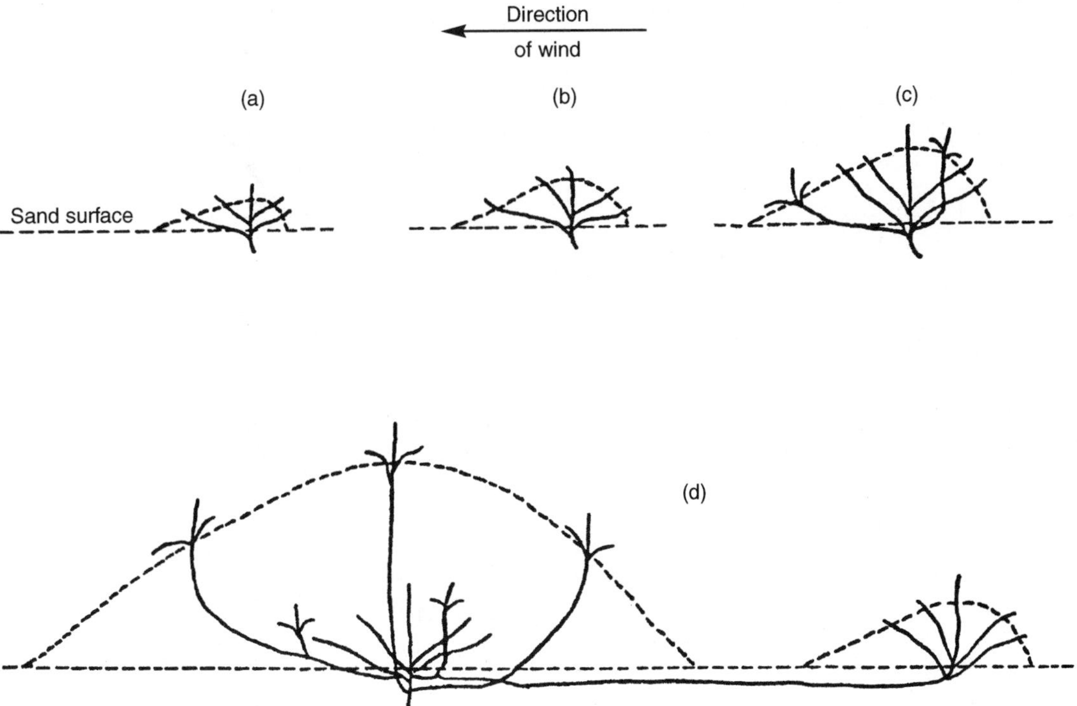

Figure 6.8 Stages (a–d) in the development of an embryonic dune round a plant of *Elytrigia juncea* (sand couch-grass). The upper broken line represents the surface of the dune. (Redrawn from Nicholson, 1952.)

sion *E. juncea* can stabilize the foreshore region, arresting blown sand and acting as a dune builder if accretion is not too great. This is seen, for example, at Balinoe, Tiree, Scotland, where *Ammophila arenaria* is locally absent (Vose *et al.*, 1957).

Multi-node rhizome fragments of *Elytrigia juncea* have an important advantage over single-node fragments and seeds in their greater ability to produce viable shoots following burial, though all three types of propagule were found by Harris and Davy (1986a,b, 1987) to be effective in colonizing bare areas. When buried experimentally at various depths, shoots from newly-germinated seedlings and single-node rhizome fragments were able to emerge from 127 mm but not from 178 mm. Shoots from multi-node fragments, which have greater reserves, grew up from the greater depth. The strandline of the study site at Holkham NNR, Norfolk, UK, was highly disturbed in comparison with adjacent foredunes, but was rapidly recolonized by seedlings and rhizome fragments following complete removal of the strandline community by the catastrophic storm surge of January 1978. Rabbit grazing was severe at this time; only rarely did strandline tillers survive long enough to flower following vernalization during the winter.

6.3 INFLUENCE OF LAND FORM, SOIL AND WATER REGIME ON DUNE PLANTS

Figure 6.9 shows one of several rather isolated parabolic dunes which arose in the southern

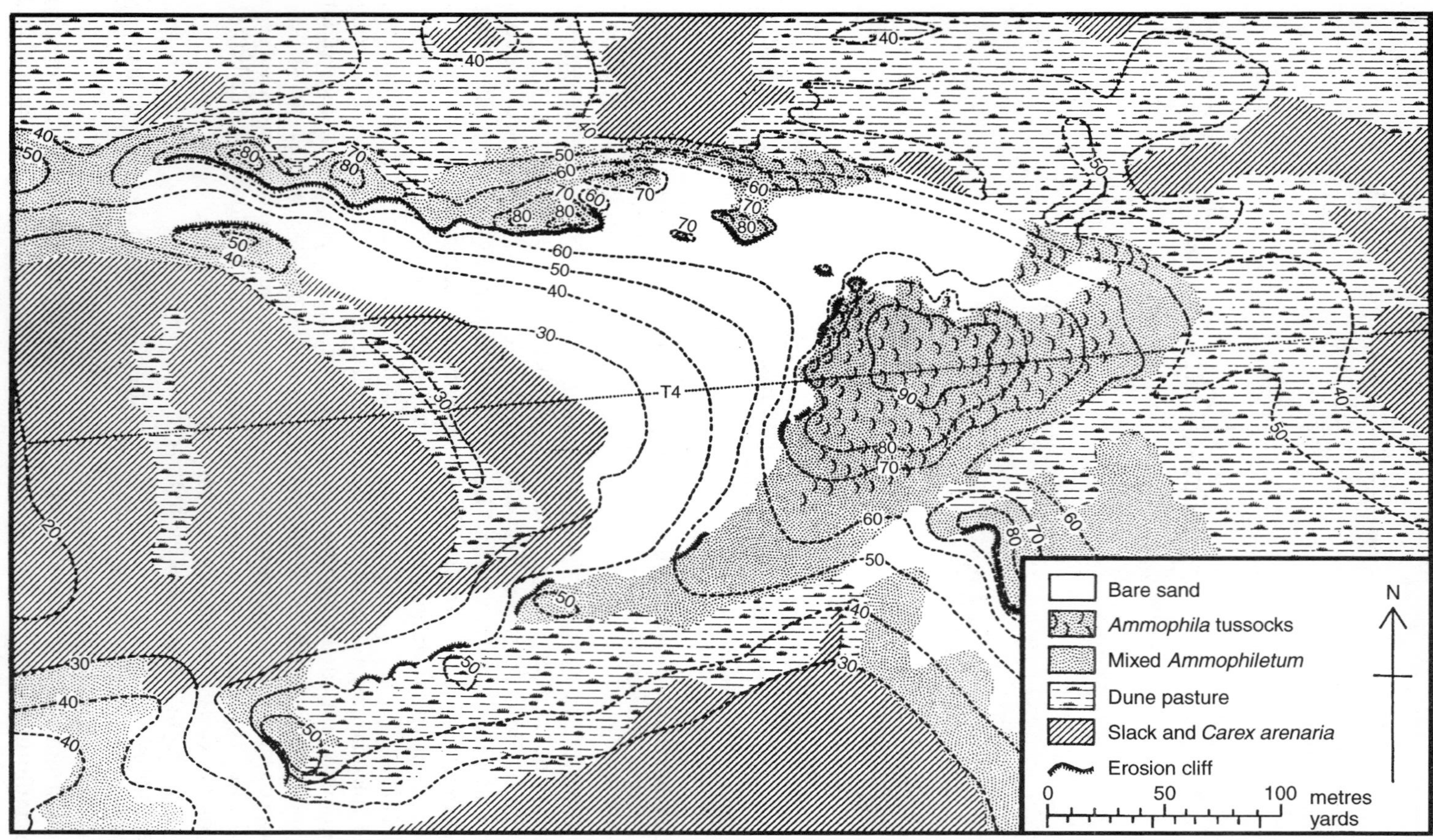

Figure 6.9 The topography and vegetation of a parabolic dune at Braunton Burrows, North Devon. Heights are in feet OD. (From Willis *et al.*, 1959b; courtesy *Journal of Ecology*.)

part of Braunton Burrows, North Devon, as a consequence of exposure to high winds and a reduced sand supply. The highest point of the dune had an altitude of almost 30 m (100 ft) in 1948, and was along the line of Transect 4 (see also Figure 6.18). The ridge was breached a little to the north of the central high region of the parabola, with the result that sand spilled through onto the dune pasture and slack in the lee of the ridge. The breach led to accelerated erosion on the windward faces of the dune, and small erosion cliffs (1.5 to 3 m high) are a prominent feature of most of the ridges in the area, as they are also in parts of the Ynyslas dunes. Much of the structure of this parabolic dune system is strongly influenced by *Ammophila arenaria*, which partially stabilizes it. Steep-sided hummocks and degenerating marram are prominent near the erosion cliffs, where the rhizome systems of marram are exposed.

Typical rounded and building marram tussocks occur in regions receiving wind-blown sand on leeward slopes flanking the area of bare sand. *Carex arenaria* plays an important role in stabilizing the floor of the slack between opposite arms of the parabola, and the two small ridges of dune pasture running transversely across it may well represent previously stabilized areas first colonized by sand sedge at the foot of the windward face of the dune in previous phases of its irregular migration towards the east.

In such systems, where most of the mobile sand is derived from the erosion faces of the ridges, considerable sorting of the sand grains is evident with the coarsest, and least easily transported, fractions remaining, partially stabilizing some of the windward slopes and the crest of the breach. Such stabilization, associated with the formation of relatively large sand ripples, is a prominent feature of the lower windward slopes of parabolic dunes (cf. Bagnold, 1984, p. 153 *et seq*.). Fine particles, which in contrast are readily moved by the wind, are deposited in the sheltered lee of the dune (Figure 6.10). The finest mineral particles of all may remain suspended in upward air currents and be carried considerable distances.

The distribution of vegetation across a high dune where there is substantial erosion on the steep windward face, with inundating sand to the lee, and scarcity of water supply almost everywhere, is shown in Figure 6.11 (note the high proportion of bare ground except for the lee slope and the three different measures employed). Species which contributed little to

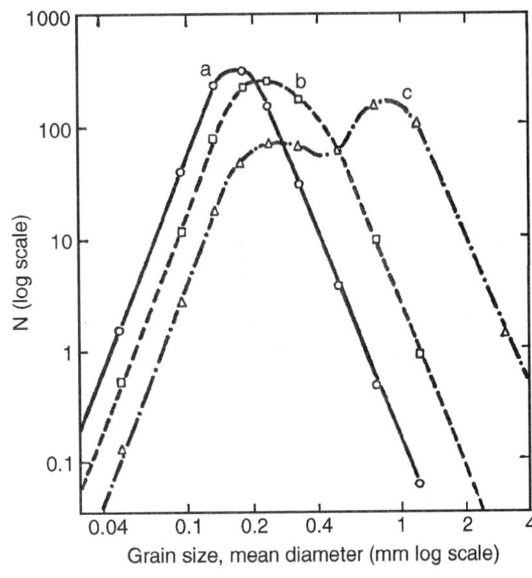

Figure 6.10 Grading diagrams for sand samples from three different situations at Braunton Burrows, North Devon. For each of the three samples the graphs show the proportion of sand having a particular grain size. This proportion is expressed by the parameter N, which is the percentage of the total sample which falls in a given fraction, divided by the difference between the logarithms of the maximum and minimum particle sizes in that fraction (see Bagnold, 1941, p. 113). The samples illustrated are as follows: (a) fine sand from among *Ammophila* shoots on the lee slope of the main line of dunes; (b) medium sand from the shore above high water mark; and (c) sand from the erosion face of a blow-out and containing two components, a coarse fraction left on the surface by wind sorting and a medium sand from below. (From Willis *et al*., 1959a; courtesy *Journal of Ecology*.)

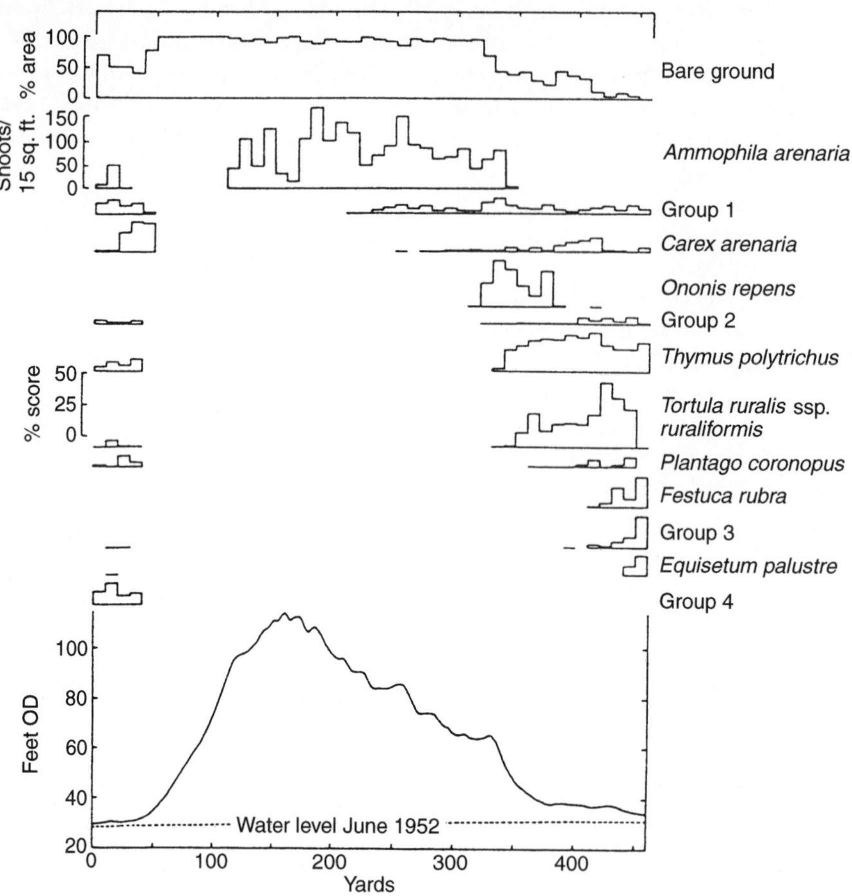

Figure 6.11 The distribution of vegetation across a typical high dune at Braunton Burrows. The seaward side is on the left; the dune profile is at the bottom of the diagram. Estimates of vegetation are shown by histograms; figures are plotted for each 9.14 m (30 ft) length of transect, and are derived from scores in 10 frames of 90 × 15 cm (3 × 0.5 ft). The figures for *Ammophila* give numbers of shoots. The scores for all the other species are expressed as percentages. The species which individually contribute little to the total bulk have been divided into four groups. (From Willis *et al.*, 1959b, courtesy *Journal of Ecology*.)

the total bulk were divided into groups 1–4 (respectively Mixed Ammophiletum 15 species; Dune pasture (high level) 16 species; Dune pasture (low level) 15 species; and inter-dunal slack 12 species); species lists are given by Willis *et al.* (1959b).

6.3.1 COMPOSITION OF DUNE SOILS AND CHANGES WITH TIME

Dune soils alter markedly as they age. Freshly blown sand from the beach is often low in mineral nutrients, many sand-dune systems having limiting levels of major nutrients, notably of nitrogen, phosphorus and potassium (section 6.6). The content of shell fragments may range from very high to very low, leading to calcareous and acidic conditions respectively. The ability of dune soils to retain water and mineral nutrients rises as organic matter content increases. As Salisbury (1922) demonstrated long ago at Blakeney Point, Norfolk, organic matter content is greater in fixed than in mobile dunes, though calcium carbonate content and soil pH drop along transects running inland from the shore (Figure 6.12). At Blakeney, almost all the sands are acidic and low in nutrient content but concentrations of calcium, magnesium, sodium and nitrate are higher in the mobile dune sands than in those of the grey dunes; the opposite is true of soluble inorganic phosphate and potassium (Gorham, 1958).

Chemical analyses from a range of sites at Braunton Burrows show low values of several major mineral nutrients (Willis *et al.*, 1959a). Here, the calcareous soils have a high pH, being alkaline even in the oldest parts of the dune system. Carbonate, mainly of calcium, is highest in dunes near the sea, where fragmented mollusc shells are common. Leaching results in somewhat lower values in the older, inland parts, but even so these remain distinctly calcareous (Table 6.2). Magnesium, highest in the foredunes, is at low levels further inland. Organic carbon is extremely low in the seaward dunes and slacks, but substantially higher where scrub develops and in old well-vegetated slacks. Total nitrogen is also very low, except for areas bearing scrub and slacks of considerable age. As expected, sodium and chloride are appreciable in the foredunes, but at low levels elsewhere. Potassium, which is readily leached, is in scant supply, particularly in the older, drier parts of the system.

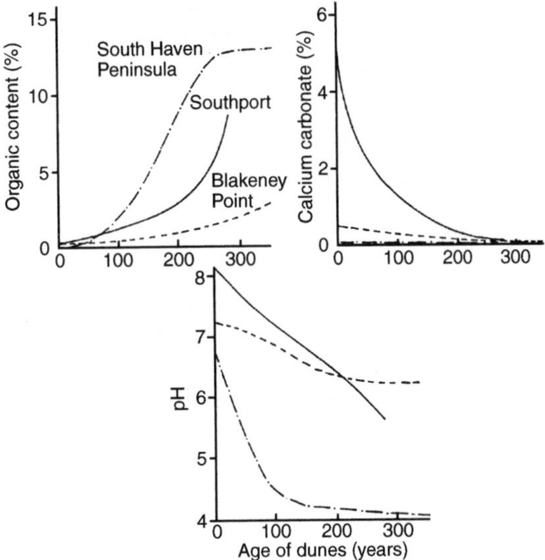

Figure 6.12 Changes in the surface soil over time in organic matter, calcium carbonate content and pH in three coastal sand dune systems (South Haven Peninsula, Dorset; Southport, Lancashire; Blakeney Point, Norfolk). Values are based on those given by Salisbury (1952) and Wilson (1960). (From Willis, 1989; courtesy Royal Society of Edinburgh.)

6.3.2 DOMED WATER TABLES IN DUNE SYSTEMS

Where dunes are formed on raised ground, perhaps over an elevated rocky substratum,

Table 6.2 Chemical composition* of soils at Braunton Burrows. (After Willis *et al.*, 1959a)

Soil component/ property	*Foredunes* (Ammophila)	*Main dunes* (Ammophila)	*Dry dune pasture*	*Old dune pasture (lichens)*	*Dune scrub*	*Slack (near sea)*	*Slack* (Carex nigra)	*Slack* (Salix repens)
pH	9.05	9.06	8.70	8.66	8.18	8.99	8.12	8.11
Organic C (mg)	0.52	0.19	0.74	2.44	12.60	0.41	22.93	13.55
Total N (mg)	0.18	0.11	0.23	0.41	2.15	0.15	2.38	1.38
Calcium (mg)	70.4	69.5	57.3	60.6	33.9	64.4	46.8	50.4
Magnesium (mg)	2.27	0.99	1.06	0.88	0.22	1.57	0.57	0.28
Potassium (μg)	50	6	7	7	13	11	26	15
Sodium (μg)	528	14	6	11	55	15	50	26
Carbonate (mg)	119.6	115.4	92.8	98.2	51.0	109.7	73.5	78.3
Phosphate-P (μg)	109	110	107	59	148	112	133	110
Chloride (μg)	845	14	3	3	10	10	29	22

* Results expressed per g dry soil.

and also where there is substantial drainage into the dune system, the water regime is complex. At Braunton Burrows, and in other sand-dune systems developed on low ground with their own catchments, the hydrology is that of a virtually isolated granular deposit. Measurements at Braunton Burrows, based on well-equilibrated water holes, show that the water table of this system is some 6 or 7 m higher in the centre of the system than on the shoreline or at the inland boundary drain (Willis *et al.*, 1959a; see Figure 6, p. 18). The steepest gradients are near the inland boundary drain where, with a silt admixture to the sand, permeability is lowest. Domed water tables are known also at Newborough (Ranwell, 1959) and at Berrow, North Somerset (A.J. Willis, unpublished data).

There is a highly significant correlation between the rainfall of the preceding 3 months and the water level at Braunton Burrows, with a greater seasonal change in water level in sites near the top of the dome than in those near the margins. Gradients of groundwater were consistent with laboratory determinations of sand permeability and estimates of the volume of water discharged if 25–35% of the rainfall reaches the groundwater (the rest of the rainfall being lost by transpiration and evaporation).

6.3.3 DEW FORMATION

The occurrence of dew in the summer months is of considerable significance for the survival of shallow-rooted plants on dune slopes during drought. The increase in moisture per night when there are clear skies in the summer may be as much as 0.9 ml per 100 ml soil (Salisbury, 1952, p.187); Monteith and Unsworth (1990, p.193) give 0.2 to 0.4 mm per night as a maximum for dewfall (i.e. 2–4 ml in an area of 10 cm^2). Such amounts may be sufficient to meet the transpiration needs of small annuals which are rooted near the surface of the sand, which is often hot and very dry in periods of drought. Moisture-laden air from the sea may condense on the cool surfaces of plants and sand grains; dew may be seen, for example, running down the leaves of marram grass at dawn after a clear summer night during which the sand surface may be some 20°C colder than during the day. Besides condensation from the atmosphere, however, a substantial contribution to the water content of the sand in the rooting zone may be made at night by the upward movement of water vapour, under a temperature gradient, from

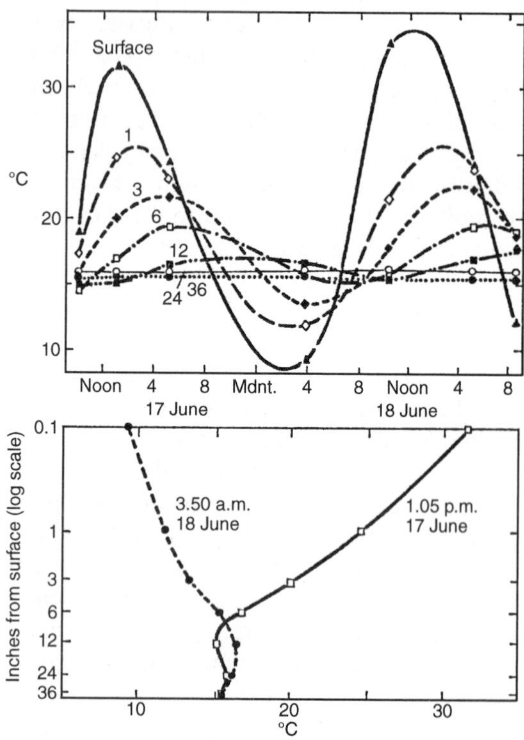

Figure 6.13 Diurnal fluctuations in temperature down a sand profile at Braunton Burrows, North Devon, UK. An undisturbed profile 40 inches (c. 1 m) deep was exposed by digging a pit in the top of a bare dune. Resistance thermometers were inserted without delay into the profile at 36, 24, 12, 6, 3 and 1 inches from the surface, and the excavated sand was restored as far as possible in its original position. After 2 days, periodic measurements of temperature were made (upper diagram). The temperature at the surface was recorded by means of a mercury-in-glass thermometer. The maximum extent of the fluctuation was plotted against a logarithmic scale of depth (lower diagram). Times are given in GMT. (From Willis *et al.*, 1959b, courtesy *Journal of Ecology*.)

warm moist sand some distance below the surface to the cooler drier upper layers. The existence of a marked temperature gradient has been clearly demonstrated at Braunton Burrows (Figure 6.13). The greatest increase in water content was recorded by Salisbury at a depth of about 1 m, where temperatures are very unlikely to fall below the dew point. However, the temperature gradients observed at Braunton Burrows, and by Salisbury, seem to preclude the possibility of increased water content within the dune by condensation from the external atmosphere, the upward movement of water vapour from wet sand below appearing more important. More detailed measurements both of temperature and of moisture gradients, especially under conditions of dew formation, are necessary before the movement of water and its availability to dune vegetation is fully elucidated.

6.3.4 WET SLACKS, DRY SLACKS, DUNES AND THE WATER TABLE

Ranwell (1959) referred to three physiographic units: **wet slacks** where the water table in summer remained within 1 m of the surface, **dry slacks** where it was between 1 and 2 m from the surface, and **dunes** where the summer water table was more than 2 m from the surface. The absolute annual range of the water tables at Newborough and Braunton Burrows is around 1 m at most sites, though at Winterton, Norfolk where the climate is much drier, it is only about 0.5 m. Soil moisture profiles (Figure 6.14) show that the sand is saturated for some 10–15 cm above the free water table, while capillary action causes high moisture contents for some 45 cm above it. The soil moisture profiles indicate that the water table has no significant influence on soil moisture contents more than 1 m above its free level in the soil.

In wet slacks as defined above, plants growing on the surface have their roots within reach of relatively high moisture zones throughout the year. Some slacks of this type are too damp for *Salix repens*, which cannot survive in permanently waterlogged habitats. Ranwell (1959) recorded that the extreme upper limit of occasional winter flooding occurred at the junction between wet and dry slacks, and was frequently marked by an abrupt change in *S. repens* from a dwarfed growth form 2–3 cm high to a taller form 10 cm or more high in the dry slack. Thus, the lower limit of *S. repens*, and probably of other species, appears to be correlated with the extension of waterlogging into the growing season.

Where the water table drops to between 1 and 2 m from the surface during June–August, shallower-rooted species of the dry slack are beyond its influence during the period when potential evapotranspiration is at a maximum, while plants of the dune associes are always well above the water table during summer droughts when they receive only occasional rain and dewfall.

At Braunton Burrows in periods of dry weather, sand more than 50 cm above the water table has a low water content (<5% gravimetric). In prolonged drought, the sand near the surface is effectively air-dry, holding only 1–2% water. There may be a fairly uniform change of water content down the soil profile, but heavy spells of rain influence the pattern of this change. In dry periods, patches of bare sand on dunes are often moister below the surface than comparable vegetated areas. At Braunton Burrows in August 1955, sand at a depth of 15 cm beneath *Tortula ruralis* ssp. *ruraliformis* had a water content of 1.6%, whereas that of sand at the same depth on a nearby bare dune crest was 3.6% by weight (Willis, 1964). At a depth of 60–90 cm the values beneath dry dune pasture and *Ammophila arenaria* on a high dune were 1.7% and 4.9% respectively. Willis *et al.* (1959b, p.265) suggested that these differences might result from rapid absorption and exploitation of water by shallow-rooted species of the dune pasture.

In some situations capillary water is of considerable significance. Sand-dune plants such as *Euphorbia portlandica* (Portland spurge), *Hypochaeris radicata* (cat's-ear) and *Thymus polytrichus* (wild thyme) have root systems longer

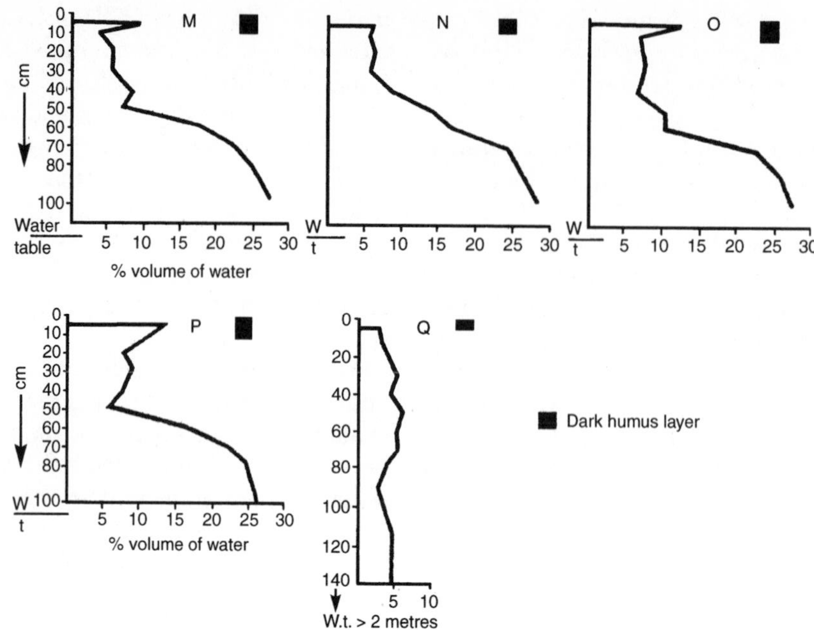

Figure 6.14 Soil moisture profiles, Newborough Warren, North Wales. M and N, beneath dry slack *Agrostis capillaris–Festuca rubra* turf associes. O and P, beneath dry slack *Salix repens* associes. Q, beneath semi-fixed dune *Salix repens* associes. (From Ranwell, 1959; courtesy *Journal of Ecology*.)

than 0.5 m, while others have roots reaching depths of 1.5 m; those of *Ononis repens* sometimes reach 2–3 m (C.H. Gimingham, personal communication).

6.3.5 EFFECTS OF WATER ON PLANT DISTRIBUTION

The distributions of many sand-dune species, including those of damp slacks, appear to be controlled by the water regime, showing very close correlation with particular conditions. Whereas some species are totally confined to wet sites subject to flooding, e.g. *Anagallis tenella* (bog pimpernel), *Juncus articulatus* (jointed rush) and *Ranunculus flammula* (lesser spearwort), others are almost always found on ground above the level of the water table, e.g. *Ammophila arenaria, Ononis repens* (restharrow), *Pteridium aquilinum, Tortula ruralis* ssp. *ruraliformis* (Figure 6.15). Yet others, such as *Carex arenaria* (Figure 6.16) and *Festuca rubra,* can grow in both dry and fairly wet sites, while certain species, e.g. *Juncus maritimus, Lotus corniculatus* (which may have roots nearly 1 m long), occur typically within the capillary fringe. Some plants, e.g. *Salix repens* and the rare *Scirpoides holoschoenus*, can establish only in wet sites, but once established may build up their own sand hummocks above the water table.

The wetness of a site may be quantified by an '**index of flooding**' – the average number of months in a representative year for which a site is under water, from 0 (free from flooding) to 12 (permanently flooded) (Willis *et al*., 1959b, p.271). Studies of the performance of plants relative to the index of flooding show distinctive differences between species, their abundance being affected in characteristic ways. Whereas typical dune slack plants, such as *Hydrocotyle vulgaris* (marsh pennywort) and *Juncus articulatus* are most successful in the wettest sites, *Carex arenaria* and *Leontodon saxatilis* (autumn hawkbit), though tolerant of wetness, are sparse in very wet parts. Transpiration rates and stomatal control in slack and dune slope species are discussed in section 3.5.

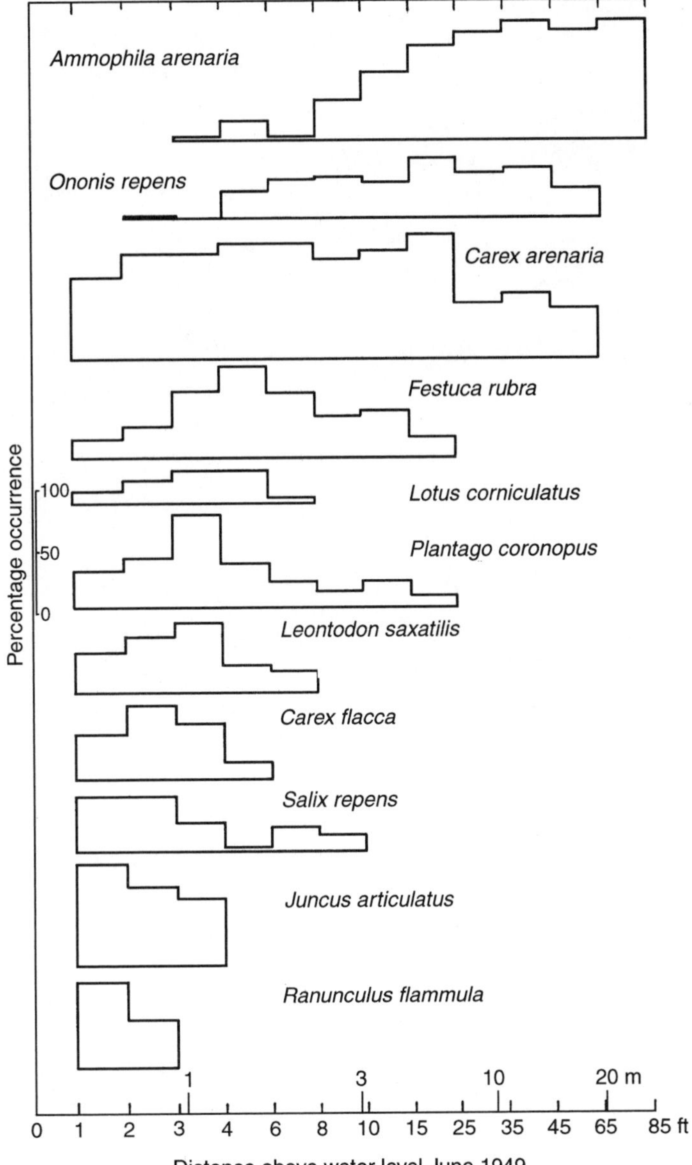

Figure 6.15 The distribution of some important species in relation to water level at Braunton Burrows, North Devon. Histograms are based on study of the vegetation at 270 surveyed sites, grouped into convenient categories of height above the water level of June 1949. Percentage occurrence shows the percentage of sites in each height category occupied by each species. (After Willis *et al.*, 1959b.)

6.3.6 INFLUENCE OF WEATHER

At Oostvoorne, The Netherlands, regularly recorded permanent plots have been used to investigate fluctuations in the abundance of sand-dune species correlated with changes in the weather. On the basis of abundance records made in 1972 and 1979, when the sum-

Figure 6.16 Lines of vigorous and flowering sand sedge (*Carex arenaria*) shoots colonizing loose sand at base of blowout, *Ammophila arenaria* tussocks on skyline, Braunton Burrows, North Devon. (Photograph by M.C.F. Proctor.)

mers were markedly wetter than average, and in 1976 and 1977 when they were much drier, some differences were seen in species response, as found also in experiments conducted in 1978–1979 when rainfall was left to nature, raised or lowered on a number of plots. There were complicating factors, such as incomplete prevention of rabbit grazing in the experiment and the weather of the preceding years, but *Anthoxanthum odoratum*, *Centaurium erythraea*, *Luzula campestris*, *Thymus pulegioides* (large thyme) and *Rhytidiadelphus squarrosus* tended to decrease under dry conditions and increase under wet ones (van der Maarel, 1981). *Cerastium semidecandrum* (little mouse-ear) did well during 1972 and 1979 in both the mesosere and xerosere, but under the manipulated reduced rainfall showed a strong increase attributed to the opening up of the herb and moss layer and activities of ants.

6.4 THE ROLE OF NON-VASCULAR PLANTS IN DUNE SYSTEMS

Mosses and lichens are the most conspicuous non-vascular plants of dune systems; both help to stabilize the dune surface. Pioneer mosses can withstand moderate accretion and lichens containing Cyanobacteria (formerly Cyanophyceae, blue–green algae) play a significant role in nitrogen fixation. At Blakeney Point, Norfolk, Richards (1929) traced a succession in the dunes in which pioneer mosses colonized bare sand, the ensuing intermediate lichen-dominated phase lasting until the cover of higher plants closed. After this, bryophytes again assumed dominance in the cryptogamic layer where the number of species gradually increased.

6.4.1 DUNE BRYOPHYTES AND THEIR GROWTH FORMS

Richards (1929) noted four primary moss colonizers on loose but immobile sand between marram tussocks on the lee side of yellow dunes at Blakeney Point. Of these, *Tortula ruralis* ssp. *ruraliformis* normally attains a height of at least 2 cm and forms a Tall Turf with erect branches; its leaves contract on drying and straighten so sharply on rewetting that sand is flicked off them. The other acrocarps (with upright, little-branched stems), *Ceratodon purpureus* and *Bryum* spp., form Short Turfs with erect branches. The fourth, *Brachythecium albicans*, is a mat-forming pleurocarp (prostrate and much-branched) whose ability to adopt a pseudo-erect habit allows it to compete effectively with the turf-formers on unfixed sand.

The same four mosses occur in similar situations on Scolt Head Island, Norfolk, where they again consolidate the surface of the sand and raise its organic content, making it easier for vascular plants, lichens and more demanding mosses, notably *Hypnum cupressiforme*, to establish, so initiating the grey dune phase. The rather poor bryophyte floras of Blakeney and Scolt are very similar, and reflect the low lime content of east coast dune sands. *Homalothecium lutescens* (= *Camptothecium lutescens*) is found in considerable abundance in the much richer floras of the more calcareous dunes on the west coasts of England and Wales, where it occurs as a fifth member of the bryophytic pioneers of yellow dune. This species is absent from Blakeney and rare on Scolt.

Of the five growth forms recognized in British terrestrial bryophytes (section 2.1) three – turfs, mats and wefts – were found by Birse and Gimingham (1955) to be the main types of sand dunes, although the dendroid form *Climacium dendroides*, a **canopy former**, occurs occasionally and **cushions** may be present on large stones or boulders. **Turfs**, whose parallel upright stems may have an abundant development of rhizoids, have a growth form able to resist burial by sand in mobile dunes. As stabilization proceeds, **mats** – generally dense and interwoven systems extending horizontally over the substratum – become increasingly important, with *Hypnum cupressiforme* assuming dominance over much of the young and old dunes. At this stage in the succession *Brachythecium albicans* has a prostrate habit, but soon grows upwards again if sand is blown over it. **Wefts** – systems developed as a result of the loose intertwining of straggling shoots and branches, often ascending and luxuriant – tend to be last in the succession and are represented at Scolt Head Island, Norfolk, by *Hylocomium splendens* and *Rhytidiadelphus triquetrus* on the grey dunes. Page *et al.* (1985) state that *R. triquetrus* is an indicator of increasing acidification on older dunes in the generally calcareous system at Ynyslas, North Wales, though elsewhere (Watson, 1981) it also occurs on calcareous substrates.

Birse (1958) studied the relationship between the growth form of mosses and ground water supply in a dune slack at Foveran Links, Aberdeenshire. Tall turfs with branches erect, Short Turfs, and also Rough and Smooth Mats in their typical development grew where they were never within reach of the water table. Wefts and the Dendroid form *Climacium dendroides* were luxuriant where the surface sand received an additional supply of water from the water table in summer. Where the water table was constantly near the surface Tall Turfs with divergent branches of limited growth, e.g. *Campylium stellatum*, and Tall Turfs (rhizoidal), e.g. *Rhizomnium punctatum*, were present.

Even pioneer mosses of British dunes appear unable to sustain upward growth through more than 4 cm of sand, as Birse, Landsberg and Gimingham (1957) showed by both experimental burial and observations on transplanted turfs under natural conditions. However, *Bryum algovicum* var. *rutheanum* (= *B. pendulum*), *Ceratodon purpureus* and *Brachythecium albicans* can emerge from burial by 4 cm of sand, and *Tortula ruralis* ssp. *ruraliformis* from a depth of 3 cm, recovering fairly rapidly from shallow coverings of sand. In most species, slender shoots bearing rhizoids reach the surface, then producing normal growth. In dunes in Victoria, Australia, *Barbula torquata* and *Tortula princeps* were also found to survive burial to a depth of 4 cm under field conditions, recovering by monopodially renewed growth, innovations at the apex or secondary protonemata from rhizoids (Moore and Scott, 1979). Figure 6.17 shows the production of leafy shoots by *Bryum algovicum* at the sand surface after burial. Even more striking examples of this phenomenon of '**flotation**', whereby accreting sand stimulates the shoot system to grow to successive dune surfaces, are seen in vascular plants; Salisbury (1952, p.217) illustrated a plant of *Eryngium maritimum* which had grown at five successively higher surfaces.

The importance of shade in controlling the distribution of dune bryophytes was demonstrated by Robertson (1955), who transplanted a number of species to a sunny site. All except two species died out in the order appropriate to levels of shading present in the sites in which they normally grew. In dune slacks mosses are an important component of the flora and contribute substantially to the biomass. Frequently abundant are *Bryum pseudotriquetrum*, *Calliergon cuspidatum*, *Campylium stellatum*, *Drepanocladus lycopodioides* and *Drepanocladus sendtneri*. Liverworts are much less evident but in moderately vegetated slacks may include the thalloid forms *Aneura pinguis*, *Moerckia hibernica*, *Petalophyllum ralfsii* and *Preissia quadrata* and a few leafy species such as *Leiocolea turbinata*.

6.4.2 DUNE LICHENS

Rabbits play an important part in maintaining short open vegetation by feeding and scratching, so retarding higher plant succession and the growth of bryophytes, while favouring persistence of the lichen phase. After myxomatosis in Britain there was a substantial increase in growth of grasses on dunes. Bryophytes at first increased in formerly lichen-rich areas, but then diminished as shade increased and litter isolated them from the surface. The striking sequence of lichens developed on dune surfaces of known age at the South Haven Peninsula, Dorset (Alvin, 1960) is described in section 6.5. Differences between this and another lime-deficient system – that of the island dunes at Laeso, Denmark (Böcher, 1952) – are ascribed to climatic factors.

Besides wood-encrusting species, some of which grow on stems of old *Suaeda vera*, other bushes and wooden posts, lichens at Scolt Head Island, Norfolk occur mainly on grey dune, eroded dune and stable shingle. On old dunes *Cladonia arbuscula* (= *C. sylvatica*) is char-

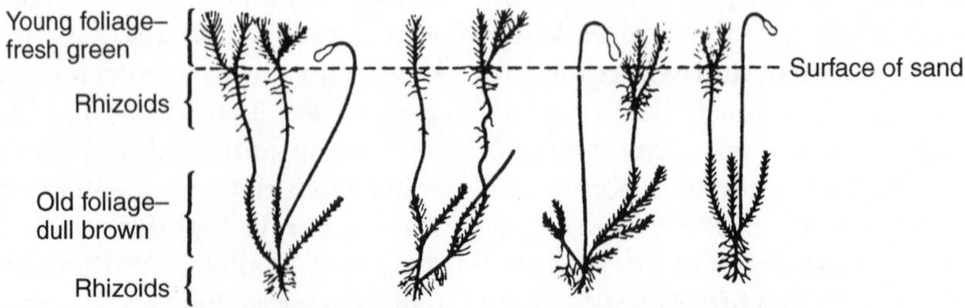

Figure 6.17 Excavated *Bryum algovicum* var. *rutheanum* (= *B. pendulum*) showing re-establishment of leafy shoots at the new dune surface following sand accretion. (After Gimingham, 1948.)

acteristic of sheltered slopes and hollows, often with mosses and *Carex arenaria*. *Peltigera membranacea* is found in damp mossy spots, while several species of *Cladonia* and *Peltigera* develop on more open ground, especially where trampling or grazing by rabbits reduces competition from vascular plants. Characteristic grey dune lichens are also present on the stable eroded dune in which some shingle is present, but *Parmelia saxatilis, Hypogymnia physodes, Coelocaulon aculeatum* (= *Cetraria aculeata*), *Rhizocarpon richardii* and *Buellia punctata* are typical of this habitat. *Lecanora dispersa*, a calcicole, grows mainly on oyster shells. The main saxicolous lichens of the stable shingle ridges are *Tephromela* (*Lecanora*) *atra, Xanthoria parietina, Caloplaca flavovirescens, Buellia punctata, Rhizocarpon richardii* and *Verrucaria maura* (Ellis, 1960). A much richer assemblage of lichens is present on the shingle ridges of Dungeness (Laundon, 1989; section 8.4), and on the Chesil Beach, Dorset (Watson, 1922; Tansley, 1949, p.893).

The succession and floristic composition of lichen communities in a coastal dune area in south-east Skane, southern Sweden, have been elucidated by Magnusson (1982). Lichens dominate the pioneer cryptogamic vegetation in open dry sites, with sparse cover of higher plants and little competition from mosses. The ground layer of the outer dunes, which are stabilized by *Ammophila arenaria*, consists of a *Cladonia glauca–C. chlorophaea* community associated with *Ammophila* debris. Further inland, a *Calluna–Empetrum* heath develops and the pioneer *Cladonia* community is partly overgrown by *C. portentosa*, the moss *Dicranum scoparium* entering the succession. On the secondary inner dunes *Corynephorus canescens* (grey hair-grass) colonizes, and *Coelocaulon aculeatum* is the pioneer cryptogam, rapidly invading by wind-borne thallus fragments and able to withstand the exposed conditions. A final lichen stage is dominated by *Cladonia* spp., notably *C. mitis* and *C. uncialis*.

This lichen zonation sharply contrasts – as expected from the very different substrata – with that demonstrated by a quantitative study of the lichen flora of siliceous rocky shores of Anglesey, North Wales. Here, Fletcher (1973a,b) showed that the distribution was of overlapping bands low on the shore but more diffuse in the supralittoral zone. Species of *Verrucaria* and *Lichina* were important in the littoral zone; in the supralittoral, a band dominated by *Caloplaca marina* was followed inland by a band including *Xanthoria parietina* and then by the 'Xeric–Supralittoral' supporting *Ramalina siliquosa*. Determining factors include topography, wave-action, aspect, slope and the light, nutrient and water status of the shores.

6.4.3 ECOPHYSIOLOGY OF TWO DOMINANT LICHENS IN DUTCH COASTAL DUNES

Cladonia furcata var. *furcata* has been described as a calcicolous species, occurring on soils whose original calcium carbonate content by weight was 1–2%, whereas the calcifuge *C. portentosa* (= *C. impexa*) grows on sands with a low lime content. The ion content of the thallus of *C. furcata* greatly exceeds that of *C. portentosa*; this is reflected by nutrient levels in the top 5 cm of the soils on which they were growing (Rozema and Spruyt, 1985). Overall mineral status appears to be a major factor controlling the distributions of these two species. Soil pH immediately beneath *C. furcata* was only 4.7, but rose rather rapidly with depth, reaching 8.3 at the bottom of a 13-cm soil profile. Soil pH beneath *C. portentosa*, initially slightly higher than that beneath *C. furcata*, rose only slightly with increasing depth, remaining close to 5.

6.4.4 FUNGI OF DUNE SYSTEMS

A great diversity of macro-fungi can frequently be found on sand-dune systems. Ellis (1960) provides an annotated list of fungi from Scolt Head Island, including many that are parasitic on sand-dune and salt-marsh plants. Pugh (1979) reviews the distribution of fungi in coastal regions, pointing out that in the decomposition of plant remains fungi are the main organisms which can decompose cellulose. Thus, in dunes and salt marshes, as elsewhere, fungi play a vital role in the **decomposition processes** upon which higher plant life is dependent. Additionally, mycorrhizal species importantly aid the mineral nutrition and growth of green plants. Indeed, a sequence of mycorrhizal types can be recognized across some sand-dune dune systems from facultatively and obligately vesicular–arbuscular mycorrhizal, to ectomycorrhizal and finally, in the oldest, most inland parts, to ericoid mycorrhizal (Read, 1989).

Watling and Rotheroe (1989) show the richness of macro-fungi in British sand dunes, giving records based on two intensive studies – of Ynyslas National Nature Reserve, North Wales and Bettyhill, northern Scotland – as well as for dunes elsewhere in Britain. Their list includes about 150 members of the Agaricales, 15 'Gasteromycetes' and eight Ascomycetes. Rather few fungi colonize yellow dunes, but dune grassland has a substantial mycoflora similar to that of grasslands in the near vicinity. Dune slacks support many species and high levels of fruit body biomass at phenological peaks, associated with willow, abundant litter and protection under the canopy.

6.5 TWO CONTRASTING DUNE SYSTEMS

The dunes at Braunton Burrows, North Devon, cover some 800 ha, are over 2000 years old, show a great diversity in water regime, are constantly changing under the influence of the wind, and are built of sands which are almost everywhere quite strongly calcareous and, on the foredunes, relatively high in magnesium. The dune system on the South Haven Peninsula, Dorset, with an area about one-quarter that of Braunton Burrows, has soils distinctly low in calcium (and also in some micronutrients), and supports an acidiphilic vegetation on its older dunes. Its origin is as recent as the early 17th century and it is still **prograding** (building towards the sea), while the shoreline at Braunton has changed very little for hundreds of years. Both have highly diverse floras, were damaged by military activity during 1939–1945, and in some places are tending to become covered with woodland scrub.

6.5.1 BRAUNTON BURROWS, NORTH DEVON

These coastal dunes, which extend for 5 km northward from the joint estuary of the Rivers Taw and Torridge, form one of the highest dune systems in the British Isles, rising to over 30 m OD in places (Figure 6.18). On the seaward side the dunes are fully exposed to the prevailing westerly winds and at low tide Saunton Sands provide a plentiful supply of wind-blown sand. The western part of the Burrows is a complex of dunes whose irregular ridges run approximately north–south parallel to the coastline; the highest crests are mostly to landward. Three or four irregular ranks of dunes can usually be recognized in the central sector, separated by low-lying slacks. Dunes are lower in the north which is partially protected by Saunton Down. Bare sand associated with conspicuous tussocks of marram on Figure 6.18 indicates the positions of many of the main crests. On the exposed southern flank only two ranks of dunes can be distinguished in places, the landward rank consisting of a series of parabolic dunes, running back alongside the estuary. The inland part of the dune system is irregular but generally fairly low-lying, with prominent sand hills only locally; the low areas of this region are flooded in wet seasons, as are numerous slacks between the high dunes. The gentle seaward slope of the interdunal slacks is stabilized by the water table (wind does not move wet sand).

Transect data for dune slacks in the southern part of the system show a clear distinction between the pioneer species established before *Salix repens* achieved dominance after about 10 years, and the secondary species

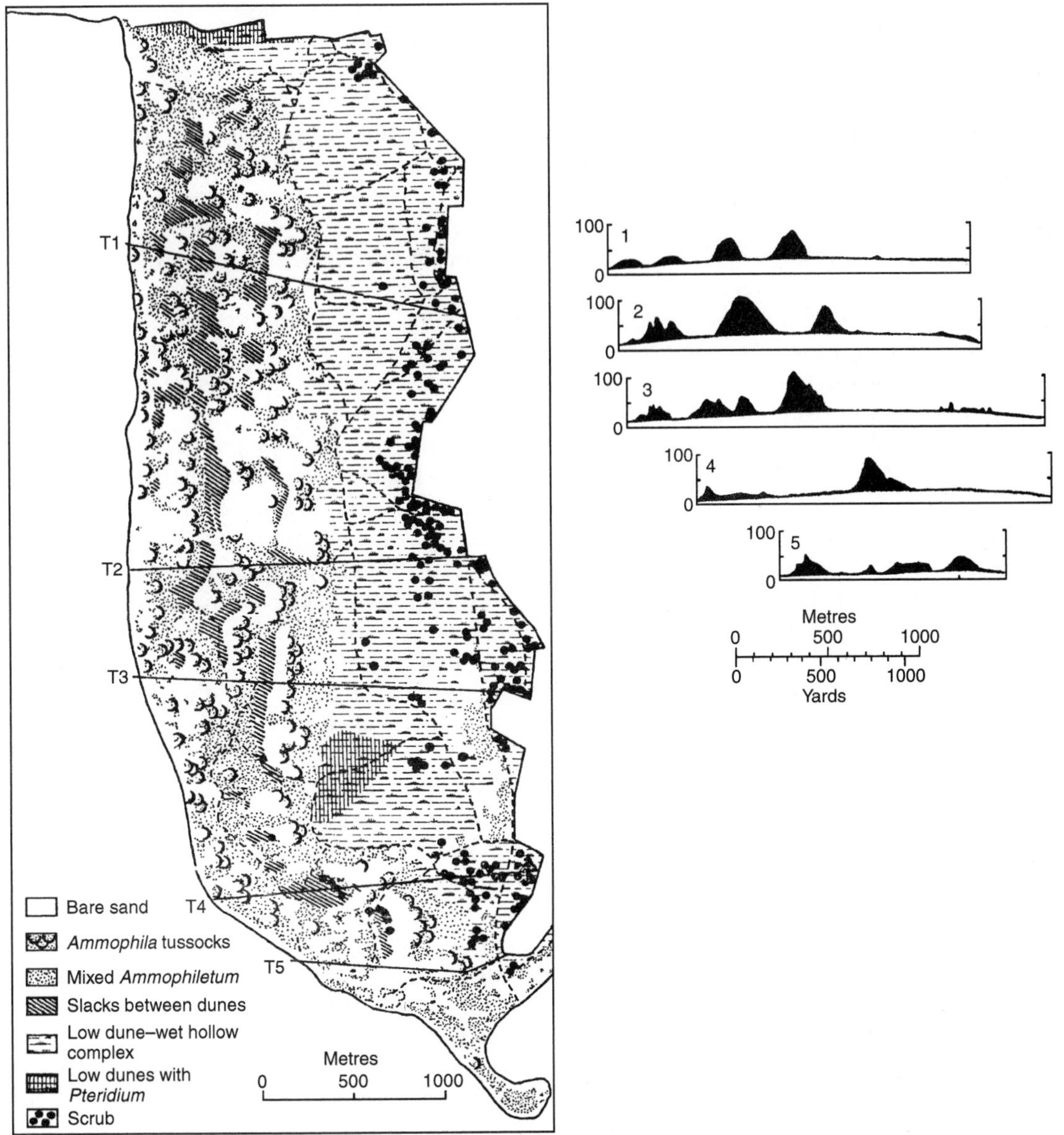

Figure 6.18 The distribution of the vegetation at Braunton Burrows in 1947, together with five profiles across the dune system. Heights, obtained by survey, are given in feet OD. In the profiles the sand above the water table of June 1952 is shown in black. (From Willis *et al.*, 1959a; courtesy *Journal of Ecology*.)

which came in after this event (Hope-Simpson and Yemm, 1979). Shoreward slacks quite often develop **crusts**, either of hardly recognizable moss debris with a large algal population or of calcium carbonate. The two forms of crust may sometimes be mixed; both are unfavourable for colonization by seed but rhizomatous plants (notably *Carex flacca*) penetrate from below, leading to gradual alteration of crusted ground.

Since 1947, about 340 taxa of angiosperms, some 10 species of pteridophytes, over 70 species of bryophytes and a substantial lichen flora have been recorded from the dune system. The large number of vascular plants in the southern and middle sectors is about 80 more than would be expected from the Arrhenius regression $\log y = \log a + n\log x$, where y = number of species, x = area, a is a constant and n an index of regression (Dony, 1977; see also Connor and McCoy, 1979) for continuous exponential increase of number of species with area (Willis, 1985a). Naturalized and alien species form an interesting but minor element of the flora, while a considerable number of flowering plants rare or uncommon in Britain, such as *Teucrium scordium* (water germander) and *Scirpoides holoschoenus* (round-headed club-rush), occur on the Burrows, some in substantial quantity.

There are numerous reasons for the species richness of Braunton Burrows (Willis, 1985a). As in other dune systems its dynamic nature ensures the constant provision of habitats ranging from the bare sand favourable to the growth of annuals to more stable terrain suitable for perennial herbs or shrubs. The influence of the water regime, including the domed water table illustrated in Figure 6.18, on niche differentiation has already been described in section 6.3, while features such as aspect, slope, exposure, and protection further widen the variety of environments available. On the Burrows the high soil pH and calcium levels have prevented the development of the rather monotonous dune heath seen in the South Haven system. Moreover, rabbit grazing and the low levels of soil nitrogen, phosphorus and potassium (discussed in section 6.6) considerably diminish the competitive ability of large species of high RGR and allow the continued survival of many 'stress-tolerant' forms. The distributions of *Juncus acutus* and *S. holoschoenus*, which are discussed in section 2.4, appear to be controlled by climate. Some flowering populations, such as those of *Dactylorhiza praetermissa* (southern marsh-orchid) and *Dactylorhiza incarnata* (early marsh-orchid) vary substantially from year to year, perhaps owing to 'runs' of weather.

6.5.2 THE SOUTH HAVEN DUNE SYSTEM AND ITS DEVELOPMENT

The numerous studies made of the South Haven Peninsula, Dorset, are outstanding – especially for the information which they provide regarding the time factor in soil development. Diver (1933) made a physiographical survey of the dunes, using early maps and charts to trace their development. The vascular plants have been investigated by Good (1935), the fungi by Brown (1958), the lichens by Alvin (1960), and the soils by Wilson (1960) and Willis (1985b, section 6.6).

The development of the peninsula is outlined in Figure 6.19. In the map of 1607 only the narrow and slightly arcuate strip of the Bagshot Beds was present. Its upper surface still forms a gently undulating plateau rising gradually towards the south. The plateau had previously been eroded to form low cliffs facing Poole Harbour to the west and Studland Bay to the east. Since 1607 the present extensive dune system has built up as a result of wind acting on a submarine accretion of sand in Studland Bay, much of the east-facing cliff being concealed as a result.

The dunes for long developed as two separate regions north and south of the present Little Sea lagoon (Figure 6.20), becoming unified only with the establishment, at the beginning of the 20th century, of a continuous line of foreshore dunes facing Studland Bay. The expansive and rhythmic succession of dune ridges and slacks of the northern region, which extends some 820 m seawards of the pre-1600 shoreline, are described here rather than the more condensed (320 m) southern sequence.

During the 17th century, blown sand accumulated against the east-facing cliffs and formed shallow dunes on the cliff top. This system developed into the long continuous Third Ridge complex that began the enclosure of the Little Sea lagoon, and was more or less complete when the 1721 map was drawn. This also shows the early hummocks of the Second ridge complex, which grew considerably by

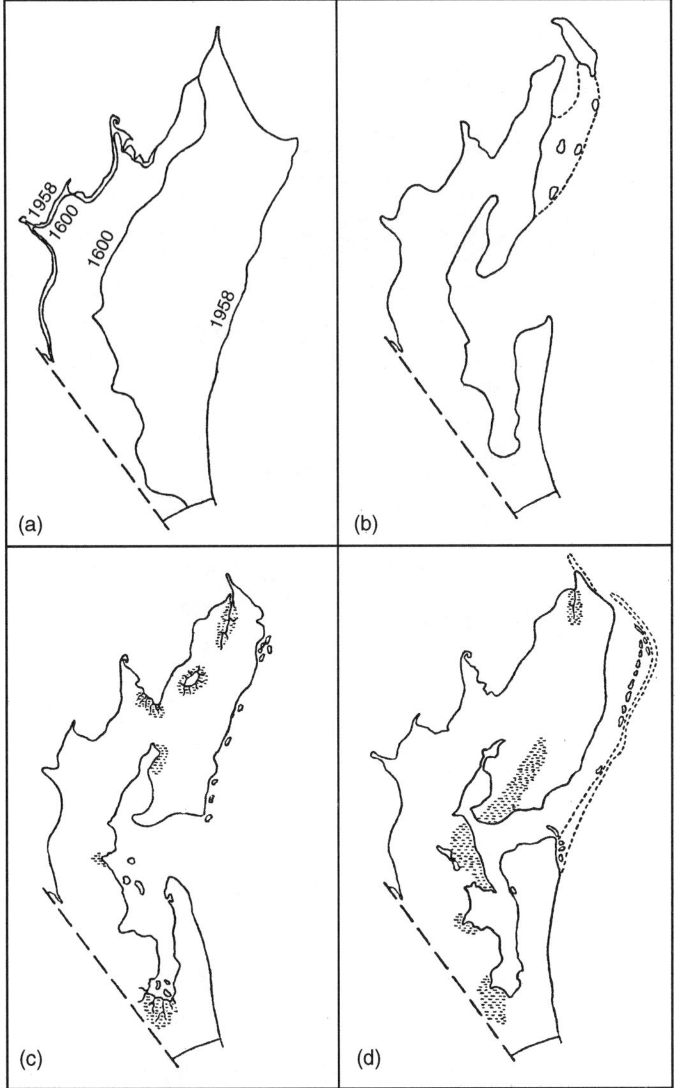

Figure 6.19 The historical development of the South Haven Peninsula dunes, Dorset: (a) c. 1600, with outline for 1958 superimposed; (b) 1721; (c) 1785; (d) 1849. (From Wilson, 1960; courtesy *Journal of Ecology*.)

1785 and must have been near its full development by 1849, when early hummocks of the First Ridge complex were mapped. The slack between the Third and Second ridges is now known as Central Marsh, while that between the Second and First Ridges is Saltings Strip. The hummocks of First Ridge had fused by 1900 and also linked with the foreshore dunes to the south. The last opening for tidal water, to seaward of Eastern Lake, closed shortly after 1900; First Ridge continued to grow towards the sea during the 1930s. After 1945 a new ridge some 2 m high developed on the foreshore east of First Ridge, from which it is separated by an intervening slack.

The marshy tract between Eastern Lake and Little Sea lagoon, formerly much influenced by flows of tidal water to the latter, was later

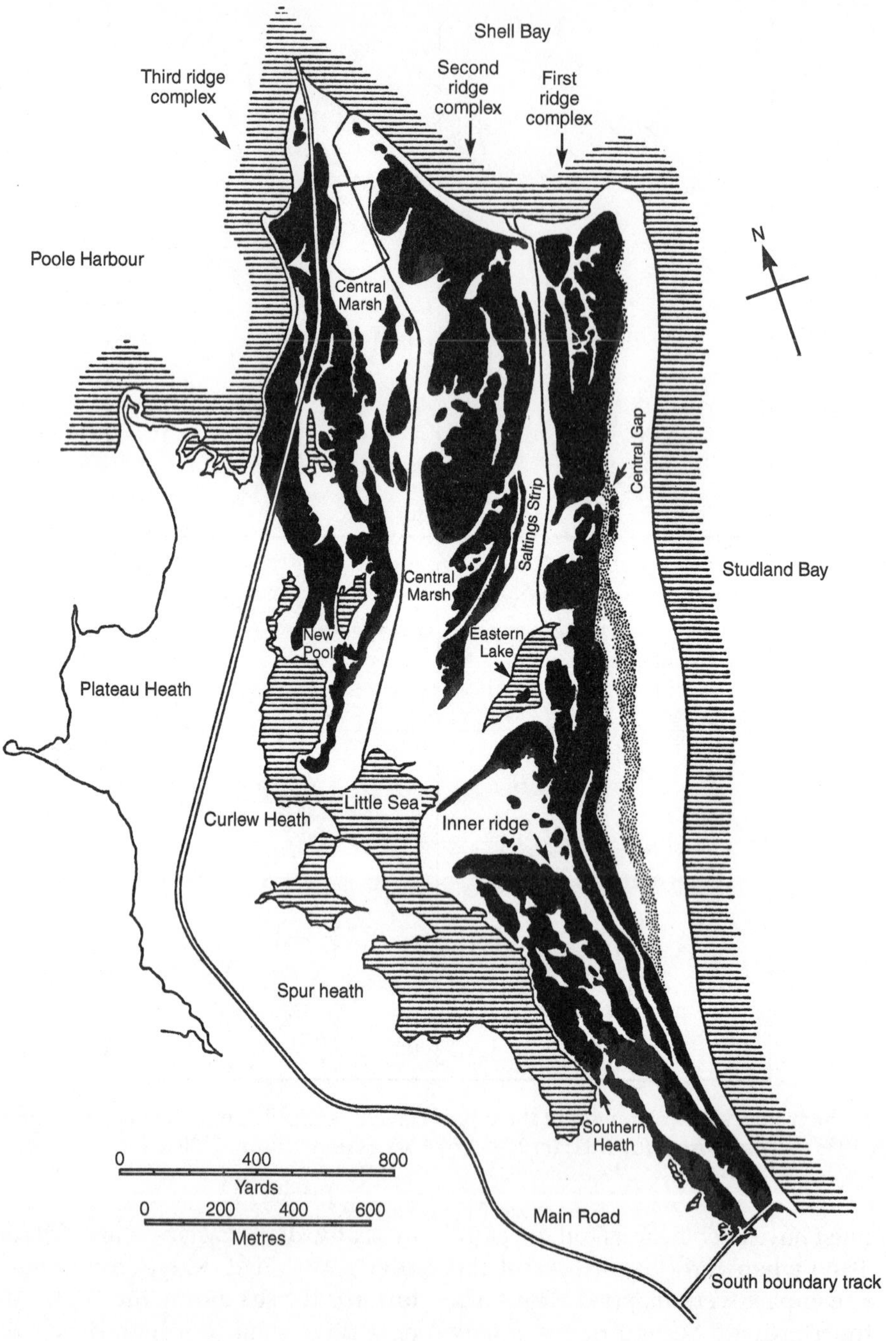

Figure 6.20 The South Haven Peninsula dune system, Dorset. Old-established dunes in black (cf. Diver, 1933): more recent growth in stipple. The width of the beach is only approximately indicated; no specific tide-line is implied. (From Wilson, 1960; courtesy *Journal of Ecology*.)

flooded by water travelling in the reverse direction. Little Sea lagoon receives much surface water from the Western Plateau and would have overflowed into the slacks had not the Central cut (shown by a single line on Figure 6.20) been dug to allow it to drain northwards. The southern dune system is less easy to interpret. Slacks between its ridges have always been much narrower than those in the north, while the whole system has been moulded by alternate, or even contemporaneous, phases of erosion and accretion. In the north the Second Ridge particularly was truncated by lateral erosion at Shell Bay; in 1894 the central region of this shore extended some 180 m further north than it did in 1935. A fringing dune parallel to the shore of this bay developed subsequently.

The sands of the dune system and of the Western Plateau both have an appreciable amount of iron, a high proportion of pure silica, a low mineral nutrient content and – except where there is much molluscan shell debris as at Shell Bay – little calcium. This striking similarity can be explained in terms of the parent material. On the Western Plateau the Bagshot Beds have weathered *in situ*, whereas the sands of the dune system have been eroded from the same formation elsewhere and then swept into place by the tides and winds. An important difference is that a thin layer of gravel, generally overlain by a very thin peat layer, extends over much of the surface of the plateau.

Realizing that the construction of the motor road to the end of the peninsula in the later 1920s rendered its wildlife liable to heavy pressure, Good (1935) mapped the vegetation of the area in 1932–1933, listing more than 15 vegetation types and providing numerous species lists, that for Wood Marsh including the rare royal fern (*Osmunda regalis*). The alternating ridges and slacks result in two main vegetation series, a psammosere and 'a reversed hydrosere commencing with saline sand and passing gradually by increase of fresh water to an open aquatic condition'. The making of the Central Cut also initiated a conventional hydrosere in the area of the Little Sea and Central Marsh.

The psammosere commences with Dune-Grass dominated by *Ammophila arenaria* with varying amounts of *Elytrigia juncea*, *Festuca rubra* and other herbs. This is followed by Dune-heath with *Calluna vulgaris* and *Erica cinerea* as pioneers and ultimate dominants; finally comes the edaphic climax Dry Heath – with or without *Pteridium aquilinum* – from which dune species are absent. Where soil moisture is high Dry Heath is replaced by Damp Heath dominated by *Erica tetralix*. In Dune-Gorse, *Ulex europaeus* almost entirely replaces the ericoids of Dune Heath.

The beginnings of the Reversed Hydrosere are less definite. Normally an open turf community arises after colonization by numerous small herbs. Such communities vary greatly in appearance and constitution but were placed by Good in Damp Sand Communities. Locally, relatively pure *Juncus maritimus* with characteristic salt-marsh associates passes into less saline marsh or into Damp Sand Communities according to subsequent edaphic developments. The dominant species of Acid Marsh are *Molinia caerulea*, *Myrica gale* and *Erica tetralix*, with *Juncus*, *Carex* and *Eriophorum* among the more conspicuous associated genera. Acid Marsh can be formed directly from Damp Sand Communities or from this type via Damp heath. It can also result from the gradual leaching of *Juncus maritimus* marsh, flooding of heath or Dune-Heath, and from the drying out of Swamp. In the 1930s, Acid Marsh was nearly always associated with increasing soil water so that swamp usually developed. This was either a true reed-swamp dominated by *Phragmites australis*, or swamp from which this species was almost entirely absent, dominance being shared by several Junci, *Bolboschoenus maritimus*, *Schoenoplectus tabernaemontani* and *Eleogiton fluitans* with *Myrica gale* (which has N-fixing root nodules) being important among them. As a considerable depth of standing water accumulated, a true aquatic community developed.

The remaining three vegetational conditions described by Good are Sallow-Birch Thicket, communities owing their present constitution to a partial overwhelming of former vegetation by blown sand, and Acid Bog. This last is a *Sphagnum* bog with *Drosera, Hypericum, Hydrocotyle* and *Eleocharis* which develops from heath where stagnant water accumulates. Writing in 1960, Wilson notes that the most major changes since the 1930s were the great extension of *Betula* scrub in more sheltered parts of the older dunes, the extension of *Betula–Salix cinerea* scrub in the slacks and growth of marginal reed-swamp so extensive that it was threatening to block narrower parts of Little Sea. These trends subsequently became so extreme as to demand active management.

6.5.3 LICHENS OF SOUTH HAVEN DUNES

Lichen distributions at South Haven are closely related to the general psammosere sequence – strandline vegetation : Dune Heath : Dry Heath – and to edaphic changes over a 300-year period since the creation of Third Ridge (Wilson, 1960; Figure 6.21). Vegetational succession in time parallels the developmental zonation encountered on the dunes going inland from the beach and is closely related to soil conditions. The steady fall in soil pH with increasing distance from the sea is illustrated in Figure 6.21, an east–west section across the mobile foredunes and the First Ridge complex. As there is normally a complete absence of mosses and lichens on loose sand, Alvin (1960) started his distribution records, based on frequency rather than cover, in the *Ammophila*-dominated Zone 1 behind the foredunes, where pioneer lichens are normally found with moss or the dead remains of higher plants. In this zone quadrats were restricted to the low hillocks where *Ammophila* was often the only vascular species present, but where there were extensive patches of moss. Even so, nearly half the quadrats contained no lichens. The unsampled and completely loose sand between the hillocks had probably been trampled.

Zone 2, the steep seaward-facing slope of First ridge, was in contrast almost completely stabilized apart from the tracks; the 17 quadrats from which lichens were absent were occupied by dense *Ammophila*. The first six species in Figure 6.22 (Group I) are the main constituents of the lichen flora of Zones 1 and 2, but *Cladonia cervicornis* and *C. floerkeana* also occur commonly in Zone 2 besides being so common on older surfaces associated with *Calluna* that Alvin allocated them to the heathland species (Group II). All the more important species of *Cladonia* inhabiting younger surfaces – Group I – characteris-

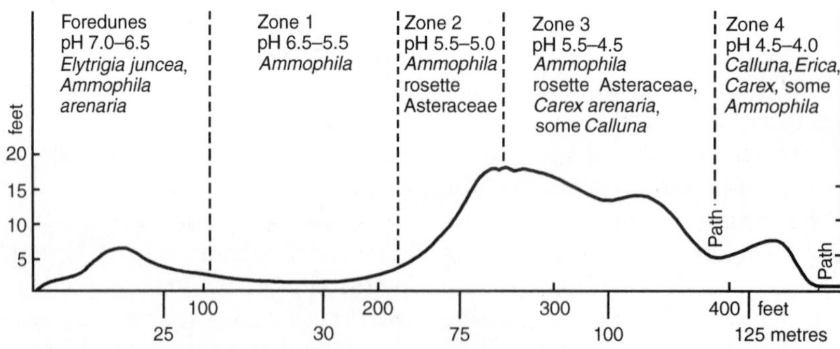

Figure 6.21 Section across First Ridge of the South Haven Peninsula, Dorset, from east to west showing the four quadrat zones in relation to the distribution of the main flowering plants. Approximate pH values are based on the data of Wilson (1960). (The pH of the soil on both Second and Third Ridges is about 3.8–4.5). (After Alvin, 1960.)

tically produce abundant soredia, a feature found in only two other common species, *Cladonia coccifera* and *C. floerkeana*. This characteristic apparently increases both the rate of reproduction and the speed of establishment of younger plants, so conferring important biological advantages on species growing in places subject to slow inundation by sand. (**Soredia** are multicellular structures containing both the fungal and algal partners. They develop faster than the purely fungal ascospores discharged from apothecia, which themselves develop rather later in the life of an individual lichen.)

There is an intergrading of Groups I and II in Zone 3. In Zone 4, the extreme landward side of first Ridge, *Ammophila* is still frequent but the area is dominated by discrete clumps

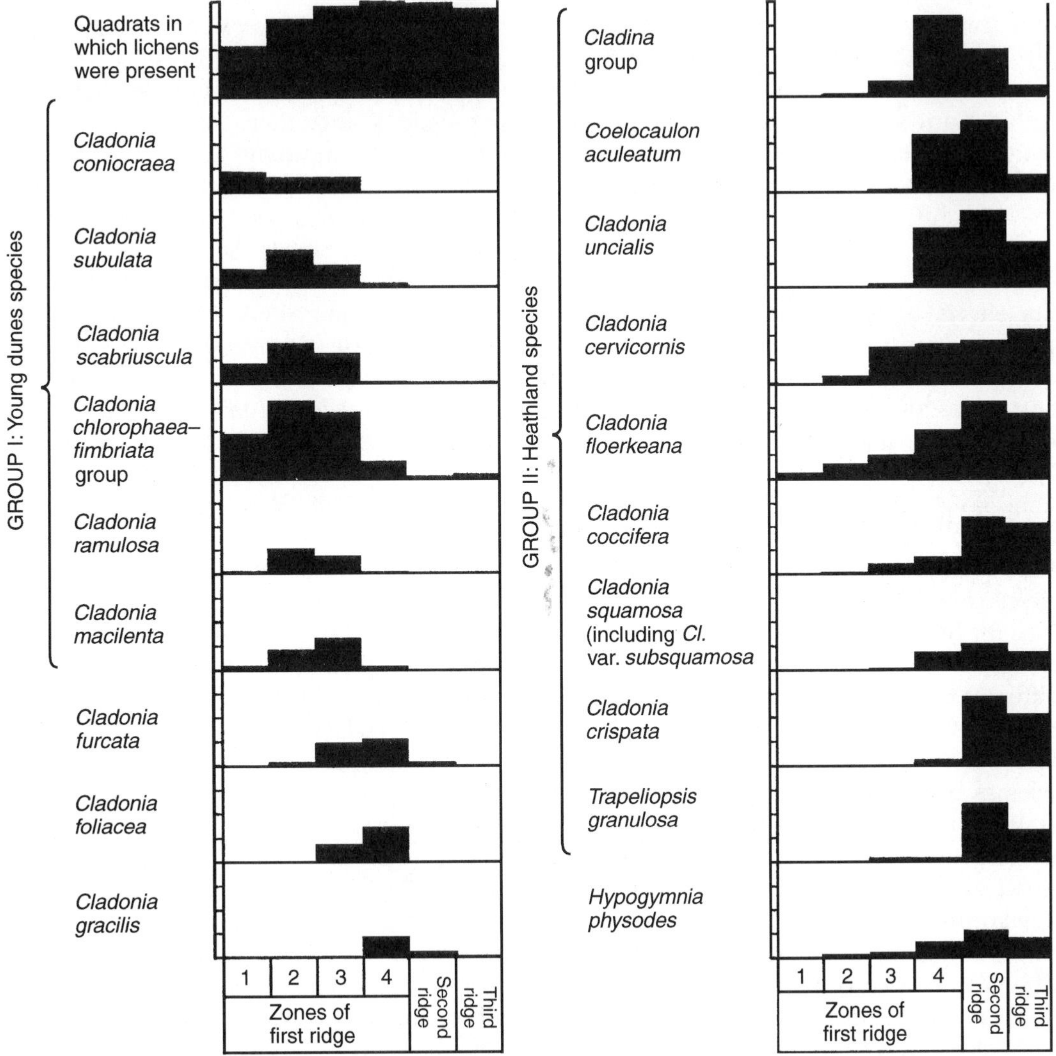

Figure 6.22 Histograms representing the frequencies of the main lichen species in six zones of the South Haven, Dorset, dunes as determined in 100 quadrats of 30.5 cm × 30.5 cm in each zone. One division of the vertical scale represents 23 occurrences. The top left histogram shows the number of quadrats in which lichens occurred. (After Alvin, 1960.)

of *Calluna vulgaris* and *Erica cinerea*, with the heath-like lichen vegetation luxuriantly developed between them often forming a continuous sward 10 cm deep. On the *Calluna*-dominated Second and Third Ridges, the soil is covered by a crust-like mor bearing a mat of lichens where sufficient light is available. *Hypogymnia physodes* occurs mainly on the older stems of *Calluna* but sometimes spreads to the ground and other lichens; it also occurs rather rarely on the basal parts of *Ammophila*. The virtual absence of *Peltigera* from dune surfaces at South Haven is in great contrast to the Culbin Sands, where before afforestation *Peltigera membranacea* frequently covered newly stabilized sand between phanerogams.

6.6 GRAZING, NUTRIENTS AND DIVERSITY

The effects of biotic and abiotic factors on species diversity and vegetation structure in sand dunes have been gradually revealed by a long series of investigations involving field observations and experimental work in Britain, Holland, Japan and elsewhere. The pioneer work of Farrow (1916, 1917) on the Breckland, where large populations of rabbits were shown to transform *Calluna*-heath to grass-heath on sandy soils, had direct implications for British dune systems which supported great numbers of these herbivores. Different types of vegetation vary in their tolerance of grazing and trampling; Farrow described the zone surrounding isolated rabbit holes as bare with concentric zones of grass heath, *Carex arenaria* and finally Callunetum. Similar concentric zones surround isolated rabbit holes at Braunton Burrows, Newborough Warren and Ynyslas dunes today, though the species involved are usually different. This early work, with its carefully observed exclusion experiments with rabbit-proof fences and cages, clearly established the role of herbivores in reducing plant biomass, in diminishing flowering and fruiting (often selectively) and in promoting opportunities for less common species to colonize gaps in the vegetation. Plants of high RGR tend to crowd out small, slow-growing species where conditions, including an adequate supply of mineral nutrients, are most favourable. Grazing and trampling militate against this effect. Other species of herbivore are often involved; sheep grazing of dunes and salt marshes has been practised for centuries and is an important management tool (section 9.3).

Since the Second World War we have seen the great natural experiment provided by the advent of myxomatosis in British dune systems (1954) and the partial recovery made by rabbit populations since then. Field experiments on the effects of nutrient additions have been made in both the absence (Willis, 1963) and the presence (Boorman and Fuller, 1982) of large populations of rabbits. There have also been a number of related glasshouse experiments (Willis and Yemm, 1961; Willis 1965), which in at least one case (Kachi and Hirose, 1983) yielded results apparently at variance with those found in the field. A further illustration of the importance of mineral nutrient status in influencing vegetation structure has come from Western Holland. Here the infiltration, from 1955 onwards, of coastal sand dunes with polluted surface water from elsewhere in order to store and purify it to meet drinking water standards has so eutrophicated the system that nitrophiles grow vigorously over large areas, a situation very different from natural dunes where the soils tend to be low in major nutrients (section 9.2; van der Werf, 1974).

6.6.1 RABBITS AND THE INFLUENCE OF GRAZING ON PRIMARY PRODUCERS

Intensive grazing not only reduces the standing crop of the primary producers but, as can be seen in the case of **rabbit grazing**, operates differentially on species of various life forms. This lagomorph, now so characteristic of British sand-dune systems, is in the present interglacial a western Mediterranean species. It appears to have been deliberately introduced into England by the Normans around

1100 AD, and was at first a delicate animal which only later developed its present hardiness and burrowing ability. Until myxomatosis so greatly reduced its population on the sand dunes of Newborough Warren, Anglesey, in 1954 the amount of woodland scrub was low. Today it is very appreciable, still increasing, and apparently also influencing soil fertility. Hodgkin (1984), who recorded 19 tree and shrub species here, found evidence of nitrogen and phosphorus enrichment beneath *Crataegus monogyna*, the main scrub species. *Hippophae rhamnoides*, which is dioecious, bird-dispersed and possesses N-fixing root nodules, is also invading the dunes, moving out from the planted pine forest. It is controlled by cutting (Gibbons 1994) but were it allowed to form the kind of dense, almost impenetrable scrub found in other Welsh sites, including Cefn Sidan and Tenby, the loss of habitat for rare dune species of low RGR and the encouragement of such aggressive species as *Urtica dioica* and *Chamerion angustifolium* would be serious. Similar impacts, which also involve changes in the soil flora and fauna, have been caused elsewhere by sea buckthorn and by other nitrogen-fixing trees and shrubs (Binggeli *et al.*, 1992). The role of alien species, which may be important in the formation of dune scrub, is considered in section 9.4.

6.6.2 THE IMPACT OF MYXOMATOSIS

There were around 100 million wild rabbits in Britain at the beginning of the 1950s, but the introduction of myxomatosis in 1953 wiped out 99% of this population within a couple of years. Recovery since then has been very uneven; the largest concentrations now occur in East Anglia and some parts of north-east Scotland.

At Newborough Warren the most intense rabbit-grazing occurred in stabilized dry slack turf which formed a mosaic with patches of *Salix repens*. When myxomatosis began to spread into Wales in 1954, Ranwell (1960b) made measurements in this vegetation before the disease reached the area in June of that year. In May 1955 rabbit pellets still lay thick in the turf areas and the grass was only 1–2 cm high; mosses were very evident. In that summer, when rabbits were virtually extinct, the most noticeable feature was the greatly increased flowering of most plants, but especially of *Festuca rubra* and *Agrostis capillaris*. By May 1956, rabbit pellets had almost disappeared and the turf was 5–6 cm high. In that year there was a large increase in seedlings of *Agrostis capillaris* and *Festuca rubra* in the semi-stable dune areas, presumably as a result of the greatly increased seed output of the previous season. *Calluna* had also grown well. By the summer of 1958 a deep mat of turf 15–20 cm high had developed. Mosses were much less evident than formerly, but their abundance had hardly changed. Acrocarpous mosses such as *Dicranum scoparium* and *Bryum* sp., as well as the dendroid moss *Climacium dendroides*, decreased, and prostrate pleurocarpous mosses such as *Pseudoscleropodium purum* increased, following the removal of rabbit grazing. An increase in *Carex flacca* and a decrease in *A. capillaris* were also statistically significant. Ranwell considered the anomalous behaviour of the latter, in relation to other grasses, to result from an inability to compete effectively with ungrazed *F. rubra* for available moisture on dry sandy soils. There was also a decrease in low-growing dicotyledonous species and in the very rich therophyte flora (originally over 30% of total species; Figure 2.1).

The main myxomatosis outbreak led to the development of scrub at Newborough Warren (Hodgkin, 1984) and in ungrazed grassland generally, while over long periods of time mosses and therophytes decreased in pastures previously grazed by rabbits (Watt, 1957). At Braunton Burrows loss of the rabbit population was followed by a great increase in vegetative growth and flowering of *Festuca rubra*, particularly near to nitrogen-fixing legumes such as *Lotus corniculatus*; growth of *Carex arenaria* also improved. The annual grass *Vulpia fasciculata*, which appears to have been highly preferentially grazed, increased remarkably quickly to dominate large areas of some dune slopes, flowering abundantly; its hybrid with *F. rubra* ssp. *rubra* (*Festulpia hubbardii*) was first seen here in 1963. *Anthyllis vulneraria*, known to be much eaten by rabbits, was first noticed in the 1960s and is now well established (Willis, 1985a).

The feeding habits of rabbits on Ynyslas dunes were studied by Oldham, who used rate of accumulation of faecal pellets as an index of activity, during the early spring of 1971 (Wootton and Sinclair, 1976, p.86). The pellets were also analysed to identify major food sources. *Festuca rubra*, *Poa* sp., *Holcus lanatus* and *Ammophila arenaria* formed over half the diet in the first 2 weeks of the investigation, gradually increasing to at least two-thirds. *Festuca* was the most important part of the diet, of which *Ammophila* formed a small but significant fraction. In a cafeteria experiment *F. rubra* was preferred, closely followed by *Poa trivialis*; *Agrostis capillaris* and *Anthoxanthum odoratum* were intermediate, with *Lolium perenne* apparently unpalatable. At Ynyslas, *Elytrigia juncea* suffers far more severely from rabbit grazing than *Ammophila* (which is more at risk from erosion following burrowing); selective grazing severely restricts its seed production.

6.6.3 NUTRIENT ADDITIONS IN THE ABSENCE OF GRAZING

Chemical analyses showed the sandy soils of the dune system at Braunton Burrows to be deficient in the macronutrients nitrogen, phosphorus and potassium, though levels of calcium – mainly derived from broken mollusc shells – were high. The influences of these shortages upon plant growth and distribution were studied in the glasshouse by adding additional nutrients to turf transplant cultures, and to dune sand in which the tomato was grown as an indicator plant (Willis and Yemm, 1961). A second approach was to add nutrients, in various combinations, to areas of contrasted vegetation – particularly of the slacks and dry dunes – in the Burrows themselves and to observe their effects on the vigour, distribution and relative abundance of the plant species present (Willis, 1963). This study was made during the period June 1957 to June 1961 when the rabbit population was very small: grazing seemed not to influence the results whereas attempts to conduct similar investigations in 1948–1949 had to be abandoned because it proved impractical to fence out the very abundant rabbit population.

In the glasshouse study, the effects of added nutrients on the growth of tomato were compared in three soils from Braunton Burrows: A, freshly-blown coarse sand from an *Ammophlia* dune; B, less coarse sand from a partly vegetated dune with abundant *Festuca rubra* and *Tortula ruralis* ssp. *ruraliformis*; and C, finer sand from an old, stable, less mossy turf with plentiful lichens, including *Cladonia* spp. and *Peltigera* spp. Sand cultures were treated with complete (all essential) nutrient solutions or similar solutions lacking individual elements. Figure 6.23 shows the effects of

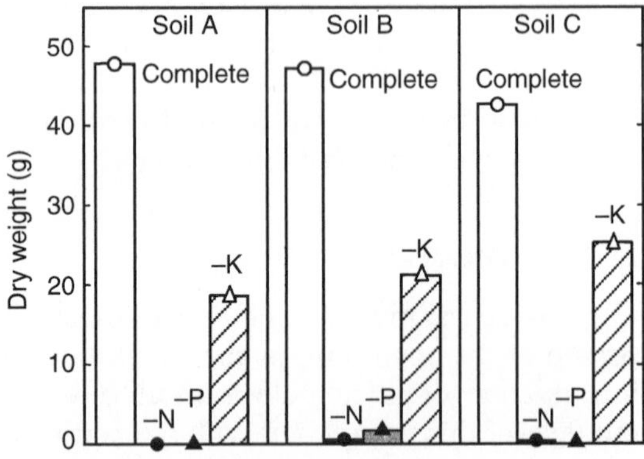

Figure 6.23 The effect of deficiencies of nitrogen, phosphorus and potassium on the growth of tomatoes on three dune soils from Braunton Burrows, North Devon, described in the text. The histograms show the dry weights of the plants, after growth for 111 days, on the three types of dune soil treated with a complete nutrient solution (○) or one lacking nitrogen (●), phosphorus (▲) or potassium (Δ). (From Willis and Yemm, 1961; courtesy *Journal of Ecology*.)

complete nutrients and of solutions deficient in nitrogen, phosphorus or potassium. Nitrogen deficiency severely limited growth; even in soils B and C, plants with no added nitrogen had a dry weight of only about 1% of those receiving complete nutrients. Phosphorus deficiency was almost as extreme, but potassium less limiting, growth (–K) being about half that made when all nutrients were supplied. Similar investigations on minor elements (Fe, B, Mn, Zn, Cu, Mo) using rigorously purified river sand for comparison, showed that all the Braunton Burrows soils contained sufficient of all the micronutrients to support vigorous growth.

In transplant experiments turfs from damp pasture at Braunton Burrows placed in plastic bowls and given complete mineral nutrients showed a three-fold increase in fresh weight of shoots and a four-fold increase in height after growth for 38 weeks in a glasshouse. There was a large reduction in the number of species present, while the grasses *Agrostis stolonifera* and to a lesser extent *Poa humilis* (= *P. subcaerulea*) increased to make up over 95% of shoot biomass (Willis and Yemm, 1961). When only nitrate was added to the cultures, diversity remained high, and height and fresh weight of the vegetation were doubled, *Carex flacca* showing the greatest enhancement. In further studies of transplants from a damp slack (Willis, 1963), little change was found when the levels of all nutrients except nitrogen were increased, but overwhelming dominance by *Agrostis stolonifera* occurred when all nutrients were made good (Figure 6.24). With only phosphorus limitation, however, *C. flacca* and *Anagallis tenella* became abundant, although there were

Figure 6.24 Turf transplants from a dune slack at Braunton Burrows, showing the effect of different treatments with mineral nutrients. The transplant on the left received complete nutrients, that in the middle complete except for phosphorus, and that on the right complete except for nitrogen. The photograph was taken after the treatments had been applied for 31 weeks (1 December 1960). Note the overwhelming dominance of *Agrostis stolonifera* in the transplant receiving complete nutrients, the abundance of *Carex flacca* and of overhanging *Anagallis tenella* in the –P culture, and the short turf, with many species, of the –N culture. The scale in the background is in inches. (Experiment by A.J. Willis)

visual symptoms of P deficiency and the level of P in the shoot system of *C. flacca* was less than a quarter of that found under complete culture conditions on a dry weight basis.

Dry dune transplants to which nutrients were supplied (Willis and Yemm, 1961) showed an overwhelming increase in *Festuca rubra*, whether the nutrients were complete, N+P+K, or nitrate only. In turf transplants from lichen-type pasture there was a marked die-back of *Cladonia* spp. in all the treated cultures. *Hypnum cupressiforme*, *Barbula convoluta* and *Peltigera* spp. were completely eliminated from cultures having added nutrients, whereas these species persisted in the control transplants. The responses of *Ammophila arenaria* to nutrient additions, which were positive (Willis, 1965), are discussed in section 7.1.

In the field nutrient experiments at Braunton Burrows (Willis, 1963) there were six additions of powdered salts per plot, three per year over a 2-year period, usually in mid-June, October and late April. The six sites studied were a dry dune with 'relict' *Ammophila*, typical dry dune with *Festuca rubra*, *Thymus polytrichus* and *Homalothecium lutescens*, lichen-type dry dune pasture, a belt ranging from dry dune pasture to slack vegetation, and two samples of slack vegetation.

In most of the sites, replicate treatments with mineral nutrients were given, the layout was based on a Latin Square and access strips were sufficiently wide to avoid lateral movement of nutrients into other treatments. Field additions of nutrients at Braunton Burrows showed:

1. There was no limitation of vegetation by minor mineral nutrients, even when deficiencies of major elements (NPK) were made good.
2. When complete nutrients were added to dry dune pasture the composition of the vegetation changed very considerably: *F. rubra* and in some sites *Poa humilis* ultimately became strong dominants. Figure 6.25 shows the results obtained in a dry dune with 'relict' marram. The histograms give the mean of six treated areas (three NPK additions and three NPK plus minor nutrient additions). The biomass of *Ammophila arenaria* increased during the period, but appears relatively less because of the much greater increase in *F. rubra*. A few of the larger dicotyledonous species, e.g. *Crepis capillaris*, *Leontodon saxatilis* and *Senecio jacobaea*, like *Phleum arenarium* and *Arenaria serpyllifolia*, were temporarily favoured, but after treatment for 2 years were of very minor importance compared with the grasses. Many small plants (e.g. rosette forms, bryophytes) were completely eliminated.
3. In the dune slacks, *Agrostis stolonifera* dominated when complete nutrients were added and, as on the dry pasture, the variety of species was much reduced and the yield and height of the vegetation substantially increased. (In one site after 2 years there was a 15-fold increase in height and a seven-fold increase in fresh weight.)
4. When nutrients complete except for nitrogen were added to the vegetation of the slacks, there was very little improvement in growth, and indication of severe N deficiency. Limitation of vegetation by potassium, though important, was much less pronounced. Addition of nutrients complete except for P led to some improvement in growth and distinctive changes in the floristic composition of the vegetation, some sedges and rushes increasing.

The studies as a whole show that the sparse and open character of Braunton Burrows, apart from areas of scrub, is largely a result of severe deficiencies of nitrogen and phosphorus. If these deficiencies were not so intense a less varied and taller vegetation, with grasses strongly dominating, might be expected.

6.6.4 CALCIUM AND MICRONUTRIENT DEFICIENCIES IN DUNE SOILS

Figure 6.26 shows the results of some nutrient experiments with two soils from the dunes of the South Haven Peninsula, Dorset, where the sand of the older dunes is markedly acidic (pH 3.8–4.4) and the dunes have, in contrast to those of Braunton Burrows, a very low calci-

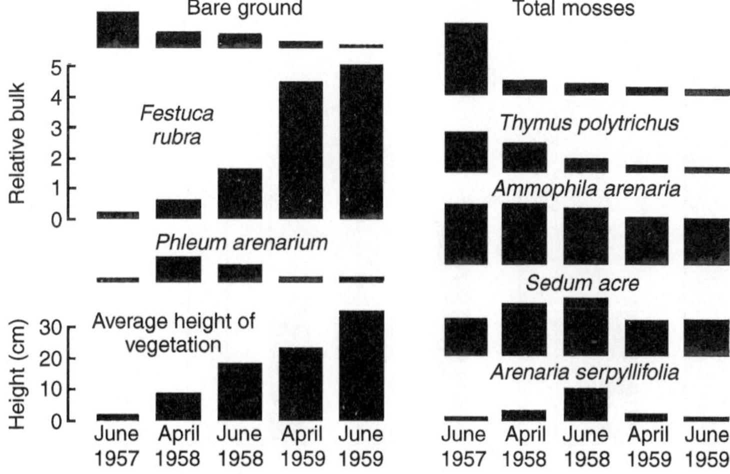

Figure 6.25 The chief changes in the vegetation of dry dune pasture at Braunton Burrows resulting from the application of mineral nutrients. The results, given for five successive occasions, are based on the average assessments of the six treated areas of the Latin Square. (Redrawn from Willis, 1963.)

um carbonate content. Besides acidification (the foredunes have a surface reaction near pH 7), there has, as usual, been a build-up of organic material with time. The organic content of the soil of the first dune ridge is about 0.2%, rising to about 13% in the inland *Calluna* heath of the third ridge where the dunes are some 300 years old. The culture experiments showed some limitation in respect of micronutrients, confirmed the deficiency of calcium, and demonstrated even stronger limitation of growth resulting from very low levels of N, P and K. The highly leached inland dunes are more deficient in nutrients than the foredunes, notably in respect of B, Mn, Cu, Mg, K and N. Investigations on turf transplants from South Haven again showed that particular limitations of nutrients resulted in distinctive combinations of species. With additions of complete nutrients, *Agrostis capillaris* dominated strongly (Willis, 1985b).

6.6.5 NUTRIENT ADDITIONS IN THE PRESENCE OF GRAZING

Boorman and Fuller (1982) report the results of adding different combinations and concentrations of N, P and K to two nutrient-poor, rabbit-grazed dune swards at Holkham, Norfolk, using the same experimental design as Willis (1963). The plant communities included a wide range of annual species with some perennials, bryophytes and lichens. Minor nutrients were not added because it was thought that these would be present already in adequate amount; changes in abundance were recorded as frequency values. In February 1976, at the end of the experiment, soil samples were taken. They generally showed that phosphorus and potassium, when added, had accumulated in the soil, in contrast to the behaviour of nitrogen.

Perennial grasses did not come to dominate the sward, any increased production being cropped by the large rabbit population for which *Festuca rubra* forms the main diet (Boorman and Fuller, 1977). Rabbit grazing at Holkham increased during the experiments; other factors which may have limited response to nutrients are the low rainfall and high evaporation in this area. Species diversity was reduced when fertilizer (especially N) was added, with diminished cover of bryophytes and lichens, and lower frequency of annual plants including the winter annuals *Aira praecox*, *Cerastium semidecandrum*, *Erophila verna*,

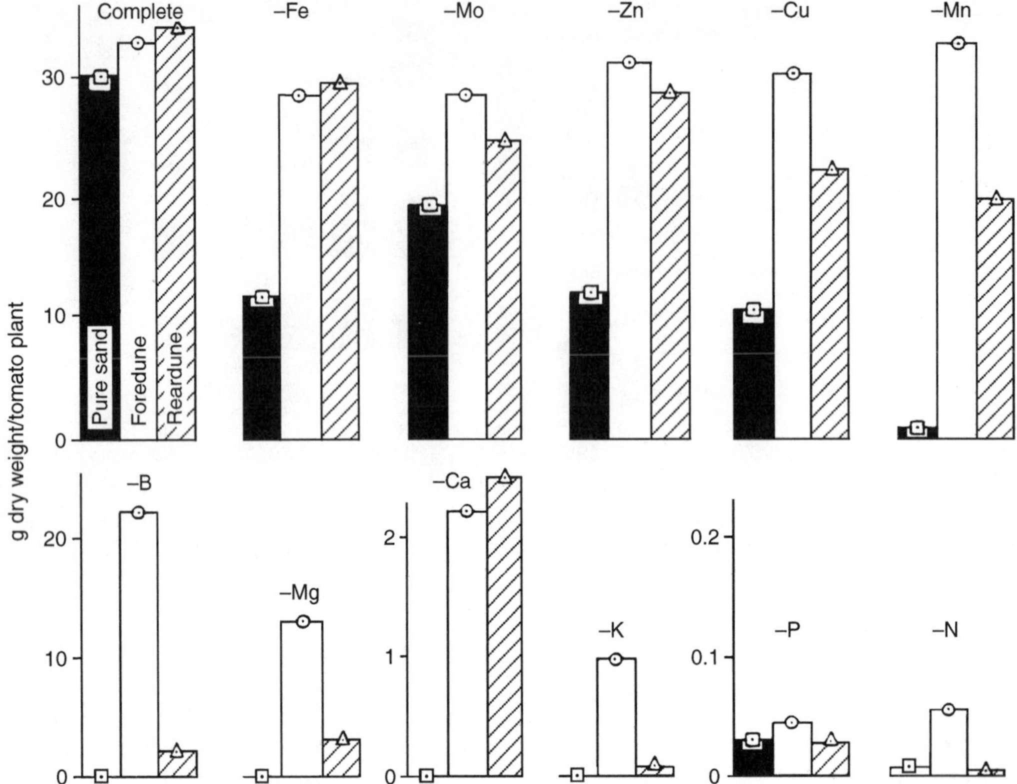

Figure 6.26 The mineral nutrient status of soils of the South Haven Peninsula, Dorset. Tomato plants were grown in glasshouse conditions for about 3 months on purified sand and on sands from the foredunes and inland dunes (for procedure, see Willis and Yemm 1961). Nutrient solutions deficient in one element were added, complete nutrients serving as a control. Note the use of three different scales. (From Willis, 1985; courtesy of Nature Conservancy Council.)

Phleum arenarium and *Valerianella locusta*. Annual species largely depend on open conditions for survival and although rabbit grazing prevented perennials from smothering them, nutrient addition may have accentuated competition for moisture. Added nitrogen was implicated in the decline of summer annuals (e.g. *Crepis capillaris*, which may be biennial), short-lived perennials (*Leontodon saxatilis*) and longer-lived perennials (*Rumex acetosella*, *Sedum acre* and *Viola canina*), all of which may have been competitively excluded from the relatively open short turf in which they normally grow. Reactions to P and K varied according to species.

Boorman and Fuller (1982) concluded that the relatively diverse communities at Holkham owe their existence both to low nutrient status and the relatively high grazing intensity exerted by rabbits. Local variations in nutrient supply provide niches for species with different nutrient requirements; sea bindweed (*Calystegia soldanella*) apparently behaves as a nitrophile.

6.6.6 USE OF *OENOTHERA* AS AN INDICATOR SPECIES IN A JAPANESE DUNE SYSTEM

Kachi and Hirose (1983) used the commonly biennial evening primrose (*Oenothera glazio-*

viana = *O. erythrosepala*) as an indicator plant on well-drained, semi-fixed dunes in the Azigaura dune system, Japan, which has an area of about 12 km^2 and faces the Pacific Ocean. The study site had less than 20% vegetation cover and its calcareous sand (pH 8.2–8.7) includes small amounts of organic matter, inorganic nitrogen and available phosphorus. In both field and pot experiments *O. glazioviana* showed that potassium was not limiting, but in the field experiments nitrogen was. Phosphorus, rather than nitrogen, caused limitation in the pot experiments, where the higher concentration of nitrate seems to result from greater mineralization and nitrification without the leaching of nitrate which occurs under field conditions. The decreased phosphorus levels found in incubated soils compared with fresh soils may result from assimilation of readily available phosphate, from formation of insoluble phosphate, or both.

This work indicates that laboratory results may not always apply directly in the field, because the nutrient status of potted soils may be changed by microbial activities. The use of rosette diameters as an indication of performance could profitably be extended to British dune systems, such as Crymlyn Burrows and Cefn Sidan, where evening primrose is abundant.

Eutrophication of dune soils by surface waters with a high load of nitrogen and phosphorus is considered under environmental impacts (section 9.2), where its effects and methods of reducing the problem are discussed.

SAND-DUNE DYNAMICS AND COMMUNITIES 7

7.1 MARRAM: THE OREGON DUNES, MACHAIR AND DUMPED SEA SAND

This section commences with a résumé of the ecology and biology of *Ammophila arenaria* (European beachgrass) whose major influence on the form and development of European dune systems has already been described. It then discusses the sand dunes of the Pacific Northwest coastal region and the way in which their dynamics were changed by the introduction of *A. arenaria*. Finally, it considers machair communities in which the influence of marram is gradually lost – but to which it returns after blowouts – and also an inland dune community with marram developed upon dumped sea sand.

7.1.1 MARRAM: FORM, DISTRIBUTION AND GROWTH

Ammophila arenaria is commonly regarded as a protohemicryptophyte, with uniformly leafy stems, but with the basal leaves usually smaller than the rest. On the classification of Gimingham (1951a) marram is a large tussock plant, the tussocks being dome-shaped and over 60 cm in diameter (section 2.2). This species plays a major role in stabilizing mobile and semi-fixed dunes in the British Isles, continental Europe and – as an exotic – in some parts of North America; on mobile dunes it may be the only vascular plant present. It occurs along all European coasts south of 63°N (Huiskes, 1979), is widespread bordering the Mediterranean and also present on the Black Sea coast (its southern limit is approximately 30°N). In more northerly parts of Europe its role is progressively taken over by lyme-grass (*Leymus arenarius*).

The leafy shoots of *A. arenaria* substantially slow the wind speed above ground level (Willis *et al*., 1959b). In consequence, the saltating grains fall to the ground, accreting the dune which tends to grow higher as the marram expands upwards through the fresh sand (section 6.1). Under conditions of appreciable sand supply, dense tall growth is made (mostly about 75 cm, but leaves may exceed 1 m), with often 100–200 tillers of living marram per m^2, apparently the carrying capacity of the sand (Huiskes, 1979). The majority of the rhizomes and roots which help to stabilize pioneer dunes ramify in the top 1 m of sand, but some extend to depths of 2 m or more; tillers of this plant reach high densities when moderate amounts of sand are being deposited round it (Willis, 1989). Marram can just withstand accretion rates of 1 m yr^{-1}, but may be overwhelmed by sudden heavy deposition. Growth is very slow when accretion of sand is small, with root failure occurring.

7.1.2 ESTABLISHMENT, OCCURRENCE AND VIGOUR OF MARRAM

Ammophila arenaria often establishes from rhizome fragments on mobile dunes not far from the coastline, but may also arise from seedlings, especially in more inland dunes. Though tolerant of moderate salinity – up to almost 1% sea salt in the substratum (Huiskes, 1979) – it is not such an effective halophyte as sand couch (*Elytrigia juncea*) with which it often occurs in the foredunes. Marram is most vigorous in mobile (yellow) dunes, frequently forming large tussocks in pure Ammophiletum. Extensive tillering leads to tussock formation (Gemmell, Greig-Smith and Gimingham, 1953), with a pattern at a scale of 20–40 cm (Greig-Smith, 1961); horizontally growing rhizomes extend the area of dune stabilized, producing a pattern at a large scale from the outgrowth of buds.

In semi-fixed dunes, with more evenly spread 'pasture-type' marram, species such as sand sedge (*Carex arenaria*), restharrow (*Ononis repens*) and the moss *Tortula ruralis* ssp. *ruraliformis* are found in mixed Ammophiletum, together with short-lived species such as ragwort (*Senecio jacobaea*), sand cat's-tail (*Phleum arenarium*) and mouse-ears (*Cerastium* spp.) which help to stabilize the surface.

In fixed dunes, with only a little sand supply, the vigour of marram is low, and it may later die out as the vegetation becomes more closed. Many reasons have been advanced for this well-known decline of vigour in old dunes (Willis, 1989). **Loss of vigour** has been attributed, for example, to poor aeration, mineral content deficiency, competition for nutrients and water, toxicity and natural senescence (Willis *et al.*, 1959b; Wallén, 1980; Eldred and Maun, 1982). Marshall (1965) showed that, in the grass *Corynephorus canescens*, the efficiency of the root system falls with ageing, with loss of cortex from the roots, good growth depending on the production of new roots higher on the morphological axis. This root development needs moist conditions at the site of production of adventitious roots. The behaviour of roots of *A. arenaria* is similar; old roots become decorticated but new roots may arise from the nodes of newly buried shoots (often three or four roots at a node). Vigour of relict marram can be renewed by substantial sand accretion; in pot cultures, considerably increased growth resulted from the addition of sand, partly burying the plants, which formed many new roots (Willis, 1965). In the last decade attention has been focused on the importance of soil biota (especially mycorrhizal fungi and nematodes) in influencing the vigour of marram. In the Netherlands and France the nematode *Longidorus kuiperi* was found to be exclusively associated with the roots of *A. arenaria* (Brinkman *et al.*, 1987) but whether it significantly debilitated the plant was not established. Read (1989) suggested that mycorrhizal association may be important for seedling establishment of the pioneer grasses *Ammophila* and *Leymus*.

In glasshouse experiments, van der Putten, van Dijk and Troelsta (1988) showed that seedlings of *A. arenaria* grew significantly better in sea sand than in rhizosphere sand and that sand sterilization by gamma-radiation increased biomass significantly. They attributed the higher vigour to elimination of harmful soil biotic factors, and degeneration to accumulation of pathogens in the rhizosphere. In factorial glasshouse studies, Little and Maun (1996) grew seedlings of *Ammophila breviligulata* buried in sterilized sand, with rhizosphere inoculation with arbuscular mycorrhizal (AM) fungi and the plant–parasitic nematodes *Heterodera* and *Pratylenchus* (treatments were separate and in combination, with organisms from natural populations). Burial enhanced the foliar growth rate in all combinations except when the plants were nematode-infected before burial. It was concluded, however, that burial did not allow *A. breviligulata* to escape from harmful nematodes, which are also the cause of decline of *Hippophae rhamnoides* (section 9.4). AM fungi, in contrast, were shown to reduce susceptibility to the nematodes.

Detailed investigation of the nature of soil-borne disease of *A. arenaria* and its relationship with sand deposition (de Rooij-van der Goes, 1996) showed that burial with unsterilized root zone sand was less beneficial than with sterilized sand. Fungi were found to colonize the freshly deposited layer of sand faster than plant–parasitic nematodes. Enhanced populations of the semi-endoparasitic nematode *Telotylenchus ventralis* (cultured from Dutch coastal foredunes) led to reduced growth of *A. arenaria* (de Rooij-van der Goes, 1995), and rejuvenation of stands along accumulating edges of blowouts was attributed to reduced inoculum pressure of plant–pathogenic organisms (notably nematodes and potentially pathogenic fungi) in the deposited sand (de Rooij-van der Goes, 1996). Management practices aimed at increased burial with wind-borne sand are consequently recommended to maintain vigour of *A. arenaria.*

7.1.3 SAND STABILIZATION WITH MARRAM

Marram can regenerate even from dormant rhizome buds; it is the most effective dune-fixing species known, and has been widely planted for this purpose, with varying success, on the coasts of Britain, continental Europe and North America. Establishment is most successful on moderate slopes where there is a fair amount of incoming sand, and when the management adopted takes into account the major ecological and morphological features of the plant. Some slopes are too steep and unstable for successful establishment. If the sand supply is too small the plants may die after failing to produce adequate roots.

If a moderate length of shoot – effectively a buried vertical rhizome – is present in the propagules used, new roots can be formed if the sand is not too dry, and successful growth will be made. Where burial by sand is not extensive, it is advantageous to place the rhizomes horizontally; this gives more rapid vegetation cover, with more horizontal rhizome development , than when rhizomes are placed vertically. Foliage should not be cut off or damaged by knotting and, where planting is horizontal, the foliage should be brought above ground (Hobbs *et al.*, 1983). Although marram has substantial requirements for the major nutrients nitrogen, phosphorus and potassium (Willis, 1965), addition of fertilizer in some sites is of little advantage. Marram grows best in calcareous dunes but has a quite wide pH tolerance, in Britain occurring within a pH range of about 6.0–9.0; it is absent from soils more acid than pH 4.5 and of limited value in the management of strongly leached old landward dunes.

In dune systems on the Atlantic coast of North America, a stabilizing function similar to that of marram is played by *A. breviligulata*, which is often considered to be a subspecies of *A. arenaria*. This sand-binder extends from North Carolina to Newfoundland and the shores of the Great Lakes. Its role in dominating the pioneer stage was recognized by Cowles (1899), whose classic description of the sand dune vegetation of Lake Michigan was a major early contribution to the concept of plant succession.

7.1.4 INFLUENCE OF MARRAM ON THE OREGON DUNES

The raw material of the Pacific Northwest coastal dunes is derived from an enormous reservoir of sand lying on the continental shelf off the Oregon coast. This sand was formed by erosional breakdown of soft sedimentary rocks of the coast range. It is moved shorewards by the waves and tides before being blown inland by the wind; the beach sand consists of grains of fine and medium size, mainly of quartz but with some feldspar.

At the beginning of the 20th century European beachgrass (*A. arenaria*) was planted on the estuaries of the Siuslaw and other rivers in the Oregon Dunes National Recreation Area (**NRA**), whose sands stretch – unbroken by rocky outcrops – for 58 km (36 miles) from north to south. This plant, intended to stabilize dunes that threatened to close river channels to boat traffic, has entirely altered the dynamics of a sand dune system which has existed for over 10 000 years. It quickly colonized sand pockets amidst the driftwood, derived from timber-cutting on the Coast Range, which forms a stable foredune base in this area. *Ammophila arenaria* then spread very rapidly and formed a high and apparently permanent foredune barrier in the path of the sand which had so long been blown inland from the seashore. The onshore winds now deposited sand within the foreshore dunes and then scoured sand from the area to the east, which was lowered to the level of the water table and became a **deflation plain**, which was swiftly vegetated and continues to widen. The area has an average annual precipitation of 1575 mm (62 inches). Much of this water is stored within the dune system; the hydraulic pressure of the groundwater is sufficient to produce numerous areas of unstable 'quicksand' within the wet hummocks and transverse dunes.

Figure 7.1 shows a profile across the Oregon Dunes from the sea to the transition forest. There is an extensive flora and the general successional relationships of sand-dune plant communities of the Pacific Northwest, of which the Oregon Dunes NRA is such an important part, are indicated by Figure 1.16, which gives Latin names. American common names are used for the numbered plant communities given below. The most important tree is lodgepole pine (*Pinus contorta*), here often called the shore pine.

The native pioneer foredune vegetation of American dunegrass–yellow sandverbena (community 1), along with such species as *Ambrosia chamissonis*, *Calystegia soldanella* and *Tanacetum douglasii*, used to be widespread just above high tide line, but is now found in a pure form only in a few places. The European beachgrass community (2) has largely taken over the upper beach and active sand zone (Wiedemann, 1984), giving rise to large hummocks which undergo a hummock cycle as in Europe (Figure 1.16). Maritime pea (*Lathyrus japonicus*) occurs in northern areas of this community and tree lupin (*Lupinus arboreus*) in the south. Behind the foredune and on sand plains where the substrate is less active the seashore bluegrass–beach pea community (3) develops. This has distinctive associates including *Carex macrocephala*, *Glehnia leiocarpa*, *Poa confinis*, *Erigeron glaucus* and *Eriogonum latifolium*. This community may invade European beachgrass deteriorating because of a diminution of accreting sand; it may itself be invaded by *Ammophila* if its plant cover is low and there is some accretion of sand. The red fescue–golden rod community (4) develops on sheltered sand plains, old blowouts and other areas where movement of sand is restricted.

Plant communities develop rapidly on the damp, stable surfaces of the deflation plains. Where there is ample moisture but no standing water at any time of year the typical meadow red fescue–seashore lupin community (6) develops; this probably has more species than any other but appears not to occur in northern California. Standing water on the surface of this area leads first to the species-rich sickle-leaved rush–springbank clover (7) and then to the slough sedge–Pacific silverweed (8) community. The herbaceous plant communities

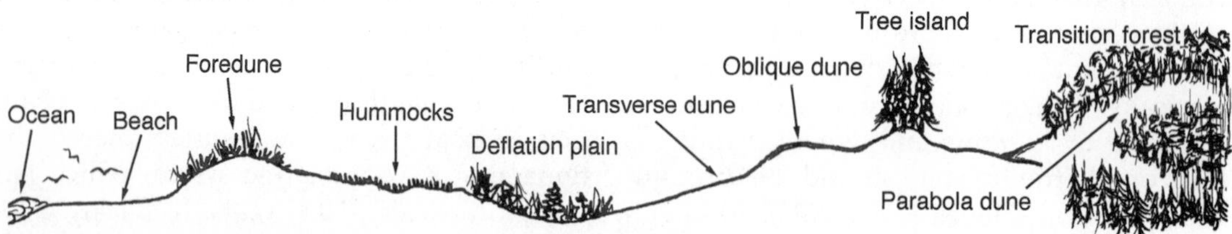

Figure 7.1 Lifezones of the Oregon Dunes, Pacific Northwest. Note particularly the foredunes colonized by *Ammophila arenaria*, at which sand carried by onshore winds is now deposited. Features further inland receive far less sand than formerly and the winds have scoured out a moist deflation plain. The transverse ridges formed at right-angles to the prevailing wind are 1.5–6 m high. The oblique dunes reach up to 55 m and are often over 1 km in length. Crowns of the trees on the nine tree islands within the National Recreation Area (NRA) are wind-pruned and relatively small. Parabolic dunes are formed when unidirectional winds blow into unstable vegetated regions at the margin of the transition forest. (Redrawn from diagram supplied by USDA Forest Service.)

frequently occur in a **complex mosaic** over the landscape. If the environmental gradient from dry unstable sand to very wet stable sand is gradual, the communities intergrade. Where the local environment changes rapidly, however, contrasting communities are adjacent to each other.

Shrub and tree seedlings develop in the communities described above from a quite early stage, and this leads to the formation of shrub communities. The salal–evergreen huckleberry community (11) occurs on sand plains, slopes and ridges, while the kinnikinnik/silver moss community (13) of undisturbed sand plains and inactive blowouts, often damaged by off-road vehicles (**ORVs**), is becoming increasingly rare. In deflation plains and swales where water stands for part of the year the Hooker willow–Pacific wax-myrtle community (14) is most common. *Pinus contorta* seedlings are nearly always present in these three community types; many of the wetter ones also have *Picea sitchensis*.

The most common forest community is (15) lodgepole pine/western rhododendron which develops on high, well-drained sites and develops an open canopy when mature. In northern California, rhododendron is replaced by *Garrya elliptica*. The lodgepole pine/bristly manzanita community (16) is found on dry ridge tops and other dry sites; the forest canopy is open and *Cladonia rangiferina* is often present. The Sitka spruce–lodgepole pine/Hooker willow/Oregon beaked moss (17) communities of the deflation plains contain varying proportions of the two conifers. Lodgepole pine is frequently dense and provides protection for Sitka spruce which is not tolerant of salt spray, but once the latter reaches 4 m or so it tends to pull away from the pine, having a faster growth rate and a greater maximum height. Lodgepole pine/slough sedge communities (18) commonly develop where water stands for much of the year, sometimes with *Sphagnum* spp. Western red cedar/mountain Labrador–tea (19) is found where acid bog conditions have developed through long accumulation of organic matter. *Sphagnum* is frequent; much rarer is the California pitcher plant (*Darlingtonia californica*), whose insectivorous habit supplements the meagre supplies of available nitrogen and phosphorus.

Relatively small areas of **climax forest** occur on the present-day dunes. The Douglas fir/western rhododendron community (20) is found on higher, drier sites; western hemlock is a frequent associate and grand fir (*Abies grandis*) occurs with Douglas fir in northern California. On moister areas there is a little of the western hemlock–Sitka spruce/salal/deer fern community (21) which is the climatic climax for the region.

7.1.5 SCOTTISH SAND DUNE SYSTEMS

Sand dune soil profiles often reveal repeated layers of organic matter. This is particularly common in the very dynamic coastal systems of the Outer Hebrides, where 'machair stratification' is a characteristic feature of calcareous dunes whose plant communities are frequently buried as a result of adjacent agricultural disturbance. Survival of the predominantly perennial species of periodically buried vegetation is clearly related to their ability either to grow through the sand or to re-establish from propagules. Inundation is succeeded by a period of competition whose outcome is decided by the ecophysiological characteristics, notably growth rate, of the species involved.

Gimingham (1964) shows that Scottish dune pasture and machair communities are, apart from the reduction of *Ammophila arenaria* to a small number of scattered degenerating groups of shoots, not greatly different from the later stages of surface fixation in marram dunes. *Agrostis capillaris* and *Holcus lanatus* are important in acid pastures on siliceous sands where the numbers of associated species, including *Carex arenaria*, *Campanula rotundifolia*, *Galium saxatile*, *Rumex acetosa* and *Rumex acetosella*, are relatively low. Among the mosses and lichens are several characteristic of hs – *Hylocomium splendens, Pleurozium* ~~beri~~, *Cladonia* spp. and *Peltigera* sp.

Much richer floras, including many species often regarded as **calcicoles** (calcium-loving plants), occur on pastures on 'base-rich' sands, where the calcium carbonate content may be only little over 1% but the mineral ion status of the soil is high. Where grazing is light a *Carex arenaria*-grassland often develops. Heavy grazing produces a short turf, which though dominated by grasses, contains species such as *Anthyllis vulneraria, Bellis perennis, Botrychium lunaria, Centaurea nigra, Coeloglossum viride, Daucus carota, Erodium cicutarium, Euphrasia* spp., *Pimpinella saxifraga, Plantago lanceolata, Primula veris, P. vulgaris, Prunella vulgaris, Ranunculus acris* and *Thymus polytrichus*.

Pastures developed on very lime-rich shell sands, in which the calcium carbonate content can be up to 80% by weight (Dickenson and Randall, 1979), are the richest in species and known as **machairs**, especially where the sand has blown inland over low-lying flat areas of peat or other substrate. Found in the Hebrides and the north and west of Scotland, a distinctive feature is the profusion of orchids, including *Coeloglossum viride, Dactylorhiza fuchsii, Dactylorhiza purpurella, Epipactis atrorubens, Gymnadenia conopsea, Listera ovata* and *Orchis mascula*. Numerous other species include *Antennaria dioica, Gentianella amarella, Trifolium medium, Trollius europaeus* and many Cyperaceae. A number of arctic–alpine species are present in the unusual communities at Bettyhill, Sutherland. Here, *Saxifraga aizoides* occurs in the slacks and *Saxifraga oppositifolia* grows with marram. *Dryas octopetala* occurs on base-rich sands that have been blown against cliff slopes, and onto cliff tops, where it forms a sward with *Empetrum nigrum* and other species. *Oxytropis halleri* occurs on shingly sand on old river terraces and among rocky outcrops in fairly open vegetation.

Scottish machair has a long association with humans, as the excavation of long-buried settlements reveals, and is of international ornithological, botanical, physiographic and conservation interest (Angus and Elliott, 1992; Boorman, 1993). A proportion of it is kept under cultivation, often for 3 years followed by 3 years under grass. Its soils are derived from comminuted shells and from the redeposition of fluvioglacial sediments washed inshore. Onshore winds and fluctuating sea levels caused these sediments to be carried inland, so contributing to the dune, machair and blown sand regions shown in Figure 7.2. The blown sand can overlie bedrock, glacial till, old raised beaches, or peat bog. The sequence – dune → plain (to which the term machair should strictly be restricted) → loch → blackland (a transitional habitat bordering the peatlands) is characteristic. When machairs were brought into cultivation seaweeds were used to improve the tilth and to act as a fertilizer; severe erosion occurred in the main period of kelp manufacture (1735–1822) for the chemical industry. This was further compounded when shell sand was used to bring peatland into cultivation, but later landlords introduced regulations to protect the vegetation. The rate of erosion is now lower than in the 16th and 19th centuries, but still a cause for concern (section 11.3).

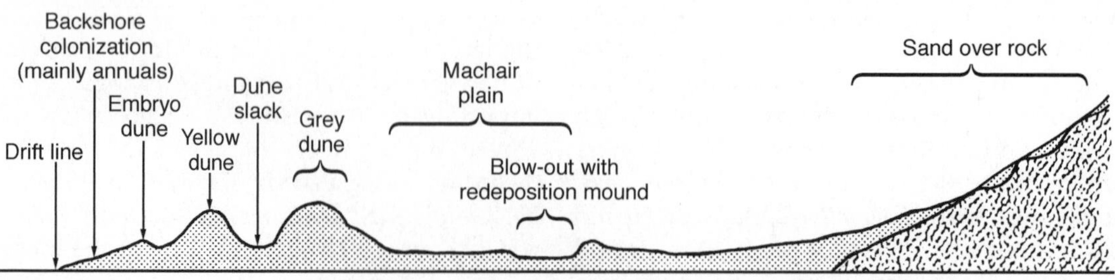

Figure 7.2 Relationship between dunes, machair and blown sand in Scotland. (After Boorman, 1993.)

7.1.6 INLAND DUNE FLORAS ON DUMPED SEA SAND

Such communities frequently encounter severe competition from local mesophytes. They are usually too small to escape eventual extinction but may persist for a surprisingly long time. That at Emscote, near Warwick, UK, was first reported in 1964 and arose when large quantities of sea sand from Margam, Glamorgan, were used to extinguish a long-burning fire in a railway embankment, probably in 1958. Bowra (1995) reports that *Ammophila arenaria*, *Carex arenaria*, *Corispermum leptopterum* (Chenopodiaceae, a casual from southern Europe), *Echium vulgare*, *Erodium lebelii*, *Oenothera* sp., *Ononis repens*, *Phleum arenarium*, *Rubus caesius*, *Salix repens*, *Saponaria officinalis*, and *Viola canina* still persist. *Cakile maritima*, *Suaeda maritima*, and *Trifolium arvense* disappeared early; *Elytrigia juncea* was not seen in 1993–1994 and *Diplotaxis muralis* last appeared in 1990.

7.2 PHENOLOGY AND PLANT POPULATION BIOLOGY

Phenology is concerned with the time of onset and duration of the seasonally determined activity phases of animals and plants throughout the year. These phases are largely adapted to long-term climate but are influenced also by weather of particular seasons, the dates on which they occur differing from year to year, although the order in which the various plant species unfold their buds, flower, fruit and senesce is much the same. Animals are similarly influenced by changes in the weather, both directly – often through temperature- and light-mediated hormonal changes – and indirectly, because of their ultimate dependence, whether herbivores or carnivores, upon seasonally influenced plant structures for food.

7.2.1 SEASONAL ACTIVITY OF REPTILES AT STUDLAND HEATH NNR, DORSET, UK

All six native British reptiles occur within the confines of this reserve, a comparatively large area of the sadly diminishing European heath, and their seasonal behaviour affords a good illustration of phenology in animals. Like amphibians they are **poikilothermic**, or cold-blooded, a rather misleading term. Reptiles in particular have high body temperatures when active; indeed, these are necessary for successful mating, reproduction, digestion and growth. Unlike warm-blooded animals, they require the heat of the sun in order to raise and maintain body temperature. As cold weather reduces them to a state of torpor, their loss of activity in winter is much more marked than in birds and non-hibernating mammals, and one of their first significant actions after the winter is to slough off the dead, but protective, stratum corneum which would otherwise limit growth.

Of the three snakes, the adder (*Vipera berus*) uses its rapidly acting poison to kill small mammals, frogs and other prey which it swallows whole. Adders have been sighted in every month in the year, but spend the depth of winter in hibernacula from which the non-breeding females and juveniles emerge later than breeding adults. Courtship and mating continue for about a month, ending in mid-May. The males indulge in ritual combat in which they both rear up, coil round each other and attempt to push the opponent to the ground. The adder population in the Wyre Forest (near Birmingham, UK) has been studied for many years by Sheldon (1995); living young are produced from mid-August onwards. Although often found away from water, the much larger grass snake (*Natrix natrix*) is an excellent swimmer and spends much of its time hunting newts and frogs through the pools and *Sphagnum* bogs. During early summer the females each lay up to 20 eggs in rotting vegetation, in which individual nests may ultimately contain several hundred eggs. The much rarer and smaller smooth snake (*Coronella austriaca*) also lays eggs.

The common lizard (*Lacerta vivipara*) is a very hardy reptile which feeds on insects, notably grasshoppers, and other small animals, especially spiders, and can frequently be observed basking in sunny places. It resembles the adder in retaining its 5–10 eggs until midsummer, when the young are capable of an independent existence. Females of the larger sand lizard (*Lacerta agilis*) (Figure 7.3) lay 8–12 eggs in a hole in the soil and leave them to hatch. The flanks and underparts of male sand lizards turn bright green during the mating season of May and June. The legless slow worm (*Anguilla fragilis*) which, like other lizards, has movable eyelids, feeds mainly on earthworms, slugs and snails. Up to 20 young are born live in late summer; if they survive the attentions of numerous predators they will become adult at the age of 6 years and may live for well over 20 years. The seasonal behaviour of all these reptiles is strongly influenced by temperature, and the availability of food and adequate cover. The need to lay down enough body fat to survive the winter is particularly crucial for the young which are themselves prey items for many other species.

7.2.2 PLANT POPULATION BIOLOGY OF WINTER ANNUALS, A RHIZOMATOUS SEDGE AND A TUSSOCK-FORMING GRASS

Sand dune habitats are, to a greater extent than most others, continuously in a state of change, including succession. Dune formation or remoulding is continuous and in large systems most phases from the bare sand of blowouts to mature dunes (and slacks) can generally be found. Such habitats are unusual in that, for plant populations already established in the area, new sites suitable for colonization are almost invariably close at hand. Consequently the dispersal of propagules in space, or in time (as buried seed which remains viable for long periods), is less important for successional species of sand dunes than it is, for example, for those of woodlands.

Watkinson *et al.* (1979) analysed the demography of a group of short-lived winter annuals, which are generally characteristic of the

Figure 7.3 Sand Lizard (*Lacerta agilis*). (Drawn by P.R. Hobson.)

drier parts of semi-fixed and fixed dunes, of *Carex arenaria*, a rhizomatous perennial sedge capable of extensive vegetative growth, and of the large tussock-forming grass *Ammophila arenaria*. All these species were studied in the Aberffraw and Newborough Warren dune systems on the south-west coast of Anglesey, UK. The three main life-cycle patterns were each considered as a strategy of meristem replacement and proliferation.

The **demographic approach**, strongly advocated by Harper (1977), involves counting numbers of genets, shoots, buds, seeds or other entities and establishing measures of population behaviour such as 'birth rates', 'death rates' and 'expectation of life'. (A **genet** represents an original zygote. It may be a tiny seedling or a clone, of a grass for example, whose ramets extend for hundreds of metres.) Attempts can then be made to discover the mechanisms determining numbers and rates of change, in this case by studying the responses of populations of these sand dune species to various **experimental perturbations**. These may include the application of nutrients, controlled alterations of plant density and the weeding out of species growing with the one under study.

In the winter annuals of sand dunes, particular interest attaches to the influence of low temperatures on floral induction, and of high temperature and long day length on rapid flower production. Annuals are rare or absent from wet stands in the slacks and their abundance is negatively correlated with the vegetation cover of perennial species (Pemadasa, Greig-Smith and Lovell, 1974). They may be considered 'opportunists', occurring where drought or some other factor inhibits a full cover of perennials. All five species shown in Figure 7.4 require low temperatures (<10°C) for floral induction, a requirement normally met under field conditions by the end of December. The behaviour of the different species once vernalized is, however, not the same. *Cerastium diffusum* (= *C. atrovirens*) and the grass *Mibora minima* become almost neutral to day length; their failure to flower in midwinter seems to result from a need for temperatures of at least 10°C. Additional chilling delays flowering in these two species (and also in *Catapodium rigidum* and *Saxifraga tridactylites*); this effect occurs but is less marked in *Aira caryophyllea*, *A. praecox* and *Vulpia fasciculata* (= *V. membranacea*). In *Aira praecox* and *Aira caryophyllea*, inflorescence development is considerably delayed under short day conditions. Though induction of development is produced by very similar conditions in the two species, inflorescence emergence is considerably faster in *A. praecox*. *Vulpia fasciculata* is an obligatory long day plant which does not begin to flower until May (Pemadasa and Lovell, 1974b). *Myosotis ramosissima* (= *M. hispida*) is a day-neutral, early flowering species common on many dunes.

In sand-dune systems these winter annuals are characteristically depauperate. Many plants of *Cerastium diffusum* produce only one mature capsule containing 3–4 seeds, often the annual grasses bear only a single shoot, and many mature plants of *Vulpia fasciculata* produce only one or two seeds. Poor performance is not surprising in plants living in mineral-deficient soils, having so short a period to gather resources from the environment, and spending most of their lives at low light levels and temperatures, though usually with adequate water. Only when cultured in fertile soils or when they grow in naturally fertilized patches near rabbit burrows do dune annuals show a capacity for increase (r_m) approaching that characteristic of typical *r*-species.

Though dune annuals are small, they strongly interfere with the growth of each other. This was demonstrated by sowing seed of *Cerastium diffusum*, *Mibora minima*, *Phleum arenarium*, *Saxifraga tridactylites* and *Vulpia fasciculata* at random on 'flats' of sterilized sand from Aberffraw in various mixtures and at different densities (Mack and Harper, 1977). The position of every plant was mapped and all were grown to maturity. Up to 69% of the dry weight and seed production of the individual

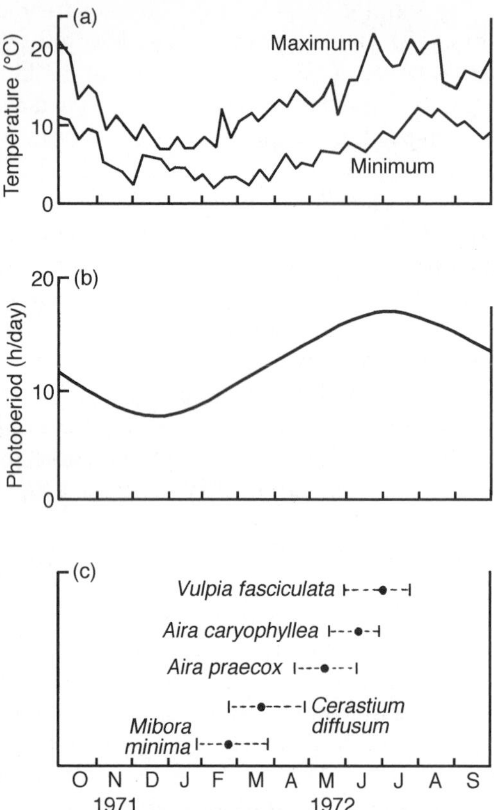

Figure 7.4 (a) Mean weekly temperature (maximum and minimum) and (b) photoperiod for the period October 1971 to September 1972. (c) Diagrammatic representation of the time of flowering in the field of five species studied experimentally (mean time of flowering and range shown). (After Pemadasa and Lovell, 1974b; courtesy *Journal of Ecology*)

plants could be accounted for in terms of the number, arrangement and species of neighbours within 2 cm. The order of aggressiveness of the species (as measured by dry weight or seed output) was *Vulpia* > *Phleum* > *Mibora* > *Cerastium* >*Saxifraga*. Plants in a crowded natural population of *Cerastium diffusum* produced 44 seeds per plant. Seedlings from the wild transplanted into dune sand in January and grown to maturity 5 cm apart in an unheated glasshouse yielded a mean number of 388 seeds per plant (Harper, 1977, p.532). The chances of an emerged seedling of an annual surviving to maturity are considerably reduced by the presence of *Festuca rubra*; this reduction is particularly significant in nutrient-enriched sand at medium and high plant densities (Pemadasa and Lovell, 1974a).

Neighbour effects among annuals can also be elucidated by manipulating natural field densities as in studies on *Vulpia fasciculata* on the fixed dunes at Aberffraw and Newborough Warren (Watkinson and Harper, 1978). The field densities obtained varied from 100 to 10 000 plants per 0.25 m^2. There was a negative correlation between seed production and density of flowering plants in the field, but no evidence of density-dependent mortality. The fate of dispersed seed was followed by labelling the seeds with radioactive scandium-46 applied in nail varnish. The survivorship curve of *V. fasciculata* is similar to the curves of small desert annuals which also have a very low seed output but a high probability of survival. Vulnerability to **self-thinning** is strongly influenced by plant geometry. The narrow, erect leaves of grasses improve the efficiency of light interception; Lonsdale and Watkinson (1983) suggested that this allowed a closer packing of plant material into a given space or the development of a greater stand height before self-thinning commences.

The **low reproductive capacity of dune annuals** makes the viability and germination performance of their seeds factors of major importance; Pemadasa and Lovell (1975) found that in *Aira caryophyllea, Aira praecox, Cerastium diffusum, Erophila verna, Mibora minima, Saxifraga tridactylites* and *Vulpia fasciculata* a high proportion of viable seeds could germinate in autumn. Only seeds lying near the surface of the soil are capable of rapid germination. The inability of seeds lying deeper in the soil (>3 cm) to germinate leads in some of these annuals to a limited **seed bank**, which can contribute to the population of the following year, by which time the seeds may have been exposed by sand movement. Seeds of all these species can remain viable for more than two years. Because they germinate in autumn and flower in spring, these **winter annuals** receive only limited light and experience low temperatures (5–10°C) during the over-wintering period which covers some two-thirds of the life span. Experimental increases in light level improved the performance of all seven species considerably. Vegetative growth of *Aira caryophyllea, Aira praecox, Mibora minima* and *Vulpia fasciculata* was restricted by short days and low temperatures, and was arrested by flowering. Pemadasa and Lovell (1976) concluded that, even under high nutrient conditions, the timing of the life-cycle of these annuals restricts their vegetative growth.

Though the **density of rabbits** at Aberffraw and Newborough Warren stayed very low – compared with pre-myxomatosis levels – for many years after 1954, it remained an important factor in the population dynamics of *Vulpia fasciculata*. The effect is selective; rabbits graze a larger proportion of the plants with several tillers and large spikes, which are often those which germinate early. When rabbits are absent the fittest plants are those which germinate early, grow rapidly and produce several tillers and a large inflorescence. When rabbits are present the plants safest from predation germinate late and are depauperate single-tillered forms.

The main vegetation of the Anglesey dunes is perennial and dominated by rhizomatous species, such as *Carex arenaria* and *Ammophila arenaria*, in which single genetic individuals form expansive clones which regenerate and proliferate from buds on the rhizomes. *Salix repens*, the only woody species present in significant amount on the dunes, behaves somewhat similarly. Single genets growing clonally may form extensive stands. Figure 7.5, which shows the fate of cohorts in dense populations of marram seedlings at the foot of a dune slope, supports the view that at least some stands of *A. arenaria* may be genetically diverse. At the end of 3 years a dense stand had developed to flowering and there were 36 genets per m^2.

Rhizomatous perennials produce relatively short-lived aerial shoots from the more permanent rhizome system. The production of each aerial shoot involves the loss of a bud from the rhizome, but the forward-growing rhizome adds new buds to the subterranean bank of dormant meristems. In *Carex arenaria* new

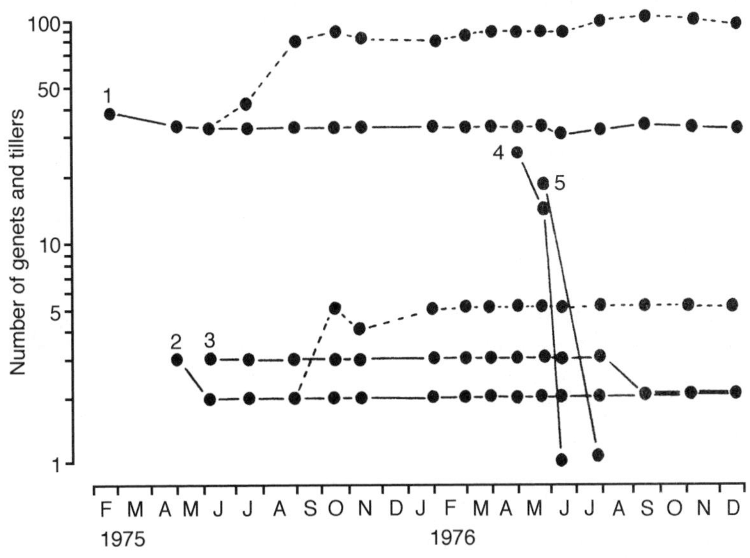

Figure 7.5 Survivorship curves of five successive cohorts of seedlings of *Ammophila arenaria* established naturally at the foot of a dune slope. ●—●, numbers of genets; ●---●, numbers of aerial shoots. (From Huiskes, 1977.)

shoots are usually produced at intervals of four internodes; each of these shoots contributes a basal bud which replaces it in the 'bud bank'. In some populations of sand sedge the number of viable dormant rhizome buds may reach 1400 m^{-2} and exceed the number of aerial shoots by 11 : 1 (Watkinson *et al.*, 1979).

Making estimates of cover or repeated counts of shoots fails to record the flux which occurs in the birth and death of the shoots and their parts. *Carex arenaria*, like *Pteridium aquilinum*, is a clonal plant which has well-defined growth phases. The leading edge of the clone is (a) the **juvenile phase**, which has behind it (b) the **adolescent phase** in which shoot density approaches the carrying capacity, **K**, of the site; in (c) the **mature phase**, the shoot population has reached carrying capacity and most shoots, which are **monocarpic** (i.e. flower once only), are flowering or about to flower; (d) is a **senile phase** with fewer shoots and (e) a 'slack phase' equivalent to the **hinterland phase** in *Pteridium* (Watt, 1947). Leaf weight ratio (LWR) rises steeply from the leading edge of the clone to the early adolescent phase and then falls again, becoming steadily more adverse as the proportion of the biomass present as the respiratory burden of rhizome increases through the mature, senile and slack phases of the continuum.

Noble (1976) repeatedly mapped shoots of *Carex arenaria* in replicated 0.5-m^2 quadrats for a period of 2 years. Shoot density was higher in summer than winter but the numbers of shoots present in the same months of succeeding years were similar; this tended to conceal the very rapid flux in shoot births and deaths. It appears that either the death of old tillers promotes the birth of new tillers (perhaps by reduced shading, reduced demand for nutrients, or reduced correlative inhibition of dormant buds), or that the birth of new tillers causes demands on resources in old tillers, which die when these resources are withdrawn. The latter explanation is analogous to the progressive transfer of nutrients from leaf to leaf in developing annuals on nutrient-poor sands.

When *Carex arenaria* sites were perturbed by the addition of NPK fertilizer, the number of sedge shoots increased when *C. arenaria* was the main species present. Expectation of life declined as a consequence of the very rapid increase in birth and death rates of shoots caused by addition of fertilizers in all but the slack phase population. Where it was a less important component of the vegetation *C. arenaria* showed very little reaction to nutrient application. In the dune slacks the associated species rather than the sand sedge grew vigorously, suggesting that these species had absorbed most of the additional nutrients. Expectation of life of individual tillers declined as a consequence of this change in flux rate.

After fertilizers had been applied the population became dominated by young tillers and the bank of dormant rhizome buds was temporarily depleted. In the mature phase the shoot to dormant bud ratio was 1 : 2; this fell to 1 : 1 in the fertilized plots. The breaking of bud dormancy is probably a nitrogen effect. Heavy doses of fertilizer profoundly altered turnover of growth modules as well as appearing to raise carrying capacity somewhat.

Studies of leaf and shoot dynamics of *Ammophila arenaria* at Newborough Warren (Huiskes, 1977) involved (a) addition of NPK, (b) repeated additions of sand, or (c) repeated removal of the aboveground parts of associated species. Birth and death rates of leaves are strongly synchronous throughout the year, being low in winter and high in summer. As in *Carex arenaria* tillers, the period of rapid leaf birth is the time when existing leaves are most at risk. Fertilizer application increases turnover of leaves; again this effect is marked only in areas where other species are rare.

When *Ammophila arenaria* was grown with *Festuca rubra* in pots of dune sand 50 cm high, the more superficially rooted fescue suppressed the marram when nutrients were supplied to the sand surface. The species were in much closer balance when nutrients were applied to the bottoms of the pots. *Ammophila* has a deep-lying horizontal system of rhizomes, with rather few buds, from which vertical rhizomes develop. The latter may show an extensive development of buds which give rise to the tussocks characteristic of this plant. Relatively few marram shoots flower at any one time and the individual shoots live considerably longer than those of the sand sedge, whose horizontal growth tends to confine it to stable and eroding dunes. The ability of the vertical rhizomes of *Ammophila* to elongate rapidly and to stabilize accreting sand may be eventually disadvantageous to the plant (sections 6.1 and 7.1), which commonly declines in vigour when the stabilized sand is colonized by the more superficially rooted species (which cannot survive very rapid accretion of sand) responsible for concentrating recycled nutrients in the surface layers of the soil.

The lives and growth patterns of the dune annuals, and the perennials *Ammophila arenaria* and *Carex arenaria*, illustrate several important features of dune plants, which exist in a habitat where open sites are relatively common. Establishment from seed is uncertain and unusual on mobile dunes, while perennial clonal growth of genets is the most important way in which plant populations are maintained in both accreting and eroding areas. Given the continuous presence of open sites suitable for colonization, the fact that almost all released seed of *Vulpia fasciculata* germinated within a few months, with only 2–15% being eaten by predators and a virtual absence of a buried bank of viable seed, is not surprising; such features might not, however, be expected in habitats where the early stages in succession occur only rarely. Only in stabilized but still open areas are therophytes generally successful; even here most of the plants involved are small short-lived winter annuals whose entire life-cycles are completed in the period when perennial activity is low (Watkinson, 1978a).

The nutrient deficiency of the site appears to be overwhelmingly important for these species; competitive interactions for nutrient resources may to some extent be responsible for the limitation of even the vigorous rhizomatous *C. arenaria* and *A. arenaria*. The ability of certain annuals and perennials to move to uncolonized areas seems to be of a similar order. An advancing front of *C. arenaria* shoots can move at a rate of 2 m per year, whereas the maximum initial distance of dispersal from the parent in *Vulpia fasciculata* was found to be 36 cm and even on a very open site no seed moved more than 92 cm from its point of landing.

Vulpia fasciculata is an annual that is primarily inbreeding, while the perennials are capable of rapid clonal growth. Both life-cycle patterns tend to perpetuate a **narrow range of genotypes** in a system involving **rapid turnover** of generations (in the annuals) or of growth modules (in the perennials). Watkinson *et al.* (1979) considered that the low seed output, the genetic systems, limited dispersibility of seed and dominance of species with persistent banks of vegetative meristems in the soil, suggested that dune plants have evolutionary and ecological tendencies which are suitable for an ecosystem that consistently contains all (or at least most) of the phases of succession in its vegetational mosaic.

This view still holds in the light of more recent work. Indeed, a 9-year population study of *V. fasciculata*, reported by Watkinson (1990), emphasizes the **role of instability** if dune systems are to retain a full complement of annuals and short-lived perennials so maintaining **maximum species diversity**. The plots involved were established in the Aberffraw and Newborough Warren dune systems in Anglesey, UK. During the first 4 years the numbers of individuals varied very little; but then declined rapidly to extinction. This decline, which coincided with an increase in abundance of perennials, was associated with a decrease in local movement of surface-blown sand and an increase in mortality in seedlings which germinated on the sand surface. A computer model predicted that *V. fasciculata* populations can be expected to persist only where the percentage of bare sand remains greater than 50%.

7.2.3 SALT MARSH AND SAND DUNE ANNUALS

The population biology of annuals in these two habitats, both of which are characterized by periods of low water availability, marked spatial and temporal heterogeneity and a zonation which – within limits – reflects successional change, is discussed by Watkinson and Davy (1985), who also review the factors favouring annual versus perennial reproduction. The distributions of **annuals** in these apparently dissimilar habitats also show marked resemblances; they are usually **dominant in the pioneer stages**. *Salicornia*-dominated low marsh is comparable with strandline ephemeral populations of species such as *Cakile maritima* and *Salsola kali* of embryo dunes in being at the bottom of the shore (section 6.2), although *Atriplex* spp. are more characteristic of salt-marsh strandlines (section 5.1). They also occur in small **gaps in the matrix of perennials** in mature vegetation; at least some of these gaps arise from drought or disturbance. In terms of **phenology**, however, there is a major contrast; only summer annuals occur on salt marshes, whereas winter annuals predominate on dunes (apart from strandline annuals and the hemi-parasitic annuals *Euphrasia* spp. and *Rhinanthus* spp., which are summer annuals that track the phenology of their perennial hosts).

Dune annuals are also characteristically depauperate: Watkinson and Harper (1978) found that on the Anglesey dunes *Vulpia fasciculata* and *Cerastium diffusum* produced an average of only 1.7 and 7.3 seeds per plant respectively. (Many plants of *V. fasciculata*, however, formed several spikelets with at least one seed per spikelet.) These low numbers contrast strongly with values given by Salisbury (1942) of 39 to 176 000 seeds per plant for a range of annuals of various habitats. Also contrasting is an average of c. 1000 seeds per plant found for *Salicornia europaea* in parts of the lower marsh at Stiffkey, Norfolk, UK (Jefferies, Davy and Rudmik, 1981), though isolated individuals of the same species produced an average of only 33.8 seeds on the upper marsh. Seed production within a species varies in time and space; density and interference – notably in terms of nutrient acquisition – from perennials or other annuals are two of the most striking influences identified by Watkinson and Davy (1985). For *Cakile edentula*, *Cakile maritima*, *Salicornia europaea* and *Vulpia fasciculata*, plant performance is always higher among the sparse vegetation of the pioneer zone than in denser vegetation. This is exemplified by the *Cakile edentula* populations in Nova Scotia investigated by Keddy (1981, 1982). Here there is a net flow of seeds landward from the much larger plants on the beach; levels of reproduction and survival of the plants on the densely vegetated dunes are so low that these populations would become extinct if a large annual flux of seed did not occur. *Cakile edentula* on open shingle beaches exhibited density-dependent reproductive output but density-independent mortality (as in *Vulpia fasciculata*). Where shingle was covered with decaying mats of *Zostera marina*, however, both mortality and reproductive output varied with density.

Synchronous germination of annuals in hazardous environments may put the entire population at risk; seed longevity and a large buried seed population or **seed bank** may be expected in such habitats. However, large banks of seeds persisting for more than the year following their production are not found in populations of dune annuals in North Wales (Mack, 1976, Watkinson 1978a) or of *Salicornia europaea* populations of English salt marshes (Jefferies, Davy and Rudmik, 1981). In contrast, those of populations of *S. europaea* in Hudson Bay, Canada, at the northern limit of its distribution, are very large indeed (Jefferies, Jensen and Bazely, 1983). This would act as a safeguard against a shortfall in seed in areas where, in contrast to the relatively predictable habitats of the UK, there is considerable variation in the length of the growing season.

7.3 PLANT AND ANIMAL COMMUNITIES

Community interactions, as long realized, can most readily be studied in relatively simple systems. Because sand-dune ecosystems are so complex there is still much to learn about the relationships between the plants, animals, fungi, bacteria and viruses which exist within them. However, many important processes, such as the modification of vegetation by the dunging, grazing and, trampling pressures exerted by mammals including rabbits, sheep, horses (Oosterveld, 1985) and cattle are now well understood as also, for example, are the feeding requirements of many insects which require particular food plants. This section is largely concerned with those features of sand-dune systems which favour animal diversity; it starts by considering the range of animals present in particular systems.

Conditions in the large coastal dunes of the Pacific Northwest are far more natural than those in Europe and they support extensive bird and mammal faunas. The Oregon Dunes NRA (section 7.1), which forms a transition zone between the ocean and the coastal mountain forest, consists of open sand, estuaries, salt marshes, beachgrass communities, other grasslands, shrub thickets and predominantly 'shorepine' *Pinus contorta* forest (Wiedemann, 1984). Some 247 species of birds (118 aquatic, 108 song birds, 21 birds of prey), 50 mammals, eight salamanders, three frogs, one toad, two snakes and the Northern alligator lizard (*Gerrhonotus coeruleus*) are recorded for the NRA. The red crossbill (*Loxia curvirostra*), chestnut-backed chickadee (*Parus rufescens*) and Steller's jay (*Cyanocitta stelleri*) are abundant here, and the characteristic large mammals include the elk (*Cervus elephus*), black-tailed deer (*Odocoileus hemionus*), bobcat (*Felis rufus*) and black bear (*Ursus americanus*). Deer mice (*Peromyscus maniculatus*), snowshoe hares (*Lepus americanus*) and Douglas squirrels (*Tamiasciurus douglasii*) are very common. In grasslands common garter snakes, the Pacific tree frog (*Hyla regilla*), the striped skunk (*Mephitis mephitis*) and the Beechey ground squirrel are often seen; the latter eats the seeds of marram and burrows in its hummocks. The open dune habitat, now being encroached upon by *Ammophila arenaria* and also subject to disturbance by off-road vehicles (ORVs), is utilized by many invertebrates including numerous bee species. Snowy plovers (*Charadrius alexandrinus*) nest at the windward bases of foredunes.

In contrast, European dune systems are both smaller and more influenced by humans. In Britain, where the extinction of large mammalian carnivores is a major contrast with dune systems on the North West Pacific coast of the USA, the fox (*Vulpes vulpes*) – present in almost all UK dune systems – is the largest predator. Rabbits formed an appreciable proportion of the diet of the fox before the myxomatosis epidemic of 1954; it now has to be controlled by culling when its predation of nesting birds becomes excessive. The bird populations harassed by foxes on the Sands of Forvie Reserve, Scotland, quoted as an example in section 2.8, include species which roost and nest in coastal sites while obtaining food elsewhere.

Although small by North American standards, many British nature reserves support important animal populations, some of which are well known. Foster (1989) and Seago (1989), for example, give full descriptions of the insect and bird populations of Blakeney Point and Scolt Head Island, Norfolk, where brown rats are regular scavengers and also take eggs and nestlings from the terneries. Populations of stoats here fluctuate roughly in parallel with those of the rabbit, their main prey; weasels are uncommon. At Braunton Burrows, North Devon, the rabbit population has fluctuated considerably since 1954, but is no longer large enough to maintain the close-grazed sward of former years (section 6.6). Insectivores – common and pygmy shrews, moles and hedgehogs – and rodents (short-tailed and bank voles and woodmice) are prominent among the small mammals. Larger carnivores include the fox, weasel and mink. The common lizard is frequently sighted; palmate newts and common toads are more abundant now that the provision of ponds allows their tadpoles to mature. Snails are common, notably *Cepaea nemoralis* and also e.g. *Cochlicella acuta*, *Helicella* spp. and *Helix aspersa*. Insects present include the scarlet leaf beetle (*Chrysomela populi*), thousands of whose larvae feed on *Salix repens* in summer and which exude blobs of distasteful liquid to deter predators. Butterflies, notably the common blue (*Polyommatus icarus*), the marbled white (*Melanargia galathea*) and the fast-flying dark green fritillary (*Argynnis aglaja*), are conspicuous in summer. Also frequent are small heath (*Coenonympha pamphilus*), small copper (*Lycaena phlaeas*), small skipper (*Thymelicus sylvestris*) and grizzled skipper (*Pygus malvae*). Later in the season meadow browns (*Maniola jurtina*) and gatekeepers (*Pyronia tithonus*) are present on the flowerheads of mint and hemp-agrimony (*Eupatorium cannabinum*). A common beetle in the marram tussocks of many dunes is the carabid (*Risophilus monostigma*). Another, the scarab beetle (*Anomala dubia*), flies like a cockchafer and often crawls over bare sand. Braunton Burrows is on the west coast migration route of many birds of passage, while large flocks of waders feed on the adjacent Taw/Torridge estuary. Magpies, kestrels and the occasional buzzard are conspicuous among the more local population. Elliston Wright (1932), who also provides an extensive list of insects, noted that shelduck nest in the Burrows and dig up cockles from sand ridges uncovered at low water; the cockles are swallowed whole, but the shells crushed in the gizzard and disgorged in small heaps on the sand.

7.3.1 ENVIRONMENTAL CHANGE, DISEASE AND BIODIVERSITY

The decline of many rare species is often linked to alterations in the ecology of the communities to which they belong, a point illustrated by the changing status of the natterjack toad (*Bufo calamita*) over the past 60 years. Beebee (1977) demonstrated that this species had declined much more extensively on inland heaths than on coastal dunes, its other major habitat. This occurred at a period when increased forestry activity and a cessation of grazing allowed encroachment of taller vegetation on the inland heaths. The resulting shade was unfavourable to the natterjacks and, more importantly, encouraged the entry of its successful competitor, the common toad (*Bufo bufo*). On coastal dunes, SSSI or nature reserve status has helped protect the species from many threats, but inadequate management leading to scrub invasion has caused declines at some of the largest extant natterjack sites.

The natterjack has been widely publicised in recent years, as has work by the British Trust for Conservation Volunteers (**BTCV**) aimed at protecting and extending the Sefton Dunes, Formby Point, Merseyside, at the northern end of its distribution. Much has been done to provide it with an appropriate habitat and suitable breeding sites (section 11.2). The number of natterjack sites documented by conservationists, excluding those resulting from translocations, rose from 21 to 40 between 1970 and 1990 (Banks *et al*., 1994), but the actual overall number of sites has probably changed very little. Over 90% of British natterjacks now live in the vicinity of just five estuaries, those of the rivers Alt, Duddon, Esk, Ribble and Solway; breeding conditions for these animals would be markedly altered if these rivers were affected by barrage schemes.

Biodiversity and the continued existence of rare species may also be strongly influenced by the effects of disease upon a single **key species**, such as the rabbit (*Oryctolagus cuniculus*), whose British populations were greatly reduced by myxomatosis (section 6.6) but later increased again. British sand-dune systems are still influenced by grazing, burrowing, trampling and dunging by this lagomorph (Figure 7.6), which is now threatened by Rabbit Viral

Figure 7.6 Recent burrowing damage caused by rabbits (*Oryctolagus cuniculus*) on grey dunes at Ynyslas, North Wales. Plant species present include *Agrostis capillaris, Ammophila arenaria* (a few scattered plants), *Carex arenaria, Erodium cicutarium, Euphorbia paralias, Festuca rubra, Galium verum, Ononis repens, Plantago lanceolata, Rosa pimpinellifolia* and *Sedum acre*. (Photograph by John R. Packham, June 1995.)

Haemorrhagic Disease (**RVHD**). First noted in domestic rabbits in China in 1984, this has since spread widely: the first confirmed deaths of wild rabbits from RVHD in Britain occurred in 1994 in Devon and Kent. Mortality from RVHD among wild rabbits in Spain – where the breeding success of some 40 vertebrates is related to the abundance of this herbivore – has exceeded 80% (Sutherland, 1995). A further semi-permanent diminution in rabbit populations (which may show varying degrees of immunity to RVHD) of British coastal systems would, through the reduction in grazing pressure and herbivore biomass, reduce the diversity of both the plant and vertebrate populations. Other examples of the reduction of key species include the wasting disease of *Zostera* and inland parallels caused by Dutch elm disease and American chestnut blight.

7.4 PRODUCTIVITY, ENERGY FLOW AND NUTRIENT CYCLING

Studies of biomass and productivity in sand dunes are not as numerous as those on salt marshes; that of Deshmukh (1979) at Tentsmuir NNR, eastern Scotland, was concerned with *Ammophila arenaria*, the most important pioneer in many sand dune successions. **Net above-ground primary productivity** was shown to decrease along a transect running inland from the shore (Table 7.1). Litter bags full of marram leaves, placed at the sand surface and at a depth of 10 cm, were analysed at 3-month intervals to determine overall loss of energy. Loss of energy in the first year of **litter decay** was estimated at 39.8 ± 3.5(SE)% (ash-free basis); there were no clear differences between bags at different zones and depths.

Table 7.1 Estimates* of annual above-ground net primary production (*Pn*) of *Ammophila arenaria* for 1971 as the sum of biomass increment (ΔB) and loss by death (L) in each zone. (From Deshmukh, 1979; courtesy Blackwell Scientific Publications.)

	Zones		
	Mobile	*Semi-fixed*	*Fixed*
ΔB	3367.2 ± 412.8	2342.2 ± 183.8	939.6 ± 145.3
L	1360.8 ± 115.6	969.7 ± 125.2	498.3 ± 52.3
P_n	4727.3 ± 415.8	3311.9 ± 400.7	1437.8 ± 150.3

* Values are mean ± SE, kJ (ash-free) m^{-2}.

Microbial respiration rates were measured as rates of oxygen uptake per unit weight of litter at temperatures corresponding to those encountered under field conditions. This involved placing material from the litter bags into a Gilson respirometer for 24-hour periods. On the basis of these results it was calculated that microbial respiration released 42.8 ± 2.3(SE)% of the energy content in one year of the litter originally placed in the bags, a result statistically indistinguishable from that obtained by the litter bag method. The only animal found in the soil that consumed *A. arenaria* litter in feeding trials was the millipede *Cylindroiulius latestriatus*; its highest population levels (100 individuals m^{-2}) occurred in the semi-fixed dunes where it consumed less than 0.25% of litter production annually. At Tentsmuir, where faecal analysis showed that rabbits were not feeding on marram grass, microbial respiration was responsible for almost all the energy released from *A. arenaria* litter during the first year of its decomposition.

The graphical model constructed by Deshmukh (1979), in which the ages of the mobile, semi-fixed and fixed dune zones of this rapidly accreting system were taken as 13.5, 18.9 and 24.3 years old, showed a good fit of **litter accumulation predictions** with independent harvest estimates. Additional factors must come into play outside the study zones as *A. arenaria* tussocks, admittedly moribund, occur in dunes more than 100 years old, whereas the model suggested cessation of marram production after 29 years and elimination of its litter after 33 years.

7.4.1 BIOMASS AND NUTRIENT ACCUMULATION IN A DUTCH DUNE SYSTEM

The long-term studies of Olff, Huisman and van Tooren (1993) reconstructed the species dynamics and pattern of nutrient accumulation which occurred during early primary succession in the Dutch island of Schiermonnikoog. Observations on permanent transects across a topographic profile from a moist plain, over a slope and on to the dry dune, followed the progress of a succession which began on bare sand following construction of a sand dike. By 1990 the lower plain, which is flooded by seawater every winter, was dominated by *Juncus gerardii*, *Bolboschoenus maritimus* and *Phragmites australis*; the small dunes and their slopes had become densely covered by *Hippophae rhamnoides*.

Figure 7.7 shows standing crop for all three sites at 12, 20 and 28 years after commencement of succession. Soil analyses and a fertilization experiment showed that above-ground biomass production was limited by nitrogen. Over some 16 years the total amount of nitrogen in the organic layer of the soil increased from 7 to 50 g N m^{-2} in the Plain (with similar figures for the Slope), and from 1 to 15 g N m^{-2} in the Dune. **Shading** became an increasingly important factor governing successional

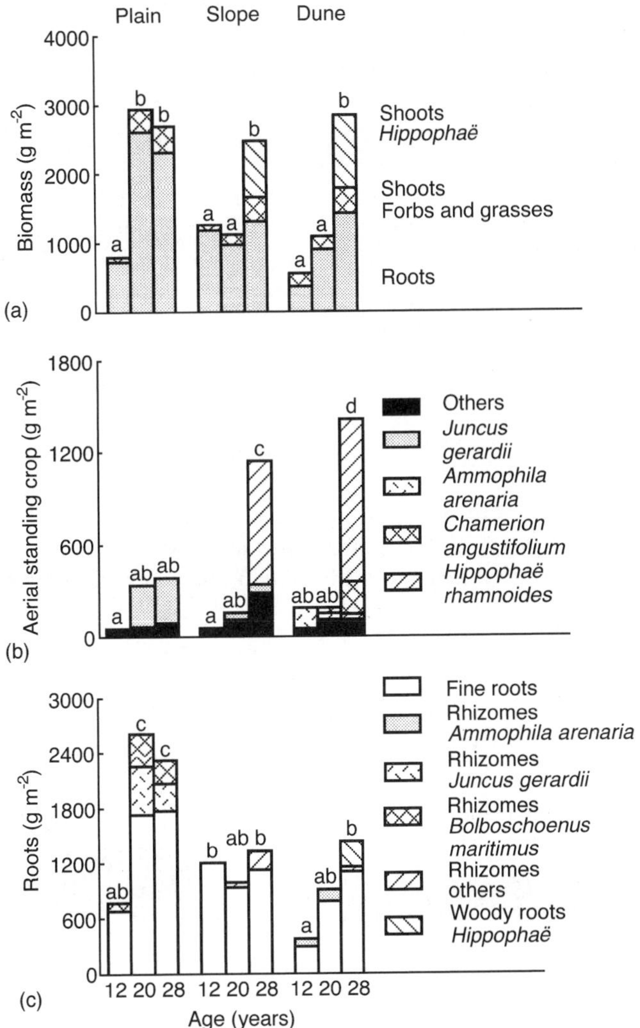

Figure 7.7 Reconstruction of (a) total biomass; (b) above-ground biomass; and (c) below-ground biomass of different plant species in Plain, Slope and Dune at three stages of primary succession at the Dutch Island of Schiermonnikoog. Totals with the same letter within each subfigure were not significantly different. (Redrawn from Olff, Huisman and van Tooren, 1993; courtesy *Journal of Ecology*.)

processes once nutrients had begun to accumulate in the ecosystem.

Strong successional trends were shown by the changing species composition on all three sites as the **life form spectra**, which are weighted for probability of species occurrence, indicate (Figure 7.8). The phanerophytes, which entered last, became important on both slope and dune. The proportion of therophytes, initially high, diminished less on the plain than on the slope. Decrease in occurrence of several grass and forb species on both slope and dune was correlated with an increase in *Hippophae*, a formidable competitor once it over-tops species present earlier in the succession. On the Dune, small species, especially chamaephytes (*Cerastium fontanum*, *Sagina nodosa* and *Sedum acre*) and therophytes (*Aira praecox*, *Cerastium semidecandrum* and *Euphrasia stricta*), were most common in the intermediate successional stages. *Agrostis stolonifera*, *Ammophila arenaria*, *Festuca rubra* and *Sonchus arvensis* were abundant at an early stage, diminishing later as *Hippophae rhamnoides*, *Chamerion angustifolium*, *Poa pratensis*, *Calamagrostis epigejos* and *Elytrigia atherica* became established. The glycophytic annuals *Odontites vernus* and *Arenaria serpyllifolia* occurred on the dune in years with heavy spring rainfall.

7.4.2 CATION FLUXES AT THE AINSDALE DUNES, UK

Nutrient fluxes in sand-dune systems differ from time to time and from place to place and are, as in salt marshes (section 4.5), difficult to measure and interpret. James *et al.* (1986) describe a **cation budget analysis** for the Ainsdale, Merseyside, coastal dune system and also provide a table giving a wide ranging comparison with other published budget studies. Ainsdale resembles many other dunelands in that no drainage channels appeared to operate, even when water-table levels were at maximum height. Such systems, unlike those of river catchments, lack a focus for the outflow of nutrient-bearing water and output is by discharge through the ground. Nutrient losses are therefore calculated from mean ionic concentrations in ground water and from estimated ground water discharge. Moreover, cation input by aerosols impacting plant and ground surfaces, though far higher than in most inland locations, is very difficult to estimate accurately.

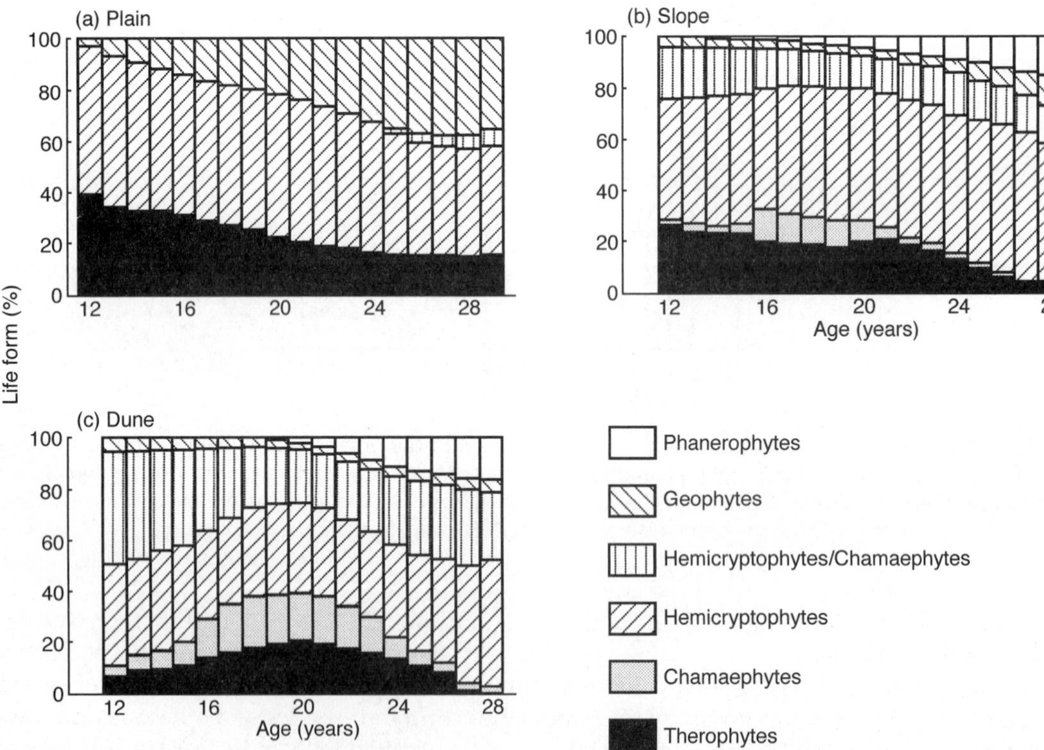

Figure 7.8 Changes in the life-form spectra in: (a) the Plain; (b) on the Slope; and (c) on the Dune during primary succession. Each species is weighted for probability of occurrence, Schiermonnikoog, The Netherlands. (Redrawn from Olff, Huisman and van Tooren, 1993; courtesy *Journal of Ecology*.)

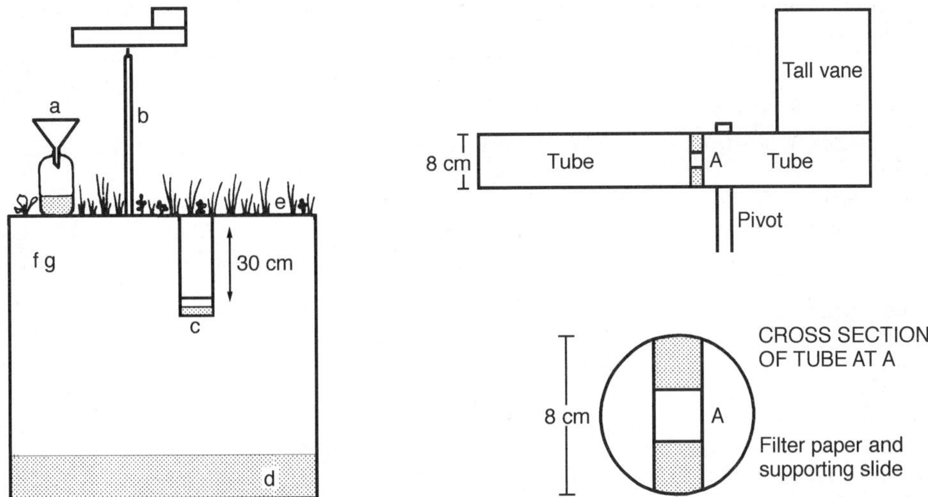

Figure 7.9 Measurement of cation concentration for the non-afforested duneland of Ainsdale, Merseyside, UK, in plant and soil stores, in atmospheric inputs and in leaching and groundwater outputs: a, deposition in bulk precipitation; b, deposition by aerosol impact (on filter paper targets; detail shown on right); c, leaching output (30 cm core); d, groundwater output (groundwater discharge estimated using a water balance approach); e, store in plant cover; f, 'available' cations extractable in ammonium acetate; g, 'slowly available' cations digested in concentrated nitric acid. (From James and Wharfe, 1989; courtesy SPB Academic Publishing.)

Methods used in the measurement of cation concentrations in atmospheric inputs and in soil and plant stores, together with leaching and groundwater outputs, are shown in Figure 7.9. **Aerosol impaction** was measured on a filter paper mounted on a slide within a plastic tube directed into the wind by means of a tail vane and pivot. [Note, however, that aerosol deposition, which is discussed by Monteith and Unsworth (1990), depends on the nature of the surface and the size and shape of the collecting structure, impaction efficiency on, for example pine needles being much greater than on larger objects.] Aerosols are naturally deposited on vegetation and the dune surface but the complexity of the system involved, in which deposition is greatly influenced by topography, led to the presentation of aerosol data in terms of weight of the four elements per unit area of filter paper. During the period January to December 1981, the relative values of cations measured in aerosol impaction (Na > Mg > Ca = K), with the weight of Na exceeding that of Mg by a factor of 9.5, reflected their concentrations in seawater (James *et al.*, 1986). During the year November 1980 until October 1981, the order of cation influx in **bulk precipitation** was Na > Ca > Mg > K (15.5 : 2.7 : 1.5 : 1.0). Differences in the cation deposition as bulk precipitation in rainwater for this period and the 2-year run considered by James and Wharfe (1984) are related to the exceptionally wet period of 1980–1981 (the high value for calcium may result from the leaching by rainfall of calcareous sand retained by the glass wool filter of the funnel neck of the collecting bottles).

Table 7.2, which uses two methods to calculate the balances, shows the summary of the annual flux of bases through soils of the non-afforested region of Ainsdale given by James and Wharfe (1989). They conclude that the positive Na balance is dominated by high inputs of aerosols in sea spray. Ca and Mg

Table 7.2 Cation balance* for the non-afforested duneland of Ainsdale, Merseyside, UK, during 1980–1982 (kg ha^{-1} a^{-1} and kg ha^{-1})

		Na	*K*	*Ca*	*Mg*
1.	Bulk precipitation	159.7	10.2	29.5	17.7
2.	Aerosol impaction	86.8	4.6	8.1	9.5
3.	Total input	246.5	14.8	37.6	27.2
4.	Ground water outflow	152.8	11.6	421.1	45.1
	Balance 1–4	+6.9	–1.4	–391.6	–27.4
	3–4	+93.7	+3.2	–383.5	–17.9
	Biomass (kg ha^{-1})	45.2	199.9	194.5	48.8
	Soil store:				
	'Available'	132.5	69.7	4669.4	134.7
	'Slowly available'	464.8	757.0	20270.6	5180.0
	'Unavailable'	5455.2	7319.1	1339.7	625.9

* Two balances are presented because of the difficulty of estimating true input by aerosol impaction. (From James and Wharfe, 1989; courtesy SPB Academic Publishing.)

have lower atmospheric inputs and are gradually being leached from the soil store, which in the case of Ca is very high. K is initially lost by leaching from raw sand, but accumulates a positive balance in pararendzina soils and the associated biomass.

COASTAL SHINGLE 8

8.1 LOCATION AND FORMATION OF SHINGLE STRUCTURES

This chapter concerns the communities and geomorphology of coastal features made of **shingle**, which consists of particles whose equivalent diameter is predominantly greater than 2 mm, the upper limit for sand, but smaller than the lower limit for boulders (200 mm). The locations of major coastal deposits in Britain are shown in Figure 8.1, and the geomorphology of coastal structures in general is outlined in section 1.2. Of the five categories of shingle structures commonly recognized (Oliver, 1912; Randall, 1989), **fringing beaches, shingle spits** and **bars** (= barriers) tend to make the coastline more regular, and are in essence foreshores regularly washed by spray and storm waves which have limited or ephemeral vegetation over much of their area. In contrast, **cuspate forelands** and **barrier islands** are larger structures much of whose vegetation is of a more terrestrial and permanent nature.

Shingle structures will form only when sedimentary material is available during periods when the waves, winds and tidal currents are suitable for its movement. Constructive waves build small foreshore ridges on the beach at the limit of the swash. During a period of change from neap to spring tides (which have a greater range), there will be a single ridge at the limit of the latest tide. In contrast, on a series of falling tides – from spring to neap – there are several recent ridges, each representing the top of a single tide. Conditions at a particular site may vary over time, in some cases leading to deposition, in others to erosion.

The powerful backwash of destructive storm waves causes the rapid removal of shingle from the upper foreshore to deeper water beyond the break-point of the waves. This material may be carried back again later but large quantities may be lost from the system temporarily: Randall (1973) records that in August 1964 storm waves at Shingle Street, Suffolk, cut back over 2 m of shingle within 4 hours.

Storm waves often throw shingle up to the top of the beach crest above the reach of ordinary waves, and because the structure is permeable the backwash cannot remove it. The ridges of large shingle structures such as Orfordness and Dungeness consist of pebbles brought to the area by longshore drifting and then thrown up by storm waves; the fact that they contain a much higher proportion of fine and medium particle grades than the lows between them is of great importance to the establishment of vegetation (Fuller, 1987).

Fringing beaches, well seen in the UK at Brighton (Figure 8.2) and Hove in Sussex, and between Shingle Street and Bawdsey in East Anglia, consist of a strip of shingle in contact with the land at the top of the beach. Shingle spits are most common on shores with frequent changes of direction. Bars are effectively spits which run right across the mouths of estuaries or indentations in the coast thus enclosing a lagoon or other water body: their lee sides have a much less maritime environment than those of spits.

Shingle, instead of being carried along the coast, may pile up in front of a fringing beach or spit. If this is then driven landwards by storm waves an **apposition beach** is formed.

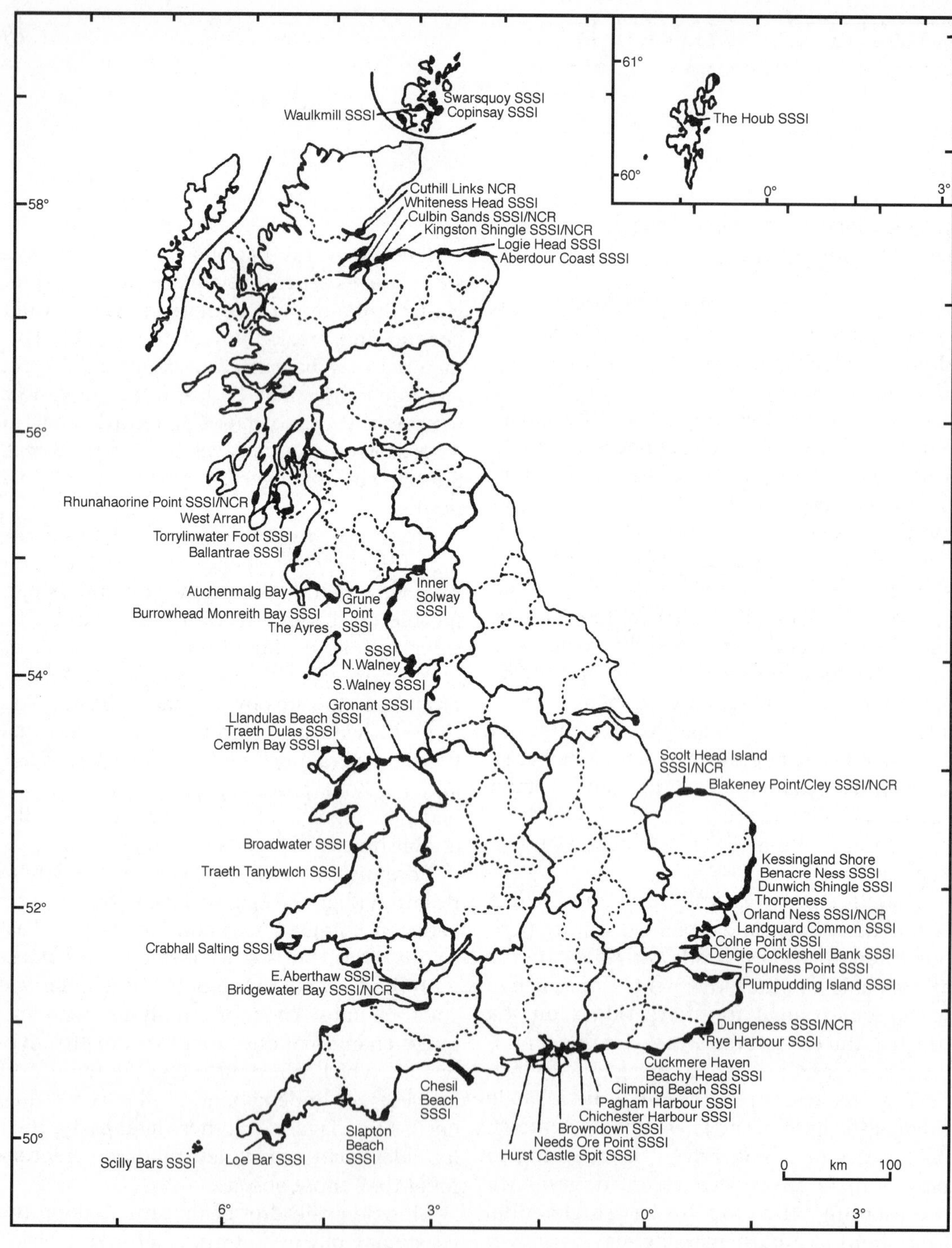
61°
60°
0°
3°
The Houb SSSI
Swarsquoy SSSI
Copinsay SSSI
Waulkmill SSSI
58°
Cuthill Links NCR
Whiteness Head SSSI
Culbin Sands SSSI/NCR
Kingston Shingle SSSI/NCR
Logie Head SSSI
Aberdour Coast SSSI
56°
Rhunahaorine Point SSSI/NCR
West Arran
Torrylinwater Foot SSSI
Ballantrae SSSI
Auchenmalg Bay
Burrowhead Monreith Bay SSSI
The Ayres
Grune Point SSSI
Inner Solway SSSI
SSSI
N.Walney
S.Walney SSSI
54°
Gronant SSSI
Llandulas Beach SSSI
Traeth Dulas SSSI
Cemlyn Bay SSSI
Scolt Head Island SSSI/NCR
Blakeney Point/Cley SSSI/NCR
Broadwater SSSI
Traeth Tanybwlch SSSI
Kessingland Shore
Benacre Ness SSSI
Dunwich Shingle SSSI
Thorpeness
52°
Orland Ness SSSI/NCR
Landguard Common SSSI
Colne Point SSSI
Dengie Cockleshell Bank SSSI
Foulness Point SSSI
Plumpudding Island SSSI
Crabhall Salting SSSI
E.Aberthaw SSSI
Bridgewater Bay SSSI/NCR
Dungeness SSSI/NCR
Rye Harbour SSSI
Cuckmere Haven
Beachy Head SSSI
Climping Beach SSSI
Pagham Harbour SSSI
Chichester Harbour SSSI
Browndown SSSI
Needs Ore Point SSSI
Hurst Castle Spit SSSI
Chesil Beach SSSI
Slapton Beach SSSI
50°
Scilly Bars SSSI
Loe Bar SSSI
0 km 100
6°
3°
0°
3°

Repetition of this process produces a series of roughly parallel shingle ridges which grow out from the shore as a **cuspate foreland**. The location and modes of formation of **offshore barrier islands** are considered in section 1.2.

Randall and Doody (1995) provide a habitat inventory for European shingle and list the significant vegetated shingle structures of Great Britain. Their outline description of the shingle vegetation of Europe emphasizes the importance of *Mertensia maritima* in the coarse shingle beaches of the low Arctic (Iceland, North Norway and Russia). These have rapid drainage, low nutrient status, and experience intense wave action and scouring; oysterplant is the only common species on these shingle shores and on gravel banks in estuaries. In contrast to *Polygonum maritimum*, which is probably advancing northwards, it is retreating at the southern edge of its range in the UK – apparently as a result of recreational pressure and shingle removal (Figure 8.3).

Other species influence the geomorphology of the structures on which they occur. Waterborne seed of *Suaeda vera* (shrubby seablite) often accumulates along the driftline on the lee shore of travelling beaches in Norfolk, UK. After germination, long tap roots rapidly grow down and enable the plants to establish. Although these can grow to over 1 m high they are repeatedly overwhelmed by shingle fans driven from the sea towards the land. On each occasion the now horizontal shoots put out new roots and new vertical shoots and help to stabilize the shingle beach (Chapman, 1947, 1976). *S. vera* is a C-3 plant formerly called *S. fruticosa* (a related C-4 species occurs in the Middle East).

8.2 CHARACTERISTICS OF SHINGLE VEGETATION

No more than two dozen species of flowering plants are especially characteristic of British shingle, and most of these also occur in the vegetation of cliffs or other coastal habitats (Randall, 1989). Some of these plants are locally abundant on shingle in particular parts of their geographical range; several of the examples given below are typical of such regions rather than exclusive to them. Shingle species with a southern or southeastern distribution in Britain include *Beta vulgaris* ssp. *maritima*, *Crambe maritima* (Figure 8.4(a)), *Glaucium flavum*, *Lathyrus japonicus* and the low evergreen shrub *Suaeda vera*; often on stabilized sandy shingle at the upper margin of salt marshes are *Frankenia laevis*, *Limonium binervosum* and the grasses *Parapholis strigosa*, *P. incurva* (more rarely) and *Elytrigia atherica*. The small tufted grass *Corynephorus canescens* and the spiny umbellifer *Eryngium maritimum* are both characteristic sand-dune species and the well-naturalized Mediterranean shrub *Tamarix gallica* also occurs in sandy places.

Figure 8.1 Distribution of coastal shingle deposits in Great Britain. British shingle communities are described by Sneddon and Randall (1993a) with appendices for Wales (1993b), Scotland (1994a) and England (1994b). The shingle community classification employed was derived from independent TWINSPAN analyses, but Table 3 of the main report lists the NVC categories (Rodwell, 1998) with which they are best matched. **Literature sources for important coastal shingle features listed below are given in the References**. Sites are listed in clockwise order starting from the top right: North Sea coastal margin: Doody, Johnston and Smith (1993); Culbin Spit: Fuller, (1975); Scolt Head Island and Blakeney Point: Allison and Morley (1989); Orfordness: Fuller and Randall (1988); Shingle Street, Suffolk: Randall (1973); Dungeness Shingle Beach: Lewis (1932), Ferry, Waters and Jury (eds) (1989); Solent: Brewis, Bowman and Rose (1996). (From Randall, 1989; courtesy of Linnean Society of London and by permission of the publisher, Academic Press Limited, London).

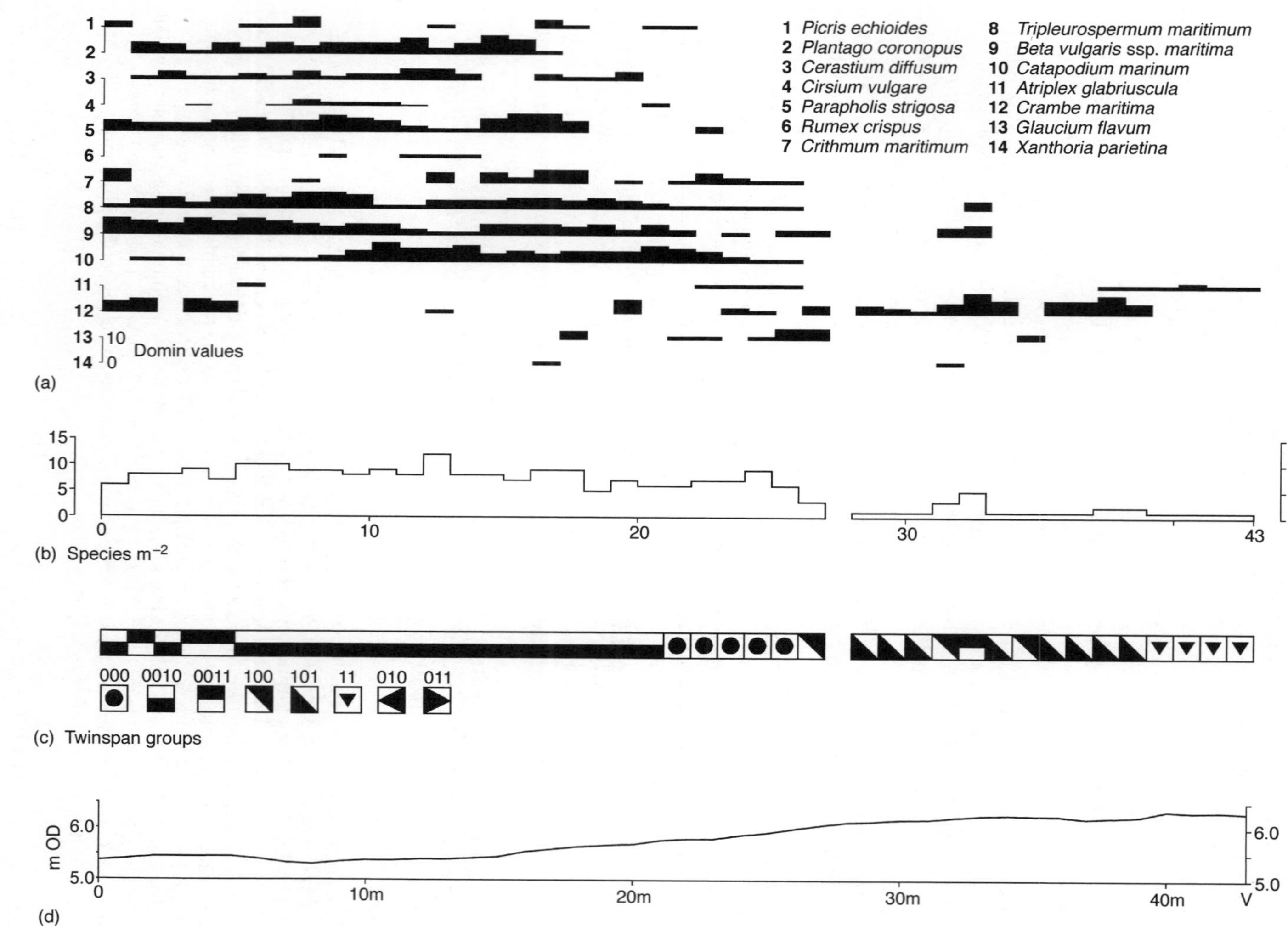
1 Picris echioides
2 Plantago coronopus
3 Cerastium diffusum
4 Cirsium vulgare
5 Parapholis strigosa
6 Rumex crispus
7 Crithmum maritimum
8 Tripleurospermum maritimum
9 Beta vulgaris ssp. maritima
10 Catapodium marinum
11 Atriplex glabriuscula
12 Crambe maritima
13 Glaucium flavum
14 Xanthoria parietina
10
0
Domin values
(a)
15
10
5
0
0
10
20
30
43
(b) Species m^{-2}
000
0010
0011
100
101
11
010
011
(c) Twinspan groups
6.0
5.0
m OD
0
10m
20m
30m
40m
V
(d)

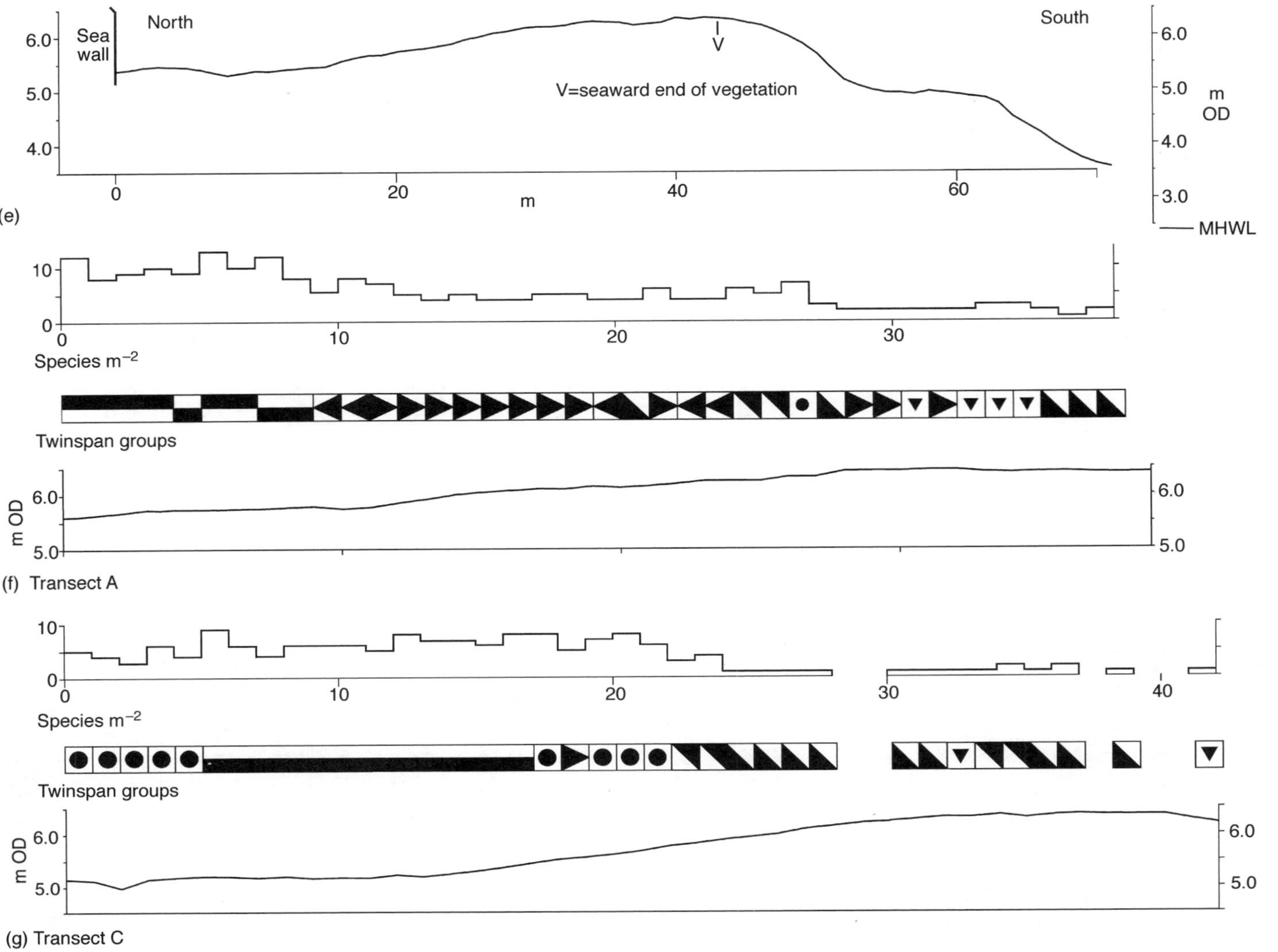

North
South
Sea wall
V
V=seaward end of vegetation
m
m OD
MHWL
6.0
5.0
4.0
3.0
0
20
40
60
(e)
10
0
30
Species m^{-2}
Twinspan groups
(f) Transect A
(g) Transect C

Figure 8.2 Black Rock Beach, Brighton, East Sussex, UK, May 1995. For **Transect B** the diagram shows: (a) Domin values for 14 of the more important of the 25 species present; (b) species density per m^2; (c) Twinspan groups to which the stands belong (based on analysis of the 34 species present in 117 stands in three transects parallel to, and west of, the western breakwater of Brighton Marina); (d) the topographic profile; and (e) an extended profile. The sparsely vegetated region nearest to the sea is higher than the area of stable vegetation adjoining the sea wall, where species density and plant cover are much greater. In the landward part of the beach the pebbles are frequently admixed with coarse sand and gravel, and the species present include ruderals. *Honckenya peploides*, absent from Transect B, formed large patches elsewhere, especially along Transect A. Species densities, Twinspan groupings and topographic profiles for **Transects A and C** are shown in (f) and (g). The eastern margins of Transects A, B and C were, respectively, 5.0, 10.85 and 23.2 m west of the western breakwater. The Twinspan analysis indicates a strong environmental gradient. Only the therophyte *Atriplex glabriuscula* was present at the seaward margin; it re-establishes from seed every year and grows especially well where seaweeds have been deposited along the strandline. Although this plant grows above the level of normal tides (Mean High Water Level here is +2.45 m OD and MLWL is –2.15 m OD), the pebbles here are disturbed by major storms. *Crambe maritima,* the largest species established on the beach, has a wide distribution and occurs nearer the sea than any other perennial. A biological spectrum for this community is shown in Figure 2.1. (Data of J.R. Packham and A. Spiers.)

Among shingle species with a south-western distribution are *Geranium purpureum* (local and closely resembling *G. robertianum*), *Polygonum maritimum, Raphanus raphanistrum* ssp. *maritimus* and the typically cliff plants *Crithmum maritimum* and *Lavatera arborea.* Characteristic of western shingle deposits are *Polygonum oxyspermum, Sedum anglicum* and *Coincya monensis* ssp. *monensis*. In the north are *Mertensia maritima* (Figure 8.4(b)) and *Cochlearia scotica* (a close relative of *C. officinalis*). All round the coast of Britain are the chenopod *Atriplex glabriuscula* (but rare in places) and *Silene uniflora,* the latter often present on cliffs and occasional on dunes and in inland habitats.

Predominantly inland species with shingle ecotypes include a small form of *Arrhenatherum elatius, Cytisus scoparius, Galium aparine,* a dwarf non-straggling variant of *Geranium robertianum, Rumex crispus* (Figure 8.4(c), prostrate plants of *Solanum dulcamara,* and also *Tripleurospermum maritimum.* Annuals, especially ephemerals, are not usually a major component of the vegetation, perhaps reflecting the often unfavourable conditions for establishment.

8.2.1 INFLUENCE OF ABIOTIC FACTORS

Although the geographic range of many species is largely determined by climate, species composition on shingle features is affected very strongly by the stability of the beach, the nature of the substrate, particle size distribution, and hydrology. Scott (1963a) showed that **stability** influences species composition more than any other factor, arranging shingle into five stability classes on this basis:

1. Vegetation is absent on very unstable beaches, including many fringing beaches on high-energy, west coast shores.
2. Summer annuals such as *Galium aparine* and various oraches (*Atriplex* spp.) grow on beaches that are stable between spring and autumn.
3. Short-lived perennials, e.g. sea fern-grass (*Catapodium marinum*) and *Sedum acre* (longer-lived in other habitats), develop where beaches are stable for 3–4 years. *Suaeda vera* also grows on such beaches in the south-east.
4. Long-lived perennials are found on shingle that has been stable for 5–20 years; such species include *Crambe maritima, Rumex crispus* and *Silene uniflora.*

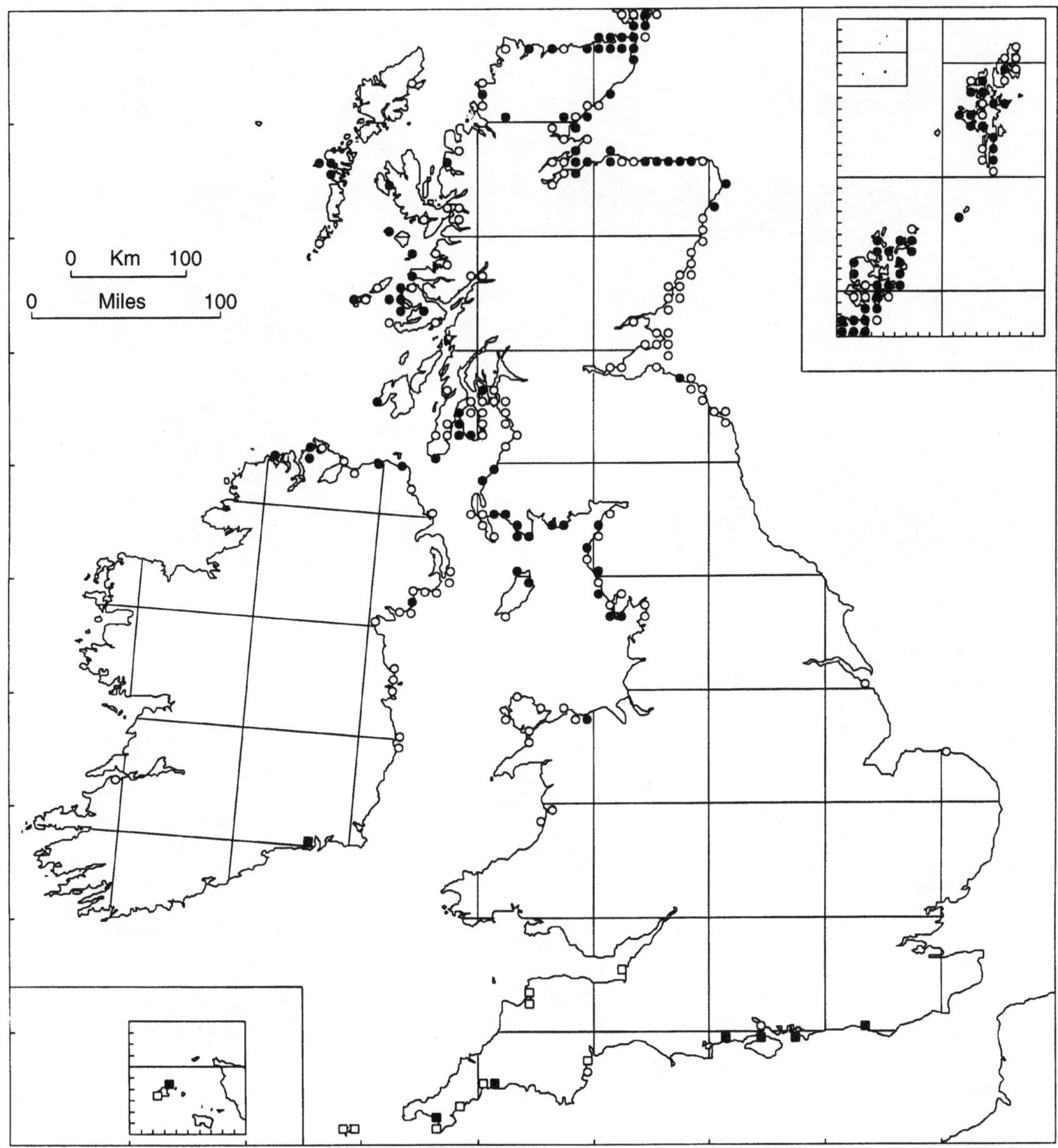

Figure 8.3 The distribution of two characteristic plants of shingle beaches. In the UK, *Mertensia maritima* (oysterplant) is retreating from the southern edge of its range: ●, 1970 onwards; ○, before 1970 (see also Stewart *et al.*, 1994, in which a series of maps indicates the former sporadic occurrence of *Mertensia* along the southern and Welsh coasts). In contrast, *Polygonum maritimum* (sea knot-grass) is likely to advance northwards with continued global warming: ■, 1970 onwards; □, before 1970. (Prepared by Mrs J.M. Croft, Biological Records Centre, Monks Wood.)

Figure 8.4 Characteristic shingle plants. (a) Sea kale (*Crambe maritima*) and rock samphire (*Crithmum maritimum*) at Black Rock Beach, Brighton, UK; (b) Oysterplant (*Mertensia maritima*) on Orkney, Scotland; (c) Curled dock (*Rumex crispus*) and minute plants of sea rocket (*Cakile maritima*) on Borth Beach, North Wales; (d) Yellow horned-poppy (*Glaucium flavum*) on Slapton Sands, South Devon, UK. (Photographs (a–c) by John R. Packham; (d) by M.C.F. Proctor.)

5. Heath or grass-heath vegetation develops on shingle that has remained stable for very long periods. *Arrhenatherum elatius, Festuca rubra, Rubus fruticosus* agg., *Cytisus scoparius,* and ling/bell heather (*Calluna/Erica cinerea*) in northern Britain, are typical of such communities.

Beach composition is a further major factor which Randall (1989) employs to divide shingle beaches into four major types, dependent on the nature of the fine particles present as a **matrix** within the shingle:

1. Beaches which consist **entirely of shingle** and lack any kind of matrix are relatively uncommon. Their vegetation is limited to encrusting lichens and a small number of pioneer angiosperms, e.g. *Lathyrus japonicus*.
2. Those beaches which consist of **shingle with a sand matrix** have the hydrological conditions of foredunes and the stability of reardunes. *Lotus corniculatus*, *Plantago lanceolata*, *Honckenya peploides* and *Sedum acre* are common here. Other species may include *Plantago coronopus*, *Plantago maritima* and *Armeria maritima*.
3. **Shingle with an organic matrix**. Rotting seaweed is nutrient-rich and its presence, subject to local conditions of drainage and salinity, may enable many species to occur. *Beta vulgaris* ssp. *maritima* and a number of *Atriplex* spp. tend to arrive as seeds in tidal debris. If drainage on the inland side is poor, *Phragmites australis* or *Iris pseudacorus* is likely to establish. In excavated lows, as at Dungeness, ranker soils with *Salix* scrub may be present.
4. On beaches where the **shingle has a silt/clay matrix** the vegetation is ecologically related to that of salt marshes, although drainage is freer. *Atriplex portulacoides*, *Glaux maritima* and *Seriphidium maritimum* (sea wormwood), all common on salt marsh levées, are prominent.

8.2.2 HYDROLOGICAL CONDITIONS AND PARTICLE SIZE

Shingle beaches have high porosity combined with low losses through evaporation (Randall, 1989); moreover, during the day upper dry shingle may form a **mulch** between dry air above the beach and damp, lower shingle layers. **Water-holding capacity** is closely related to the **particle size** of the shingle. Fuller (1987) found that fine shingle (taken as ≤10 mm diameter) retains four times more moisture than coarse shingle (>10 mm diameter) and that 15% fine shingle doubled the water-retention capacity of coarse shingle at field capacity.

The characteristically **vegetated ridges** of large shingle formations, such as Dungeness and Orfordness, have a considerable proportion of fine shingle and are formed during storms which pile up shingle that has steadily collected on the foreshore under calm conditions. Fuller (1987) postulates that, of the mixed particles involved, fine material has come from low-water, medium grades from the mid-tide zone, and coarse from the high-tide mark. Backwash in the crest of the storm ridge occurs as downward percolation so no fine shingle is lost. After the storm is over the ridge is out of reach of the ensuing lower waves which build up the shingle, consisting almost entirely of coarse material, which constitutes the **low** that is formed before another storm ridge develops.

Besides increasing moisture-retention capacity, the presence of humus also prevents seeds sinking too deeply for establishment, and germination is more successful. This is particularly important in promoting the growth of annual species (**therophytes**), which are adversely affected by instability, and also do better when there is an adequate matrix of sand or organic matter (as is present in parts of Black Rock Beach, Brighton, Figure 2.1(d)).

The water table of many shingle formations is at least 2 m beneath the surface, while the tap roots of the largest perennial plants, such as *Cytisus scoparius*, rarely penetrate to more than 1 m. Only where the ground level is within reach of the water table, as around the borrow-pit lagoons at Shingle Street and the Open Pits at Dungeness, can this water be exploited by plants on shingle. The marshes which develop under these circumstances are frequently co-dominated by *Phragmites australis*, *Epilobium hirsutum* and *Lythrum salicaria*. On smaller shingle beaches and spits **fresh water** floats as a **lens** on the denser salt water beneath, but even here borings have shown that the water table is very frequently out of the reach of the longest-rooted beach plants. With lack of appreciable capillarity in shingle there is very little upward movement of liquid water and the smaller or short-lived plants are even further from the water table. How then do they survive?

Glaucium flavum, a plant with a reported high transpiration rate (despite having leaves with a thick cuticle), has a substantial capacity to withstand drought (Scott, 1963b); like other species it may be partially dependent on dew.

8.2.3 SOURCES OF WATER AVAILABLE TO SHINGLE PLANTS

Rain is a primary source of moisture for shingle plants, many of which utilize the **pendular water** that has permeated to the lower layers of shingle or matrical sand, but which is well above the **water table**. Soil water content is strongly dependent on moisture held in the finer fraction within the shingle; without these small particles the amount of available water may be insufficient for seeds to germinate and seedlings to develop.

Two other important sources are **ordinary dew**, in which water vapour comes from the atmosphere, and **internal dew** whose origin is described below and which in summer may be the main source of water above the water table. Pronounced temperature gradients can be expected in shingle in periods of hot sunshine, with much cooler night temperatures. With high porosity, extensive diffusion of water vapour from lower to higher levels in the shingle may be expected in daytime; lower night temperatures may mean that saturation point is reached, condensation of water vapour taking place. Water would be rapidly absorbed in the effective rooting zone by plants suffering a water deficit. Replenishment of water in the upper layers from below by internal dew formation as well as ordinary dew may be critical in times of drought.

The possible importance of water derived by upward diffusion from the water table, which was indicated long ago (Olsson-Seffer, 1909; Hill and Hanley, 1914), appears often to be overlooked. Oliver (1912) was in no doubt of the ability of shingle plants to survive long periods without rain, noting that 'The water of the shingle is astonishingly copious and suffers no diminution during periods of prolonged drought'. He refers to sheep grazing on fresh *Silene uniflora* on the Chesil Bank, Dorset, in September 1911 after the prolonged drought which resulted in the vegetation of the adjoining mainland being completely parched (see also Tansley, 1949, Plate 156). Further study on the relative importance of direct rainfall, ordinary dew and internal dew formation in affecting the establishment and water relations of the vegetation of shingle beaches is desirable.

Transplant experiments at Sizewell, Suffolk, are also relevant to this question (section 11.1). Walmsley and Davy (1997c), who describe the shingle-beach habitat as 'both xeric and nutrient poor', recorded high survival rates despite the fact that the plants were watered only twice (on 4 and 29 May 1990), the temperature being high and precipitation during May very low. Rainfall between May and August was only 73% of the 5-year mean for 1989–1993. Wilting in May and June 1990 was more marked in coarse substrates; mortality was insufficient to determine statistically whether wilting led to higher mortality in the seaward sites. Growth of established seedlings of *Glaucium flavum* and of container-grown plants of several shingle species was greater in substrates dominated by coarse particles.

Though climate is a major influence on the macro-scale, the species composition of the vegetation of individual shingle deposits is largely controlled by their stability, matrix and hydrology.

8.2.4 SUCCESSIONAL RELATIONSHIPS ON SHINGLE

Randall (1989) considers that external environmental factors influence the vegetation of shingle foreshores so strongly that only physical changes will produce alterations in the species composition of the vegetation, though chance plays an important role in determining which seeds are available for germination. **Allogenic factors** – influences not involving living organisms – may largely control **vegetational change** here, although the gradual build-up of humus is also important. In **autogenic succession** plants and other organisms are the major influence on the processes of change. Dungeness is one of the few British shingle formations large enough for allogenic activity to be replaced by autogenic succession. Other examples include the development of a woody cover of *Betula/Salix/Prunus* on raised shingle or cobble beaches in west Scotland, and the varied vegetation with extensive scrub on the Kingston Shingles of the Moray Firth.

Elsewhere, the exposed maritime community of the foreshore may be replaced by less open herb-rich grasslands somewhat similar to those of stabilized dunes or cliff tops. In such places inland weed species may become abundant. Vegetation fails to develop at all in some parts of shingle beaches simply because there is no available soil material. Most 'lows' of the Dungeness shingle which – as already explained – are characterized by coarse particles, and parts of the raised beaches of the Island of Jura, Scotland, support no plant life apart from occasional lithophilous lichens.

The colonization of bare dry surfaces is of major importance on shingle, but there is also a **hydrosere succession** from open water lagoons or freshwater pools to terrestrial vegetation. **Lagoons** are semi-isolated from the sea at several places, especially Slapton Ley, south Devon, and Shingle Street, Suffolk, UK. Contact with the sea in such lagoons is usually only via seepage and the water is normally brackish to fresh.

8.2.5 SHINGLE RIDGE SUCCESSION AT DUNGENESS

The shingle vegetation of the foreland was described by Scott (1965), who emphasized the role of a prostrate form of broom (*Cytisus scoparius*) in what he considered was an '**autogenic xerosere**' leading to climax holly (*Ilex aquifolium*) woodland. The term 'xerosere' with respect to shingle vegetation, suggesting a dry habitat, however, may be rather misleading (cf. Oliver, 1912, p.99) in view of the copious water which appears to be available in at least some shingle banks. Hubbard (1970b) listed the major communities of Dungeness (section 8.4), noting especially the holly wood on Holmstone beach which is known to have been a major feature in 1539 and appears to have existed for hundreds of years before that. He also described the previous use of Dungeness by humans, the whole area serving as a military base from 1939 to 1946. Peterken and Hubbard (1972) considered the Holmstone holly wood to be of natural origin; their interpretation of the successional relationships is shown in Figure 8.5. Others regard the holly wood as planted and not of critical importance as far as plant succession is concerned.

Ferry, Barlow and Waters (1989) confirmed the role of fine shingle in promoting plant establishment. Figure 8.6 outlines, in terms of the National Vegetation Classification (NVC), their conclusions concerning the processes of succession and vegetational change which occur at Dungeness in which the **cyclic build-up of humus** is undoubtedly a key process. The MC8 community, which is strongly dominated by *Festuca rubra* and frequently contains much *Silene uniflora*, *Armeria maritima* and *Cladonia rangiformis*, is mainly found on ridges along the southern margin that are being returned to sea by erosion.

8.2.6 GENERALIZED PATTERN OF PLANT SUCCESSION ON SHINGLE SITES IN BRITAIN

The successional models discussed above appear inappropriate to British shingle as a whole, for which Figure 8.7 provides a series of pathways for autogenic succession leading to **local edaphic/climatic equilibria with vegetation** rather than a definite climax at any particular time. (This anastomosing pattern is

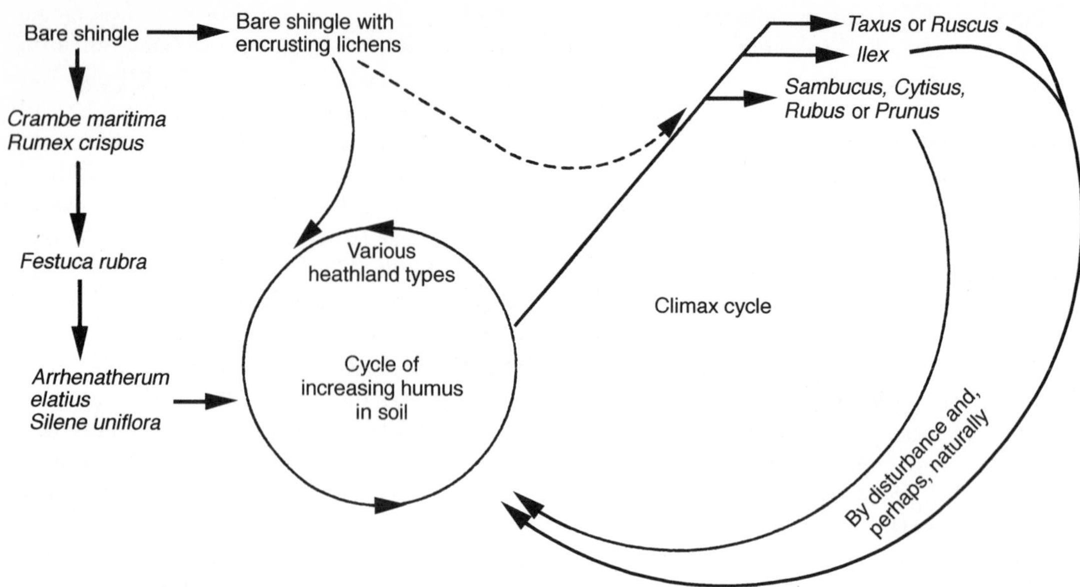

Figure 8.5 Summary of presumed successional relationships of communities on the shingle at Dungeness. (After Peterken and Hubbard, 1972; courtesy of *Journal of Ecology*.)

similar to that of Figure 8.9, which applies to New Zealand shingle.) However, it differs from this and previous models, all of which emphasized the importance of humus accumulation resulting from the presence of *Cytisus scoparius*, mesotrophic grassland or ericaceous communities, in showing two starting points. These involve pioneer maritime herbs where the maritime influence is extreme, and lichens and mosses where the initial pioneer stage is more terrestrial (Randall and Doody, 1995).

The importance of the latter starting point is emphasized by a general lack of soil development across the majority of shingle sites (average depth <5 cm) and the continuing presence of bryophytes in succeeding successional stages. Mosses such as *Dicranum scoparium* occur on bare shingle and throughout more mature communities of 'vegetation islands' at sites such as the Kingston Shingles, near Spey Bay, Moray Firth, Scotland. Under some circumstances bryophytes may negate the need for increased humic material in the soil, perhaps even acting as a mulch, reducing water loss from the shingle by evaporation. In other instances, mature grasslands seem to facilitate further vegetational development more effectively than *Cytisus scoparius*.

As previously emphasized, more research, particularly on the hydrology of vegetated and non-vegetated shingle, is urgently required. Is internal dew formation reduced when the vegetation cover is complete? Does the presence of a bryophyte layer cause a greater loss by swift evaporation of intercepted rain than is prevented by its operation as a mulch? These questions can be settled only by direct measurement.

8.3 SHINGLE VEGETATION OF A BOULDER BANK IN NEW ZEALAND

The development of coastal shingle, which on a world basis is rare but widely distributed, is dependent upon an appropriate supply of rock fragments. In Britain, many shingle deposits are formed from flints eroded from the chalk, while in South Island, New Zealand, the Tasman Intrusives weather to provide

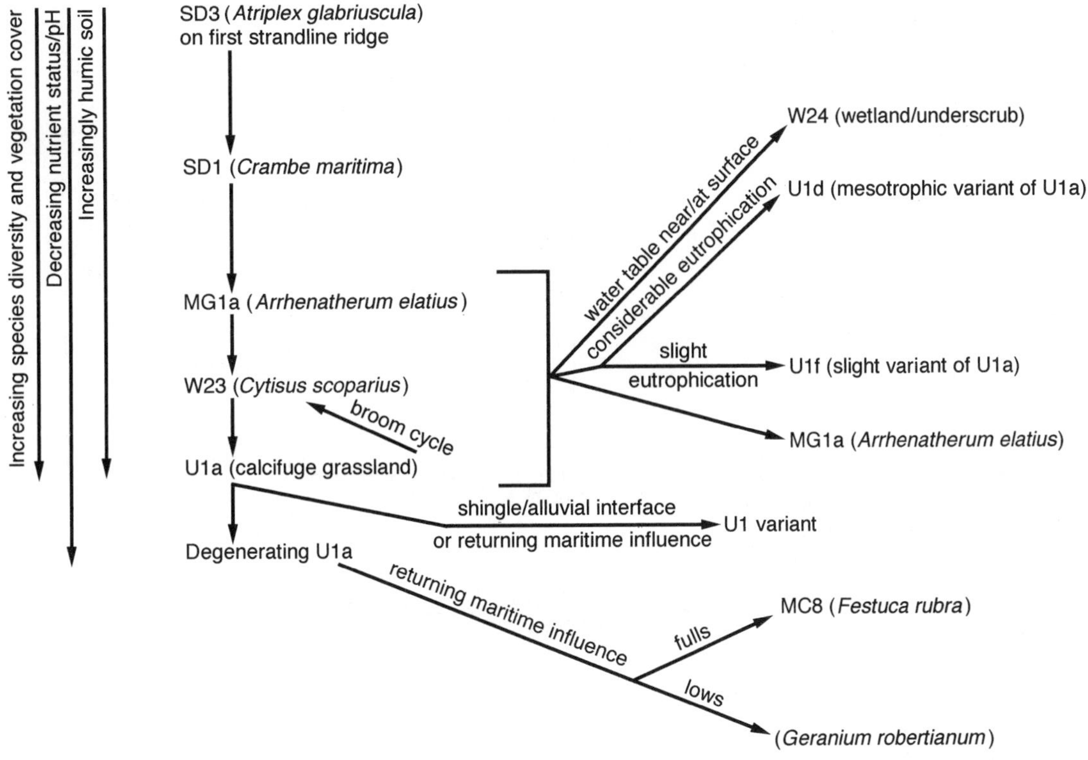

Figure 8.6 Suggested interrelationships between National Vegetation Classification (NVC) communities from both intact shingle ridges and areas of disturbed and damp shingle at Denge Beach, Dungeness. MG1a, mesotrophic grassland dominated by *Arrhenatherum elatius*. W23, *Ulex europaeus–Rubus fruticosus* agg. underscrub in which *Cytisus scoparius* replaces *U. europaeus*. U1a, which replaces W23 on the inland side, calcifuge grassland dominated by *Festuca tenuifolia* and *Rumex acetosella* in mosaic with patches of *Dicranum scoparium* and *Cladonia* spp. W24, *Rubus fruticosus–Holcus lanatus* underscrub. (After Ferry, Barlow and Waters, 1989; courtesy of Linnean Society of London and by permission of the publisher, Academic Press Limited, London).

fragments many of which are within the shingle range (2–200 mm). Randall (1992) describes the shingle vegetation of the Boulder Bank enclosing the Haven at Nelson in Tasman Bay, South Island, which runs south-west for over 13 km and for most of its length is 1.5 km offshore. This account of a southern hemisphere shingle system describes how introduced plants and animals, together with direct human interference since 1841, have modified the native vegetation and the form of the bar.

The bank consists of rock fragments swept by longshore drift from the coastal exposure of Tasman Intrusives at Mackays Bluff. As the boulders are swept towards the south-west they become increasingly smaller and more rounded. For most of its length the bar at high tide is 50–55 m wide. It overlies sand and silt and reaches a height of 6 m above High Water Mark (HWM) in the north and 2.5 m at the southern end near Haulashore Island where it is breached by the artificial entrance known as 'The Cut', made in 1906. Sheep and goats introduced at Mackays Bluff since colonization by Europeans largely destroyed the natural vegetation, leading to increased erosion and

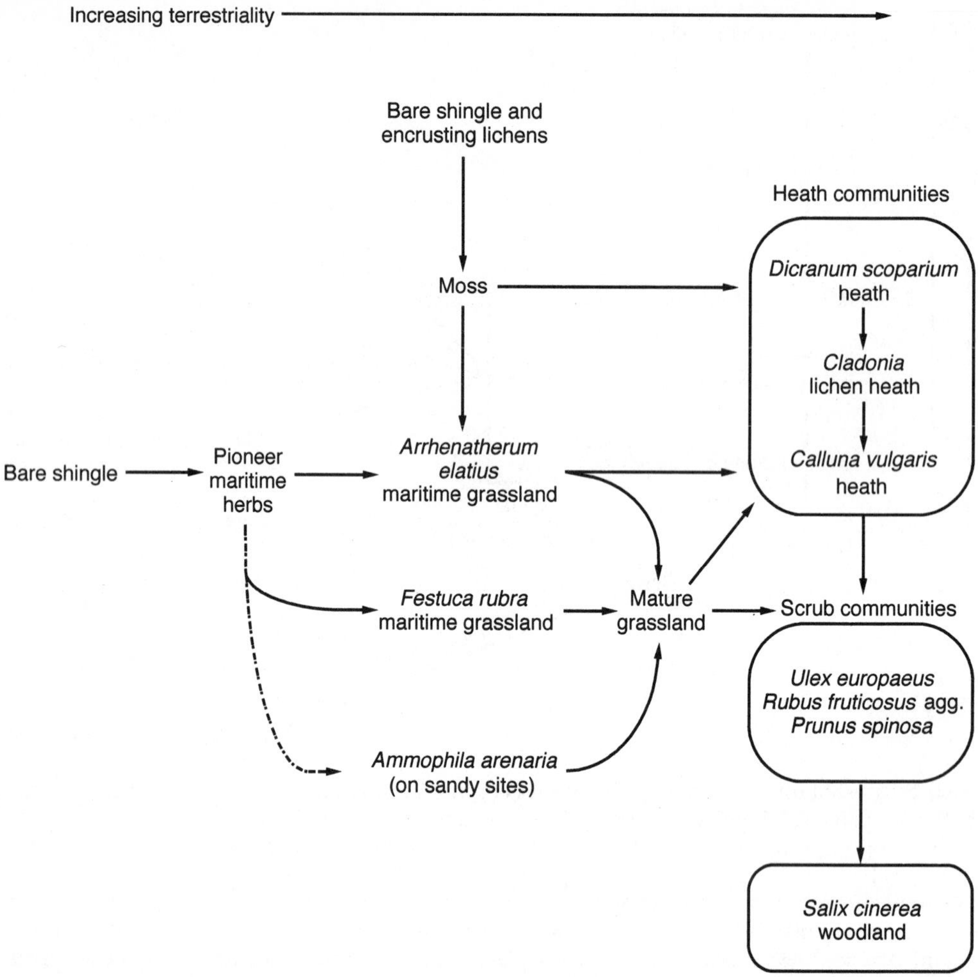

Figure 8.7 Proposed sequence of vegetation succession on coastal shingle sites in Britain. (After Randall and Doody, 1995.)

enhancing the growth rate of the bank, though the use of material from the bank as road metal partially offsets this trend.

Whiri (*Asplenium flabellifolium*), a salt-tolerant fern, often occurs low on the seaward slope, but above the drift-line the bank forms an asymmetric double ridge in section (Figure 8.8) and has a **salt marsh** with *Spergularia marina*, *Suaeda nova-zelandiae* and *Salicornia australis* on the inner bank which faces the Haven. The crest of the inner ridge is occupied by the native shrub pohuehue (*Muehlenbeckia complexa*), on the seaward side of which is an area badly scarred by ORVs. Over 1000 southern black-backed gulls breed on the bank, being concentrated in the hollow between the two ridges where their guano, and that of a community of red-billed gulls, greatly enriches the soil. This area contains numerous adventive species including *Medicago arabica*, *Trifolium repens* and *Geranium molle*, as well as the nitrophilous weeds *Stellaria media*, *Coronopus didymus*, *Veronica persica*, *Anagallis arvensis*, *Poa annua*, *Rumex crispus*, *Senecio sylvatica* and *Urtica urens*.

When surveyed in August–September 1981, a total of 60 exotic and 25 native grasses, forbs

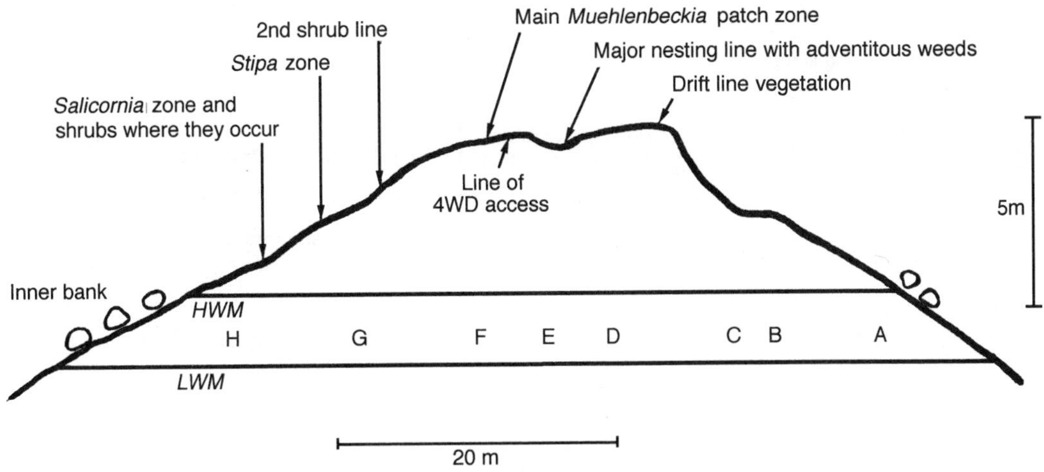

Figure 8.8 Typical cross-section of Nelson Boulder Bank, New Zealand. A, large boulders 45 cm diameter; B, mixed shingle and boulders, 10–30 cm; C, mainly 30 cm with some 5–10 cm; D, mainly 5–8 cm shingle with some to 15 cm; E, mainly 30 cm with some 15 cm; F, 5 cm pebbles; G, mix of 5–30 cm shingle and boulders; H, gravel 2–5 cm. (After Randall, 1992.)

and shrubs were recorded, together with four indigenous ferns, and another four native species known to have been planted. Allan (1936) considered the success in New Zealand of plants from Europe was largely related to the presence of the introduced fauna, and that the removal of the alien fauna would lead to a resurgence of the indigenous plants. Rabbits ate palatable native forbs and grasses and ring-barked the shrubs on the Bar until controlled by ferrets in the 1970s.

Nelson Boulder Bank, which Randall (1992) compares with the only other extensive area of coastal shingle in New Zealand, that of **Kaitorete Spit** which encloses Lake Ellesmere to the south of Christchurch, South Island, is an essentially open community. The very patchy distribution of the introduced species may be ascribed largely to chance, with seed availability being a major factor. Native species have a more uniform distribution and there is a successional sequence related to improved environmental conditions on the leeward side of the Bar.

Randall (1992) presents a successional scheme with local variations (Figure 8.9) as being applicable to extensive New Zealand shingle deposits. He concludes that evidence from New Zealand coastal shingle vegetation shows increasing **stability and the presence of organic matter** are the only requirements for succession to occur, as maritime influences decline. Once established in the sequence, shrub hummocks greatly accelerate humus production, emphasizing the mosaic pattern characteristic of stabilized shingle. **Low shrub vegetation** appears to be the climax for such areas; dense woodland does not occur in this habitat. These conclusions are similar to those reached from studies of the *Betula*/*Salix*/*Prunus* woody cover on the raised shingle or cobble beaches of western Scotland, and the *Calluna vulgaris* heathland of the Kingston Shingles on the Moray Firth, North Scotland, two areas which – unusually for shingle (but like Dungeness) – are large enough for autogenic processes to play a major role in succession.

8.4 TWO OUTSTANDING BRITISH SHINGLE BEACHES

Many **cuspate forelands**, e.g. Cape Hatteras, North Carolina, (Figure 1.7(b), develop where two curved and active spits meet. A different

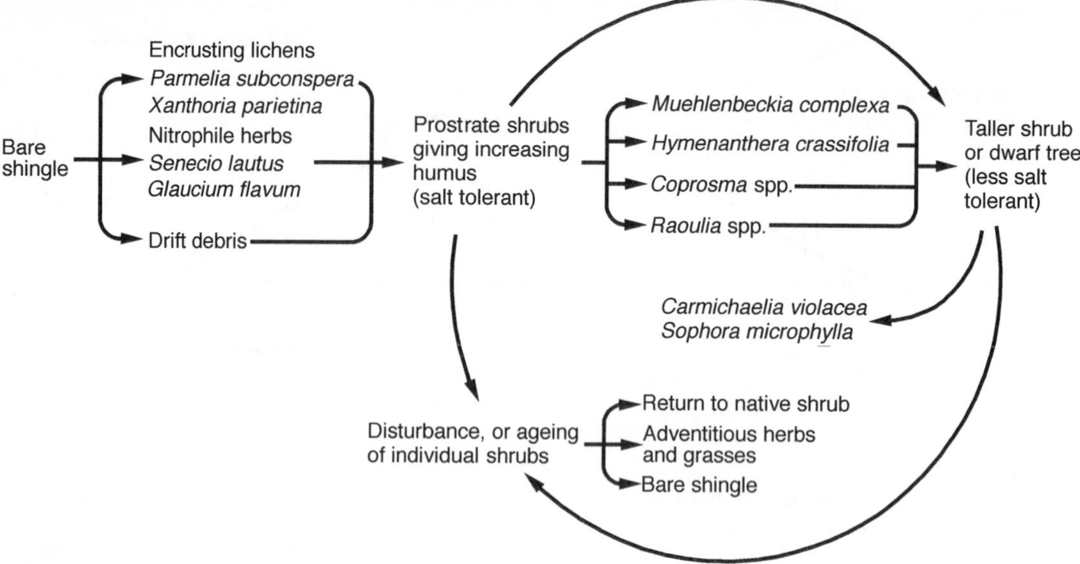

Figure 8.9 Anastomosing sequences in New Zealand shingle vegetation. (From Randall, 1992; courtesy *New Zealand Journal of Geography*.)

mechanism is involved at Dungeness, which has two seaward sides of which the southern is retreating before the attack of the waves, while the eastern has been built out so effectively that the foreland has advanced 2.4 km since the beginning of the 13th century. Spits linked to the mainland tend to continue parallel to its coast, or to curve inwards towards the shore if the water shallows in that direction. It is thus unusual for wave-built structures to move away from the general trend of the coastline: when they do so it is often in response to a shallowing of the water seawards – perhaps towards an island or shoal (Holmes, 1965). In some instances, however, the change in form is at least in part a reaction to the average direction of the ridge-building storm waves, which is itself dependent on the directions of the dominant winds and of the maximum fetch. At Dungeness the proximity of France prevents waves approaching from the south-east or south. This huge cuspate foreland was first shown by Lewis (1932) to have evolved, via a series of apposition beaches that gradually changed direction, from a simple spit that grew from the cliffs near Fairlight east of Hastings, and which existed in Roman times.

8.4.1 ORFORDNESS AND DUNGENESS

Orfordness and Dungeness are fine examples of vegetated shingle beaches. Both consist almost entirely of flint pebbles, retain features 2000 or more years old, are in the drier south-eastern region of Britain, and have been colonized by animals and plants adapted to the distinctive conditions of the shingle. The two structures may be considered to complement each other (Fuller, 1989). Orfordness (385 ha) is a relatively young, highly dynamic **coastal shingle spit** with a well-developed flora of pioneer and early successional species. In contrast, Dungeness has been established for a much longer period, has been frequently disturbed by humans, and is six times larger (2179 ha), being the most important area of vegetated shingle in Britain (Doody, 1989a). Moreover, the area away from the coast has been subject to autogenic succession and is very stable, supporting shingle-heathland communities with scrub development.

Figure 8.10 compares Orfordness as it is now with Dungeness as it is believed to have been 700–800 years ago, when it ran as a shingle spit across a bay which occupied the present site of the Romney Marshes and terminated somewhere between New Romney and Hythe. In the 12th and 13th centuries, Dungeness had a form very similar to that of Orfordness today; the shingle spit prevented the Rivers Rother, Tillingham and Brede from flowing directly into the sea at Rye. Instead they probably joined and flowed 15–24 km to the north-east behind the spit. In Figure 8.10(b) the apparent fragility of the shingle barrier at Aldeburgh, where it deflects the River Alde, matches that which formerly existed at Rye.

8.4.2 PLANT COMMUNITIES OF DUNGENESS

The vegetation of Dungeness varies with the **substrate, distance from the sea and availability of water**; salt marshes, freshwater marshes, dunes and shingle beaches are all well represented. The plant communities have been greatly influenced by long continued grazing, surface disturbance, the introduction of various materials and the excavation of shingle. Denge Beach has the greatest diversity of habitats and includes the largest area of English shingle bearing acidic heathland, the series of freshwater hydroseres known as the Open Pits, a series of man-made habitats, and a coastal plant zonation which changes in relation to accretion or erosion along the shore line (Hubbard, 1970b). The inland areas of Holmstone Beach were described by Peterken and Hubbard (1972) as being occupied mainly by an ancient holly wood capable – at least until recently – of natural regeneration. *Ilex aquifolium* occurs as discrete wind-shaped thickets intermixed with *Sambucus nigra* and *Ulex europaeus*, the thickets being surrounded by bare shingle. Individual holly bushes are of coppice-like structure with flat stools up to 1 m in diameter and their horizontal branches frequently root.

8.4.3 LICHEN COMMUNITIES AND LICHENICOLOUS MITES AT DUNGENESS

Some 150 species of lichen occur at Dungeness, among the most notable being *Cladonia mitis, Lecanora helicopis, Lecidea insidiosa, Parmelia soredians, Rinodina aspersa* and *Usnea glabrata*. Three main communities are present (Laundon, 1989), the first consisting of **crustose lichens colonizing bare pebbles**. The dominant species adjoining the drift line is *Rhizocarpon richardii* which blackens the pebbles of a zone 5 km long and up to 300 m wide. Away from the drift line this species is replaced by other encrusting forms including *Buellia aethalea, Rhizocarpon obscuratum, Aspicilia caesiocinerea* and *Parmelia glabratula*, together with a number of local and less common species.

The second community is that of the *Cladonia* turf, which forms the **lichen heath** and plays a major role in the vegetation cycle involving *Cytisus scoparius* (Figures 8.5 and 8.6). The most common species near the sea, where the shingle is less acid, is *Cladonia rangiformis*, a whitish-grey species which is associated with *C. foliacea* (yellow) and *C. cervicornis* (grey) in a community in which *Sedum anglicum* is prominent. Elsewhere the vegetation is largely controlled by the depth of the water table. *Cladonia gracilis*, though constantly present in the lichen heath, has low cover values. *Cladonia mitis*, an Arctic lichen, is a post-glacial relic found in England only at Dungeness.

The third group comprises **epiphytes on woody scrub**, including gorse, broom, holly, and the prostrate *Prunus spinosa* which occurs in transition zones from shingle to alluvial soils on the landward side of all the beaches. In the 19th century, a number of oceanic lichens very sensitive to air pollution flourished here. Forms such as *Lobaria pulmonaria* and *Lobaria scrobiculata* became extinct proba-

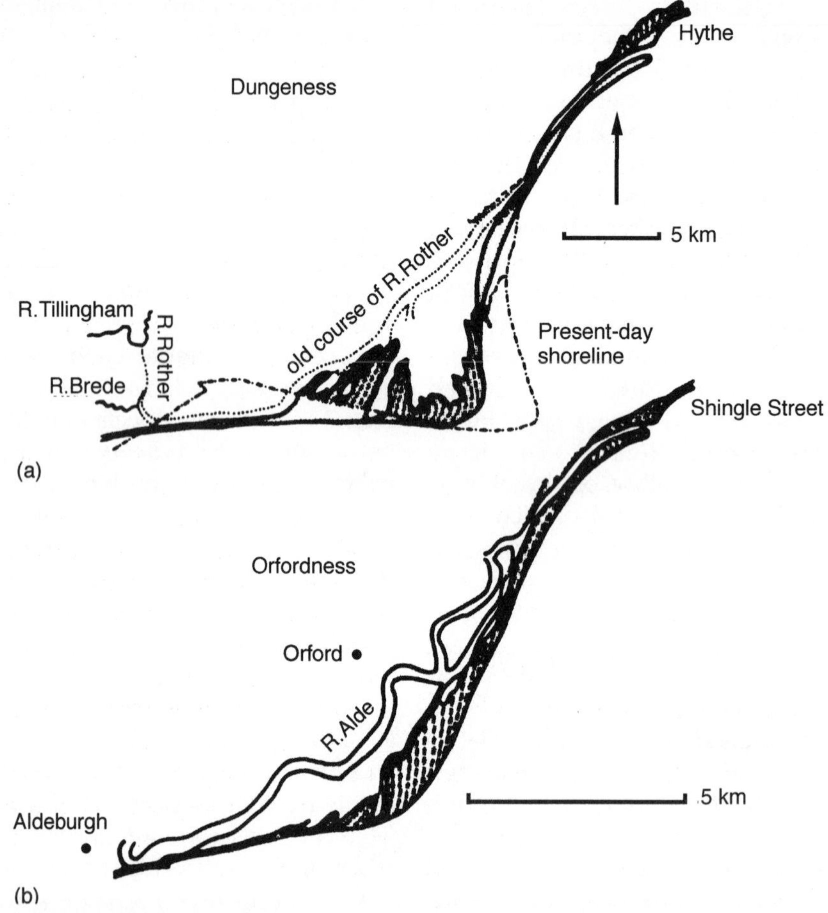

Figure 8.10 (a) The probable form of Dungeness, before the breach by the River Rother near Winchelsea, in the 13th century. The exact course of the River Rother and the length and breadth of the Dungeness spit are unknown. They would have varied throughout the life of the spit but the existence of counter-spits from Hythe indicates that the Dungeness spit once extended to this point. (b) Present-day outline of Orfordness seen in mirror-image and orientated to assist comparison. (After Fuller, 1989; courtesy of Linnean Society of London and by permission of the publisher, Academic Press Limited, London).

bly early in this century because of the rise in background air pollution indicated by the spread of *Lecanora conizaeoides* on to Denge Beach. The lichen flora of the old willow (*Salix cinerea*) scrub of the damp hollows has been described by Ferry and Waters (1984).

Also present are lichens growing on **artificial substrates** such as concrete (common species of *Caloplaca, Candelariella, Lecanora, Physcia s.l.* and *Xanthoria*), wooden posts (*Cyphelium inquinans, Lecanora conizaeoides, Lecanora varia, Lecidea insidiosa*), and railway clinker.

The population dynamics and feeding preferences of mites have been examined within soil/litter habitats of many geographical regions. This interest arises from the role that mites play in decomposition, energy flow and nutrient turnover (Packham *et al.*, 1992, pp. 259–262). Energy flow in such systems attributable to mites is small, but by physically comminuting the litter they greatly increase the surface area which can be attacked by other animals, fungi and bacteria. At Dungeness the population dynamics of lichenicolous mites has been investigated on seven late successional species – *Cladonia portentosa, C. ciliata, C. foliacea, C. gracilis, C. rangiformis, Coelocaulon aculeatum* and *Hypogymnia physodes* at Denge Beach. Barlow and Ferry (1989) list 30 species of mites for the Dungeness shingle (one astigmatid, 15 cryptostigmatids, 11 mesostigmatids and three prostigmatids). Of these, *Carabodes minusculus, Oribatula tibialis, Trichoribates incisellus, Trhypochthonius tectorum, Camisia spinifer* and *Camisia horridus* are the most common of the mycophagous mites. Lichens are consumed by many species of animals but the Acari are probably, in terms of numbers and consistency of occurrence, the most important group.

During 1981–1982, bimodal seasonal peaks in mite numbers occurred; low numbers in the summer were associated with high temperature. This may not be a direct effect, but related to the hydration state of the lichen thalli. After a September peak, the winter decline in mite numbers may well be directly related to the low temperatures.

8.4.4 OTHER DUNGENESS INVERTEBRATES

Dungeness has more species of bumble bees (13 out of 16 British *Bombus* species) than are known in any other locality in the British Isles. Williams (1989) compared a 72-ha survey area from Dungeness with a 50-ha sample area with only seven species of bumble bee near Shoreham, Kent, UK, in terms of three theories concerning the number of species occurring in a particular locality. His results indicated that the high diversity of bumble bees at Dungeness is a consequence of the high density of nectar-rich flowers, especially wood sage (*Teucrium scorodonia*) and viper's bugloss (*Echium vulgare*), on which bumble bees can feed. *Bombus* spp. forage most successfully when the bees have learned to visit flowers whose corolla depths are similar to their own proboscis lengths, and when the distances between such flowers are relatively short. There was no correlation between diversity of bumble bees and diversity of food plants at the two sites, nor could bee diversity be accounted for in terms of particular food plant species.

The populations of the medicinal leech (*Hirudo medicinalis*) in the Lade Pit at Dungeness are of international importance to the conservation of this species (Wilkin, 1989). This animal is now protected from being killed, injured, taken, or sold under the Wildlife and Countryside Act 1981, which also makes intentional damage or destruction of the habitat of such a species an offence. Though this blood-sucking ectoparasite lacks aesthetic appeal, it can be used to reduce bruising and swelling after surgery and produces a range of important biochemicals, including anticoagulants, and is now farmed commercially at Swansea.

Studies of comparative mortality in the brown-tail moth (*Euproctis chrysorrhoea*), whose minute larval hairs cause intense skin irritation to humans, have also been made at Dungeness. Sterling and Speight (1989) found the Dungeness population to be infected by *Parasarcophaga uliginosa*, a dipterous parasitoid not previously found in Britain or indeed recorded as a parasitoid on the browntail in continental Europe.

8.4.5 ENVIRONMENTAL DEGRADATION AT DUNGENESS AND ORFORDNESS

Dungeness today is still a magnificent sight from the sea or the air, but has over the centuries been much altered by the hands of humans. The road along its eastern shore, with its associated ribbon development running south to the Pilot Inn, reduces the natural grandeur of the system, but is perhaps no more damaging than the developments described by Laundon (1989), and by Findon (1989). These began with the compulsory purchase in 1958 of the area now occupied by

Dungeness nuclear power stations A and B, and the almost simultaneous development of large-scale gravel winning. The most controversial of these developments was the digging of the large Burrowes pit within the RSPB reserve next to the Open Pits. The RSPB received a substantial income from the sale of gravel royalties, and now has the use of the flooded pit, which has been subjected to 'creative management', making it suitable for a number of bird species which benefit from the islands, graded shorelines and other features.

Though the RSPB reserve is undoubtedly valuable, and other gravel pits have also been excavated nearby, conflict between two conservation considerations, one much more powerful than the other, set in train the reduction of the former magnificent lichen heath to a series of isolated pockets. In contrast, the present RSPB policy of controlling predation by the increased fox and carrion crow populations (Hill and Makepeace, 1989) in order to re-establish the once very large breeding seabird colony seems entirely justified. The gravel extraction pits, railway, roads, housing, airport, military use and the nuclear power station have all modified a complex system renowned for its semi-vegetated ridges; future policy should protect what remains of the basic natural pattern of the Dungeness foreland.

In contrast, Orfordness has few buildings and no nuclear power station. Fortunately, Steers (1926) described it before much damage had been done to the three main groups of shingle ridges north of the present lighthouse, beyond which the main spit swings sharply to the southwest. By 1993 English Nature was responsible for the south-western area, where much of the vegetation and many of the dune ridges were still intact. The northern area, whose ownership has recently passed to the National Trust, presents numerous management problems including the control of ORVs, which damage shingle ridges and their vegetation, travelling southwards down the spit from Aldeburgh and which it is intended to exclude by physical barriers. Remodelling of the natural seaward shingle barrier and the extraction of huge amounts of shingle and gravel have considerably obscured the original ridge patterns. The National Trust area is still littered with the remains of a military airfield, a battle area, buildings used by the Atomic Weapons Research Establishment and the extensive 'Cobra Mist' early warning site (Bacon and Bacon, 1992). There is also the problem of returning abandoned pasture and arable land to a more natural vegetation which it is intended to graze.

8.4.6 A SANDY CUSPATE FORELAND IN THE BALTIC

The Darss peninsula, on the Baltic coast between Rostock and Rugen, has considerable similarities to Dungeness in form, scale and geomorphology, but not in terms of its ecology and in the detail of developmental processes involved in its formation (Waters, 1989). The Darss foreland is considered to have developed by transport of coastal material, principally from west to east, over a century or so. Like Dungeness, it possesses a set of ridges that form a pattern, appearing to represent the positions of former shorelines. The parent material is almost entirely sand, carried by water to the foreshore and subsequently transported and moulded into a 'ridge and low' pattern by the wind. The coastal vegetation is largely dune grasses, while most of the foreland is covered by pinewood which regenerates freely. *Sphagnum* dominates some areas between old dunes; in other low-lying areas *Alnus* carr has a field layer rich in ferns and sedges.

8.5 SPITS, BARS, BARRIER ISLANDS, LAGOONS AND SALINAS

Spits, bars and barrier islands (introduced in section 1.3) are frequently associated with lagoons and salinas. Wetland ecosystems of great importance are discussed here after a description of two major sand and shingle structures of the Norfolk coast.

8.5.1 BLAKENEY POINT AND SCOLT HEAD ISLAND, UK

The 40-km stretch of North Norfolk coast from Hunstanton to Weybourne consists of an extensive series of sandflats, shingle ridges and salt marshes whose evolution under the action of sea, wind, climatic change and the resident plant and animal communities has also been greatly influenced by humans. An informative background to the area is provided by Jefferies (1976), who emphasizes that the coastal system must be considered as a unit if the functioning of its individual components, including Blakeney Point and Scolt Head Island (Figures 8.11 and 1.2(d)), is to be understood. The action of **wind** in transporting sand from the sand flats exposed at low tide is obvious; its influence on the transport of sand by wave movement is also important. Many waves approach the North Norfolk coast obliquely from the north-east; their action moves stones and fine material along the shore in the process of **beach drifting** which here results in a predominantly westerly movement. Steers (1960) observed that at Scolt Head Island there is an eastward flow of the ebb current for about 3 hours after high water. After a short interval the flood tide then runs west until about 2 to 2.5 hours before the next high water. The westward deflection of the harbour channel from Brancaster in the west, and the easterly turn of that from Overy Staithe in the east, is caused by sand deposited at both ends of Scolt Head Island as a result of these **tidal currents**.

Blakeney Point, a spit joined to the land at its eastern end, contains a far higher proportion of shingle than any other feature on this coast (Steers, 1969). The eastern region is a simple, high, fairly broad shingle ridge facing relatively deep water and is reminiscent of such features as Orfordness, Chesil Beach and Slapton. To the west, about a mile beyond Cley, arise the first laterals which are short shingle ridges (the Marrams) whose extremities usually show a marked eastward bend and which are interspersed by salt marshes. The extent of sand exposed at low tide is much greater here; this western region clearly evolved differently from the eastern end, is typical of the shallow water areas of the Norfolk coast, and terminates in some sweeping ridges of which the most westerly runs from the Headland to Far Point.

Scolt Head Island has been studied by numerous specialists over a very long period; the account edited by Steers (1960) represents one of the most thorough treatments of coastline anywhere in the world. The essential skeleton of this feature is formed by the sand and shingle ridges around which the dunes and salt marshes have developed. Though the major east–west ridge is liable to be pushed landward, the island has generally grown towards the west as a result of beach-drifting, the western point of the major ridge being deflected from time to time, thus forming the present series of laterals. Much development of Scolt Head Island has been the result of slow continuous processes, but sudden catastrophic storms have also brought about significant physiographic change. The storm of 12 February 1938 breached the seaward hills and a large fan of shingle was pushed over part of Hut Marsh. The great storm of 31 January 1953 spread a huge shingle and sand flat over the marshes behind another breach originally opened at the western end of the Norton Hills in 1938. This breach was reclosed by August 1953 as a result of sand accumulation along three lines of wire and brushwood fencing erected in the spring of 1953.

A concise account of Blakeney Point and Scolt Head Island (Allison and Morley, 1989) indicates the great diversity of the animals and plants present. At Blakeney, *Spartina anglica* spread very rapidly after 1945, growing on even the sloppiest and most mobile mud where annual species of *Salicornia* did not readily establish. *Spartina* meadows began to develop and for a time it looked as if the character of the salt marshes would be permanently changed. The spread of *Spartina* has now

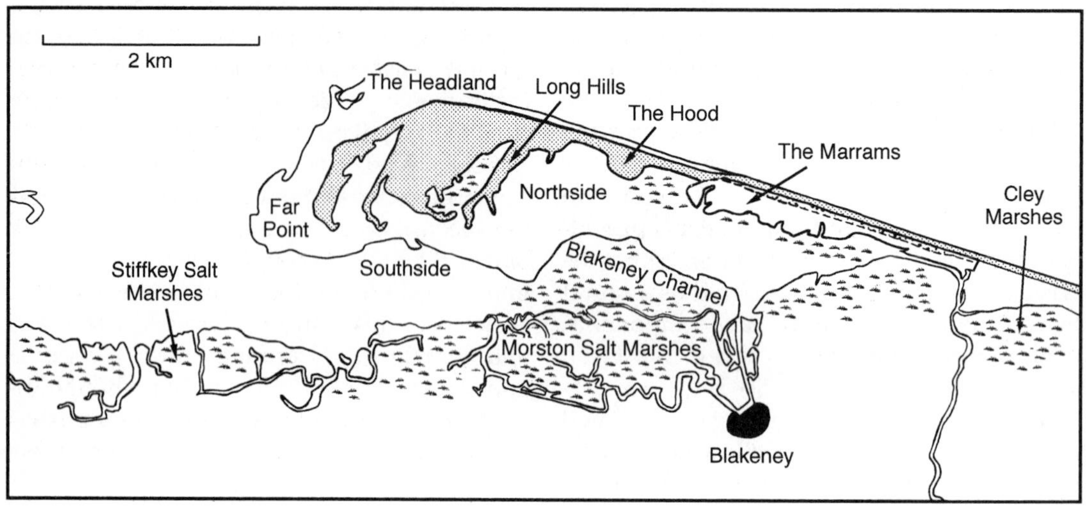

Figure 8.11 Blakeney Point, Norfolk, UK. The coastal area shown here, of great ecological, ornithological and geomorphological importance, is largely owned by the National Trust and by the Norfolk Naturalists' Trust, established in 1926.

largely halted and in many places it is being replaced by *Salicornia* and *Aster tripolium*, which grow well on areas of marsh which were raised and made firmer as a result of accretion induced by *Spartina*. Other changes include the outbreak of canine distemper which in 1988 ravaged the 700 common seals and about a dozen grey seals found along this coast in 1987; subsequent recovery took some years. The bird populations of this area are most important; their annual rhythms are discussed in section 2.8.

8.5.2 LAGOONS

Coastal lagoons are shallow, often almost tideless, pond or lake-like bodies of coastal saline or brackish water, which although partially isolated from the adjacent sea by a sedimentary barrier, receive an influx of water from the sea. They occupy 13% of the world's coastline, and are most characteristic of **microtidal regions**, in which the tidal range does not exceed 2 m. (Large tidal ranges lead to water movements potentially capable of damaging sedimentary barriers and also to extensive drainage during low tide.) Four main types of coastal lagoon may be recognized (Barnes, 1994a):

1. **Estuarine lagoons** formed when coastal barriers move onshore and partially block existing drowned river valleys. Their long axis is usually perpendicular to the coastline, and the river or stream may be completely blocked by the sedimentary barrier. Lagoonal status continues if seawater can enter by overtopping the barrier during high water of spring tides. Many former estuarine lagoons have become completely

freshwater habitats because landward movement and raising of the barrier now prevent entry of seawater.

2. **Bahira lagoons**, such as Poole Harbour, Dorset, are pre-existing, partially land-locked coastal embayments which were drowned by postglacial rise in sea level only to have their mouths later almost completely blocked by the development of sedimentary barriers.
3. Typical **coastal lagoons** have become isolated from the parent sea by the development of a spit, as in the Vistula and Kursk lagoons of the Baltic coast (Figure 1.3), or by one or more offshore or longshore barrier islands, as in the majority of the world's lagoons, including that of Venice. More rarely, lagoons are enclosed by multiple tombolos connecting an offshore island to the mainland. In all these instances seawater enters or leaves the lagoon through natural or man-made channels, such as Small Mouth, the opening of the Fleet, Dorset, the largest non-bahiral British lagoon.
4. **Percolation lagoons** develop where low-lying land occurs behind longshore barriers through which seawater can percolate. Reclaimed former salt marsh behind longshore shingle ridges frequently receives an influx of water from the barrier water table. This water, which collects in depressions such as the remnants of former salt marsh creeks, consists of seawater which soaks into the barrier at high tide but is diluted by rainwater falling on the barrier.

Despite their diverse form and origin, all coastal lagoons are partially or completely enclosed bodies of largely non-tidal saline or brackish water. They usually support meadows of submerged macrophytes such as *Ruppia* or *Zostera* which root in the soft sedimentary floor, typically possess one or more isolating sand or shingle barriers, and are often fringed by salt marshes or reed beds (*Phragmites*).

Tidal range is of great importance in determining the amount of lagoonal habitat. Microtidal Italy, for example, possesses over 150 000 ha of typical coast lagoon habitat. There are also substantial lagoon areas in southern France (Languedoc) where 'étangs' of varying salinity support large populations of brine shrimp (*Artemia salina*) and are visited by flamingos. Macrotidal Britain has only 600 ha of non-bahiral lagoon habitat and the Fleet accounts for most of this. Most British lagoons are of the estuarine or percolation types (Barnes, 1994a), but despite their small total area they are of great importance in conservation terms (section 11.4).

Coastal lagoons are major feeding and over-wintering areas for waders, wildfowl, grebes and other marshland birds. Lagoonal invertebrate and fish faunas contain three major elements: (i) species which are essentially freshwater; (ii) those which are essentially marine/estuarine; and (iii) specialist lagoonal (**paralic**) animals. The last are found also in Eurasian inland seas as well as in brackish water ponds and drainage ditches, particularly in reclaimed coastal areas of Britain and the Netherlands. They are clearly related to the marine/estuarine group as shown by such species pairs as the lagoon and edible cockles *Cerastoderma glaucum* and *C. edule*. Barnes (1994a,b) provides examples of all three groups and an identification guide to the brackish water fauna of north-western Europe.

8.5.3 THE WADDENSEE

The **Waddensee** – a tidal wetland shared by the Netherlands, Germany and Denmark – includes the whole coastal area from Den Helder in the west to the Skallingen peninsula in Denmark. It is a strip of tidal flats, sandbanks and barrier islands with an area of almost 8000 km^2 (Figure 1.7(a)), making it the largest region of tidal wetlands in Europe and one of the most important habitats for shore birds and coastal waterfowl in the world. It is the chief staging area for millions of birds using the East Atlantic Flyway, and is the moulting place for almost all the shelduck in north-west Europe; over 100 000 are present there every year, being unable to fly for about 4 weeks while they renew their wing and tail feathers. The Waddensee is under pressure from embankments and land claim, over-exploitation of renewable resources such as mussels (*Mytilus edulis*) and cockles (*Cerastoderma edule*), disturbance of wildlife through recreation, hunting, and military activities, and through input of pollutants and of nutrients which cause eutrophication. The three governments are cooperating in an integrated policy designed to protect this unique area. Hunting of migratory species is being phased out in the Waddensee itself; mussels will no longer be collected in certain areas, and recreational activities will be forbidden in certain ecologically sensitive sites (Enemark, 1993).

8.5.4 CHESIL BEACH AND THE FLEET

Chesil Beach, Dorset, varies considerably in width – much of it being 150–200 m across – and for half its length protects the Fleet, the largest saline lagoon in Britain, whose organisms and communities have been surveyed by the Fleet Study Group (Ladle, 1981). It is essentially a simple, **linear shingle storm beach** which commences at Bridport, where it is joined to the land as a fringing beach, and runs south-west for 29 km before culminating as a **tombolo** at the Isle of Portland (Figure 8.12). Brunsden and Goudie (1981) consider that at one time it may have extended as far west as Golden Cap. Its exceptional geomorphological and conservation importance is indicated by the SSSI citation sheet for Chesil and the Fleet (1986).

Chesil Beach consists mainly of flints derived from the Chalk; other rock types include chert from the Upper Greensand, Jurassic limestones from Portland, and a relatively small number of rocks from further afield. It shows remarkable grading of the pebble size, with pea-sized stones in the west and cobbles in the east (Figure 8.13b). Pebble extraction from the foreshore and the lee slope of the bar poses a threat to its stability.

Borehole investigations and pollen analysis have thrown light on the origin of Chesil Beach and the Fleet (Brunsden and Goudie, 1981). It is suggested that the first stage in their evolution occurred in the last (Ipswichian) interglacial and glacial (Devensian) periods, when the last raised shorelines were cut and the last solifluction deposits laid down. Peat deposits recovered from boreholes suggest that sea level was –45 m approximately 10 000 years ago. When the ice caps melted, average sea level rise was 1.5 m every 100 years, and Chesil was driven onshore (along the lines of Figure 10.2(d)) as a bar of material from the river gravels of the enlarged glacial coastal plain. The forces involved during storm episodes are very great; material is often thrown right over the beach and slabs of peat may be left along its crest (as in April, 1995). The predominantly flint shingle barrier of Slapton 'Sands', which encloses Slapton Ley, South Devon, UK, also resulted from the same Flandrian rise in sea level.

Several open pioneer plant communities may be recognized along the top of the lee slope of the beach and are given 'Shingle Community Classification' (Sneddon and Randall, 1993a) notation below. Much of the cover in these communities is of *Silene uniflora* (sea campion)(SH7); *Atriplex* spp., *Beta vulgaris* ssp. *maritima*, *Glaucium flavum* and *Sonchus arvensis* are less common associates. An even more open pioneer community (SH3) is characterized by the constant presence of *Rumex crispus*, usually with *S. uniflora*, and sometimes *Geranium robertianum* or *Geranium purpureum*. In another open pioneer community, *Silene uniflora* occurs with *Cochlearia danica* (SH5); *Arrhenatherum elatius*, *Beta vulgaris* ssp. *maritima* and *Raphanus raphanistrum* ssp. *maritimus* are occasional associates. Where *Rumex crispus* is an important component of the vegetation (SH4) *Cochlearia danica* is also constant. *Silene uniflora* may also be present, with *Geranium robertianum* and *Glaucium flavum* as additional

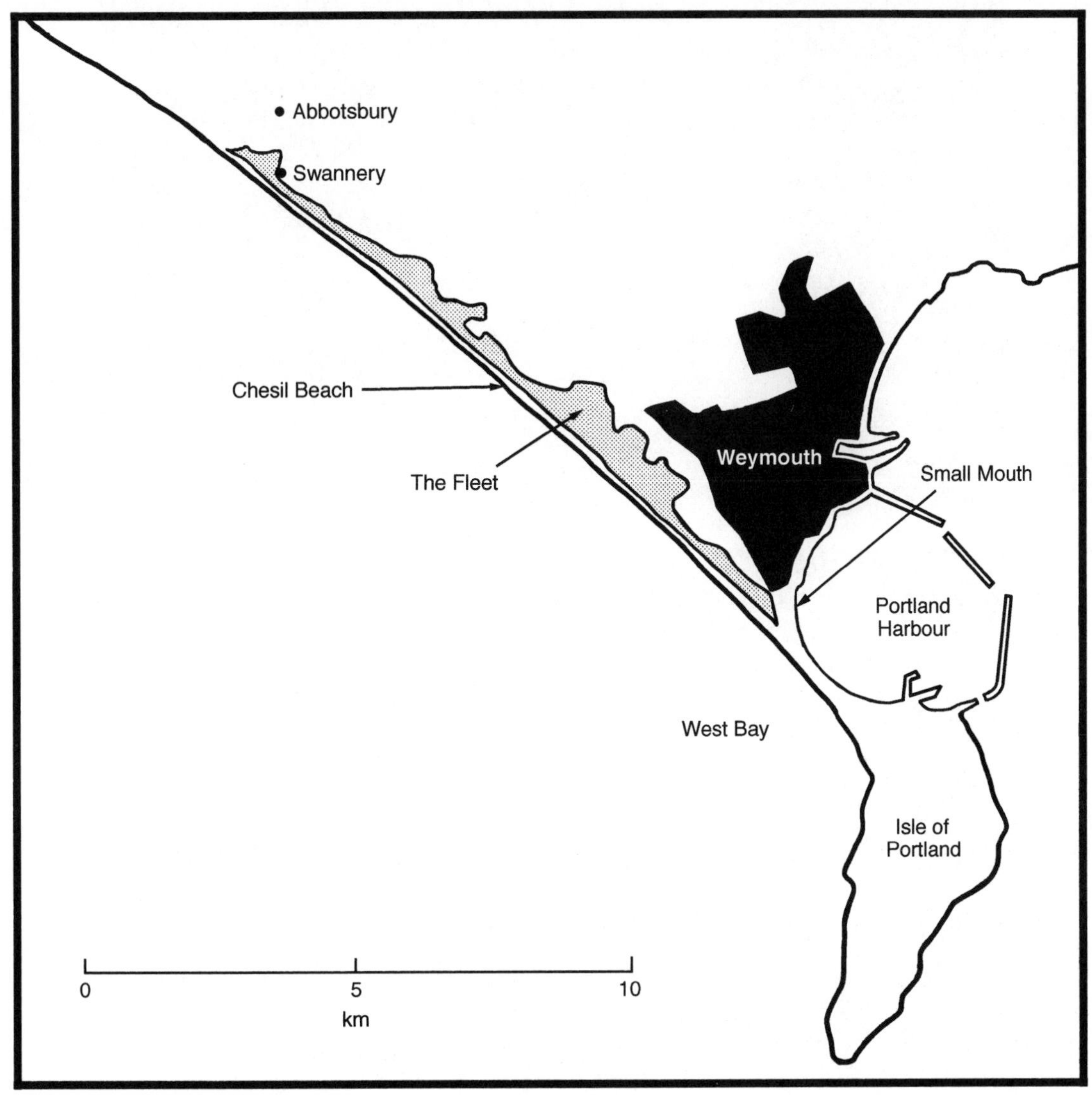

Figure 8.12 Chesil Beach and the Fleet, Dorset, UK. The arrow shows the position of Small Mouth (←) where a channel 75 m wide continued to allow communication with the sea after construction of the Ferry Bridge.

(a)

(b)

Figure 8.13 (a) Walkway to the crest of Chesil Beach at Abbotsbury, Dorset, UK. Most of the coastal walk is arranged inland of the ridge, avoiding damage and disturbance. Information is provided by the display at the extreme left. (b) Shingle from the main ridge, Chesil Beach. The shingle on the left was towards the west at Abbotsbury; that on the right – on which a *Buccinum undatum* (whelk) egg mass is resting – is from the east near the Portland Causeway. The diameter of the lens cap is 5.7 cm. The contrast in particle size between the extreme western and eastern ends of Chesil Beach is even greater than that depicted here. (Photographs by John R. Packham.)

associates. Sneddon and Randall (1994b) consider all the communities listed above to be largely restricted to Chesil Beach; some of the distinctions made are rather fine and may not long persist.

A *Silene uniflora–Crambe maritima* pioneer community (SH6), in which *Glaucium flavum* and *Sonchus arvensis* occur only occasionally, is largely confined to the western end of the Beach. *Lathyrus japonicus* communities are often monospecific (SH11), but *Silene uniflora* is a locally important component (SH11a). The *Silene uniflora–Rumex crispus–Arrhenatherum elatius* SH25 assemblage found to the west of the car park at West Bexington is typical of many shingle sites; here, disturbance has led to the introduction of *Leontodon hispidus*, *Solanum dulcamara* and *Sonchus arvensis*. A grassland assemblage rich in maritime herbs (SH70) appears to occur where there is a high silt content in the shingle matrix. It is characterized by the constant presence of *Festuca rubra*, *Lotus corniculatus* and *Silene uniflora*; key associates include *Armeria maritima*, *Cochlearia danica* and *Ononis repens*.

A narrow strip of *Suaeda vera* salt marsh (NVC : SM25) occurs along the edge of the Fleet (Figure 8.14). *Atriplex portulacoides*, *Beta vulgaris* ssp. *maritima* and *Xanthoria parietina* are minor constants in a community in which shrubby seablite responds vigorously to burial by fresh shingle.

Cladonia furcata, *Festuca rubra* and *Cochlearia officinalis* are constant, and major, components of the SH51 grassland on stable shingle on the lee slope, often at its base and where the slope angle is less steep. Many other *Cladonia* species are present; these lichens take a long time to establish and are easily disturbed, particularly by vehicles. Herb associates commonly include *Plantago coronopus*, *Plantago lanceolata*, *Sedum acre*, *Silene uniflora* and *Cerastium glomeratum*. *Ceratodon purpureus* and *Hypnum cupressiforme* are the major components of the well-developed bryophyte flora (Sneddon and Randall, 1994b).

Some 50 pairs of little tern (*Sterna albifrons*) and 30 pairs of ringed plover (*Charadrius hiaticula*) – the only sizeable populations of these species in the south-west – nest along the beach; human access is accordingly prevented during early summer.

The Fleet is 13 km long, 75–900 m wide and has a surface area of 4.9 km^2 at high tide. Its waters average only 1.5 m or less in depth, deepening to 5 m in the Narrows towards the east. Tidal flow decreases from Small Mouth, the opening into Portland Harbour, where it has a height of c.2 m, to almost nil midway along the lagoon. Salinity grades from marine in the east to mildly brackish – and at times almost fresh – at the Abbotsbury end, where up to 1200 mute swans (*Cygnus olor*) overwinter at the Swannery whose duck decoy, built in 1655, is the oldest still in use in Britain. The lagoon is well known for its underwater meadows of *Zostera noltii* (dwarf eelgrass) and *Zostera angustifolia* (narrow-leaved eelgrass) where fully saline conditions occur; *Ruppia cirrhosa* (spiral tasselweed) and *Ruppia maritima* (beaked tasselweed) grow where salinity is reduced. Some 23 species of fish have been recorded from the Fleet; common eels, smelt and grey mullet are abundant and the greater pipe-fish (*Sygnathus acus*) occurs where upright algae are present. The intertidal mudflats and associated vegetation provide important wintering grounds for wigeon, Brent geese and numerous waders, including dunlin and lapwing.

8.5.5 SALINAS

The nature conservation values of **salinas**, artificially controlled coastal basins used to produce common salt by the evaporation of seawater, are frequently high. Documentary proof of salt production at the Slovenian Ramsar site of Secovlje Salina on the Mediterranean coast of Slovenia dates from the 13th century, and in the next century the quality of the salt produced there was improved by the introduction of blue–green algae to the bottom of the pools. These, together with calcite and gypsum, formed a carpet-like basement layer which prevented

Figure 8.14 Two mute swans (*Cygnus olor*) at the western end of the Fleet Nature Reserve, Dorset. Abbotsbury Swannery lies to the left and Chesil Bank to the right of the view shown. The northern shore of the lagoon is edged by reedswamp (*Phragmites australis*). Shrubby sea-blite (*Suaeda vera*), sea purslane (*Atriplex portulacoides*), red fescue (*Festuca rubra*), sea aster (*Aster tripolium*) and sea plantain (*Plantago maritima*) occur in salt marsh in the lee of Chesil Bank. (Photograph by John R. Packham.)

the common salt deposit from mixing with the muddy sediment, and by the 15th century the Secovlje salt-pans were well recognized for their production of high-quality white salt. Halophytes are prominent in the protected area and include *Atriplex portulacoides* in the drier areas, *Limonium angustifolium* in damper places, and *Salicornia europaea* in shallow water, including that used for saline concentration (Beltram, 1994).

In the Secovlje Salina terns, waders, plovers and gulls nest on the dikes between the saline pools, while the halophyte meadows provide breeding grounds for Kentish plover, tawny pipit, black-winged stilt and terns. The great reed warbler, Cetti's warbler, moorhen and water rail are included in the birds breeding in the reed beds. Kestrels, hen harriers, goshawks, buzzards, Scops owls, little owls, eagle owls, and long-eared owls, all hunt in the area. These birds of prey sometimes build their nests in abandoned houses on the reserve and often feed around the salt-pans. Twelve species of small mammal live in the saline area, of which the pygmy white-toothed shrew (*Suncus etruscus*), that weighs no more than 2 g, is the only typical Mediterranean species. Secovlje Salina is subject to multiple demands on its slender resources and many of its communities benefit from the activities involved in production of moderate quantities of salt. It is just one of the many Mediterranean salinas whose wildlife management is discussed by Walmsley (1995), who emphasizes their importance in affording resting places for birds migrating between

Palaearctic breeding grounds and their winter quarters in Africa.

8.6 DYNAMICS OF A US NORTHERN BARRIER SPIT

Nauset Spit, Cape Cod, Massachusetts, is one of many barrier beaches on the east coast of the United States which are retreating towards the land as the sea level rises. Quantitative information about the vegetation going back over 100 years was used by Zaremba and Leatherman (1986) for the first detailed study in north-eastern USA to consider the interaction between physical processes – mainly overwash – on plant communities and physiographic features of barrier beaches. A series of 15 levelled transects was surveyed, each perpendicular to the north-to-south trend of the spit, on which dunes commonly border the open sea to the east while salt marshes occur along the western edge of the spit, most of which adjoins the quieter waters of Chatham Harbour and Pleasant Bay (Figure 8.15).

The transects were 30 m wide and starting from the north they were designated X, Y, Z, A–L. Numbers of plant species within each complete belt and within the five quadrats sampled at each 10-m interval along the belt were determined. In 1978, four of the belts (X,Y,J,L) were less than 10 years old, in terms of time since overwash, two (G,K) were approximately 26 years old, three (D,H,I) were 40–92 years old and six (Z,A,B,C,E,F) were older still.

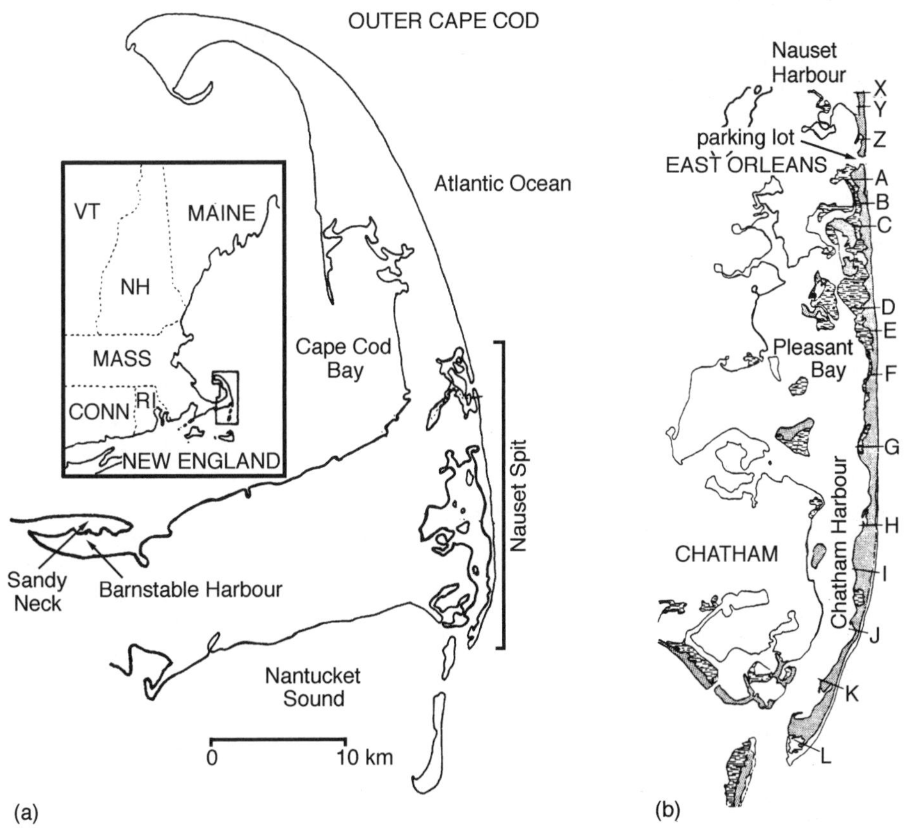

Figure 8.15 (a) Cape Cod Peninsula, Massachusetts, showing position of the predominantly sandy Nauset Spit and the Great Marshes at Barnstable. (b) Positions of the 15 belts of vegetation along Nauset Spit described by Zaremba and Leatherman (1986).

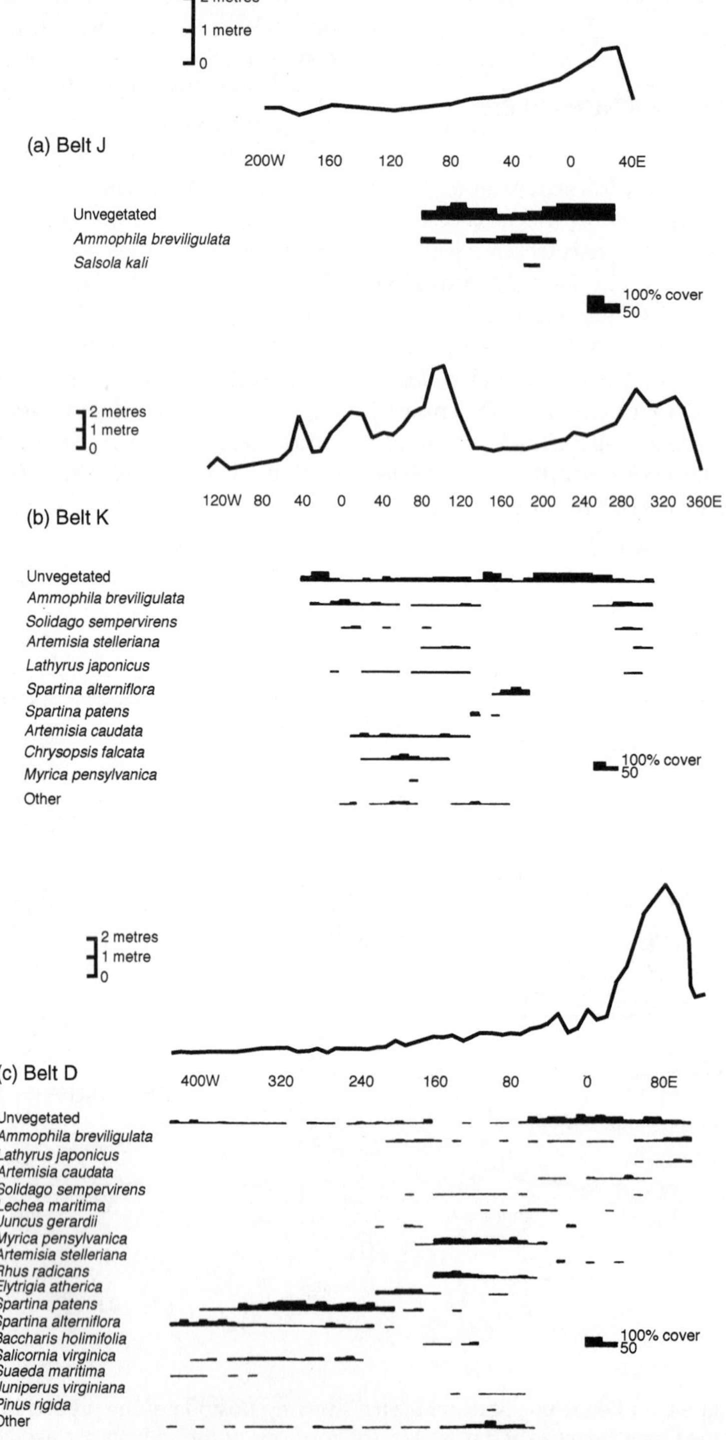

Figure 8.16 Altitudinal profiles and plant cover records at Nauset Spit for: (a) Belt J (3, Group I); (b) Belt K (31, Group III); and (c) Belt D (38, Group II). The first figure inside the brackets represents the total number of species found within the belt concerned. (After Zaremba and Leatherman, 1986.)

On ordination, the two planted areas (X,Y) were placed with those which had been most recently overwashed (J,L) in Group I, which was to the right of axis 1, had the lowest species diversity and whose vegetation was principally dominated by *Ammophila breviligulata*. Axis 1 was related to species richness; of the oldest belts all but Z were placed in Group II at the opposite end to Group I. Belts B and F both have well-developed *Hudsonia tomentosa* (beach heath: Cistaceae) communities, as well as *Spartina patens* and *Elytrigia atherica* grasslands which are characteristic of marginally supratidal substrate and absent from young areas. Belt D is in Group II in spite of having been barren sand in 1938; all the major barrier communities developed in only 40 years. The salt marsh in this belt is highly diversified, the *Spartina alterniflora* biomass is low and this grass has many associates (Figure 8.16).

Group III (G,K,Z,H) lies between the two preceding groups. Belt Z crosses a brackish pond that is the last remnant of the waterway that until the mid-19th century connected Pleasant Bay to Nauset Harbour; no significant changes occurred here between 1938 and 1978. Belts G and K developed after 1952 while Belt H, though initially formed in the 1930s, was partially overwashed in 1978.

The major dune species, particularly *Ammophila breviligulata*, can recover from overwash burial. Thus dune-building may be rapid in areas which in aerial photographs appear unvegetated. The greatest *A. breviligulata* biomass on Nauset Spit occurs on recently overwashed belts, where it exceeds that even of most building foredunes.

New dunes frequently develop in response to spit elongation and to massive washovers; the first dunes often start to grow along drift lines at the western (bayward) margin of the spit. Such dunes often fail to survive once overwash and continued sand supplies are reduced or eliminated. If they persist as slightly elevated sites they may be colonized by *Elytrigia atherica*, but frequently deflation is rapid and there is a transition to high marsh vegetation.

Development of physiographic features and plant communities on the spit is extremely rapid, dunes forming within 3 years after overwash and salt marshes within 10 years. Once dunes cease to increase in height, several plant species rapidly invade less vigorous stands of *A. breviligulata*; shrubs may colonize the area if seeds are available and foredunes are high enough to reduce sea spray. In sites which are stable for many years *A. breviligulata* declines, yielding to *Hudsonia tomentosa*, *Chrysopsis falcata*, *Artemisia caudata*, *Lechea maritima* and *Cladonia* spp. These species can be used as **indicators of substrate stability**, but senescent dune zones can develop very rapidly as is indicated by Belt K, which although less than 26 years old had a broad stable dune with extensive stands of dead and dying *A. breviligulata* in the interior of two large spit recurves.

Most of the species important on the barrier can grow in bare sandy substrates. Plant succession to a climax appears not to occur on the spit, which is gradually migrating as a result of a cyclic series of events including inlet dynamics, overwash processes and dune building.

ENVIRONMENTAL IMPACTS 9

9.1 POLLUTION, RECLAMATION AND MINERAL EXTRACTION

The soft coastlines of the world are increasingly polluted by ocean-borne material of all kinds, of which plastic rubbish is a major constituent (Figure 9.1(a)). This contamination has a particularly marked affect on salt marshes where long-lying plastic and other debris along the shoreline can kill the vegetation beneath it, and where water-borne oil impedes gas exchange in plants and causes death or metabolic disturbance in animals. Gaseous pollutants may arise from nearby industry and an important input from the land is often freshwater, affecting the salinity regime.

Pollutants discharged into rivers and reaching estuaries often spill over salt marshes. These pollutants include fertilizers from agricultural areas, heavy metals, and organic compounds from industry and sewage waste which is often channelled directly into the sea. Such discharges lead to considerable eutrophication of coastal waters which enrich salt marshes with mineral nutrients, increasing especially the levels of inorganic nitrogen and phosphorus. The concentration of inorganic nitrogen in the Rhine, for example, increased six-fold between 1930 and 1986 and discharged sewage sludge in Long Island Sound near New York has led to very high concentrations of ammonium (Rozema and Leendertsee, 1991). This eutrophication enhances primary production. Nutrient enrichment has in recent decades resulted in large increases of macro-algal species, notably of *Ulva* and *Enteromorpha*, in salt marshes in the south of England, the Netherlands and parts of the Mediterranean. These algae can utilize the ammonium-nitrogen prevalent under the anaerobic conditions of salt marshes. Sulphurous gases from decomposing algae in the Waddensee, Holland, give noxious smells; the high respiration rates of phytoplankton and other algae may cause anaerobiosis severe enough to kill large populations of fish (Beukema and Cadee, 1986).

Heavy metals may be adsorbed on small particles. Sedimentation of particulate matter on salt marshes and mangrove swamps makes them a 'sink' for these pollutants as well as of adsorbed organic substances. Uptake of heavy metals by salt-marsh plants is mainly through the root system and is affected chiefly by the oxidation state of the soil, pH, salinity, particle size and organic matter content (Otte, 1991). The uptake of heavy metal ions is species-specific, monocotyledonous salt-marsh plants tending to be excluders and dicotyledonous plants accumulators; however, acute toxicity symptoms have not been reported (Adam, 1990).

The effects of oil spillages on vegetation have been studied for many years; while many of the basic principles which would reduce their incidence have long been known (Packham, 1971) they continue to occur with monotonous regularity, sometimes on a very large scale. Oil pollution of salt marshes affects plant community structure, the fauna and environmental conditions. In general it seems that a single large spill causes less permanent damage to shore organisms than numerous small spillages or continuous low-level discharges, though damage to birds through oiling can be very great. Oil is unsightly and in tourist areas there is a strong tendency to 'tidy up' by bulldozing, burning or cutting oiled vegetation. Burial of oiled sand prevents its continued breakdown by photo-oxidation.

(a)

(b)

Except for very thick persistent layers of oil, these procedures are biologically more destructive than simply leaving the oil to be broken down by natural processes. Attempts to accelerate such breakdowns by introducing cultures of the appropriate bacteria came to the fore when the Persian Gulf was badly oiled during the Gulf War of 1991. The use of chemical oil dispersants after the *Torrey Canyon* disaster of 1967 caused more damage than did the oil itself; modern lower-toxicity dispersants are better but still cause finely dispersed oil to penetrate more deeply into sediments where they take longer to degrade. The release of 83 700 tonnes of light crude oil when the *Braer* sank off Shetland in 1993 was used by Ritchie (1995) as a case study in discussing oil spill contingency plans and the importance of on-site scientific expertise in making the major decisions required soon after a major spill. In this instance the oil was rather rapidly dispersed by natural means, although birds were severely oiled. In contrast the 37 000 tonnes of oil released in the *Exxon Valdez* incident of 1989 caused very extensive damage. Some 72 000 tonnes of oil were released when the *Sea Empress* ran aground off St Ann's Head, Pembrokeshire, Wales, in 1996. Most – perhaps all – of this spillage could have been avoided if the ship had been double-hulled and powerful tugs available at the outset.

The damage to plants from heavy crude oils is chiefly a smothering effect, whereas aromatic compounds from lighter and refined oils penetrate the plant, causing tissue damage (Scholten and Leendertse, 1991). These effects, together with deposition and absorption of oil by leaves, lead to decreased gas exchange, water loss and light absorption, causing death of green tissues. Oil-polluted sediments usually become anaerobic, with mortality of the benthic fauna, conditions under which degradation of oil components is very slow.

Long-term studies under the auspices of the Field Studies Council, which has centres adjacent to the oil port of Milford Haven, south Wales, show that oil pollution affects species of varying growth form and physiology differentially (Baker, 1975, 1979). When oily films kill the shoots of short-lived plants, such as *Salicornia* spp. and *Suaeda maritima*, they have no large underground systems from which to regenerate; population recovery is delayed and is from seed. In contrast, some grasses, including *Agrostis stolonifera* and *Elytrigia atherica*, and some rosette perennials, including *Armeria maritima*, *Plantago maritima* and *Triglochin maritimum*, are resistant. *Spartina anglica*, an important plant of the lower salt marsh, has a large underground system of roots and rhizomes. Light oils disrupt its plasma membranes; heavy oils tend to smother it. When an oil spill kills the aerial shoots, growth resumes from buds just below the soil surface. *Spartina* is not resistant to chronic pollution, however; by blocking the stomata, oil interferes with oxygen diffusion pathways from the shoots to the roots whose surroundings are often anaerobic (Baker *et al.*, 1990). Cessation of the normal outward diffusion of oxygen from the roots subsequently allows toxic reduced ions to accumulate in the soil around them. As *Spartina* commonly occurs in monocultures its death can result in large expanses of bare mud.

The ecological effects of repeated small oil spillages and chronic discharges are discussed by Dicks and Hartley (1982). In 1951, effluent

Figure 9.1 Contrasting shingle beaches. (a) Beach contaminated by sea-borne rubbish at Brighton marina, Sussex, UK. Angled concrete blocks break the force of the waves. (b) Driftline near the Neolithic settlement of Skara Brae, Bay of Skaill, Orkney, UK. The drift is dominated by dead stipes of *Laminaria*, among which grow a few orache plants (*Atriplex* sp.). The community growing on the sand at the top of the beach includes *Ammophila arenaria*, *Elytrigia juncea*, *Festuca rubra*, *Galium aparine* and *Tripleurospermum maritimum*. (Photographs by John R. Packham.)

from a refinery–petrochemical complex began discharging to the dendritic creek system of the Fawley Marshes, a Spartinetum near Southampton, UK, and by 1970 c. 0.6 km^2 of marsh had been de-vegetated. An improvement programme greatly reduced the scale of the discharges and the subsequent re-colonization was monitored twice yearly from 1972. The most rapid recolonization was by those species which seeded well, notably *Salicornia* spp. and *Aster tripolium*. Re-invasion by the formerly dominant *Spartina anglica* was initially slow and from vegetative fragments, but this species subsequently seeded and spread more rapidly (Baker *et al.*, 1990). An important finding was that, by 1980, sediments heavily contaminated with weathered oil supported apparently healthy marsh plants and oligochaete worms.

Upper salt marshes often have mosaics in which *Juncus maritimus*, *Festuca rubra* and *Agrostis stolonifera* occur together with various dicotyledons, including the umbellifer *Oenanthe lachenalii*. A single spill is often sufficient to kill most of the *Juncus*, in which case *A. stolonifera* colonizes the dead *Juncus* patches with its fast-growing stolons. Chronic pollution of this community will gradually eliminate various species according to their **tolerance ranking**; *Oenanthe* is very resistant and can survive at least 12 successive monthly oilings. In contrast to a *Spartina* monoculture, the more complex upper marsh community has a greater ability to resist total destruction. The effects of oiling on various animal and plant communities differ considerably, and there are various post-pollution reactions between organisms, some being reduced or eliminated while tolerant species gain a competitive advantage. Growth stimulation following initially deleterious effects of oil pollution can occur, one possible cause being fertilization from oil decomposition products (Baker, 1979).

Mechanical beach-cleaning, particularly when it involves the complete removal of algal debris, wood and leaves from the strandline, is very damaging to invertebrate populations. Amphipods are particularly vulnerable as their juveniles cannot bury themselves in sand and remain among newly deposited seaweeds, where relative humidity is high even during exposure at low tide. Llewellyn and Shackley (1996), who investigated this problem on the shores of Swansea Bay, South Wales, advocate a compromise in which a limited area is mechanically cleaned and the remainder of the beach is hand-picked – avoiding sand compaction – to remove objectionable man-made items.

Sand dune vegetation is vulnerable to hydrological changes resulting from drainage, water abstraction and infiltration schemes employed to purify river water: these last schemes are described in section 9.2 and in Holland have caused eutrophication of dunes previously low in mineral nutrients.

9.1.1 RECLAMATION AND MINERAL EXTRACTION SCHEMES

Much of the world's salt-marsh area has been reclaimed for commercial, residential or agricultural use, while the impact of tourism and an insatiable demand for holiday homes along the margins of dunes and shingle beaches also contribute to the coastal squeeze described in section 10.3. Whereas the tendency is to enclose, raise and drain salt marshes, many dunes and shingle beaches have been adversely affected by mineral extraction schemes. These are often exploited for aggregates for use in the construction industry, but sometimes because of their chemical nature – as in the RTZ claims on the titanium-bearing ilmenite sands of the unique coastal forests of Madagascar. In the Outer Hebrides, lime-rich machair sands have often been exploited as a source of calcium carbonate when attempts were made to bring peatland into agricultural use.

9.2 DRAINAGE AND WATER ABSTRACTION

The Dutch coastal dunes, which form an almost continuous strip 1–4 km wide along the

North Sea coast, have been an important source of drinking water for the last century and this use has had a major influence on the vegetation (van der Meulen, 1982; van Dijk, 1985, 1989). At Meijendel, freshwater at first came from the upper dune sands; after 1930 it was also extracted from the deeper layers. Ultimately, however, pumping caused the intermediate brackish water zone (or interface) that is underlain by sea water to rise towards the pumping system, a phenomenon called **interface upconing**. To reverse this trend and to meet increasing demands for drinking water the Dutch dunes have since the mid-1950s been infiltrated, via ponds 50–300 m long and 15–50 m wide, with water from the Rhine, Meuse and polder water courses, the amount of infiltrated water exceeding the natural recharge by a factor of 10–20. After storage in the dunes for 1–3 months, the water is withdrawn via drains or wells and supplied as drinking water.

Wet slacks, whose vegetation is dependent on ground water, support the most threatened plant communities in the Younger Dunes such as those at Meijendel and Berkheide. Succession here may involve the gradual drying out of the soil, changes occurring in hydrosere, hygrosere and mesosere communities of these species-rich wet slacks, where the fluctuating ground water table remains within 1–1.2 m of the surface during the growing season. The period of inundation is critical in this rabbit-grazed mosaic vegetation, where diversity is related to small topographic differences leading to complex dry to wet, acid- to base-rich, mineral to organic edaphic gradients. Plants of the dry dunes are dependent on pendular water held after rain, rather than on ground water, and the four groupings listed for the xerosere by van der Meulen (1982) represent an increasing predominance of woody vegetation.

The water works exerted a major influence on the vegetation; at Meijendel wet slack vegetation with such species as *Carex trinervis*, *Centaurium* spp., *Dactylorhiza incarnata*, *Epipactis palustris*, *Equisetum variegatum*, *Orchis morio*, *Parnassia palustris* and *Schoenus nigricans* disappeared as a result of initial extraction and transition to xerosere. Some characteristic species have returned following infiltration, among the most striking results of which is an increase in plant biomass and the appearance of tall nitrophilous species, especially in the numerous seepage areas, with a depth of 0.5–1 m, which occur between the infiltration ponds and the sea, or between the ponds and the catchment points. Together with the increase in woody forms, these have caused major loss of both species and life form diversity.

The normal dome-shaped ground water table becomes irregular when dunes are infiltrated (Figure 9.2) and the dune valleys inundated; the resulting ponds are occupied by green algae and hydrophytes such as *Myriophyllum spicatum*, *Potamogeton pectinatus* and *Zannichellia palustris*. When infiltrated, the partially purified river water responsible for eutrophication contains about 0.2 mg phosphate (PO_4^{3-}) and 15 mg nitrate (NO_3^-) per litre on average, whereas the corresponding figures for natural dune lakes in Holland are <0.07 and <0.3 mg l^{-1} respectively. The effects of eutrophication also depend on the rate at which ground water travels laterally; the faster infiltration water passes through the dunes the greater the nutrient availability to the vegetation in its path. *Urtica dioica*, *Eupatorium cannabinum* and *Epilobium hirsutum* have come to be especially common on the margins of infiltration lakes. *Calamagrostis epigejos*, *Lycopus europaeus* and *Mentha aquatica*, which are now common near seepage areas, were prominent in earlier stages of succession along infiltration lakes (van der Werf, 1974).

It was originally assumed that a layer of precipitation water would cover infiltration water in seepage areas where the latter is held for relatively long periods. The spread of nitrophilous species just discussed cast doubts on this supposition. van Dijk (1985) investigated the changing distributions of the tracer minerals chloride and potassium, and the

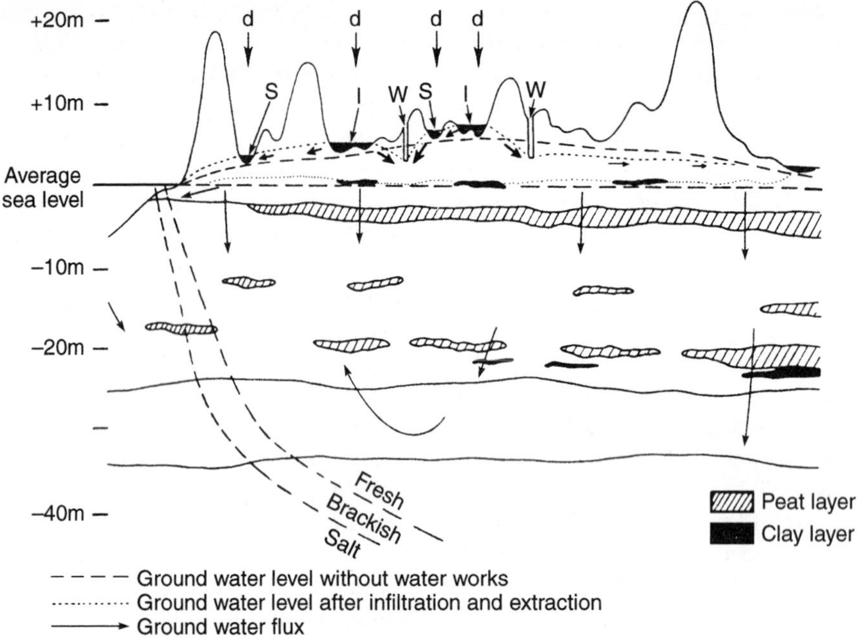

Figure 9.2 Cross-section of the Berkheide dunes, Holland, showing effect of water works on the ground water table. d, depression; S, seepage pool; W, well; I, infiltration pond. Note the dome-shape of the original ground water table. (From van Dijk and de Groot, 1987; courtesy *Water Research*.)

macronutrients orthophosphate, nitrate and, again, potassium. In the infiltrated dunes of Meijendel, Berkheide and Luchter, potassium occurred in unnaturally high concentrations in ground and surface waters up to 500 m from the infiltration ponds, thus disproving the hypothesis of a so-called 'precipitation lens' in the seepage areas. Since 1960, increases in nitrogen values of the catchment water in all three infiltration areas studied have been much greater than can be accounted for by increased contributions from infiltration water. In Meijendel two sources, a herring gull (*Larus argentatus*) colony and shrubs of *Hippophae rhamnoides*, have been responsible for considerable extra additions of nitrogen in the past few years. There may also be a reduction in the rate of denitrification in the ground water as sources of organic compounds diminish. This could result from the 'exhaustion' of easily oxidizable peat, and the decreasing concentration of organic compounds in the infiltrated water owing to improved pre-purification.

van Dijk (1985) considers phosphate to be the limiting factor for the occurrence of nitrophilous tall hemicryptophytes in infiltrated dune areas. This is fortunate since dephosphorization is a relatively uncomplicated process already employed to reduce algal blooms. High seepage pool concentrations of phosphate at Berkheide appear to be partly caused by the absence of peat in the relatively thin upper aquifer in this area, but ultimately continued infiltration leads to saturation of absorption points, even in organically rich soils.

Unless large-scale desalination is adopted, the Dutch coastal dunes will probably continue to be used for drinking water production, so management options are largely concerned with reducing the nutrient load. This can be done by cropping vegetation, removing humus, and using less eutrophic water for infiltration. Regeneration of dune slack vegetation is more complex; any measures adopted should include a 'naturally' fluctuating ground water table. It may be possible to regenerate the vegetation of old wet slacks by reducing the nutrient load, but the unnaturally high lateral velocities of the ground water militate against this. Alternatively, new wet slacks could be created in dry dunes by raising the water table, by excavation or by promoting excavation in blowouts (van der Meulen, 1982). van Beckhoven (1992) points out that *Schoenus nigricans* is likely to play a pivotal role in any attempt to restore wet slack communities by raising the water table, partly because it has a high radial oxygen loss from its roots. This is likely to facilitate colonization by such characteristic damp slack species as *Centaurium littorale*; importantly, *S. nigricans* also reduces the otherwise exuberant growth of *Calamagrostis epigejos* under wet conditions. Monitoring has been an important aspect of the whole dune project, and the availability of some 40 permanent plots studied in various parts of the Meijendel dunes shortly before infiltration began has been of great value in unravelling the complex changes since then.

9.3 MOWING, GRAZING, TRAMPLING AND DAMAGE BY VEHICLES

Mowing and grazing help prevent scrub encroachment, reduce the proportion of coarse tussock-forming species and increase species richness. Mowing or cutting has also been used in attempts to reduce the spread of particular species, notably *Pteridium aquilinum*, which was very prominent on the Oxwich Dunes, Gower Peninsula, South Wales in the late 1980s. Of four treatments employed experimentally to control *Juncus maritimus* (Packham and Liddle, 1970), cutting at ground level at the end of June (the only treatment to prevent seed formation completely) was considerably more effective than at the end of April. Cutting at a height of 15 cm at either time was less effective still.

In an investigation of the effects of cutting on *Spartina anglica* in Poole Harbour, Dorset, Hubbard (1970a) showed that repetitive monthly cutting to ground level from June to October resulted in denser growth in the following year than in the uncut control areas. The cutting treatment led to more uniform height and distinctly earlier flowering than in the control plots; the seeds from the cutting treatment showed good germination and, after storage, remained viable for at least 4 years.

The influence of rabbit grazing, discussed in section 6.6, is very important, especially on dune vegetation. Indeed, the **sheep grazing experiments** on the sand dunes of Newborough Warren, Anglesey, were a response to the coarsening of the vegetation after the rabbit decline caused by the 1954 myxomatosis outbreak. These experiments began in 1980 with a comparison of mowing and grazing by Soay sheep (Hewett, 1985). Initial findings showed that mean numbers of species per plot increased under both regimes, that the cover of *Lotus corniculatus* became greater and that of *Arrhenatherum elatius* decreased. The grazing experiments, which have been greatly extended, still continue and are briefly reviewed by Gibbons (1994). Cattle are more effective in grazing back the coarse grasses which sheep may leave untouched. Soay sheep are also used to graze areas of salt marsh, pools and carr within the Newborough Reserve. Wildfowl feed here during the winter months and geese are important salt marsh grazers in many parts of the world, particularly in Hudson Bay (section 5.4).

Controlled sheep grazing experiments were used by Ranwell (1961) at the upper limits of an area of *Spartina anglica* in Bridgwater Bay, Somerset, where salinity of the soft marsh was reduced by land drain seepage. Successional processes led to the spread of *Puccinellia maritima*, increased tillering in *S. anglica* and a reduction in the area available to the annual *Atriplex prostrata* where the vegetation was grazed, and to the formation of tall marsh vegetation with

Phragmites australis and *Bolboschoenus maritimus* where it was not.

Boorman (1989), who studied the effects of grazing on 48 British dune sites, found that those at Aberffraw, Anglesey, which are quite heavily grazed by cattle, sheep, rabbits and hares, had the highest species diversity, the mean height of the vegetation being about 5 cm. Vegetation on the completely ungrazed dunes at Alnmouth, Northumberland, was tall (over 50 cm) and dense, with a far lower species diversity. In the Netherlands, grazing of dunes by domestic animals at low stocking rates led to higher species diversity and greater occurrence of rare species than in heavily grazed areas (Oosterveld, 1985).

In the warmer, drier site at Les Quennevais, Jersey, Channel Islands, mowing has been used in attempts to restore a species-rich sward on an extensive dune plain dominated by tall, thick *Rosa pimpinellifolia* and grass-dominated communities containing few species. There was no significant difference between annual and biennial treatments but diversity was significantly greater on mown as compared with unmown plots. Increases occurred in annuals such as *Vulpia bromoides*, *Petrorhagia nanteuilii*, and species of *Euphrasia*, *Trifolium* and *Vicia*. Expansions also occurred among biennials such as *Daucus carota* and *Raphanus raphanistrum* ssp. *maritimus*, and perennials including *Armeria arenaria* (Jersey thrift), *Leontodon saxatilis*, *Lotus corniculatus* and *Poa pratensis*. Rabbit activity was encouraged by mowing once close to the warrens; greater species diversity and higher numbers of annual herbs ensued for two to three years before declining (Anderson and Romeril, 1992).

9.3.1 TRAMPLING AND PATH FORMATION

Investigations of trampling damage to various types of vegetation by many authors were reviewed by Liddle (1975a), who concluded that damage to vegetation from human trampling involved seven major features:

1. Plants possessing basal meristems tolerate trampling better than those which do not.
2. Communities of higher plants are less damaged by trampling when the ground is dry than when it is damp.
3. After an initial rise the numbers of species in a community will fall as trampling increases.
4. Tall grasses at first give way to lower-growing dicotyledons, but as the amount of trampling continues to increase broad-leaved plants are replaced by monocotyledons before they too are eliminated.
5. As plant succession continues the vegetation becomes more productive and more tolerant of trampling.
6. Light trampling stimulates primary production, but this effect is rapidly reduced as the intensity of trampling increases.
7. Plants with high potential relative growth rate are common on paths.

The soils and vegetation of the tracks and paths on the sand dune system of Aberffraw, Anglesey, North Wales, have been extensively studied by Liddle and Greig-Smith (1975a,b), together with the relative contribution of vegetation removal and soil compression to changes in the microclimate of the tracks (Liddle and Moore, 1974). Under high incident radiation, diurnal soil and air temperatures of paths in both wet and dry areas of the dune system were increased relative to vegetated areas, but the microclimatic effects of track formation are complex, being influenced by an increase in soil bulk density and thermal capacity. Compressed sand can hold more water at matric potentials within the range available to plants than uncompressed sand, but this may be offset by the occurrence of anaerobic conditions. As in other studies *Poa pratensis* and the rosette species *Bellis perennis* and *Plantago lanceolata* were characteristic of trampled areas in wetter areas of the dunes, while *Festuca rubra* survived well. Hemicryptophytes formed a major part of the 'trampled flora'. Interesting new additions to this included *Galium verum*, *Taraxacum* spp., *Juncus articulatus* and *Carex flacca*. The last two species occurred in harder wet areas and their shoots tend to leave the soil obliquely when subject to trampling.

Ground and aerial surveys at Winterton, Norfolk, resulted in a map showing 35 km of paths over 1 m wide; Boorman and Fuller (1977) recorded a further 40 km of narrower paths in the 104 ha of dune. Vulnerability to trampling of the *Ammophila* dunes was high, that of short turf, scrub, woodland and *Rhododendron ponticum* was low, with *Calluna* and rough grass communities intermediate. Sward height decreased and the amount of bare ground increased with a greater number of trampling passes across areas of the rough grass (tall *Festuca ovina–Carex arenaria*). The results obtained fitted the relationship between primary productivity of the vegetation and its ability to tolerate trampling derived by Liddle (1975b), who used the number of trampling passes required to reduce the cover or biomass of a previously unworn area of vegetation to 50% of its original value as an **index of vulnerability**.

Sand dunes are a natural target for recreational pressure which is steadily increasing; the results of survey led Boorman and Fuller (1977) to suggest that increasing visitor numbers might necessitate the institution at Winterton of laid out paths (see Figure 11.5) and fencing similar to those employed at Meijendel. In these dunes near the Hague there were 71 km of major cycle, foot and horse paths, and a further 22 km of minor paths in an area of only 105 ha.

9.3.2 DAMAGE BY VEHICLES

Coastal ecosystems based on granular deposits are particularly vulnerable to damage by the recreational use of off-road vehicles (ORVs), but conventional vehicles also cause enormous

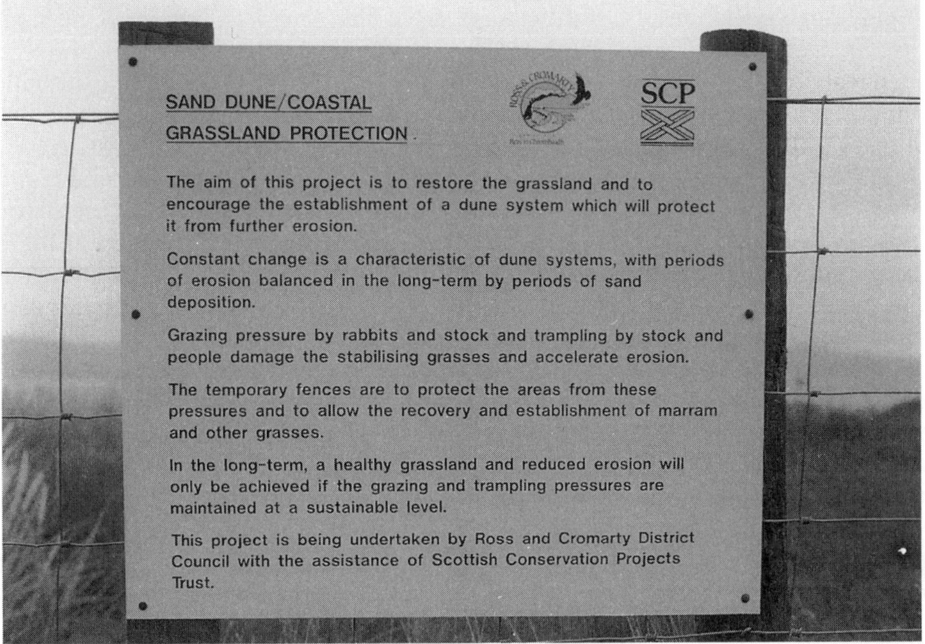

Figure 9.3 The aims of coastal dune and grassland protection communicated in a notice of a restoration scheme by the Scottish Conservation Trust and Ross and Cromarty Council, on the western shore of Loch Ewe, Ross and Cromarty, Scotland. (Photograph by John R. Packham.)

harm and have to be controlled in the interests of conservation. In 1960 the dune system at Aberffraw had 3.2 km of track and 2.2 km of footpaths; this had increased to 11.7 km of track and 16.5 km of footpaths by 1970 (Liddle and Greig-Smith, 1975a). Damage since that time has increased despite the best efforts of the Meyrick Estate. The repeated passage of cars has in many places destroyed the surface grassland, and the sand beneath has subsequently been wind-blown, leaving tracks more than 1 m lower than the adjacent turf. High-powered motorcycles are also driven over the relatively inaccessible dunes fronting the sea. These are highly dangerous to pedestrians visiting the area and so damage the vegetation as to increase the number of blowouts. The dunes fronting the sea are occasionally cut through by paths created by walkers but these breaches are relatively easy to repair.

ORVs can often travel quite rapidly along spits and over shingle ridges; the problem faced by the National Trust in protecting the rare shingle ridge vegetation of Orfordness is mentioned in section 8.4. Dunes are heavily used for recreation in many parts of the world. In California and Oregon specially modified 'dune buggies', often with balloon tyres which give low loadings per unit area, cause extensive damage when driven up and over dune crests.

On sand dunes, the main problems related to vehicles are concerned with erosion and the initiation of blowouts, but in the lower regions of many salt marshes it is longevity of the unsightly ruts they make that is the major cause for concern. These depressions are emphasized by abrupt changes in the vegetation; plants such as *Puccinellia maritima* and *Salicornia* growing within the rut contrast with less damp-tolerant species growing in the adjacent ground.

9.4 THE INFLUENCE OF INVASIVE PLANTS

'*The Ecology of Invasions by Animals and Plants*' is not only the title of the celebrated book by Charles Elton (1958), but a topic which has long interested ecologists. This section is concerned with plants; the influence upon sand dune ecosystems of the rabbit (section 6.6) and of the Canada goose upon British salt marshes (section 2.8) are considered elsewhere. As De Ferrari and Naiman (1994) point out, exotic plants not only compete with native species for light, water, and other resources, but can eventually displace them. They can interfere with successional processes, harm animals including livestock, alter disturbance regimes, and even modify geochemical processes. The type, frequency, duration and magnitude of **disturbance regimes** are most important in facilitating invasions by exotic species, while existing plant community type and a whole range of environmental and habitat factors also play a role.

Ranwell (1981) reviews examples of introduced coastal plants and comments on their influences on rare species. *Spartina anglica*, which received part of its genetic complement from the American species *S. alterniflora*, has been widely introduced by man (sections 4.4 and 11.1) and in Britain the distribution of *S. maritima* is now generally restricted to small isolated salt marsh sites in east and south-east England. At Maplin Sands at the mouth of the Thames Estuary, however, this cord-grass thrived in a pioneer situation at the seaward limit of the salt marsh. At this site silt accretion seemed to be almost counterbalanced by downward isostatic adjustment, providing a situation more favourable to *S. maritima* than *S. anglica*. More recently, *S. maritima* has declined noticeably in the Suffolk estuaries (A.J. Davy, personal communication) An experimentally introduced population of *S. alterniflora* at Maplin had by 1981 not increased significantly over a period of 5 years. *Spartina anglica* has also been shown to replace *Zostera noltii*, notably at Arne Bay, Poole Harbour, Dorset, where it first invaded the salt marsh c. 1899.

Simplifications of salt-marsh vegetation following planting of *Spartina anglica* are not necessarily permanent. Those made at Bridgwater

Bay, Somerset, in 1929 were followed by rapid expansion, but by 1980 vegetation at its former landward limits had become very diverse. Areas of open space created in the mature *Spartina* marsh by the smothering effects of its own litter had by then been colonized by many species, including the rare water parsnip (*Sium latifolium*), carried down in litter from oligohaline zones up estuary.

Arenohaline marsh is the distinctive ecotone sometimes found between salt marsh and sand dune. It is alternately affected by blown sand and thin infrequent layers of silt derived from tidal overwash, and affords a niche occupied by a number of rare species including *Carex maritima*, *Frankenia laevis* and *Limonium bellidifolium*. Ranwell (1981) noted that in the UK this zone had not so far been invaded by any well-established introduced species, but feared that *Tamarix* species might do so in the future if not depressed by rabbit grazing to which they are very susceptible. Global warming would increase this possibility as *Tamarix* is an Old World genus of warmer climates, and spread over thousands of hectares of marshland in the southern states of USA to which it was introduced early in the 20th century. The plant is deep-rooted and transpires large quantities of water in brackish areas, so the resulting increases in shading and in salinity in the surface soils could greatly change this important ecotone.

As Elton (1958) has shown, very few of the plant species which have been interchanged between different parts of the world have actually become naturalized in the areas to which they were deliberately or accidentally introduced. Coastal dunes and man-made **coastal embankments** favour such establishment; both are frequently associated with estuaries, ports and rail-heads through which many exotic species enter the country. Coastal embankments also form a refuge for such species as *Trifolium squamosum* (sea clover) and *Puccinellia rupestris* which agricultural practices have eliminated from natural transitions between salt marsh and lowland coastal grassland. The heights of taller aliens are effectively checked by salt-laden winds in such exposed sites. Ranwell (1981), who found that 22% of the 166 vascular species present along a 2 km stretch of 15-year-old coastal embankment on the Ribble Estuary, Lancashire, were aliens, concluded that management activities were potentially a greater threat to the continued survival of rare species than competition from alien plants. Introduced species can be thought of as enriching the native flora in such situations, but their presence provides a source of plants which may later invade and utterly change valuable native communities.

The low creeping forb *Acaena anserinifolia* (bronze pirri-pirri-bur) from New Zealand, which has barbed spines on its fruiting heads, has spread extensively on Holy Island, Northumberland, where it frequently establishes on the bare ground of paths in the *Ammophila* dune. In spring and summer the invasive composite *Petasites hybridus* now limits grazing on large areas of machair on the Isle of Lewis, Outer Hebrides; when it dies back the soil is exposed to erosion by wind (Angus and Elliott, 1992). *Pennisetum clandestinum* (Kikuyu grass), introduced into New Zealand from Zimbabwe in 1920, grows well on dunes and is tolerant of trampling and mowing. At Aramoana, some 70 km south of Napier, it grows with marram at the top of the beach.

The most important invasive taxa in Danish coastal habitats – *Rosa rugosa*, *Pinus mugo* and *Spartina* spp. – eliminate native endangered species through competition, and are considered by Andersen (1995) to require active control. *Rosa rugosa* (Figure 9.4) is also extensively naturalized in some British dune and shingle habitats.

Introduced trees and shrubs can greatly alter the floras of sand dunes, and conifers in particular have often been employed to stabilize dune systems in Europe. Besides shading out most of the dune flora, including rare species, the large-scale introduction of pines to the dune systems of Culbin (Moray), Tentsmuir (Fife), Newborough (Anglesey) and

Figure 9.4 *Rosa rugosa*, a species requiring active management or elimination in some European dunes and shingle beaches, establishing on Black Rock Beach, Brighton, Sussex, UK. Other species present include *Beta vulgaris* ssp. *maritima, Catapodium marinum, Glaucium flavum, Senecio vulgaris* and *Tripleurospermum maritimum*. (Photograph by John R. Packham.)

Holkham (Norfolk) has resulted in a lowering of the water table through a combination of increased transpiration and artificial drainage. On the other hand, coastal conifer plantations provide niches for a number of rare vascular plants and a range of cryptogams.

Hippophae rhamnoides, a much-branched, freely suckering shrub whose ecology has been described by Pearson and Rogers (1962), is native in England and Wales but often becomes very invasive when planted on sand dunes. In Ireland it is the most aggressive of the alien species which cause severe problems in sand dunes; others include *Acer pseudoplatanus, Lupinus arboreus, Pinus* spp. and *Rhododendron ponticum* (Binggeli *et al*., 1992). As at Newborough Warren (section 6.6), woody species on Irish sand dunes greatly increased following the diminution of rabbit grazing after the major myxomatosis outbreak in the 1950s. In the Netherlands, declining performance in populations of sea buckthorn appears to be related to parasitic nematodes (*Longidorus dunensis* and *Tylenchorhynchus microphasmis*) which attack its roots (Brinkman *et al*., 1987; Willis, 1989). This raises the possibility of restricting the vigour of *Hippophae* scrub by using nematodes as a biological control.

Impenetrable thickets of acacia, developed when the seeds of *Acacia cyclops, A. longifolia* and *A. saligna* were sown among the grasses of South African dunes as a stabilization measure, greatly increased soil nitrogen content. *Casuarina equisetifolia*, an Australian tree with low invasive potential in South Africa, has a symbiotic association with the nitrogen-fixing actinomycete *Frankia*, produces copious leaf litter and has an abundance of fibrous roots which preclude the growth of other species beneath it (Avis, 1989). In view of this, and of such problems as those associated with the introduction of *Ammophila arenaria* in Oregon (section 7.1), it is of interest that the large-scale dune stabilizations undertaken by Government agencies in South Africa now successfully employ an ecological approach using only indigenous species (Avis, 1995).

9.4.1 REALIZED NICHES IN EXOTIC POPULATIONS

Discussion so far has centred around the results of invasions by a few exotic species, but in many parts of the world particular areas have been invaded by considerable numbers of non-native species. In such instances much can be learned by comparing the behaviour of these species in their new habitats with that in their areas of origin. Investigation has frequently shown remarkably little genetic differentiation of exotic species within the invaded areas, and indeed little genetic change from the source populations. Such species appear to be pre-adapted to conditions in their new range. Is this because they find niches identical to those in their native range, or are they occupying quite different niches in the invaded areas? This is the main question considered in a comparison of realized niche relations of 24 species prominent both on the dry shingle banks of Dungeness, UK (section 8.4), where they are native, and on shallow soils in low elevation areas of the dry Upper Clutha catchment, New Zealand, where they are exotic (Wilson, Hubbard and Rapson, 1988).

Both sites are grazed, but the latter is in the rain shadow of the Southern Alps. Its soils have less large gravel and much more silt, clay and organic matter, while its climate is drier, sunnier and more extreme than that of Dungeness. The vegetation samples consisted of 10 m × 10 m quadrats and the plants recorded included such familiar British species as *Agrostis stolonifera, Arrhenatherum elatius, Cytisus scoparius, Echium vulgare, Festuca rubra, Hypochaeris radicata, Poa pratensis, Rumex crispus* and *Trifolium repens*. Habitat range within the two areas varies from very dry to aquatic; water availability is a major control on plant distribution. Ordinations of the species in the two areas show no obvious similarity; indeed agreement between inverse classifications of the same 24 species as distributed at Dungeness and in the Upper Clutha is actually slightly less than that expected on a random basis. The conclusion reached was that many of the species concerned were pre-adapted ('**exapted**') to niches in the Upper Clutha which were different from those at Dungeness, although the ecological behaviour of some of them – such as the presence of *Echium vulgare* along dry road verges – is very similar in both areas.

The multi-scale assessment of the occurrence of exotic plants on the Olympic Peninsula, Washington State, USA by De Ferrari and Naiman (1994), although largely concerned with forested areas, is a good role model for future studies of exotics in coastal ecosystems. Some 52 exotic species – 23% of the total flora – were found in this investigation. The proportions of exotic species and of the areas which they covered were greater in the riparian zones (which facilitated movement of exotic plants through the landscape) than in the upland areas.

VARIATION IN SEA LEVEL AND CLIMATIC CHANGE 10

10.1 EVIDENCE OF PREVIOUS CHANGES IN SEA LEVELS

Changes in sea level relative to a particular shore line may be either isostatic or eustatic. **Isostatic changes** result when, for example, an area of land is depressed after a thick ice sheet has formed above it, or tends to rise after such a burden has been removed. World-wide or **eustatic changes** in sea level, which have occurred many times in the Earth's history, result from the formation or decay of large ice sheets, changes in the form of ocean basins through the operation of tectonic processes, or partial infill of such basins by sediments.

Evidence of sea-level change is seen along many coasts, that of **raised beaches** cut when the sea was at a higher level being particularly convincing. The cliffs behind the Fleet lagoon, now protected by the Chesil Beach, Dorset, UK, were formerly subjected to direct marine erosion. The preservation of the 15-m raised beach to the west of the Fleet at Bexington Cliffs, which are mantled by periglacial deposits, demonstrates that they were certainly reached by the sea during the last (Ipswichian) interglacial (Brunsden and Goudie, 1981).

Sea-level rises and falls have clearly occurred many times in the past; studies of events associated with past marine transgressions and regressions, such as the recent evolution of a paraglacial estuary under conditions of rapid sea-level rise at Chezzetcook Inlet, Nova Scotia (Carter *et al.*, 1992), may afford a key to future scenarios driven by the current sea-level rise (Tooley and Jelgersma, 1993). The changes occurring along the rapidly submerging coastline of Louisiana, USA, where rates of land loss during the 20th century are estimated as being as high as 130 $km^2\,yr^{-1}$, are of particular relevance when considering policies of managed retreat (Salinas, De Laune and Patrick, 1986).

10.1.1 PALAEOENVIRONMENTAL STUDIES OF FENLAND, UK, INVOLVING SEA LEVEL CHANGE

Before drainage and use for agriculture, Fenland was an area of extensive coastal wetland, fringed at seaward localities by broad salt marshes and mud flats. All major Fenland lithological contacts – including those revealed by coring – are **diachronous** (varying in age from place to place), reflecting as they do landward and seaward migrations of palaeoenvironments with time. Stratigraphic studies have revealed widespread clay and silty clay interpolated between coastal reedswamp and salt-marsh deposits in this area. Much of the early work was by Godwin; the depositional environment of the silty clay is discussed by him (Godwin, 1978) and by Shennan (1986).

In north-central Fenland peat began to form at c. 2700 BP and continued to do so until c. 2300 BP when silty clay and ultimately, clayey silt, were deposited, the latter in a salt-marsh environment. Archaeological evidence points to cessation of salt-marsh growth about 1900 BP. There was a second phase of salt-marsh formation between 1750 and 1350 BP. Lithostratigraphic studies away from former major salt-marsh creeks at Newton in north-central Fenland, Cambridgeshire (Wheeler,

1994, 1995), revealed the occurrence of a silty clay unit separating fen peat deposits from salt-marsh clayey silt. Pollen analysis demonstrated a development from *Betula* carr, which was largely destroyed by inundation by saline water, to salt marsh (with *Plantago maritima* and *Zostera*) via an intermediate environment with pollen and spores derived from both seaward and landward facies.

Analysis and interpretation suggested that the **silty clay** unit was deposited in brackish open water which resulted from the **ponding of fresh and tidal water behind the salt marsh**. This evidence suggests the existence of a previously unrecorded successional environment between coastal reedswamp, probably dominated by *Phragmites australis*, and a salt marsh which developed rapidly on a coast

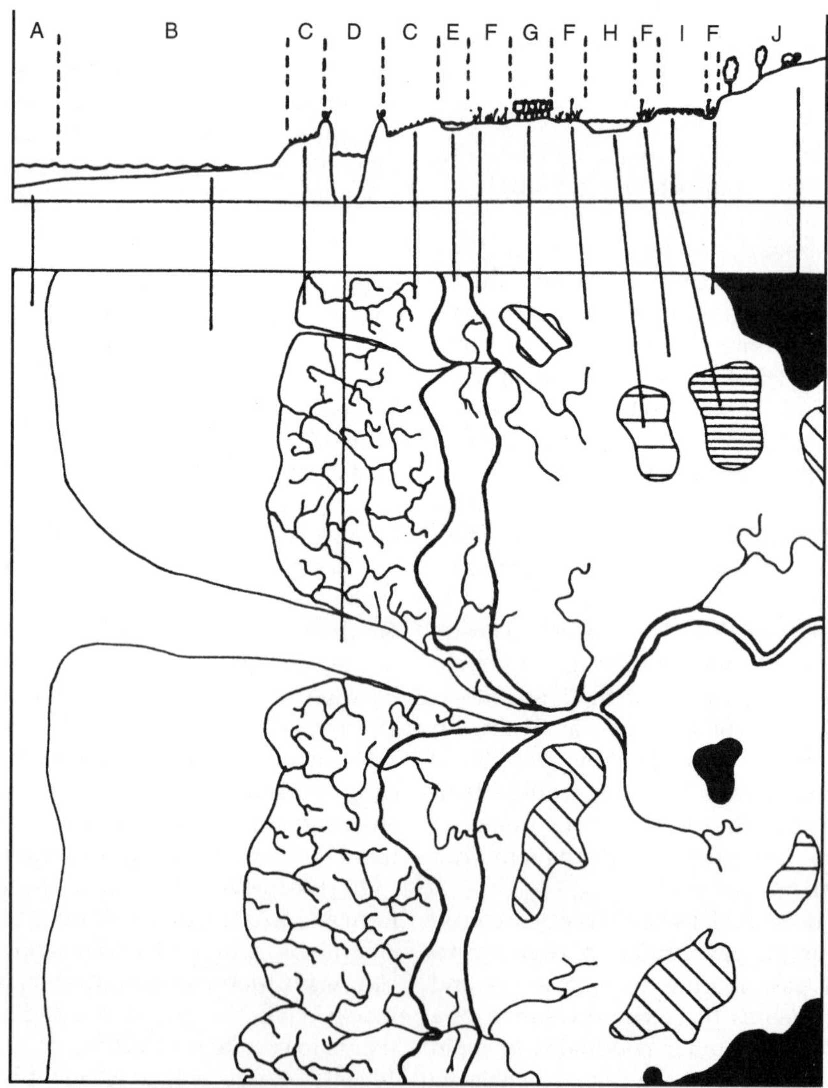

Figure 10.1 A Fenland palaeoenvironmental reconstruction in section and plan view. A, offshore mobile sands and silts; B, sandflat and mudflat; C, salt marsh; D, estuary; E, flooded transition environment; F, coastal reedswamp and sedge fen; G, carr (fen woodland); H, mere or lake; I, raised bog; J, dry land. (After Shennan, 1986.)

with a broad shallow gradient. Figure 10.1 gives a Fenland palaeoenvironmental reconstruction of the circumstances under which the silty clay unit of the Newton core is thought to have been laid down in Zone E, landward of the top of the salt marsh where accretion has been rapid enough to outstrip that in the subsequently **flooded area of the transition environment**. Largely as a result of drainage and sea defence schemes, no comparable environment exists in this area today, but Wheeler (1995) suggests that the commonly observed direct transition from coastal reedswamp to salt marsh is not applicable to extensive coastal wetland/salt-marsh systems where an intermediate environment would naturally occur.

10.2 EVIDENCE OF CONTEMPORARY CHANGE

Evidence of change in climate is drawn from a variety of sources, including direct measurement of concentrations of 'greenhouse gases' – notably CO_2 – temperature, of eustatic rise in sea level over the past half-century, and of changing patterns of ocean currents and meteorological systems. There is also evidence provided by alterations in the distribution and behaviour of various animals and plants, including those of coastal systems.

10.2.1 INCREASES IN ATMOSPHERIC CO_2

Measurements on ice cores have shown that concentrations of atmospheric CO_2 have varied between 170 and 300 ppm during the past 160 000 years. Watson *et al.* (1990), however, estimate that over the past 200 years the Industrial Revolution and increased combustion of fossil fuels have caused a rise in CO_2 concentration from about 280 to 353 ppm in 1990. They also estimate that if such anthropogenic emission levels are sustained, levels of atmospheric CO_2 will increase at a rate of about 1.8 ppm yr^{-1}, reaching some 520 ppm by the year 2100.

Carbon dioxide emissions form part of the **global carbon cycle**. Although emissions are directly to the atmosphere, this carbon will find its way to other **global C sinks**, especially the oceans and vegetation. Nevertheless, it has been estimated that c. 44% of CO_2 emissions remain within the atmosphere. Although CO_2 cannot be considered as a direct pollutant, it has several effects on global climate, which themselves directly influence plant processes. CO_2 is an important 'greenhouse gas' which absorbs and re-radiates infra-red radiation emitted by the Earth, causing temperature increases at the Earth's surface. Mitchell *et al.* (1990) give estimates from computer models that a doubling of atmospheric CO_2 will cause a global average temperature increase of between 1.5 and 4.5°C, with a 'best guess' of 2.5°C.

The work discussed above forms part of the investigations of the atmospheric and terrestrial processes of the climate system considered by the Intergovernmental Panel on Climate Change (IPCC). Rates of climatic change predicted by the IPCC (Houghton *et al.*, 1990, 1992) are now considered to be excessive; Beerling and Woodward (1994) discuss the problems of such predictions and give a more recent estimate of climatic change as it is likely to affect Britain.

10.2.2 MEASUREMENT OF SEA-LEVEL CHANGES

The importance of sea-level changes over the Quaternary Period, which has so far lasted some 2.5 million years, is reviewed by Tooley (1992), who also points to the influence on coastal processes of regional variability in sea-level behaviour over time scales of 10–1000 years. Sea level can now be measured by **satellite altimetry**, supplementing previous estimates made with **tide gauges** where it intersected a coastal surface. The **geological record** provides an indirect and less sensitive means of estimating previous changes in sea level. Tooley (1992) demonstrates that the sedimentary record provides evidence of coastal, sea-level, water-level, and water quality changes. In recent years a data-base of sea-level index points has been built up and used to quantify rates of sea-level change and vertical earth movements.

In terms of sea defence, **storm surges**, in which sea levels greatly exceed those expected from tidal predictions, are extremely important. In February 1990, for example, when the sea wall at Towyn, North Wales, was breached, a flood mark in the town was at +5.39 m Ordnance Datum and even at 2 km further south a flood mark was +5.24 m OD. The Towyn flood inundated 10 km^2 of residential and agricultural land, but very little sediment was carried landward. This is in line with the view that when storm surges breach natural defences landward transport of sediment is restricted (Tooley, 1992).

10.2.3 CHANGES IN DISTRIBUTION AND BEHAVIOUR OF COASTAL ORGANISMS

Polygonum maritimum, a species characteristic of the Mediterranean coast, is at the northern limit of its range in southern England where it has recently been recorded in East Sussex (Harmes and Spiers, 1993) and Hampshire, as well as in west and east Cornwall (Figure 8.3). This, together with records from south-east Ireland and the Netherlands, suggests that the species may be extending its range in response to the hot summers of recent years. With global warming it may be expected that other Mediterranean species may colonize southern Britain and, in general, that the distribution limits of British plants will extend northwards, as well as those elsewhere in the northern hemisphere.

Monitoring by members of the British Herpetological Society shows that a number of amphibians now start breeding earlier in the season. First spawnings of the closely monitored natterjack toads (*Bufo calamita*) in Hampshire, UK, are now at least 2 weeks earlier than they were in the late 1970s. Edible frogs (*Rana esculenta*) have gradually brought spawning times forward from mid-June to May. Spawning times of the common frog (*Rana temporaria*) appear unchanged in Sussex but are consistently earlier in counties such as Cumbria (Inns, 1995).

10.2.4 SCALE OF CHANGE

Although some meteorologists have doubted the likelihood of global warming, there is considerable evidence that global temperatures are currently rising. This appears to be leading to reduction in the areas of polar ice masses, thermal expansion of ocean waters and water from melting of polar and glacier ice giving a consequent rise in sea level, together with increasing instability in climatic patterns. Such changes will have profound impacts on salt marsh and dune species in low-lying coastal areas, some of which are considered in the next section. Many different figures have been estimated for **average rises in world sea levels** in response to global warming; although those predicted by the IPCC (Warrick and Oerlemans, 1990) – of between 8 and 29 cm by AD 2030 (with a best estimate of 18 cm) and between 31 and 110 cm (best estimate 66 cm) by AD 2100 – seemed at publication to provide a reasonable guide they may well be too large. The average rate of sea-level rise over the past 100 years has been 1.0–2.0 mm yr^{-1}; future rates of rise will be faster, but accurate prediction is difficult. Tooley (1998) presents evidence to show that sea level rises and falls during the Holocene, far from being smooth and progressive, have often been rapid. Accelerated rises of up to 40 mm yr^{-1} led to rapid destruction of salt marshes in England and Wales; similar events in the future would produce catastrophic change.

10.3 IMPLICATIONS OF FUTURE CHANGES IN SEA LEVEL AND CLIMATE

Rises in sea level will contribute greatly to **coastal squeeze**, by which salt marsh area is being lost on the landward side through reclamation for agriculture or other purposes – often referred to as **land claim** – which tends to push high water mark towards the shore, and on the seaward side through erosion or inundation by the sea. Rising sea levels cause coastal squeeze even when man-made sea defences are static; unless such defences are abandoned there is no scope for the compensatory development of salt marshes landwards and they become progressively narrower.

Besides diminishing existing dunes, salt marshes and mud flats, predicted rises in sea level and the likelihood of increased storm surges will cause losses of agricultural land, and threaten many populated areas, especially those along major estuaries. Substantial protection is needed, especially from exceptional storm surges, for nuclear power stations situated on the coast, particularly those most at risk.

10.3.1 COASTAL DUNE DEVELOPMENT IN RELATION TO SEA LEVEL RISE

Forecasts of the effects of sea level changes must take account of **coastal sediment cells**, ten of which are recognized in England and Wales (Pye and French, 1993a, p.8), and in which sand dune and salt marsh areas will be influenced differently. Each cell is a length of coast which acts as a semi-closed system along which sediment transport occurs.

A useful approach to the future evolution of sand dune systems is to examine the origins of those which now exist. Changes in sea level and rates of marine sediment supply – as well as in wind strength, rainfall, and evaporation rates – strongly influence the formation of coastal dunes. Individual cases vary widely and it is apparent that dunes can be formed during **marine transgressions** as well as in **regressions**, when the sea level is high as well as when it is low (Pye and Tsoar, 1990). Figure 10.2(a) fits the case proposed by Bretz (1960) for the carbonate dunes of Bermuda, which he concluded had been formed during a period when the sea level was equivalent to, or slightly higher than, that of the present day. Following regression, the sea level rose again and dune building resumed. Because carbonate dunes rapidly become cemented when exposed to air, Bretz argued that these dunes could not have been formed from material which migrated long distances across the Bermuda platform during a period of low sea level.

Other workers have proposed, for Bermuda and elsewhere, the model shown in Figure 10.2(c) in which dunes now exposed along the shoreline originally formed on the continental shelf during glacial **low sea-level** stands. Subsequent marine transgressions submerged some dunes, others advanced onto higher ground and became stabilized when deprived of their sand supply.

Figure 10.2(b) illustrates the theory that most dune formation occurs when sea levels fall during the switch from interglacial to glacial conditions. The argument is that the **falling sea level** lowers the wave base and increases the area of continental shelf over which sand can be transported towards the land; rising sea level raises the wave base and reduces landward sand movement. The **rising sea level model** (Figure 10.2(d)) suggests that marine transgressions initiate transgressive coastal dune development. Although rising sea levels can cause shoreface erosion and offshore movement of sand, destruction of foredune vegetation in areas of high wind energy may allow large amounts of sand to be blown inland as transgressive dunes. This has been shown to have occurred during post-glacial marine transgression on many exposed parts of the eastern Australian coast. Other smaller-scale episodes of trangressive dune activity in this area have been related to local aboriginal burning, and to fluctuations in wind and storm wave climate.

10.3.2 NON-LINEAR THRESHOLD EFFECTS

At La Pérouse Bay, Manitoba, Canada, **trophic interactions** between the herbivore population of lesser snow geese and the coastal vegetation on which they feed (sections 5.4, 5.5)

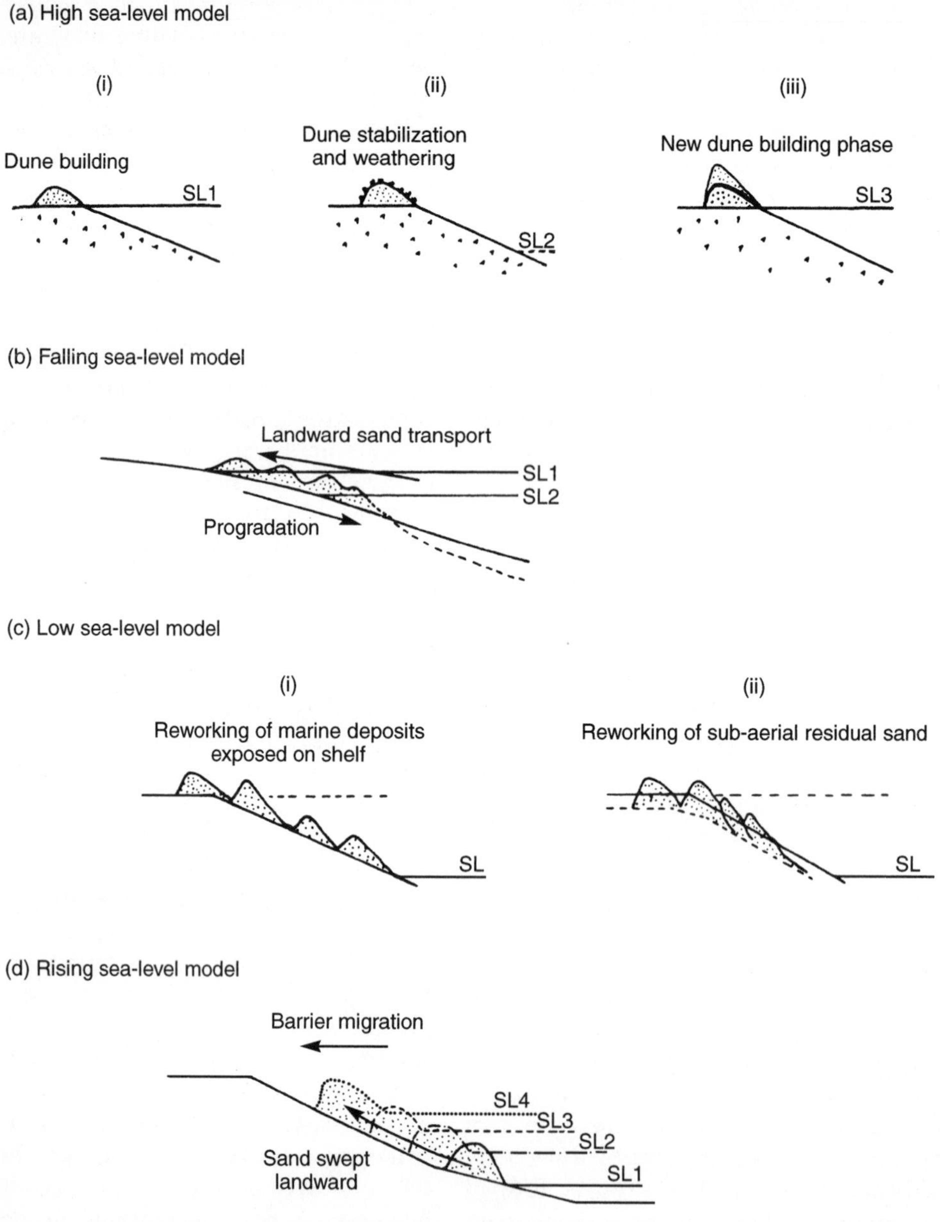

Figure 10.2 Four alternative models of coastal dune development in response to changes in sea level. (After Pye and Tsoar, 1990.)

have altered as a result of changes in the weather pattern. In contrast to the situation in the late 1970s, bad weather to the north in spring has led migrating snow geese to stage in this bay, adding to pressures exerted by the breeding goose population. The birds, which grub for roots and rhizomes of the forage plants, are destroying salt-marsh swards before above-ground growth begins, while intense grazing in summers which have frequently been dry and warm further reduces available forage. Increased evaporation has caused large areas of sparsely covered sediments, devoid of an insulating mat of vegetation, to become hypersaline. This effect is particularly marked in the upper levels of the intertidal flats, which are not inundated by the tides from the time of snow melt in early June until early August, heights of spring tides close to the summer solstice being relatively low. Much of the salt in the coastal regions of Manitoba is derived from buried marine clays which in uneroded parts of the marsh are capped by a layer of highly humified organic matter 3–8 cm deep. Tidal waters at La Pérouse Bay are low in dissolved salts and flush salts from the top soil when they cover the intertidal flats at the end of the summer.

Growth of the two most important forage species, *Puccinellia phryganodes* and *Carex subspathacea*, is reduced at higher salinities; this effect is of most importance in the latter species which grows in less-saline areas (Srivastava and Jefferies, 1995). A decline in the body size of the goslings at La Pérouse Bay has in turn adversely influenced adult size, survivorship and fecundity. Jefferies *et al.* (1995) employ this example when pointing to the importance of non-linear threshold effects in environmental conditions when predicting biological responses to global climatic change.

10.3.3 EFFECTS OF CLIMATIC CHANGE ON COASTAL SYSTEMS

There is a rapidly growing literature, some of it speculative, about such subjects as the impact of climatic change on the coastal dune landscapes of Europe (van der Meulen, Witter and Ritchie, 1991), and research in these areas will continue to develop. Climatic change will undoubtedly influence the speed and nature of pedogenic processes, both directly and by altering the nature of the vegetation; in sand dunes with freely drained soils higher temperatures and lower precipitation surpluses will lead to lower rates of acidification, less leaching and more rapid turnover of litter (Sevink, 1991; see also section 2.5).

Climatic change is largely driven by increases in 'greenhouse gases' but CO_2 itself, via photosynthesis, is central to primary plant processes, so changes in atmospheric CO_2 levels are of great interest in plant physiology. Enhanced levels of CO_2 increase photosynthetic rates of many crop species; Table 10.1 gives data for relative growth rate, photosynthesis and water use efficiency for five salt-marsh species. Clearly, increased levels of atmospheric CO_2 enhance growth in some coastal species, but may reduce it in others. In many species an increase in photosynthesis leads to an improvement in **water use efficiency** (the ratio between photosynthetic and transpiration rates). It is widely considered that increased levels of CO_2 will improve growth of C-3 species, particularly in temperate areas such as herbaceous salt marshes, but will be less advantageous to C-4 species (Welburn, 1994).

In an extended study of an Atlantic coast salt-marsh ecosystem in Chesapeake Bay, Maryland, USA, enhanced CO_2 was found to lead to significant changes in the vegetation (Curtis *et al.*, 1989, 1990). With ambient CO_2 doubled, a natural monoculture of the C-3 rush *Scirpus olneyi* showed increased photosynthesis, root biomass, number of shoots, delayed senescence and increased net carbon storage in the ecosystem. On the other hand, an adjacent stand of *Spartina patens*, at a very slightly higher elevation, showed none of these changes. After 1 year of the treatment, the estimated net primary production of *Scirpus olneyi* was 539 g m^{-2} with doubled CO_2 as compared with 345 g m^{-2} in the control. In a mixed community with the C-4 grasses *Spartina patens* and *Distichlis spicata*, *Scirpus olneyi* showed, with doubled CO_2, an even larger increase in shoot density and biomass than in monoculture. In a high CO_2 environment, *S. olneyi* may be expected to increase at the higher elevations of the marsh at the

Table 10.1 The effect of CO_2 concentration and salinity (250 mM NaCl) on relative growth rate, rate of photosynthesis and water use efficiency on four C_3 and one C_4 salt-marsh species. (After Rozema *et al.*, 1990, 1991)

	CO_2 ($\mu l\ l^{-1}$)	C_3 *monocot*		C_3 *dicot*		C_4 *monocot*
		Bolboschoenus maritimus	*Puccinellia maritima*	*Aster tripolium*	*Spergularia maritima*	*Spartina patens*
RGR	340	34.8	4.1	60.2	74.1	9.3
P_n		n.a.	n.a.	5	2.7	6.5
WUE		n.a.	n.a.	1.52	0.53	1.65
RGR	580	49.2	8	57.1	90.7	4.9
P_n		n.a.	n.a.	4.2	4.1	6.86
WUE		n.a.	n.a.	1.31	0.89	2.12

Abbreviations: RGR, relative growth rate (mg g^{-1} fresh wt day^{-1}); Pn, rate of photosynthesis (μmol CO_2 m^{-2} s^{-1}); WUE, water use efficiency (mg CO_2 [g H_2O]$^{-1}$); n.a., not available.

expense of the C-4 species (Arp *et al.*, 1993); in general, elevated CO_2 may give C-3 species a competitive advantage over C-4 species within mixed communities. Nevertheless, the full implications of increased levels of CO_2 to plant life in general are not fully understood; its interactions with other environmental factors including temperature, and availability of nutrients and water, appear to be key controlling considerations.

In providing a simple predictive model of response to climatic and atmospheric change, Long (1990) examined the factors governing primary productivity in *Spartina anglica*, a C-4 plant whose production is more enhanced by higher temperatures than that of *Puccinellia maritima*, a C-3 species with which it is often in competition. However, the latter benefits considerably from higher levels of atmospheric CO_2 which reduce its photorespiratory losses, whereas *S. anglica* does not. The model assumes a doubling of atmospheric CO_2 and a 3°C increase in air temperatures throughout the year. This increase in temperature would be sufficient to decrease the incidence of photoinhibitory damage in *S. anglica* and cause **leaf area index** (LAI) in this species (which at present leafs late) to intercept 30% of incoming radiation 50 days earlier than at present (the corresponding figure for *P. maritima* is 35 days). It would thus be able to utilize high solar inputs in May and June.

As *Spartina anglica* continues to evolve, its **realized niche** is expected to narrow owing to dieback at the lower levels and invasion by competitors at the upper limits. Gray *et al.* (1995) further discuss the effects of global warming on the distribution of this species, and utilize detailed observations of existing communities in modelling changes in the vertical limits of the *Spartina* zone likely to result from human intervention through management or tidal barrage construction.

Destruction of the ozone layer resulting in increases in UV-B irradiation will influence organisms of the coastal zone differentially. Van de Staaij, Rozema and Stroetenga (1990) showed that enhanced UV-B irradiation caused reduction in growth and photosynthesis which was more marked in *Aster tripolium* than in *Spartina anglica*. Elevated UV-B level was found to depress biomass production at ambient CO_2 concentration in the C-3 upper salt-marsh grass *Elytrigia atherica* by 31%; although doubled CO_2 level at low UV-B gave 67% increase in biomass, at elevated UV-B there was a biomass depression of 8% compared with control plants (van de Staaij *et al.*, 1993).

COASTAL MANAGEMENT AND CONSERVATION 11

11.1 RESOURCE POTENTIAL, PRESENT USE AND HABITAT CREATION

Soft coastal ecosystems based on granular deposits are immensely valuable for **sea defence** as well as in economic, scientific, **conservation** and aesthetic terms. Owing largely to their locations, they are often exploited or destroyed as natural or semi-natural habitats. Salt marshes and stabilized sand dunes have frequently been grazed, or reclaimed for **industrial, agricultural** or **forestry use**. Many dunes, barrier islands and spits have been used as residential sites – often in connection with the tourist industry – or as sources of silica, building or heavy mineral sands and shingle. The importance of these habitats in supporting highly specialized plant and animal communities, or – as with many salt marshes – acting as over-wintering sites or resting points along the migration routes of birds, has been emphasized in previous chapters. The presence of particular coastal species may have other economic consequences: for example, sea beet (*Beta vulgaris* ssp. *maritima*) acts as a host of virus beet yellows of sugar beet.

The potential of the above-ground tissues of *Spartina anglica* as **biofuel** has been investigated by Scott, Callaghan and Lawson (1990). In a trial at Southport, Lancashire, optimal harvest time was in winter from November onwards; in this treatment yield fell from 16 dry tonnes $ha^{-1} yr^{-1}$ to 8 t $ha^{-1} yr^{-1}$ in 3 years. Fertilizer application gave no increase in yield. Costs of harvesting and energy conversion of *S. anglica* are likely to be high compared with other biofuels; ash content, already high because of the presence of silica cystoliths and salt, is raised further when sediment from autumn tides is trapped in the foliage. Biomass became much lower in summer-harvested plots and *Puccinellia maritima* became increasingly dominant.

The ability of *S. anglica* to establish vigorously in mud flats whose surfaces may previously have been stabilized by organisms no larger than diatoms and filamentous algae, thus promoting rapid accretion, has been used to advantage in **sea defence** projects in many parts of the world. By 1964 its geographical limits in Europe were between 48° and 58°N. It had also been introduced successfully at 48°N on the western seaboard of the United States, but plantings had failed at 45°N on the colder eastern side. In Australia and New Zealand it was surviving, and often flourishing, between 35° and 46°S (D.S. Ranwell, personal communication). In 1963, *S. anglica* was first introduced into China, where it now covers more than 36 000 ha on coasts ranging from 21°27′N to 40°53′N, surviving in a very wide range of seasonal temperatures. It is used in reclamation, to stabilize the coast, and to ameliorate saline soils, as well as providing green manure, animal fodder, fish feed, fuel and material for paper-making (Chung, 1990).

South of Boston, Massachusetts, there is a specialist culture of cranberries (*Vaccinium macrocarpon*) in dune depressions, which are flooded during harvesting so that the berries can be skimmed off the water surface. Dune sand in high-rainfall areas is, however, rapidly leached of nutrients and is generally far less valuable for agricultural use than closer-textured soils. In arid climates this problem is much less severe and agriculture on sand can be successful provided that sufficient water is available. Under the ancient system of simple horticulture known as **mawasi** (suction in Arabic) practised in coastal dunes in Gaza and northern Sinai, vegetable and fruit crops are grown in interdune depressions (Figure 11.1). The water table approaches quite closely to the surface of these depressions, some of which are natural while others are excavated. The water here has naturally high levels of nitrogen and phosphorus; additional fertilizer is applied in the form of animal and green manure. Sand removed when plots are levelled is used to create barriers to reduce invasion by saltating sand and plot margins are further protected by the use of palm and tamarisk trees (Pye and Tsoar, 1990).

Mawasi horticulture continues, but intensive modern systems using computer-controlled trickle-irrigation are increasingly used on dunes in southern Israel and Gaza. Although uncovered crops can easily be grown in arid areas using this method, glass or polythene sheeting is necessary to prevent excessive leaching of soluble nutrients in humid areas.

The value of coastal dunes and salt marshes is increasingly recognized, as is the danger of reducing their potential through inappropriate or excessive use. A survey with these issues in mind is that made by Ritchie and Mather (1971) with respect to the conservation and use of the beaches of Sutherland, north-west Scotland. The crofting townships along this Atlantic coast are closely associated with beach and blown sand areas. Areas of arable land are very small and largely devoted to production of oats, potatoes and hay. Wind erosion following destruction of the vegetation carpet by over-grazing is fairly common, while pressures from camping and caravanning are heavy in some beach areas. Ritchie and Mather concluded that whereas some erosion and general deterioration resulting from summer tourism could be avoided, much of the dune erosion was natural and that due weight had to be ascribed to the economic benefits conferred by the holiday trade. Tourists were not deterred by damaged machairs and even saw advantages in the shelter and privacy afforded by erosion hollows.

11.1.1 HABITAT CREATION

Knowledge of the biology and germination ecology of the species involved, valuable in all attempts at habitat creation, is crucial in shingle vegetation. Walmsley and Davy (1997a–c), who experimented with both seeds and transplants in the successful restoration of shingle-beach vegetation damaged during construction of Sizewell 'B' nuclear power station on the

Figure 11.1 The mawasi horticultural system and its relationship to coastal dune slacks and the fresh groundwater table. (After Tsoar and Zohar, 1985.)

Suffolk coast, emphasize the importance of using transplants and creating the correct substrate composition. A seed mixture of *Crambe maritima, Eryngium maritimum, Glaucium flavum, Lathyrus japonicus* and *Rumex crispus* (of which all but *R. crispus* showed innate dormancy) was used in a field experiment examining the effects of proximity to the sea, particle size composition and amendment with organic matter upon seedling emergence and establishment. After two seasons vegetation cover was poor and dominated by *Glaucium flavum* when, 20 months after sowing, the entire site was buried by 15–25 cm of substrate and all the seedlings were exterminated. The use of seed in growth habitat schemes on shingle beaches appears to be less effective than in other habitats, such as woodlands.

Seeds of the species used in the sowing experiments had been gathered from the site 4.5 years before and stored at low temperature and humidity. When tested at 7 years they were found to retain much of their original viability, but had developed more stringent germination requirements, particularly with regard to salinity and temperature conditions. Aged seeds of *Honckenya peploides* lost the requirement for stratification; germination in this species was also no longer promoted by light. Appropriate pre-treatments and the use of optimal conditions for germination enable the efficient production of transplants from aged seed stock for restoration work.

Transplants of *Crambe maritima, Glaucium flavum* and *Rumex crispus* in a shingle-dominated plot showed greater growth than on a sandy plot at an equal distance from the sea. The performance of *Eryngium maritimum* was similar in both plots. At least in the short term, container-grown plants used in habitat creation of shingle vegetation show good survival and rapid growth and establishment. However, response to organic matter and fertilizer treatments in a Latin Square design field experiment with container-grown plants showed them to be both costly and ineffective; the money is better spent on creating a substrate of **suitable particle size composition** (Walmsley and Davy, 1997c; section 8.2).

Habitat creation schemes have succeeded in developing **simulacra** of a number of vegetation types, and indeed their use in areas where the natural vegetation has previously been destroyed is often worthwhile. There is, however, a danger that developers and others will use offers of habitat creation schemes in bargaining for re-allocation of land use which destroys natural habitats. Such an offer of **mitigation** was made when the owners of a hotel in Atlantic City, New Jersey, wished to fill in the saltwater cove that it surrounded, offering to replace it with 2 acres (0.8 ha) of '**facsimile**' **wetlands** further up the inlet. This example is quoted by Race and Christie (1982), who provide an evaluation of wetlands creation studies, and also discuss the role of mitigation in a number of East Coast States. The US Army Corps of Engineers has developed techniques for marsh creation to enhance and stabilize dredge spoil materials. In 1982, Race and Christie considered the available evidence to be insufficient to support the view that man-made salt marshes provide the important values of natural ones, or that they function in the same way.

On the other hand, it is clear that in the past natural soft coastal systems have developed when coastal communities migrated seawards following a fall in sea level or, as is now occurring, towards the land when sea level has risen. The economic implications of the current sea level rise have led to a whole series of research cooperations within the past decade (Nordstrom, Psuty and Carter, 1990; Allen and Pye, 1992). In conservation terms it is desirable to retain or even extend the area occupied by the evolving communities of salt marshes, dunes and shingle; moreover, the cost effectiveness of using soft coastal systems for sea defence is no longer in question. Future coast protection programmes will benefit from rapidly developing knowledge of these systems, which is increasingly based on a multidisciplinary approach involving geomorphologists, ecologists and engineers.

11.2 MONITORING CHANGE

Long-continued and systematic recording of the British flora by members of the Botanical Society of the British Isles (BSBI) is of particular value in detecting floristic change, such as the apparent increase of the southern species *Polygonum maritimum* in recent years (section 10.2).

On a smaller scale, the establishment of a baseline is also an essential prerequisite when following changes in local communities or single-species populations over long periods. This is most important when considering, for example, the role of environmental change as a cause of declines in the **natterjack toad** (*Bufo calamita*), the rarest British native amphibian. Warning given by Beebee (1976, 1977) led to effective conservation measures. In the sand-dune system at North Merseyside, Lancashire, a survey by Smith and Payne (1980) showed that most spawning took place in the shallow water of slacks or excavated scrapes; deep ponds, ditches and bomb craters were little used. Unsuccessful breeding was mainly due to shallow water bodies drying up before the tadpoles could metamorphose.

In dunes with little or no wet ground, small excavations with gentle gradients down to or just below the water table are used to diversify the system. Steep-sided excavations are to be avoided as their margins are liable to collapse and gradual slopes are more easily colonized than steep ones. This type of measure is effective in maintaining and improving the breeding sites of *Bufo calamita* (Figure 11.2), which are warm shallow pools, chiefly on coastal sand dunes (section 7.3). The excavated sand also provides sites for plant colonists. Flemming (1998) emphasizes the importance of salt marsh as a key habitat for the natterjack, which breeds in shallow pools whose water has a salinity up to 10–15% of that of seawater.

A similar instance is that of the **sand lizard** (*Lacerta agilis*) whose population on the Merseyside sand dunes dropped from 8000–10 000 or even tens of thousands during the 1920s to the 1940s, to several thousand in the 1950s and 1960s, a few hundred in the late 1970s, and then to less than 200 in the late 1980s (Corbett, 1988). During this period, 23 km^2 (42.6%) of the dune habitat between Southport and the River Allt was built over (Jackson, 1979), and much of the remainder greatly changed by the construction of golf courses, building of an airfield, abstraction of sand for building, and increased growth of woody scrub following diminution of the rabbit population after myxomatosis.

Cooke (1991) reports a monitoring exercise employing 5 m × 5 m quadrats centred around positions where individual lizards had been seen. His analysis showed that the lizards do best in dunes facing south to south-east and with a slope of 30–49° (making them convenient for basking) which are not entirely fixed but possess a mosaic of vegetation heights and some bare sand. Besides allowing habitat quality to be assessed at each site, this work also showed that the ease of sighting individual lizards – and hence total population numbers – remained unchanged during the 1980s. This was an encouraging indication of the value of protected status to a species previously thought to be on the verge of local extinction.

11.2.1 NEW COMMUNITIES ASSOCIATED WITH SEA DEFENCE WORKS

The largest European sea defence works in the mid-20th century were those associated with the Delta Plan approved by the Dutch parliament after the catastrophic storm surge of 31 January 1953, in which 1300 people were drowned. These works resulted in the formation of a series of new communities and essentially initiated a vast ecological experiment. They also largely caused the destruction of the 'green beaches' ('**groene stranden**'), which were large elevated beaches densely covered with a mosaic of salt marsh and wet and dry sand dune communities along the estuaries (van der Maarel, 1979).

Figure 11.2 Natterjack toad (*Bufo calamita*), the rarest British amphibian. (Drawn by P.R. Hobson.)

On a much smaller scale is the new plant community (Figure 8.2) which has developed on a small beach adjacent to the western breakwater of the Brighton Marina, East Sussex, UK. This structure traps a great deal of jetsam, including discarded plastic, ropes, nets, cans, and seaweed, which forms a mulch in which seeds germinate; it also tends to prevent the former eastward drift of shingle. Since the completion of this arm of the marina in the early 1970s the flint pebbles of the beach, together with patches of fine shingle and sand, have steadily built up above high tide level. During this period a plant community has arisen which has a composition found nowhere else along the Sussex coast. The most notable species is *Polygonum maritimum*; others are *Atriplex glabriuscula*, *A. littoralis*, *A. portulacoides*, *Beta vulgaris* ssp. *maritima*, *Cakile maritima*, *Calystegia soldanella*, *Catapodium marinum*, *Crambe maritima*, *Crithmum maritimum*, *Elytrigia atherica*, *Glaucium flavum*, *Honckenya peploides*, *Parapholis strigosa*, *Raphanus raphanistrum* ssp. *maritimus*, *Spergularia media* and *Tripleurospermum maritimum* (Harmes and Spiers, 1993; Packham, Harmes and Spiers, 1995). As the Channel current and the prevailing wind in summer both come from the south-west, it seems possible that the seeds of the sea knotgrass in particular may have been swept along from Cornwall, Hampshire, the Channel Islands or Northern France. It is unlikely that the seeds of all these species were brought in by birds; greenfinches and sparrows were the only seed-eating species observed on the beach.

11.2.2 LOSS OF SAND-DUNE HABITAT AND ITS COMPENSATION

Pye and French (1993a) estimate that natural and anthropogenic habitat losses in England over the next 20 years will be, in terms of area and as a proportion of the national resource where known: 240 ha (3%) for **sand dunes**, 200 ha (4%) for **shingle formations**, and 50 ha for **coastal heath**. Sand dune habitat will be lost by erosion of frontal dunes on exposed coasts and at the up-drift end of coastal cells, with further anthropogenic loss owing mainly to recreational activities. Losses are likely to be heaviest in the north-east and south-west. Compensation for these losses can be achieved by:

- encouraging frontal dune accretion where this occurs naturally, especially at the downdrift end of coastal sediment cells;
- employing beach recharge and 'soft' protection works to encourage foredune stability; and
- allowing sand drifts to move inland and cover former backbarrier deposits, pasture, or cultivated land, in areas where frontal dune erosion cannot be prevented.

11.2.3 MONITORING EROSION AND VEGETATION CHANGE IN SALT MARSHES

The salt marshes of south-east England – which protect extensive lengths of sea defences in a low-lying, highly populated region – have experienced marked erosion for many years. This has caused concern over the loss of a valuable conservation resource, the potential danger to sea defences as loss of marshes leads to destabilization of embankments, and the possible effect of sea defence maintenance work on the remaining marsh areas. Burd (1992) re-mapped the salt marshes of Essex and North Kent as part of an investigation set in train by the Nature Conservancy Council (NCC) using as a baseline a report produced in 1973 by the Institute of Terrestrial Ecology. Losses ranged from 10% on the Dengie Marsh to 44% on the River Stour; all the 11 estuaries surveyed had experienced significant losses between 1973 and 1988. The largest areas of erosion were in the sparsely vegetated pioneer zone at the seaward margin of the marshes. The relatively low losses on the Dengie (10%) and Colne (12%) Marshes were significant as both are large but with smaller marsh fronts relative to area than any of the other sites. This supports the view that erosion is wave-driven; the Dengie is also morphologically different from the other sites, being an open-coast site with very little dissection by creeks.

This example is indicative of processes influencing salt marshes in many parts of the world. It fits well with the view that the loss of salt marsh area – as well as a change to low marsh vegetation in sites which formerly supported well-differentiated low, middle and high marsh – is the result of **rising sea level,** caused in this case by a long-continued isostatic readjustment. The five-volume report by Pye and French (1993b) on erosion and accretion processes on British salt marshes, which deals with a similar topic on a much larger scale, concludes that tidal flows are of fundamental importance in controlling sedimentation on salt marshes. While waves are the main agent responsible for marsh erosion, tidal currents transport wave-eroded sediment away and may contribute to marsh dissection by enlarging tidal creeks. Figure 11.3 shows variation in marsh edge morphology resulting from accretional and erosional processes (Figure 11.4) discussed by Pye and French (1993b). The report also emphasizes the importance of monitoring change, especially by repeated aerial photographic survey and ground levelling along specified transects.

11.3 MANAGEMENT OF DUNES, SHINGLE RIDGES AND BARRIER ISLANDS

As already discussed (sections 6.6 and 9.3), the high species diversity found in dune systems is related to the widely varying water regimes, relatively low nutrient status – which permits the establishment and survival of low RGR species – and grazing, often by rabbits, which are features of so many of them. Species richness and habitat diversity in dune systems, which by their nature are highly dynamic, are greatest when a full sequence of successional stages is represented; management is desirably directed to maintain this diversity. Besides such relatively crude procedures as the removal of woodland scrub, management

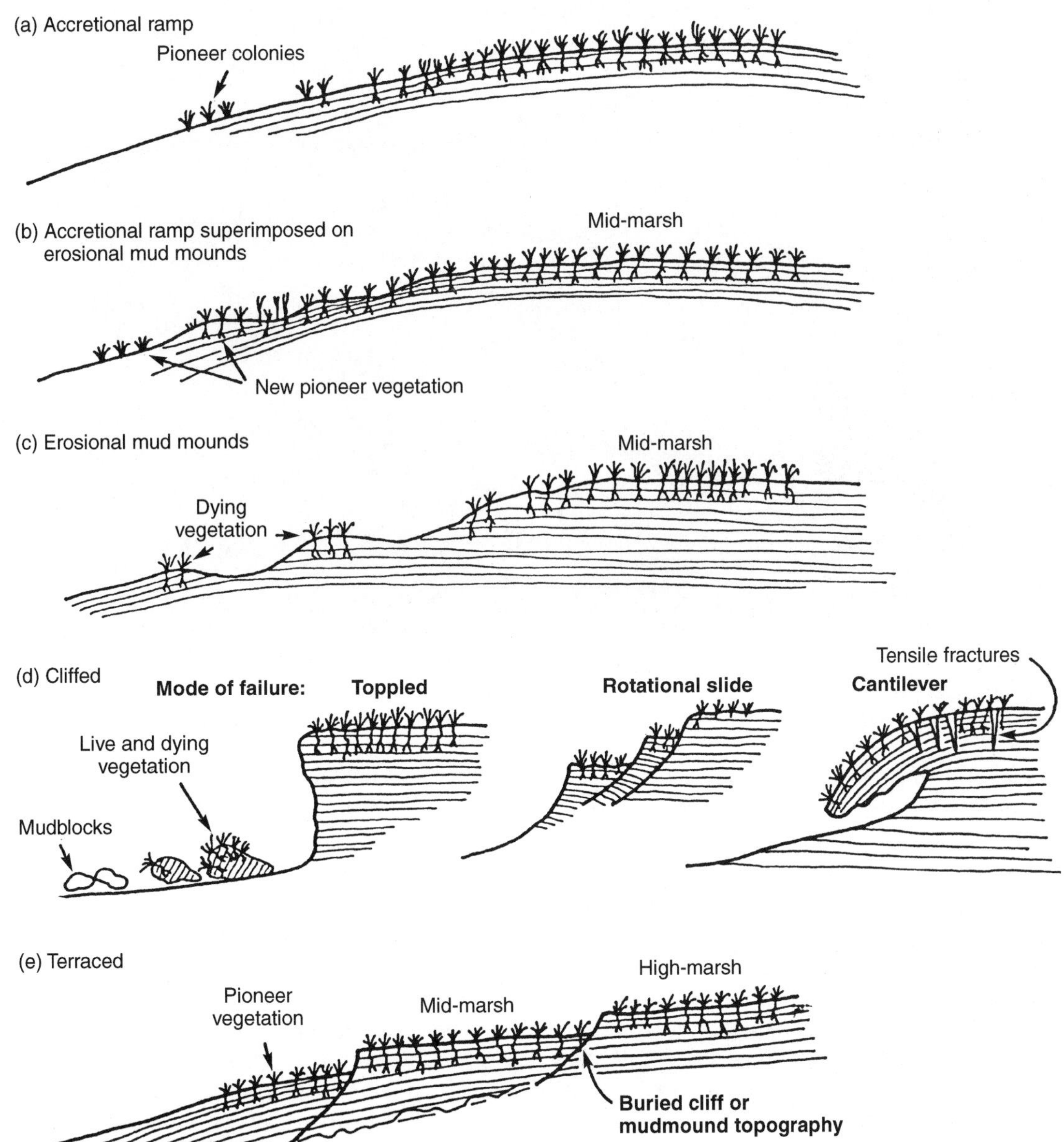

Figure 11.3 Variations in salt marsh edge morphology, in relation to formational processes. (From Pye and French, 1993b; courtesy of Cambridge Environmental Research Consultants Ltd.)

of vegetated dunes for conservation involves manipulations of the water regime, nutrient status and grazing/mowing according to the objectives involved. Where dunes are regularly subjected to trampling – as on golf courses – fertilizer treatments and regular mowing encourage a compact and resistant sward. Parts of some dune systems, such as that at Ynyslas, Wales, which may be visited by half a million people in a year, cannot withstand

Figure 11.4 Erosion on the Keyhaven salt marshes west of Lymington, Hampshire, UK. (Photograph by M.C.F. Proctor.)

constant trampling and the best solution is to institute a system of walkways and vantage points such as that shown in Figure 11.5. It is particularly important to avoid trampling vegetation when the ground is damp. However, in dune systems subject to only light recreational use, or perhaps with few or no rabbits to initiate small-scale mobility, over-stabilization may be as great a management problem as erosion. This is emphasized by Radley (1994) who identifies four main issues – the **importance of naturalness** and the **management of instability, recreation and succession** – in coastal dune management for **nature conservation**. Many organisms depend on open dune habitats, so a dynamic approach to dune conservation is advocated (Doody, 1989b).

Carter (1980) uses a comparison of slope failure on vegetated and non-vegetated dune slopes facing the sea at Portrush, Northern Ireland, to argue the case for **retaining a degree of instability** if seaward dunes are to remain effective barriers against marine erosion. Following close planting with marram, the vegetated slope had been criss-crossed with wire fences and gorse brushwood, and a dense, multi-species sward built up. High tides and waves combined to undercut and erode the base of both the vegetated and the non-vegetated dunes in February 1975. Along the base of the latter erosion scars were rapidly filled by avalanching, sand contributed to the beach and nearshore restored depleted levels and absorbed wave energy, while a near-constant angle of repose of 33.5° was maintained. On the vegetated section the basal cliff was considerably higher, little sand was supplied to replenish beach levels, and a series of small rotational slides and slumps on most of the seaward slope of the dunes gave an irregular profile, resulting in 'terracettes'. Although sediment residence time is much greater for planted as opposed to loose sand, 'over-strengthening' caused by increases in

Figure 11.5 Walkway and vantage point at Ynyslas Dunes, North Wales, where dune blowouts are been repaired by fencing and planting with marram tillers. (Photograph by John R. Packham.)

cohesive shear strength resulting from the development of root systems may cause difficulties including the risk of massive slab failure. It is essential to know the maximum stable cliff angle and the probable magnitude of storm-induced recession if stabilization is to be successful in the longer term. In an evaluation of the various restoration and reclamation techniques that have been employed in Irish dunes, Carter (1985) noted that although more recent approaches stressed scientific appraisal, sufficient financial support and effective management were generally lacking.

Dunes and machairs are dynamic systems in which natural erosion and deposition may alternate or occur concurrently. On a global basis, however, erosion of soft coasts by wind and water is currently a cause of concern. The Scottish machairs in particular are subject to scouring by fierce winds which can rapidly destabilize granular deposits whose vegetated surfaces have been damaged (section 7.1). Such damage results from rabbits and other animals (grazing, scraping, burrowing), changes in arable and pastoral agriculture, sand extraction, recreation (particularly that involving vehicles) and tourism.

Angus and Elliott (1992) discuss a number of attempts to reinstate damaged machairs in the Outer Hebrides. Such schemes usually involve the prevention of further damage to the area concerned, often by fencing it off permanently or temporarily, followed by repair of the damaged vegetation surface. Tillers of *Ammophila arenaria* and of other grasses are frequently close planted, often being covered by brushwood which slows the wind and reduces erosion. In the Scottish islands little brushwood is available but at Eoligarry, Barra, encouraging results were obtained from the use of *Ammophila* planted in holes poked through a geotextile mat. The use of solid barriers should be avoided, while the dumping of rubble and the creation of **gabions** (stone or earth-filled frameworks of wicker or metal bands) on the dune front or in blowouts simply promotes scour. **Porous barriers** in the form of fences placed at right-angles to the prevailing wind slow the air, promoting deposition close to the fence where it can be stabilized by marram planting, but allowing other sand to pass on and feed the machair behind. Cattle dung, seaweed and the planting of marram have traditionally been used to stabilize blowouts; indeed cattle dung, a rich seed source, is still used on bare areas in North and South Uist. Sheep are a problem because they shelter in blowouts and tend to shoulder into the margins of hollows, thus enlarging them, but the greatest difficulty is caused by rabbits many of whose populations have largely recovered from deliberately introduced myxomatosis. In Ireland hay bales trap sand very effectively in plantings of *Leymus arenarius*.

Leymus arenarius is the only native grass species which can be used to stabilize drifting sand in Iceland, where it is grown extensively from seed in both coastal and inland regions (section 6.2). About 30 t of seed is harvested every autumn from c. 2000 ha of wild stands of *L. arenarius* and sown in the following spring. Greipsson and Davy (1995) recommended that stands selected for harvest should have an individual seed mass (averaged for the whole spike) of at least 5 mg; such stands are most likely to be found in early-successional dune sites. These authors showed that mean seed (caryopsis) mass of 36 natural populations (34 in Iceland, one each in Scotland and England) had a six-fold variation. Large seeds germinated much faster than smaller ones and also achieved considerably greater total percentage germination.

11.3.1 EFFECT OF DIFFERENT SAND ACCUMULATORS ON RESULTANT DUNE FORM

As noted earlier, sand deposition rates can be enhanced by the planting of **vegetation**, the use of **sand fences** – which create regions of low wind velocity both in front of and behind the fence – or a combination of methods which are discussed in detail by Pye and Tsoar (1990). The systems available differ considerably in their effectiveness, and also produce a wide range of dune forms (Figure 11.6). Sand fences can produce a fairly steep high dune, while brush matting can be employed to form a broad low dune. The use of double or multiple fences can also produce broader dunes. Maximum sand deposition is achieved by fences with 36–40% porosity; a fence with less than 20% porosity behaves almost as a solid obstacle. Wooden slat fencing is most commonly employed, but brushwood and nylon mesh are also effective.

Artificial barriers have constantly to be maintained and modified: the use of vegeta-

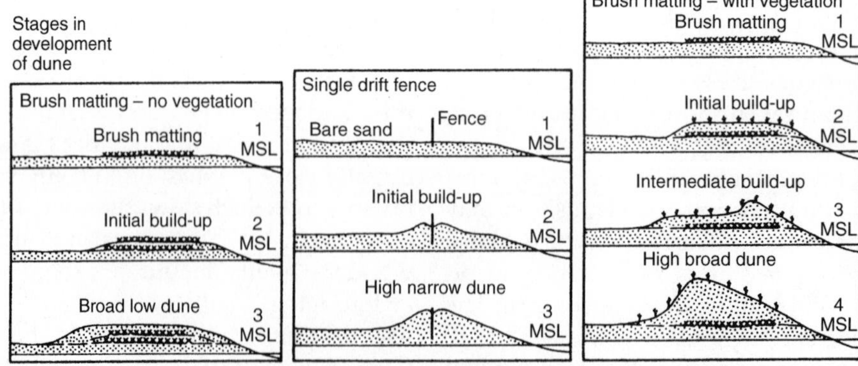

Figure 11.6 Effect of different sand accumulators on resultant dune form. MSL, mean sea level. (After Gale and Barr, 1977.)

tion undoubtedly provides the most effective and inexpensive method of dune creation and stabilization. Belts of trees and shrubs act as self-renewing fence systems; the species used should have a bushy shape to promote sand deposition and be able to survive expected rates of sand accumulation. Various species of *Tamarix* and *Eucalyptus* are very successful in Middle Eastern shelterbelts. The planting of grasses, shrubs and trees over large areas is also employed to trap sand and prevent its inundation of roads and landscapes. Pre-planting modification of the existing land form with bulldozers can promote establishment, as can the use of binding agents such as oil or latex which are sprayed onto the surface of the sand. All protection dunes should be constantly inspected if gaps are to be repaired before serious blowouts enlarge.

11.3.2 SEDIMENT TRANSPORT IN RELATION TO COASTAL PROTECTION

One method of protecting beaches is to feed them with shingle which can sometimes be removed in severe storms as, for example, in Christchurch Bay, Dorset, in 1991 when it was the larger stones which were moved first. Subsequent experiments devised at Southampton University involved the use of a platform called TOSCA (Transport of Sediments under Combined Action) which was placed on the seabed and which had electromagnetic current meters at 0.3, 0.7 and 1.2 m above the shingle surface. These were used to measure the speed of the water currents. Other systems provided estimates of wave height, average water depth, and the speed with which the shingle was moving. The ability to monitor the dynamics of such systems on an almost real-time basis is likely to provide insights of great practical value.

11.3.3 BARRIER ISLANDS

Coastal defence systems which fail to take account of natural processes are costly, ineffective or both. This is particularly true of many barrier islands and spits where the importance of 'wash-over' processes has already been emphasized (sections 1.2 and 8.6). The inner margins of the barrier islands along the mid-Atlantic coast of the USA were used as village sites by early settlers, who thus avoided many of the potentially damaging activities of storms and oceanic overwash. In the 20th century, demand for recreational land, and a desire to build at the ocean margin, have led to the siting of houses, motels, roads and public utilities in areas vulnerable to storm damage.

Sand fences were used in the late 1930s to construct a large foredune protecting the highway along much of the North Carolina Outer Banks. This dune system was extended to the southern tip of Ocracoke Island in the 1950s and stabilized by planting rear dune slopes with American beach grass *Ammophila breviligulata*, trees and shrubs in the expectation that this artificial system would prevent over-topping of the beach by storm waves and surges, and also serve as a sand reservoir to renourish the beach during storms.

Sixty years later the results of the programme can be illustrated by comparing Core Banks with Hatteras Island. Core Banks is relatively unchanged, its beach a sandy expanse varying from 100 to almost 200 m wide. The foredunes are low and the vegetation sparse, while the salt marshes on the inner sound are extensive. Storm waves expend their force on the low dunes and wash-over renews the island's inner margin with sediment. Hatteras beaches, once up to 200 m wide, have receded to little more than 30 m wide at the same time becoming much higher, up to 9 m in places. Hatteras is much greener than Core Banks, partly because greater height has protected dune grasses and shrubs from salt spray, but also because the grass plantings were fertilized. The sound side marshes, deprived of washed-over sediment, have narrowed, while stormy seas breaking with their full force on the high dunes have steadily eroded the beach sand.

Support for the dune-stabilization policy waned after the storm of Ash Wednesday, 6 March 1962, in which the physical geographer Robert Dolan, who was making observations to correlate beach changes with measurements of wave heights, currents and tides, lost his instruments when the gale hit the Outer Banks. This storm had remarkably little effect on the mainland, but devastated installations on the barrier islands, demonstrating all too clearly their essentially dynamic nature (Sackett, 1983). Groynes, breakwaters and seawalls may all protect for a time, but protection of an area by these devices invariably results in accelerated erosion elsewhere. Sand replenishment by pumping dredged material onto the shoreface of an eroding coast is usually cheaper than constructing barriers of the type just mentioned. The sand is eventually carried away by shore currents so a continuous programme is required; furthermore, dredging may well promote erosion elsewhere.

The rate and pattern of the vegetation changes resulting from the dune-stabilization programme along the Outer Banks have been determined by remote sensing, particularly of Ocracoke Island (Schroeder *et al.*, 1976). Photographs for 1958, 1962, 1968 and 1974, showed a clear delineation of five zones, whose relative areas changed during the period of study. Bare sand (zone 1) near the sea was followed in turn by prairie, woody vegetation, cordgrass/rush, with low salt marsh on the further side. Zone 1 is best regarded as an 'active sand zone'. Even the outer beach, with its drifting sand and heavy salt spray, develops a sparse vegetation of such species as *Euphorbia maculata* and spike grass (*Uniola laxa*) during calm periods. Zone 2 (coastal prairie) was defined as grassland beyond the normal reach of wash-over processes, although liable to erosion during major storms. Here the grassy dunes had a more extensive cover with sea oats (*Uniola paniculata*) and beardgrass (*Andropogon*).

On low-lying islands, such as Core Banks, storm damage and overwash cause repeated oscillation between the active sand zone and even the highest dune. Intervention can occasionally succeed, as on a section of Core Banks that was so narrow that no vegetation could establish. Protective barrier dunes eliminated overwash altogether and extensive grassy communities developed, but when dune maintenance was abandoned in the late 1960s a dynamic, overwash-adaptable vegetation system took over.

Barrier island ecosystems are so delicately balanced that significant changes result from minor alterations of the physical environment. Large artificial barrier dunes shift successional trends towards the establishment of woody communities. These are not resistant to salt spray and very extreme storms, whereas grassy dune vegetation adapts to overwash and sand burial (Dolan, 1980). Though the construction of artificial dunes may offer immediate benefits on barrier islands, the long-term implications of this form of environmental management need careful consideration.

If barrier spits or islands eventually migrate to the mainland the position is entirely altered, overwash simply driving substrate up the shore. The barriers afforded by artificially maintained or constructed dunes have been successfully used in many developed areas to prevent inland penetration of waves and storm surges; the US Army Coastal Research Centre has played an important and continuing role in this work. Situations at each site and within each coastal cell, while having many common factors, possess unique combinations of features which require full evaluation before a coast defence or management strategy is developed.

11.4 OBJECTIVES OF SALT-MARSH MANAGEMENT

The last century has seen an enormous loss of salt marsh area world-wide; the coastal squeeze described in section 10.2 has operated more severely on this coastal ecosystem than any other. Recent examples within the UK are the 858 ha of salt marsh which were lost from the Wash by land claim between 1970 and 1980, and the 14.5 ha of marsh lost from the Orwell estuary when the docks at Felixstowe, Suffolk, were extended (Pye and French, 1993a). In the first instance, therefore, primary management objectives are to **prevent further loss of salt marsh area** and to maintain or improve the quality of existing salt marshes. The achievement of these apparently simple objectives is in practice anything but easy. Commercial pressures, including those connected with transport facilities, harbours and their associated manufacturing and storage facilities, will continue to grow, while the **prevention of the pollution** associated with these activities (section 9.1) will require continuous and expert **monitoring and enforcement**.

The retention of areas of salt marsh equivalent to those which now exist will be difficult; the stabilization and preservation of all those which we have today is impossible. Salt marshes and mangrove swamps are by their very nature constantly evolving and changing; in a period of rising sea levels where tidal barriers and sea walls are employed to defend low-lying areas the tendency for these systems to migrate up the existing shore is frequently heavily constrained. The degree to which coastal defences can operate effectively is clearly related to the rate at which the sea level is rising; as already noted (section 10.3) this varies, sometimes quite sharply, from one area to another.

The long-term objectives of salt marsh management will vary according to the coastal defence strategies adopted in particular areas. Where the solution adopted is that of '**planned retreat**', or '**set-back**', management should be designed to promote the effective migration of existing species and communities. A major difficulty in assessing the extent of the total problem has been a lack of information concerning the areal extent of the various coastal habitat types. Comprehensive national inventories are, however, now becoming available; Pye and French (1993a) give estimates of combined natural and anthropogenic losses on the coasts of England for the next 20 years. These areas (and the proportion which they form of the existing resource) are for salt marsh 2750 ha (8%), for intertidal flats 10 000 ha (4%), and for saline lagoons 120 ha (10%) – estimates which should be viewed as minimum targets for habitat re-creation.

When individual cases are considered in the short term the outcome may be dictated by the species or community which it is pre-eminently desired to conserve. Tubbs (1995a) discusses an area between Lymington and Hurst Spit, Hampshire, where the sea wall was in an advanced state of deterioration and the NCC proposed to retreat the landward marshes and allow the sea to cover areas reclaimed long ago. This, or even the construction of a large new wall, would have involved the loss of major areas of the lagoons on the landward side of the wall. The lagoons were then discovered to contain major populations of the starlet sea-anemone (*Nematostella vectensis*), the lagoon sand-worm (*Amandia cirrhosa*), *Gammarus insensibilis*, and the charophyte *Lamprothamnium papulosum*. All these lagoonal species are specially protected under the Wildlife and Countryside Act 1981, and the final decision was to abandon the idea of managed retreat and to advance a new wall around the embanked marshes further down the shore. As Tubbs remarks, the rate of sea-level rise in the locality should have dictated a different outcome; we cannot ultimately defy the impact of natural change. Nevertheless, many **embanked marshes** are nature reserves, often with a special value as breeding sites for wetland birds, as high-water roosts for the waders feeding on estuarine mud flats, and as winter feeding grounds for wigeon (*Anas penelope*) and Brent geese (*Branta bernicla*). Their diverse flora includes such nationally scarce plants as *Polypogon monspeliensis* (annual beard-grass), *Alopecurus bulbosus*, *Carex divisa* and *Ranunculus baudotii*. Many existing embanked salt marshes are of no great antiquity: provided programmes of managed retreat commence in good time it should be possible to move back all the important community components along with the salt marsh.

The review of conservation of British salt marshes by Doody (1992) provides an excellent example of the information needed when devising a strategy to deal with a national resource of this nature. Starting with an evaluation of the value of salt marshes to different groups of people, existing systems are put into an historical perspective, giving the areas of salt marsh 'reclaimed' for agriculture. The distribution of the 44 000 ha of salt marsh in Great Britain is indicated (Pye and French, 1993b give a figure of c. 46 000 ha), lists are given of 28 salt marsh communities defined by the National Vegetation Classification, and trophic relationships within an estuary are outlined. Community dynamics are discussed and so, most helpfully, is the operation of the Wildlife and Countryside Act 1981 with particular regard to salt marshes. The review is designed to elucidate the general background, placing competing demands and special interests within a single framework so that best use is made of the resource.

11.5 SEA DEFENCE AND PLANNED RETREAT

11.5.1 SEA DEFENCE: THE POLICY OPTIONS

The attitudes of particular flood control authorities are strongly influenced by the nature of the areas for which they are responsible. Pethick (1996), who emphasizes that coastal shingle deposits, sand dunes and salt marshes are systems whose materials are constantly reworked, argues that – where possible – natural processes should be allowed full play; in consequence important structures should not be sited close to sea level. This philosophically attractive view is inapplicable to the Netherlands, two-thirds of which are below sea level, and where in 1990 the Dutch government decided to stop any further structural coastal recession. This 'dynamic preservation' policy aims to maintain the 1990 coastal outline almost solely on the basis of extensive **sand nourishment** and with the minimal use of hard engineering (de Ruig, 1995; Hillen and Roelse, 1995). It is hoped that using **Integrated Coastal Management** (**ICM**), alternatively known as **Coastal Zone Management** (**CZM**), to coordinate planning for coastal defence, conservation, transport, housing and other matters, will reduce the increasing stress on this dynamic and sensitive zone. Sand may be added on the seaward or landward side of the shore ridge, as a horizontal layer on the beach, or supplied to the foreshore (Figure 11.7). The ecological effects of these methods, and also of dredging for sand on benthic organisms, are discussed by Loffler and Coosen (1995).

Pye and French (1993c) conclude that in any particular coastal defence situation the five main alternatives which can be adopted, often in combination, are:

1. The '**do nothing**' option in which the coast is left to erode where unprotected, existing defences are allowed to fall into disrepair, threatened land and property are abandoned to the sea, and important installations are moved inland when necessary.
2. **Managed retreat**, in which existing defences are selectively removed or modified, and a new line of defence is established further inland.
3. **Maintained defence** involving the improvement of hard defences, including armoured embankments and concrete sea walls, or using soft engineering approaches such as beach recharge. Figure 11.8 shows an example of a **retarded defence** involving an area of seeded pseudo-salting beyond the permanent road next to the sea wall, whose toe and cap are reinforced with concrete to enhance stability. High tides over-top this wall but are contained by the embankment which forms a second line of defence.
4. **Managed advance** in which the coast is encouraged to prograde by enhancing natural accretion, artificially nourishing beaches, constructing breakwaters or reclaiming land from the sea.

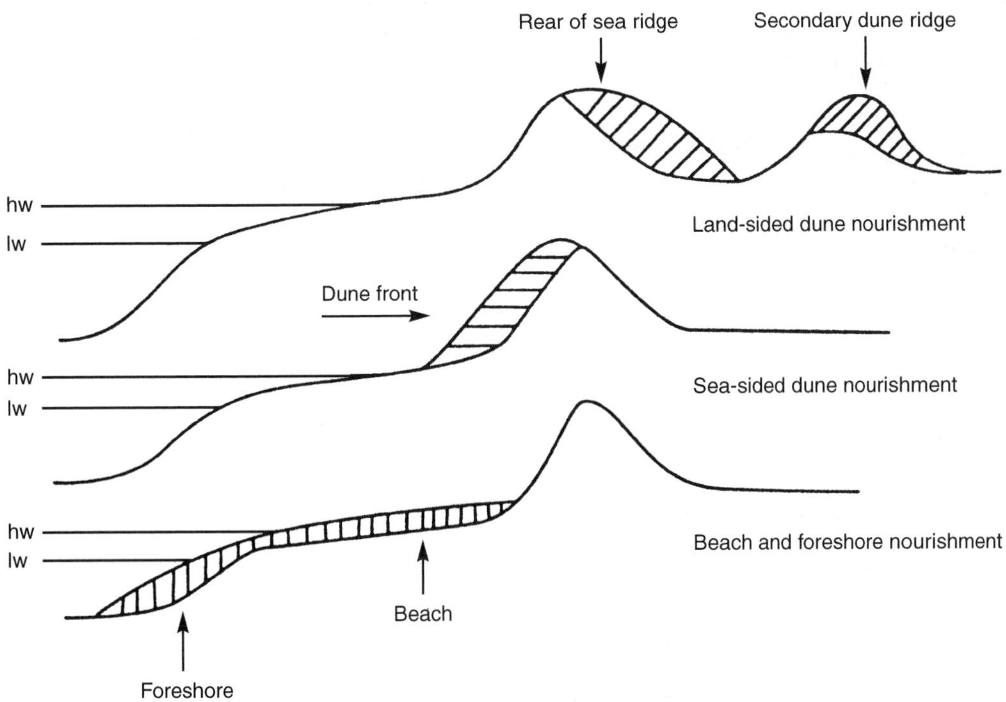

Figure 11.7 Three locations for supplying nourishment sand, a technique used widely on the Dutch North Sea coast which is 353 km long. hw, high water; lw, low water. (After Loffler and Coosen, 1995.)

5. **Large-scale hydraulic modification in estuaries** where attempts are made to modify patterns of tidal flow and sediment transport within a complete estuary by altering its morphology and hydraulic roughness.

The implications of these options for coastal protection are considered much more fully by Doody (1998). The following discussion aims to provide a backcloth for the geomorphological and ecological changes which are likely to occur around the coasts of Britain and elsewhere in the coming century.

11.5.2 PLANNED RETREAT AND SALT MARSH CREATION

Many of the issues involved in using planned retreat as a coastal defence option are discussed by Brooke (1992), who concludes that in Britain the low national value of agricultural production is increasingly resulting in rural flood-defence schemes that show marginal or negative economic benefits. Many coastal or tidal defences are continuing to deteriorate so future rises in sea level will simply compound existing problems. Unless more money becomes available as a result of major changes in grant-aiding policy, or there is a significant increase in the value of agricultural land, retreat is in her view inevitable; the only question is how to make maximum use of the environmental, engineering and economic opportunities involved.

This leads directly to a consideration of the many organizations with management inputs or responsibilities for coastal matters in the UK, whose roles are reviewed by Toft (1995). **Coastal protection** measures to prevent or diminish coastal erosion are the responsibility of the coastal Local Authorities and private land

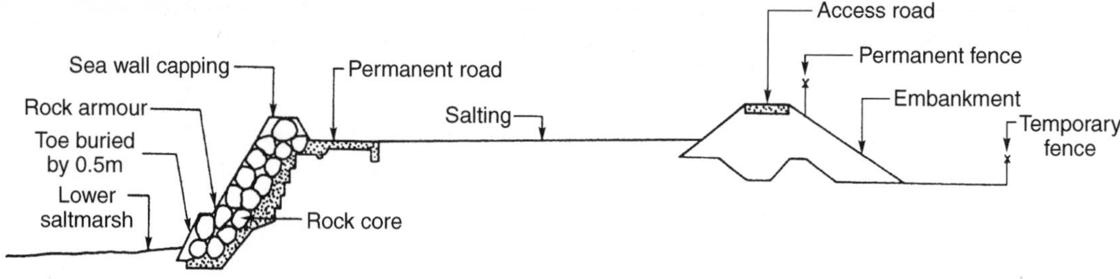

Figure 11.8 Retarded sea defence employed north of the River Parrett, Somerset. (After Toft and Townend, 1991.)

owners. Supervision of the **flood defence** of the coast and estuarine margins of England and Wales, on the other hand, was the responsibility of the National Rivers Authority (NRA) – now incorporated in the Environment Agency for England and Wales and the Scottish Environment Protection Agency – which recognized the important part played by salt marshes in protecting low-lying coastal areas from the sea. The work of the NRA in advancing and evaluating the use of saltings as sea defence (Toft and Townend, 1991) was greatly assisted by their cooperation with the Essex Saltings Restoration Project (ESRP). The Essex coast, which has been the scene of extensive reclamation and drainage schemes since Roman times, is now threatened by embankment failure resulting from loss of fronting salt marsh and lowering foreshores. It was concern about this erosion, first voiced in 1976, which led to the creation of the ESRP and its subsequent experiments, later in cooperation with the NRA, on the **restoration and regeneration of salt marsh**.

The techniques employed – derived largely from European experience – include: (i) the Schleswig–Holstein method; (ii) brushwood groynes and enclosures; and (iii) offshore breakwaters created using disused Thames lighter barges. These schemes were installed very rapidly and with minimal cost so the assessment of the dynamics, particularly of the estuarine flows, and the establishment of adequate baselines followed by monitoring has been less effective than the quantitative assessments now possible. Nevertheless, this innovative early initiative – which will take many years to mature – has provided impetus to the whole salt marsh regeneration effort in the UK and has led to much closer liaison between the numerous research and consultancy groups involved.

In the **Schleswig–Holstein method** a width of mature upper marsh is enclosed by barriers of brushwood together with a similar width of mudflat to seaward of the existing salting. The aim is to develop a new area of **salting** (mature upper marsh), which will protect the reclaimed area, and which is subdivided into several smaller enclosures or polders. The main ditches are dug perpendicular to the coast while other trenches ('grips') are parallel with it. The main trenches direct the waters of the flooding tide onto the upper areas sufficiently rapidly for them to carry the sediment up the shore instead of slowing and depositing it near the sea.

This German method built upon an earlier Dutch technique in which the mud flat areas were originally enclosed solely by earth mounds. Later the Dutch dug trenches ('grips') across the enclosed areas. New material brought in by the tide tended to accumulate in these depressions, while the material removed from the trenches remained within the embankment. Gripping was repeated each year, traditionally in late summer so that the newly dug sediment could consolidate before the winter storms, until the polder level was sufficiently high for planting and subsequent colonization to occur.

It takes many years for mature salt marsh to be formed using these methods under favourable conditions; Toft and Townend (1991) also warn that in an unfavourable environment they may destabilize the mud flat and cause a net loss of sediment.

Brushwood groynes on the Dutch model have been constructed perpendicular to the coast where they minimize wave action, slow currents along the coast and promote sedimentation. They have also been used to make small individual ungripped enclosures.

Lines of sunken disused Thames lighter barges formed effective **offshore breakwaters** which also encouraged sediment deposition. The efficiency of these structures in reducing wave energy in their lee was demonstrated by physical models and directly on site. As a result of such work, and of numerical modelling, Hydraulics Research Ltd was able to recommend that the ends of the lines of lighters at Marsh House and Horsey Island, both in Essex, be connected with the shore by groynes which encouraged accretion still further. The Marsh House wavebreak of 16 lighters was emplaced in 1984 and the groynes, a single line of stakes with geotextile cladding, added in 1986/7. The light construction of the groynes was intended to lessen trampling damage to the marsh during installation, but it required much more maintenance and was far less durable than a traditional brushwood fence.

Planned or managed retreat, which is by no means an easy option, involves allowing previously protected areas to be become intertidal, as Figure 11.9 illustrates. Experience of large-scale managed retreat with salt marsh regeneration in the United States has demonstrated that it needs to be carefully planned if stability is to be achieved and appropriate vegetation established. The hydraulic capacity of the tidal channel system must be sufficient to cope with the expected range of tidal inputs. Weirs, sluices and culverts are needed to control the tidal inundation period and tidal velocities. Temporary wave breaks should be employed during the period when vegetation is establishing, and the surface level of the proposed marsh restoration area should be raised before vegetation is initiated by planting or allowed to invade naturally (Pye and French, 1993c).

The RSPB reserve at Titchwell in North Norfolk developed as a result of subsequent management of an area of 'unmanaged retreat' following breaching of the sea defences during the 1953 storm surge. This large area of former agricultural marsh initially reverted to low diversity *Aster* marsh with a large brackish reedbed along the landward fringe. The site was purchased by the RSPB in the early 1970s and input of tidal water is now regulated by a series of sluices, banks and ditches so as to maximize the diversity of environmental conditions available to birds and other wildlife (Hollis *et al.*, 1990).

This was a fortunate outcome; wave erosion and excessive tidal scour often result when sea defences breached by storm are left unrepaired, while the stability of the tidal creek system – by which sediment enters a marsh system – is crucial in the long term. Managed retreat should be used only in circumstances that have been properly elucidated. Serious adverse consequences would result from large-scale managed retreat in the inner or middle reaches of an estuary. This would increase the tidal capacity of the area, the speed with which the tidal wave passes up the estuary and the tidal height in the inner estuary where salt marsh creeks would be heavily scoured. These effects would be most marked in an estuary whose outer regions were constrained by floodbanks or rock outcrops. If the floodbanks could be repositioned landwards along the outer estuary, permitting it to widen, and possibly become shallower, near its mouth, the increase in tidal velocities would be correspondingly negated (Pye and French, 1993c).

Managed retreat, in some ways an attractive option as sea levels rise and sea walls tend

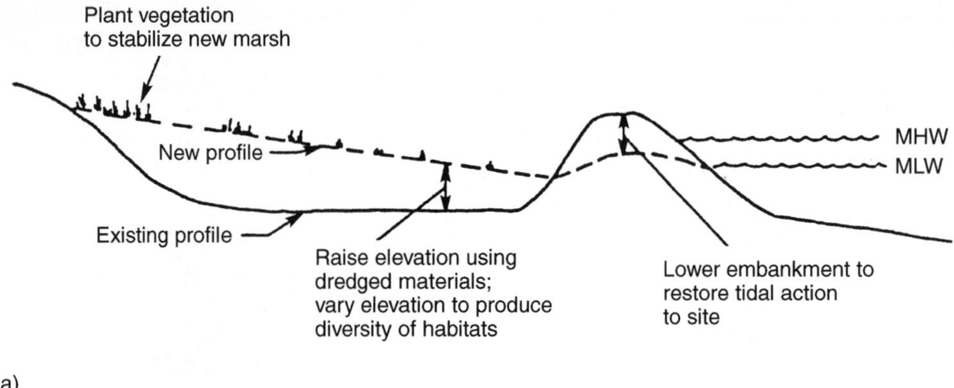

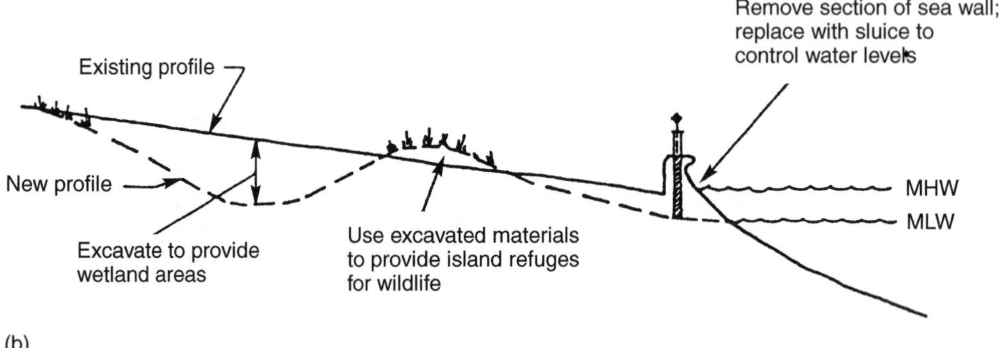

Figure 11.9 Two examples of planned retreat. (a) Raising elevation to produce intertidal areas; (b) Lowering elevation to produce intertidal areas. MHW, mean high water level; MLW, mean low water level. (After Brooke, 1992; courtesy *Journal of the Institution of Water and Environmental Management*.)

to disintegrate, is seen as having an increasing application in the UK. The flooding of another 21 ha – at Tollesbury Fleet on the Blackwater Estuary, Essex – on 4 August 1995, was a further stage in a scheme in which farmers whose arable or grazing land is restored to salt marsh will receive annual compensation.

11.5.3 RETREATS ON THE ESSEX MARSHES

When viewed at low tide on 16 March 1997, several distinct trends were already evident on the **Tollesbury experimental site**, where saline water had reduced the formerly vigorous trees to skeletons (Figure 11.10(a)). Large numbers of young *Salicornia* seedlings were present on higher levels of the incipient salt marsh distant from the breach. Areas of suitable elevation and drainage within a new, more landward, sea wall appeared to be quite rapidly acquiring salt-marsh vegetation, and monitoring supported the view that the policy of using plugs of soil and vegetation from the adjacent salt marsh (in addition to sowing seed) was fully justified. Colonization of the mud flats by intertidal invertebrates – important in the functioning of the ecosystem, notably serving as food for the fish and birds which use the site – had begun within three months of flooding and continues to be monitored by English Nature. Less encouraging were the failure of considerable areas to drain at what was a quite low tide, and the persis-

(a)

(b)

Figure 11.10 Contrasting situations on vulnerable Essex marshes. (a) Managed retreat at Tollesbury, Blackwater Estuary, Essex. Note the relatively light construction of the new 'retarded' sea wall, the dead trees and persistent vehicle tracks on the new marsh, of which extensive areas are covered by green algae. Various halophytes are establishing at higher levels. 16 March 1997. (b) Old salt marsh adjoining the managed retreat at Tollesbury on the Blackwater Estuary; scouring and prominent basins at creek terminations are evident. (c) 'Unmanaged retreat' dating from 1897 at North Fambridge, River Crouch. The former reticulate network of artificial drainage channels has been deeply scoured; in many places mud mounds derived from former vegetated ridges are now degenerating to mud flats. (d, e) Regeneration of the salt marsh landward of the old Orplands Sea Wall, Blackwater Estuary, Essex, 16 March 1997. (d) NRA notice describing the scheme, diversion of the footpath and ownership of the area. (e) View over regenerating marsh towards one of two breaches made in the old sea wall during 1995. Colonization by higher plants is proceeding well; green algae are very abundant. Designed to allow outflow from fields adjoining the new marsh, the drainage channel in the foreground was temporarily blocked by sediment brought in by the tides. (Photographs by John R. Packham.)

(c)

(e)

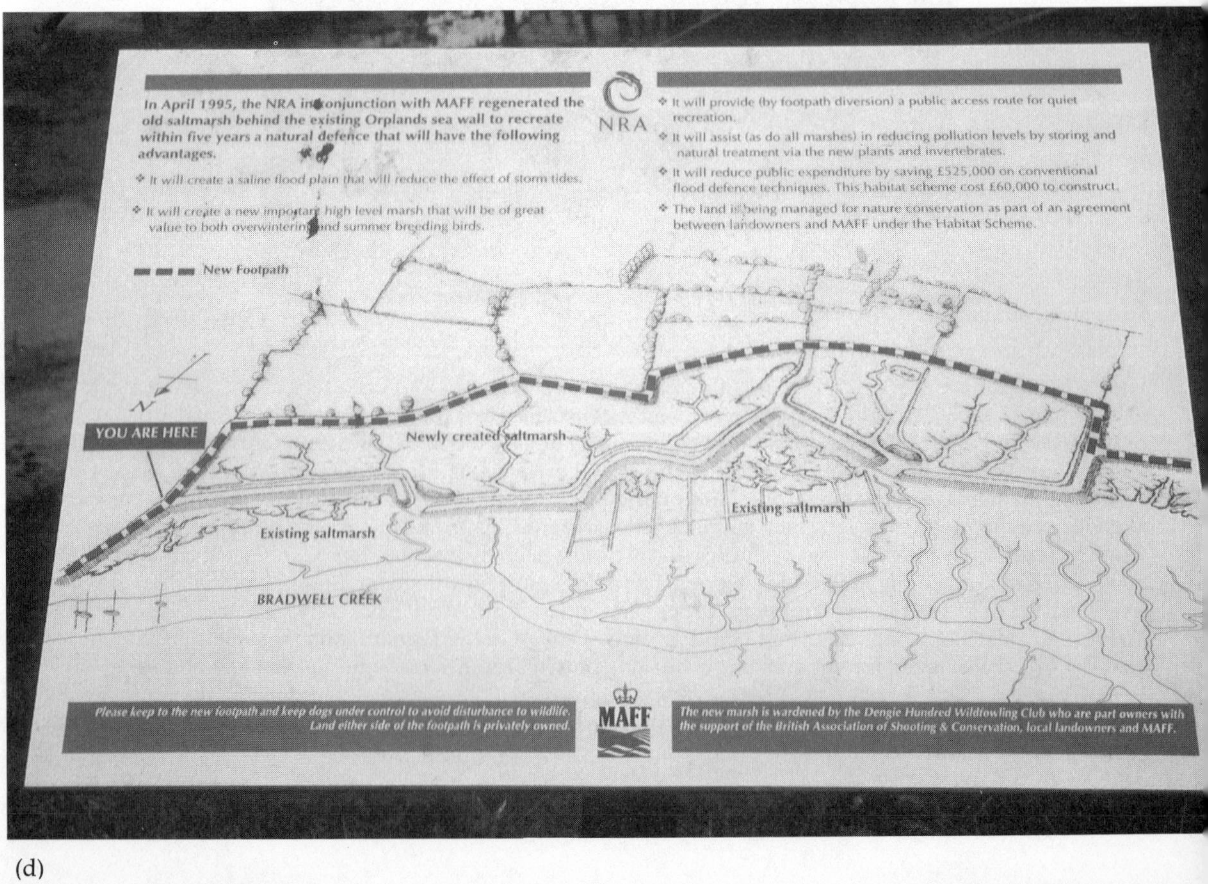

In April 1995, the NRA in conjunction with MAFF regenerated the old saltmarsh behind the existing Orplands sea wall to recreate within five years a natural defence that will have the following advantages.
NRA
It will create a saline flood plain that will reduce the effect of storm tides.
It will create a new important high level marsh that will be of great value to both overwintering and summer breeding birds.
It will provide (by footpath diversion) a public access route for quiet recreation.
It will assist (as do all marshes) in reducing pollution levels by storing and natural treatment via the new plants and invertebrates.
It will reduce public expenditure by saving £525,000 on conventional flood defence techniques. This habitat scheme cost £60,000 to construct.
The land is being managed for nature conservation as part of an agreement between landowners and MAFF under the Habitat Scheme.
New Footpath
N
YOU ARE HERE
Newly created saltmarsh
Existing saltmarsh
Existing saltmarsh
BRADWELL CREEK
Please keep to the new footpath and keep dogs under control to avoid disturbance to wildlife. Land either side of the footpath is privately owned.
MAFF
The new marsh is wardened by the Dengie Hundred Wildfowling Club who are part owners with the support of the British Association of Shooting & Conservation, local landowners and MAFF.

(d)

tence of damage caused by digger vehicles tracking over the marsh when the 60 m breach was created and a primary channel cut. Prolonged persistence of vehicle tracks on salt marshes has been noted previously (p.262); plough ridging, present when retreat occurred, is still very evident and likely to remain.

The numerous linear vegetated ridges separated by muddy channels near **North Fambridge** in the 'unmanaged retreat' on the River Crouch, Essex (Figure 10.11(c)), which dates from the high storm tide of November 1897, have a similar origin from the deepening and widening of a reticulate network of artificial drainage ditches by tidal scour. These channels may ultimately join, leading to the creation of residual mud mounds which then erode leaving bare mud flats.

As is common in Essex, the salt marsh adjoining the Tollesbury managed retreat is highly dissected with deep basins at creek terminations and numerous mud mounds (Figure 10.11(b)). Calcium content and shear strength of the sediments in this region are generally low; in contrast, active younger marsh sediments in the Severn Estuary typically contain more than 10% detrital calcium carbonate ($CaCO_3$). $CaCO_3$ influences sodium absorption ratio, moisture retention and potential dispersive behaviour of clay-rich sediments, improving their shear strength (Pye and Crooks, 1996). Many of the Essex marshes are far less robust than those elsewhere, such as the 'unmanaged' retreats on the River Severn, UK, where no great internal dissection or marsh break-up occurs (Allen, 1993, 1998). The previously calcium-saturated soils of the Tollesbury retreat site – which had mainly Ca^{2+} ions adsorbed on their particle surfaces and whose soil crumbs were relatively stable in pure water – became increasingly sodic with repeated flooding by saline water (Dexter, 1998). Soils with Na^+ ions adsorbed onto their particle surfaces are stable in sea water but unstable in fresh water. Some effects of agricultural soil management on the Tollesbury retreat site were still apparent ten months after breaching; soil shear strength was about three times that of the adjoining mature salt marsh.

Dixon and Weight (1996) describe problems of rising sea levels, declining agricultural benefits and loss of intertidal habitats in terms of managing coastal re-alignment on the southern side of the Blackwater Estuary, Essex. Here the NRA (employing a scheme designed by HR Wallingford Ltd) have recently regenerated (April 1995) old salt marsh landward of the existing **Orplands sea wall** (Figure 11.10(d, e)), creating a saline flood control zone that diminishes the effects of storm tides by reducing wave action and relieving tidal pressure. The new high level marsh is already important to both overwintering and summer breeding birds. The 'retarded' sea wall was placed to ensure the survival of an impressive row of elms running parallel to the estuary; oaks alongside watercourses on the marsh itself continue to survive – a situation similar to that on the Fal Estuary (p. 91). The new salt marsh, like others, stores sediments whose pollution levels are reduced by natural processes involving the developing plant and invertebrate communities. Retreat in the northern compartment was brought about by constructing a 50 m breach in the old sea wall; the breach, allowing tidal entry and egress in the southern compartment, was 40 m wide. Artificial feeder creeks carry silt and water-borne seeds and other propagules from the estuary onto the developing marsh, while a new pond assists land drainage and provides a fresh to brackish aquatic environment. This scheme appears to be working well, with appreciable accretion and substantial colonization by animals and plants.

'Soft engineering' is conspicuous on this coast. A kilometre or so down the estuary from Orplands **sediment recharge** using dredged material piped to an appropriate position is proving an effective control of erosion at

Peewit Island. However, the relative depression of southeast England caused by isostatic re-adjustment which followed removal – by the end of the Devensian glacial period – of the heavy ice burden that covered Scotland and northern England, has left the Essex coast in a vulnerable state; further losses of land area appear inevitable.

REFERENCES

These references, for the most part, are included in the text; additional references are given to make the bibliography as wide-ranging as possible.

Adam, P. (1990) *Saltmarsh Ecology*, Cambridge University Press, Cambridge.

Ahmad, I. and Wainwright, S.J. (1977) Tolerance to salt, partial anaerobosis and osmotic stress in *Agrostis stolonifera*. *New Phytologist*, **79**, 605–12.

Albert, R. (1975) Salt regulation in halophytes. *Oecologia*, **21**, 57–71.

Allan, H.H. (1936) Indigene versus alien in the New Zealand plant world. *Ecology*, **17**, 187–93.

Allen, E.B. and Cunningham, G.L. (1983) Effects of vesicular–arbuscular mycorrhizae on *Distichlis spicata* under three salinity levels. *New Phytologist*, **93**, 227–36.

Allen, J.R.L. (1985) *Principles of Physical Sedimentology*, George Allen and Unwin, London.

Allen, J.R.L. (1992) Tidally influenced marshes in the Severn Estuary, southwest Britain, in *Saltmarshes: Morphodynamics, Conservation and Engineering Significance* (eds J.R.L. Allen and K. Pye), Cambridge University Press, Cambridge, pp. 123–47.

Allen, J.R.L. (1993) Muddy alluvial coasts of Britain: field criteria for shoreline position and movement in the recent past. *Proceedings of the Geologists' Association*, **104**, 241–61.

Allen, J.R.L. (1998) Historical set-back on saltmarshes in the Severn Estuary, S.W. Britain, in *British Saltmarshes* (eds T.Harris and B.Sherwood), Linnean Society/Samara Publishing, Cardigan.

Allen, J.R.L. and Pye, K. (1992) Coastal saltmarshes: their nature and importance, in *Saltmarshes: Morphodynamics, Conservation and Engineering Significance* (eds J.R.L. Allen and K. Pye), Cambridge University Press, Cambridge, pp. 1–18.

Allen, J.R.L. and Rae, J.E. (1988) Vertical salt-marsh accretion since the Roman period in the Severn Estuary, southwest Britain. *Marine Geology*, **83**, 225–35.

Allison, H. and Morley, J.P. (eds) (1989) *Blakeney Point and Scolt Head Island*, National Trust, Norfolk.

Alvin, K.L. (1960) Observations on the lichen ecology of South Haven Peninsula, Studland Heath, Dorset. *Journal of Ecology*, **48**, 331–39.

Andersen, U.V. (1995) Invasive aliens: a threat to Danish coastal vegetation?, in *Directions in European Coastal Management* (eds M.G. Healy and J.P. Doody), Samara Publishing, Cardigan, pp. 335–44.

Anderson, C.E. (1974) A review of structures of several North Carolina salt marsh plants, in *Ecology of Halophytes* (eds R.J. Reimold and W.H. Queen), Academic Press, New York, pp. 307–44.

Anderson, P. and Romeril, M.G. (1992) Mowing experiments to restore a species-rich sward on sand dunes in Jersey, Channel Islands, GB, in *Coastal Dunes* (eds R.W.G. Carter, T.G.F. Curtis and M.J. Sheehy-Skeffington), A.A. Balkema, Rotterdam, pp. 219–34.

Anderson, R.S. (1989) Saltation of sand: a qualitative review with biological analogy. *Proceedings of the Royal Society of Edinburgh*, **96B**, 149–65.

Angus, S. and Elliott, M.M. (1992) Erosion in Scottish machair with particular reference to the Outer Hebrides, in *Coastal Dunes* (eds R.W.G. Carter, T.G.F. Curtis and M.J. Sheehy-Skeffington), A.A. Balkema, Rotterdam, pp. 93–112.

Archibold, O.W. (1995) *Ecology of World Vegetation*, Chapman & Hall, London.

Armstrong, W. (1982) Waterlogged soils, in *Environment and Plant Ecology*, 2nd edn (ed. J.R. Etherington), Wiley, Chichester, pp. 290–330.

Armstrong, W., Wright, E.J., Lythe, S. and Gaynard, T.J. (1985) Plant zonation and the effects of the spring–neap tidal cycle on soil aeration in a Humber salt marsh. *Journal of Ecology*, **73**, 323–39.

Arp, W.J., Drake, B.G., Pockman, W.T., Curtis, P.S. and Whigham, D.F. (1993) Interactions between

C_3 and C_4 salt marsh plant species during four years of exposure to elevated atmospheric CO_2. *Vegetatio*, **104/105**, 133–43.

Atkinson, D. and Houston, J.A. (eds) (1993) *The Sand Dunes of the Sefton Coast*. National Museums and Galleries on Merseyside with Sefton Metropolitan Borough Council, Lancashire.

Avis, A.M. (1989) A review of coastal dune stabilization in the Cape Province of South Africa. *Landscape and Urban Planning*, **18**, 55–68.

Avis, A.M. (1995) An evaluation of the vegetation developed after artificially stabilizing South African coastal dunes with indigenous species. *Journal of Coastal Conservation*, **1**, 41–50.

Bacon, J. and Bacon, S. (1992) *Orford and Orford Ness*, Segment Publications, Colchester.

Bagnold, R.A. (1941) Reprinted 1954, 1984. *The Physics of Blown Sand and Desert Dunes*, Chapman & Hall, London.

Baker, H.G. (1955) Self-incompatibility and establishment after 'long-distance' dispersal. *Evolution, Lancaster, Pa.*, **9**, 347–48.

Baker, J.M. (1975) Effects on shore life and amenities, in *Petroleum and the continental shelf of N. W. Europe* (ed. H.A. Cole), Applied Science Publications, Barking, Essex, UK, pp. 85–90.

Baker, J.M. (1979) Responses of salt marsh vegetation to oil spills and refinery effluents, in *Ecological Processes in Coastal Environments* (eds R.L. Jefferies and A.J. Davy), Blackwell Scientific Publishers, Oxford, pp. 529–43.

Baker, J.M., Oldham, J.H., Wilson, C.M., Dicks, B., Little, D.I. and Levell, D. (1990) *Spartina anglica* and oil: spill and effluent effects, clean-up and rehabilitation, in Spartina anglica *– a research review* (eds A.J. Gray and P.E.M. Benham), HMSO, London, pp. 52–62.

Bakker, J.P. (1978) Changes in a salt-marsh vegetation as a result of grazing and mowing – a five year study of permanent plots. *Vegetatio*, **38**, 77–87.

Ball, D.F. (1964) Loss-on-ignition as an estimate of organic matter and organic carbon in non-calcareous soils. *Journal of Soil Science*, **15**, 84–92.

Ball, P.W. and Brown, K.G. (1970) A biosystematic and ecological study of *Salicornia* in the Dee Estuary. *Watsonia*, **8**, 27–40.

Banks, B., Beebee, T.J.C. and Cooke, A.S. (1994) Conservation of the natterjack toad *Bufo calamita* in Britain over the period 1970–1990 in relation to site protection and other factors. *Biological Conservation*, **67**, 111–18.

Barkman, J.J. (1988) New systems of plant growth forms and phenological plant types, in *Plant Form and Vegetation Structure* (eds M.J.A. Werger, P.J.M. van der Aart, H.J. During and J.T.A. Verhoeven), SPB Publishing, The Hague.

Barlow, S.L. and Ferry, B.W. (1989) Population dynamics of lichenicolous mites at Dungeness. *Botanical Journal of the Linnean Society*, **101**, 111–24.

Barnes, B.M. and Barnes, R.D. (1954) The ecology of the spiders of maritime drift lines. *Ecology*, **35**, 25–35.

Barnes, R.S.K. (1989) Marine animals of Blakeney Point and Scolt Head Island, in *Blakeney Point and Scolt Head Island* (eds H. Allison and J.P. Morley), National Trust, Norfolk, pp. 67–75.

Barnes, R.S.K. (1994a) The coastal lagoons of Europe. *Coastline*, **3**(2), 3–8.

Barnes, R.S.K. (1994b) *The Brackish Water Fauna of Northwestern Europe*, Cambridge University Press, Cambridge.

Bassindale, R. and Clark, R.B. (1960) The Gann Flat, Dale: studies on the ecology of a muddy beach. *Field Studies*, **1**, 1–22.

Bazely, D.R. and Jefferies, R.L. (1986) Changes in the composition and standing crop of salt-marsh communities in response to the removal of a grazer. *Journal of Ecology*, **74**, 693–706.

Beebee, T.J.C. (1976) The natterjack toad (*Bufo calamita*) in the British Isles, a study of past and present status. *British Journal of Herpetology*, **5**, 515–22.

Beebee, T.J.C. (1977) Environmental change as a cause of natterjack toad (*Bufo calamita*) declines in Britain. *Biological Conservation*, **11**, 87–102.

Beeftink, W.G. (1977) The coastal salt marshes of western and northern Europe and phytosociological approach, in *Ecosystems of the World. 1. Wet Coastal Ecosystems* (ed. V.J. Chapman), Elsevier, Amsterdam, pp. 109–55.

Beeftink, W.G. (1979) The structure of salt marsh communities in relation to environmental disturbances, in *Ecological Processes in Coastal Environments* (eds R.L. Jefferies and A.J. Davy), Blackwell Scientific Publications, Oxford, pp. 77–93.

Beerling, D.J. and Woodward, F.I. (1994) Climate change and the British scene. *Journal of Ecology*, **82**, 391–97.

Belanger, L. and Bedard, J. (1994) Role of ice scouring and goose grubbing in marsh plant dynamics. *Journal of Ecology*, **82**, 437–45.

Beltram, G. (1994) Secovlje Salina: the Slovenian Ramsar site and landscape park. *Coastline*, **4**(3), 4–9.

Beukema, J.J. and Cadee, G.C. (1986) Zoobenthos responses to eutrophication of the Dutch Waddensea. *Ophelia*, **26**, 55–64.

Bhadresa, R. (1977) Food preferences of rabbits *Oryctolagus cuniculus* L. at Holkham sand dunes, Norfolk. *Journal of Applied Ecology*, **14**, 287–91.

Binggeli, P., Eakin, M., Macfadyen, A., Power, J. and McConnell, J. (1992) Impact of alien sea buckthorn (*Hippophaë rhamnoides* L.) on sand dune ecosystems in Ireland, in *Coastal Dunes* (eds R.W.G. Carter, T.G.F. Curtis and M.J. Sheehy-Skeffington), A.A. Balkema, Rotterdam, pp. 325–37.

Birse, E.M. (1957) Ecological studies on growth-form in Bryophytes. II. Experimental studies on growth-form in mosses. *Journal of Ecology*, **45**, 721–33.

Birse, E.M. (1958) Ecological studies on growth-form in Bryophytes. III. The relationship between the growth-form of mosses and ground-water supply. *Journal of Ecology*, **46**, 9–27.

Birse, E.M. and Gimingham, C.H. (1955) Changes in the structure of bryophytic communities with the progress of succession on sand dunes. *Transactions of the British Bryological Society*, **2**, 523–31.

Birse, E.M., Landsberg, S.Y. and Gimingham, C.H. (1957) The effects of burial by sand on dune mosses. *Transactions of the British Bryological Society*, **3**, 285–301.

Böcher, T.W. (1952) Lichen-heaths and plant successions at Osterby on the Isle of Laeso in the Kattegat. *Kongelige Danske Videnskabernes Selskabs (Biologiske Skrifter)*, **7**, no. 4.

Bohnert, H.J., Nelson, D.E. and Jensen, R.G. (1995) Adaptations to environmental stresses. *The Plant Cell*, **7**, 1099–111.

Bond, T.E.T. (1952) *Elymus arenarius* L.: Biological Flora of the British Isles. *Journal of Ecology*, **40**, 217–27.

Boorman, L.A. (1967) *Limonium vulgare* Mill.: Biological Flora of the British Isles. *Journal of Ecology*, **55**, 221–29.

Boorman, L.A. (1982) Some plant growth patterns in relation to the sand dune habitat. *Journal of Ecology*, **70**, 607–14.

Boorman, L.A. (1989) The influence of grazing on British sand dunes, in *Perspectives in Coastal Dune Management* (eds F. van der Meulen, P.D. Jungerius and J. Visser), SPB Academic Publishing, The Hague, pp. 121–24.

Boorman, L.A. (1993) Dry coastal ecosystems of Britain: dunes and shingle beaches, in *Dry Coastal Ecosystems: Polar Regions and Europe* (ed. E. van der Maarel), Elsevier, London, pp. 197–228.

Boorman, L.A. and Fuller, R.M. (1977) Studies on the impact of paths on the dune vegetation at Winterton, Norfolk, England. *Biological Conservation*, **12**, 203–16.

Boorman, L.A. and Fuller, R.M. (1982) Effects of added nutrients on dune swards grazed by rabbits. *Journal of Ecology*, **70**, 345–55.

Boorman, L.A. and Fuller, R.M. (1984) The comparative ecology of two sand dune biennials: *Lactuca virosa* L. and *Cynoglossum officinale* L. *New Phytologist*, **69**, 609–29.

Boorman, L.A. and Ranwell, D.S. (1977) *The ecology of Maplin Sands and the coastal zones of Suffolk, Essex and north Kent*, Institute of Terrestrial Ecology, Cambridge.

Boorman, L.A. and Woodell, S.R.J. (1966) The topograph, an instrument for measuring microtopography. *Ecology*, **47**, 869–70.

Bowra, J.C. (1995) *Corispermum leptopterum* and the sand-dune flora at Emscote near Warwick. *BSBI News*, **68**, 33–35.

Bressolier, C. and Thomas, Y.F. (1977) Studies on wind and plant interactions on French Atlantic coastal dunes. *Journal of Sedimentary Petrology*, **47**, 331–38.

Bretz, J.H. (1960) Bermuda, a partially drowned, late mature, Pleistocene karst. *Bulletin of the Geological Society of America*, **71**, 1729–54.

Brewis, A., Bowman, P. and Rose, F. (1996) *The Flora of Hampshire*, Harley Books, Colchester, UK.

Bridson, R.H. (1980) Saltmarsh, its accretion and erosion at Caerlaverock National Nature Reserve, Dumfries. *Transactions of the Dumfriesshire and Galloway Natural History and Antiquarian Society*, **55**, 60–7.

Briens, M. and Larher, F. (1982) Osmoregulation in halophytic higher plants: a comparative study of soluble carbohydrates, polyols, betaines and free proline. *Plant Cell and Environment*, **5**, 287–92.

Brightmore, D. and White, P.H.F. (1963) *Lathyrus japonicus* Willd.: Biological Flora of the British Isles. *Journal of Ecology*, **51**, 795–801.

Brinkman, H., Loof, P.A.A. and Barbez, D. (1987) *Longidorus dunensis* n.sp. and *L. kuiperi* n.sp. from the sand dune coastal region of the Netherlands (Nematoda: Longidoridae). *Revue Nématologique*, **10**, 299–308.

Brooke, J.S. (1992) Coastal defence: the retreat option. *Journal of the Institution of Water and Environmental Management*, **6**, 151–7.

Brown, J.C. (1958) Soil fungi of some British sand dunes in relation to soil type and succession. *Journal of Ecology*, **46**, 641–64.

Brown, R.A. and R.J. O'Connor (1974) Some observations on the relationships between Oystercatchers *Haematopus ostralegus* L. and

Cockles *Cardium edule* L. in Strangford Lough. *Irish Naturalist Journal*, **18**, 73–80.

Brunsden, D. and Goudie, A. (1981) *Landform Guides No. 1. Classic Coastal Landforms of Dorset*, Geographical Association, Sheffield.

Burd, F.H. (1989) *The Saltmarsh Survey of Great Britain – an Inventory of British Saltmarshes*, Nature Conservancy Council, Peterborough.

Burd, F.H. (1992) *Erosion and Vegetation Change on the Saltmarshes of Essex and north Kent between 1973 and 1988*, Nature Conservancy Council, Peterborough.

Burdick, D.M. and Mendelssohn, I.A. (1990) Relationship between anatomical and metabolic responses to soil waterlogging in the coastal grass *Spartina patens*. *Journal of Experimental Botany*, **41**, 223–8.

Burger, J., Shisler, J. and Lesser, F.H. (1982) Avian utilisation on six salt marshes in New Jersey. *Biological Conservation*, **23**, 187–212.

Burkholder, P.R. (1956) Studies on the nutritive value of *Spartina* grass growing in the marsh areas of coastal Georgia. *Bulletin of the Torrey Botanical Club*, **83**, 327–43.

Burrows, C.J. (1990) *Processes of Vegetation Change*, Unwin Hyman, London.

Cadwalladr, D.A. and Morley, J.V. (1974) Further experiments on the management of saltings pasture for wigeon (*Anas penelope* L.) conservation at Bridgwater Bay National Nature Reserve, Somerset. *Journal of Applied Ecology*, **11**, 461–6.

Cahoon, D.R., Reed, D.J. and Day, J.W. (1995) Estimating shallow subsidence in microtidal salt marshes of the southeastern United States: Kaye and Barghoorn revisited. *Marine Geology*, **128**, 1–9.

Caldwell, P.A. (1957) The spatial development of *Spartina* colonies growing without competition. *Annals of Botany, N.S.*, **21**, 203–14.

Cargill, S.M. and Jefferies, R.L. (1984a) Nutrient limitation of primary production in a sub-arctic salt marsh. *Journal of Applied Ecology*, **21**, 657–68.

Cargill, S.M. and Jefferies, R.L. (1984b) The effects of grazing by lesser snow geese on the vegetation of a sub-arctic salt marsh. *Journal of Applied Ecology*, **21**, 669–86.

Carpenter, K.E. (1993) *Nutrient, fluvial and groundwater fluxes between a North Norfolk, UK saltmarsh and the North Sea*, PhD Thesis, University of East Anglia.

Carpenter, K.E. (1997) A critical appraisal of the methodology used in studies of material flux between saltmarshes and coastal waters, in *Biogeochemistry of Intertidal Sediments* (eds T.D. Jickells and J. Rae) Cambridge University Press, Cambridge.

Carter, R.W.G. (1980) Vegetation stabilisation and slope failure of eroding sand dunes. *Biological Conservation*, **18**, 117–22.

Carter, R.W.G. (1985) Approaches to sand dune conservation in Ireland, in *Sand Dunes and their Management* (ed. J.P. Doody), Nature Conservation Council, Peterborough, pp. 29–41.

Carter, R.W.G., Orford, J.D., Jennings, S.C., Shaw, J. and Smith, J.P. (1992) Recent evolution of a paraglacial estuary under conditions of rapid sea-level rise: Chezzetcook Inlet, Nova Scotia. *Proceedings of the Geologists' Association*, **103**, 167–85.

Castellanos, E.M., Figueroa, M.E. and Davy, A.J. (1994) Nucleation and facilitation in saltmarsh succession: interactions between *Spartina maritima* and *Arthrocnemum perenne*. *Journal of Ecology*, **82**, 239–48.

Cavalieri, A.J. and Huang, A.H.C. (1977) Effect of NaCl on the *in vitro* activity of malate dehydrogenase in salt marsh halophytes of the U.S. *Physiologia Plantarum*, **41**, 79–84.

Cavers, P.B. and Harper, J.L. (1964) *Rumex crispus* L.: Biological Flora of the British Isles. *Journal of Ecology*, **52**, 754–66.

Chambers, R.M. (1992) A fluctuating water level chamber for biogeochemical experiments in tidal marshes. *Estuaries*, **15**(1), 53–8.

Chapman, V.J. (1947a) *Suaeda maritima* (L.) Dum.: Biological Flora of the British Isles. *Journal of Ecology*, **35**, 293–302.

Chapman, V.J. (1947b) *Suaeda fruticosa* Forsk.: Biological Flora of the British Isles. *Journal of Ecology*, **35**, 303–10.

Chapman, V.J. (1950) *Halimione portulacoides* (L.) Aell.: Biological Flora of the British Isles. *Journal of Ecology*, **38**, 214–22.

Chapman, V.J. (1960a) The plant ecology of Scolt Head Island, in *Scolt Head Island* (ed. J.A. Steers), Heffer, Cambridge, pp. 85–163.

Chapman, V.J. (1960b) *Salt Marshes and Salt Deserts of the World*, Leonard Hill (Books), London.

Chapman, V.J. (1976) *Coastal Vegetation*, 2nd edn, Pergamon Press, Oxford.

Chapman, V.J. (1977) Introduction, in *Wet Coastal Ecosystems* (ed. V.J. Chapman), Elsevier, Amsterdam, pp. 1–29.

Chapman, V.J. and Chapman, D.J. (1976) Life forms in the algae. *Botanica Marina*, **19**, 65–74.

Charman, K., Fojt, W. and Penny, S. (1986) *Saltmarsh Survey of Great Britain: Bibliography*, Nature Conservancy Council, Peterborough.

Cheeseman, J.M. and Wickens, L.K. (1986a) Control of Na^+ and K^+ transport in *Spergularia marina*. I. Transpiration effects. *Physiologia Plantarum*, **67**, 1–6.

Cheeseman, J.M. and Wickens, L.K. (1986b) Control of Na^+ and K^+ transport in *Spergularia marina*. III. Relationship between ion uptake and growth at moderate salinity. *Physiologia Plantarum*, **67**, 15–22.

Chung, C.H. (1990) Twenty-five years of introduced *Spartina anglica* in China, in Spartina anglica – *a Research review* (eds A.J. Gray and P.E.M. Benham), H.M.S.O., London, pp. 72–6.

Clapham, A.R., Pearsall, W.H. and Richards, P.W. (1942) *Aster tripolium* L.: Biological Flora of the British Isles. *Journal of Ecology*, **30**, 385–95.

Clapham, A.R., Tutin, T.G. and Moore, D.M. (1987) *Flora of the British Isles*, 3rd edn, Cambridge University Press, Cambridge.

Clark, J.S. (1986a) Late-Holocene vegetation and coastal processes at a Long Island tidal marsh. *Journal of Ecology*, **74**, 561–78.

Clark, J.S. (1986b) Dynamism in the barrier beach vegetation of Great South Beach, New York. *Ecological Monographs*, **56**, 97–126.

Clark, J.S. and Patterson, W.A. (1985) Development of a tidal marsh: upland and oceanic influences. *Ecological Monographs*, **55**, 189–217.

Clarke, S.M. (1965) *Some aspects of the autecology of* Elymus arenarius *L.* PhD Thesis, University of Hull.

Clements, F.E. (1916) Plant Succession. An analysis of the development of vegetation. *Carnegie Institute Washington*, No. 242.

Clements, F.E. (1936) Nature and structure of the climax. *Journal of Ecology*, **24**, 252–84.

Clipson, N.J.W. and Jennings, D.H. (1992) *Dendryphiella salina* and *Debaryomyces hansenii*: models for ecophysiological adaptation to salinity by fungi which grow in the sea. *Canadian Journal of Botany*, **70**, 2097–105.

Clipson, N.J.W., Tomos, A.D., Flowers, T.J. and Wyn Jones, R.G. (1985) Salt tolerance in the halophyte *Suaeda maritima* (L.) Dum. The maintenance of turgor pressure and water-potential gradients in plants growing at different salinities. *Planta*, **165**, 392–96.

Clipson, N.J.W., Lachno, D.H. and Flowers T.J. (1989) Salt tolerance in the halophyte *Suaeda maritima* (L.) Dum. Abscisic acid contents in response to constant and altered salinity. *Journal of Experimental Biology*, **39**, 1381–8.

Coles, S.M. (1979) Benthic microalgal populations on intertidal sediments and their role as precursors to salt marsh development, in *Ecological Processes in Coastal Environments* (eds R.L. Jefferies and A.J. Davy), Blackwell Scientific Publications, Oxford, pp. 25–42.

Conard, H.S. (1924) Second survey of the vegetation on a Long Island salt marsh. *Ecology*, **5**, 379–88.

Conard, H.S. and Galligar, G.C. (1929) Third survey of a Long Island salt marsh. *Ecology*, **10**, 326–36.

Connor, E.F. and McCoy, E.D. (1979) The statistics and biology of the species–area relationship. *American Naturalist*, **113**, 791–833.

Cooch, E.G., Jefferies, R.L., Rockwell, R.F. and Cooke, F. (1993) Environmental change and the cost of philopatry: an example in the lesser snow goose. *Oecologia*, **93**, 128–38.

Cooke, A.S. (1991) *The habitat of sand lizards* Lacerta agilis *at Merseyside*, Nature Conservancy Council, Peterborough.

Cooper, A. (1982) The effects of salinity and waterlogging on the growth and cation uptake of salt marsh plants. *New Phytologist*, **90**, 263–75.

Cooper, W.S. (1967) Coastal dunes of California. *Geological Society of America Memoir*, **104**, 131 pp.

Corbett, K.F. (1988) Distribution and status of the sand lizard *Lacerta agilis agilis* in Britain. *Mertensiella, Salamandra Supplement*, **1**, 92–100.

Cordazzo, C.V. (1994) *Comparative population studies of four dominant plants of southern Brazilian coastal dunes*, PhD Thesis, University of East Anglia.

Cosgrove, D. (1986) Biophysical control of plant cell growth. *Annual Review of Plant Physiology*, **37**, 377–405.

Cowles, H.C. (1899) The ecological relations of the vegetation on the sand dunes of Lake Michigan. *Botanical Gazette*, **27**, 95–117, 167–202, 281–308, 361–91.

Cram, W.J. (1983) Chloride accumulation as a homeostatic system: set points and perturbations. *Journal of Experimental Botany*, **34**, 1484–502.

Cramer, W. and Hytteborn, H. (1986) The separation of fluctuation and long-term change in vegetation dynamics of a rising sea-shore. Paper V in PhD Thesis of Cramer, W. *Acta Universitatis Upsaliensis*.

Crawford, R.M.M. (1989) *Studies in Plant Survival*, Blackwell Scientific Publications, Oxford.

Crawley, M.J. (1987) What makes a community invasible? in *Colonization, Success and Stability* (eds A.J. Gray, M.J. Crawley and P.J. Edwards), Blackwell Scientific Publications, Oxford, pp. 429–51.

Cresswell, W. (1994) Age-dependent choice of redshank (*Tringa totanus*) feeding location: profitability or risk? *Journal of Animal Ecology*, **63**, 589–600.

Curtis, P.S., Drake, B.G., Leadley, P.W., Arp, W.J. and Whigham, D.F. (1989) Growth and senescence in plant communities exposed to elevated CO_2 concentrations on an estuarine marsh. *Oecologia*, **78**, 20–6.

Curtis, P.S., Balduman, L.M., Drake, B.G. and Whigham, D.F. (1990) Elevated atmospheric CO_2 effects on belowground processes in C_3 and C_4 estuarine marsh communities. *Ecology*, **71**, 2001–6.

Daiber, F.C. (1977) Salt-marsh animals: distributions related to tidal flooding, salinity and vegetation, in *Ecosystems of the World. 1. Wet Coastal Ecosystems* (ed. V.J. Chapman), Elsevier, Amsterdam, pp. 79–108.

Daiber, F.C. (1982) *Animals of the Tidal Marsh*, Van Nostrand Reinhold, New York.

Dainty, J. (1979) The ionic and water relations of plants which adjust to a fluctuating saline environment, in *Ecological Processes in Coastal Environments* (eds R.L. Jefferies and A.J. Davy), Blackwell Scientific Publications, Oxford, pp. 201–9.

Dalby, D.H. (1963) Seed dispersal in *Salicornia pusilla*. *Nature*, **199**, 197–8.

Dalby, D.H. (1970) The salt marshes of Milford Haven, Pembrokeshire. *Field Studies*, **3**, 297–330.

Dale, M.F.B. and Ford-Lloyd, B.V. (1985) The significance of multigerm seedballs in the genus *Beta*. *Watsonia*, **15**, 265–7.

Dansereau, P. (1947) Zonation et succession sur la restinga de Rio de Janeiro. I. Halosere. *Revue canadien de Biologie*, **6**, 448–77.

Dargie, T.C.D. (1993) *Sand Dune Vegetation Survey of Great Britain. Part 2: Scotland*, JNCC, Peterborough.

Dargie, T.C.D. (1995) *Sand Dune Vegetation of Great Britain. Part 3: Wales*. JNCC, Peterborough.

Davies, M.S. and Singh, A.K. (1983) Population differentiation in *Festuca rubra* L. and *Agrostis stolonifera* L. in response to soil waterlogging. *New Phytologist*, **94**, 573–83.

Davies, P, Williams, A.T. and Curr, R.H.F. (1995) Decision making in dune management: theory and practice. *Journal of Coastal Conservation*, **1**, 87–96.

Davy, A.J. and Bishop, G.F. (1991) *Triglochin maritima* L.: Biological Flora of the British Isles. *Journal of Ecology*, **79**, 531–55.

Davy, A.J. and Costa, C.S.B. (1992) Development and organization of saltmarsh communities, in *Coastal Plant Communities of Latin America* (ed. U. Seliger), Academic Press, San Diego, pp. 157–78.

Davy, A.J. and Figueroa, M.E. (1993) The colonization of strandlines, in *Primary Succession on Land* (eds J. Miles and D.H.W. Walton), Blackwell Scientific Publications, Oxford, pp. 113–31.

Davy, A.J. and Smith, H. (1988) 1. Life-history variation and environment, in *Plant Population Ecology* (eds A.J. Davy, M.J. Hutchings and A.R. Watkinson), Blackwell Scientific Publications, Oxford, pp. 1–22.

Davy, A.J., Costa, C.S.B., Proudfoot, A.M., Yallop, A.R. and Mohamed, M.F. (1998) Biotic interactions in plant communities of saltmarshes, in *British Saltmarshes* (eds T. Harris amd B. Sherwood), Linnean Society/Samara Publishing, Cardigan (in press).

De Ferrari, C.M. and Naiman, R.J. (1994) A multiscale assessment of the occurrence of exotic plants on the Olympic Peninsula, Washington. *Journal of Vegetation Science*, **5**, 247–58.

de Jong, T.J., Klinkhamer, P.G.L. and Boorman, L.A. (1990) *Cynoglossum officinale* L.: Biological Flora of the British Isles. *Journal of Ecology*, **78**, 1123–44.

de Rooij-van der Goes, P.C.E.M. (1995) The role of plant-parasitic nematodes and soil-borne fungi in the decline of *Ammophila arenaria* (L.) Link. *New Phytologist*, **129**, 661–9.

de Rooij-van der Goes, P.C.E.M. (1996) *Soil-borne plant pathogens of* Ammophila arenaria *in coastal foredunes*. PhD Thesis, Wageningen Agricultural University, The Netherlands.

de Ruig, J.H.M. (1995) The Dutch experience: four years of dynamic preservation of the coastline, in *Directions in European Coastal Management* (eds M.G. Healy and J.P. Doody), Samara Publishing, Cardigan, pp. 253–66.

Deshmukh, I.K. (1979) Fixation, accumulation and release of energy by *Ammophila arenaria* in a sand dune succession, in *Ecological Processes in Coastal Environments* (eds R.L. Jefferies and A.J. Davy), Blackwell Scientific Publications, Oxford. pp. 353–62.

de Wit, C.T. (1960) On competition. *Verslagen van landbouwkundige onderzoekingen van het Rijkslandbouwproefstation*, **6**, 1–82.

Dexter, A.R. (1998) Changes in soil physical properties on the Tollesbury set-back site, in *British Saltmarshes* (eds T.Harris and B.Sherwood), Linnean Society/Samara Publishing, Cardigan.

Dickenson, G. and Randall, R.E. (1979) An interpretation of machair vegetation. *Proceedings of the Royal Society of Edinburgh, B*, **77**, 267–78.

Dicks, B. and Hartley, J.P. (1982) The effects of repeated small oil spillages and chronic discharges. *Philosophical Transactions of the Royal Society of London, B*, **297**, 285–307.

Disraeli, D.J. (1984) The effect of sand deposits on the growth and morphology of *Ammophila breviligulata*. *Journal of Ecology*, **72**, 145–54.

Diver, C. (1933) The physiography of South Haven Peninsula, Studland Heath, Dorset. *Geographical Journal*, **81**, 404–27.

Dixon, A.M. and Weight, R.S. (1996) Case study at Orplands Sea Wall, Blackwater Estuary, Essex, in *Saltmarsh Management for Flood Defence* (eds Sir William Halcrow and Partners Ltd), NRA, Bristol.

Dodds, J.G. (1953) *Plantago coronopus* L.: Biological Flora of the British Isles. *Journal of Ecology*, **41**, 467–78.

Doing, H. (1985) Coastal fore-dune zonation and succession in various parts of the world. *Vegetatio*, **61**, 65–75.

Dolan, R. (1980) Barrier islands,: natural and controlled, in *Coastal Geomorphology* (ed. D.R. Coates), Allen & Unwin, London, pp. 263–78.

Dolan, R., Godfrey, P.J. and Odum, W.E. (1973) Man's impact on the barrier islands of North Carolina. *American Scientist*, **61**, 152–63.

Dony, J.G. (1977) Species–area relationships in an area of intermediate size. *Journal of Ecology*, **65**, 475–84.

Doody, J.P. (ed.) (1984) Spartina anglica *in Great Britain*. Nature Conservancy Council, Attingham.

Doody, P. (1989a) Dungeness – a national nature conservation perspective. *Botanical Journal of the Linnean Society*, **101**, 163–71.

Doody, J.P. (1989b) Management for nature conservation. *Proceedings of the Royal Society of Edinburgh*, **96B**, 247–65.

Doody, P. (1989c) Conservation and development of the coastal dunes in Great Britain, in *Perspectives in Coastal Dune Management* (eds F. van der Meulen, P.D. Jungerius and J. Visser), SPB Academic Publishing, The Hague, pp. 53–67.

Doody, J.P.(1990) *Spartina* – friend or foe? A conservation viewpoint, in Spartina anglica *– a research review* (eds A.J. Gray and P.E.M. Benham), HMSO, London, pp. 77–9.

Doody, J.P. (ed.) (1991) *Sand Dune Inventory of Europe*, Joint Nature Conservation Committee, UK and EUCC; Peterborough.

Doody, J.P. (1992) The conservation of British saltmarshes, in *Saltmarshes: Morphodynamics, Conservation and Engineering Significance* (eds J.R.L. Allen and K. Pye), Cambridge University Press, Cambridge, pp. 80–114.

Doody, J.P. (1998) *Coastal Management for Nature Conservation*. Chapman & Hall, London.

Doody, J.P., Johnston, C. and Smith, B. (eds) (1993) *Directory of the North Sea Coastal Margin*, JNCC, Peterborough.

Drake, B.G. and Leadley, P.W. (1991) Canopy photosynthesis of crops and native plant communities exposed to long-term elevated CO_2. *Plant, Cell and Environment*, **14**, 853–60.

Dring, M.J. (1992) *The Biology of Marine Plants*, Cambridge University Press, Cambridge.

Ducker, B.F.T. (1960) The bryophytes of Scolt Head Island, in *Scolt Head Island*, 2nd edn (ed. J.A. Steers), Heffer, Cambridge, pp. 164–76.

Dupont, I.M. (1992) Salt-induced changes in ion transport: regulation of primary pumps and secondary transporters, in *Transport and Receptor Proteins of Plant Membranes* (eds D.T. Cooke and D.T. Clarkson), Plenum Press, New York, pp. 21–8.

Durell, S.E.A. Le V. dit, Goss-Custard, J.D. and R.W.G. Caldow (1993) Sex-related differences in diet and feeding method in the oystercatcher *Haematopus ostralegus*. *Journal of Animal Ecology*, **62**, 205–15.

During, H.J. (1979) Life strategies of bryophytes: a preliminary review. *Lejeunea*, **5**, 2–17.

During, H.J. (1992) Ecological classifications of bryophytes and lichens, in *Bryophytes and Lichens in a Changing Environment* (eds J.W. Bates and A.M. Farmer), Clarendon Press, Oxford, pp. 1–31.

Dyer, M.I. (1975) The effects of Red-winged Blackbirds (*Agelaius phoeniceus*) on biomass production of corn grains (*Zea mays*). *Journal of Applied Ecology*, **12**, 719–26.

Edwards, P.J. (1984) The growth of fairy rings of *Agaricus arvensis* and their effect upon grassland vegetation and soil. *Journal of Ecology*, **72**, 505–13.

Egler, F.E. (1954) Vegetation science concepts. I. Initial floristic composition, a factor in old-field vegetation development. *Vegetatio*, **4**, 412–17.

Eilers, H.P. (1979) Production ecology in an Oregon coastal salt marsh. *Estuaries and Coastal Marine Science*, **8**, 399–410.

Eldred, R.A. and Maun, M.A. (1982) A multivariate approach to the problem of decline in vigour of *Ammophila*. *Canadian Journal of Botany*, **60**, 1371–80.

Ellenberg, H. (1988) *Vegetation Ecology of Central Europe*, Cambridge University Press, Cambridge.

Ellis, E.A. (1960) An annotated list of the fungi, in *Scolt Head Island*, 2nd edn (ed. J.A. Steers), Heffer and Sons, Cambridge, pp. 179–82.

Ellison, A.M., Bertness, M.D. and Miller, T. (1986) Seasonal patterns in the belowground biomass of *Spartina alterniflora* (Gramineae) across a tidal gradient. *American Journal of Botany*, **73**, 1548–54.

Elliston Wright, F.R. (1932) *Braunton: a few Nature Notes*, Barnes, Barnstaple.

Elton, C.S. (1958) *The Ecology of Invasions by Animals and Plants*, Chapman & Hall, London.

Enemark, J.A. (1993) The Wadden Sea; a shared wetland system. Coastal management on the basis of wise use. *Coastline*, **2**(2), 30–5.

Ericson, L. (1980) The downward migration of plants on a rising Bothnian sea-shore. *Acta Phytogeographica Suecica*, **68**, 61–72.

Etherington, J.R. (1983) *Wetland Ecology*, Edward Arnold, London.

Faraday, C.D. and Thomson, W.T. (1986) Functional aspects of the salt glands of the Plumbaginaceae. *Journal of Experimental Botany*, **37**, 1129–35.

Farrow, E.P. (1916) On the ecology of the vegetation of Breckland. II. Factors relating to the relative distributions of *Calluna*-heath and grass-heath in Breckland. *Journal of Ecology*, **4**, 57–64.

Farrow, E.P. (1917) On the ecology of the vegetation of Breckland. III. General effects of rabbits on the vegetation. *Journal of Ecology*, **5**, 1–18.

Ferry, B.W. and Waters, S.J.P. (1984) The vegetation of natural freshwater pits at Dungeness – II Lichens and bryophytes. *Transactions of the Kent Field Club*, **9**, 153–8.

Ferry, B.W. and Waters, S.J.P. (eds) (1985) *Dungeness: Ecology and Conservation*, Nature Conservancy Council, Peterborough.

Ferry, B.W., Barlow, S.L. and Waters, S.J.P. (1989) The shingle ridge succession at Dungeness. *Botanical Journal of the Linnean Society*, **101**, 19–30.

Findon, R. (1989) Recent developments at Dungeness. *Botanical Journal of the Linnean Society*, **101**,125–35.

Fish, J.D. and Fish, S. (1989) *A Student's Guide to the Seashore*, Unwin Hyman, London.

Flemming, L.V. (1998) Amphibians on saltmarshes: ecology threats and conservation, in *British Saltmarshes* (eds T. Harris and B. Sherwood), Linnean Society/ Samara Publishing, Cardigan.

Fletcher, A. (1973a) The ecology of marine (littoral) lichens on some rocky shores of Anglesey. *Lichenologist*, **5**, 368–400.

Fletcher, A. (1973b) The ecology of maritime (supralittoral) lichens on some rocky shores of Anglesey. *Lichenologist*, **3**, 401–22.

Flowers, T.J. (1985) Physiology of halophytes. *Plant and Soil*, **89**, 41–56.

Flowers, T.J., Troke, P.F. and Yeo, A.R. (1977) The mechanism of salt tolerance in halophytes. *Annual Review of Plant Physiology*, **28**, 89–121.

Flowers, T.J., Hajibagheri, M.A. and Clipson, N.J.W. (1986) Halophytes. *Quarterly Review of Biology*, **61**, 313–37.

Flowers, T.J., Hajibagheri, M.A. and Yeo, A.R. (1991) Ion accumulation in the cell walls of rice plants growing under saline conditions: evidence for the Oertli hypothesis. *Plant, Cell and Environment*, **14**, 319–25.

Foster, W.A. (1989) Terrestrial animals of Blakeney Point and Scolt Head Island, in *Blakeney Point and Scolt Head Island* (eds H. Allison and J.P. Morley), National Trust, Norfolk, pp. 76–86.

Francis, R. and Read, D.J. (1995) Mutualism and antagonism in the mycorrhizal symbiosis, with special reference to impacts on plant community structure. *Canadian Journal of Botany*, **73** (Suppl. 1), S1301–9.

French, J.R. (1993) Numerical simulation of vertical marsh growth and adjustment to accelerated sea-level rise, North Norfolk, UK. *Earth Surface Processes & Landforms*, **18**, 63–81.

French, J.R. and Spencer, T. (1993) Dynamics of sedimentation in a tide-dominated backbarrier salt marsh, Norfolk, UK. *Marine Geology*, **110**, 315–31.

Frid, C. and James, R. (1989) The marine invertebrate fauna of a British coastal salt marsh. *Holarctic Ecology*, **12**, 9–15.

Fuller, R.J. (1982) *Bird Habitats in Britain*, Poyser, Calton.

Fuller, R.M. (1975) The Culbin shingle bar and its vegetation. *Transactions of the Botanical Society of Edinburgh*, **42**, 293–305.

Fuller, R.M. (1987) Vegetation establishment on shingle beaches. *Journal of Ecology*, **75**, 1077–89.

Fuller, R.M. (1989) Orfordness and Dungeness: a comparative study. *Botanical Journal of the Linnean Society*, **101**, 91–101.

Fuller, R.M. and Boorman, L.A. (1977) The spread and development of *Rhododendron ponticum* L. on dunes at Winterton, Norfolk, in comparison with invasion by *Hippophaë rhamnoides* L. at Saltfleetby, Lincolnshire. *Biological Conservation*, **12**, 83–94.

Fuller, R.M. and Randall, R.E. (1988) The Orford shingles, Suffolk, UK – classic conflict in coastline management. *Biological Conservation*, **46**, 95–114.

Funnell, B.M. and Pearson, I. (1984) A guide to the Holocene geology of North Norfolk. *Bulletin of the Geological Society of Norfolk*, **34**, 123–40.

Funnell, B.M. and Pearson, I. (1989) Holocene sedimentation on the North Norfolk barrier coast in relation to relative sea-level change. *Journal of Quaternary Science*, **4**(1), 25–36.

Gadallah, F.L. and Jefferies, R.L. (1995a) Comparison of the nutrient contents of the principal forage plants utilized by lesser snow geese on summer breeding grounds. *Journal of Applied Ecology*, **32**, 263–75.

Gadallah, F.L. and Jefferies, R.L. (1995b) Forage quality in brood rearing areas of the lesser snow goose and the growth of captive goslings. *Journal of Applied Ecology*, **32**, 276–87.

Gale, R.W. and Barr, D.A. (1977) Vegetation and coastal sand dunes. *Beach Conservation*, 28. Brisbane Beach Protection Authority of Queensland.

Gallagher, J.L. (1983) Seasonable patterns in recoverable underground reserves in *Spartina alterniflora* Loisel. *American Journal of Botany*, **70**, 212–15.

Gallagher, J.L., Reimold, R.J., Linthurst, R.A. and Pfeiffer, W.J. (1980) Aerial production, mortality and mineral accumulation – export dynamics in *Spartina alterniflora* and *Juncus roemerianus* plant stands in a Georgia salt marsh. *Ecology*, **61**, 303–12.

Garcia Novo, F. (1979) The ecology of vegetation of the dunes in Doñana National Park (South-West Spain), in *Ecological Processes in Coastal Environments* (eds R.L. Jefferies and A.J. Davy), Blackwell Scientific Publications, Oxford, pp. 571–92.

Gates, D.M. (1980) *Biophysical Ecology*, Springer, Berlin.

Gauthier, G., Hughes, R.J., Reed, A., Beaulieu, J. and Rochfort, L. (1995) Effect of grazing by greater snow geese on the production of graminoids at an arctic site (Bylot Island, NWT, Canada). *Journal of Ecology*, **83**, 653–64.

Géhu, J.M. (1985) *European Dune and Shoreline Vegetation*, Council of Europe, Strasbourg.

Gemmell, A.R., Greig-Smith, P. and Gimingham, C.H. (1953) A note on the behaviour of *Ammophila arenaria* (L.) Link in relation to sand-dune formation. *Transactions of the Botanical Society of Edinburgh*, **36**, 132–6.

Gerdol, V. and Hughes, R.G. (1993) Effect of the amphipod *Corophium volutator* on the colonisation of mud by the halophyte *Salicornia europaea*. *Marine Ecology Progress Series*, **97**, 61–9.

Gerdol, V. and Hughes, R.G. (1994) Effect of *Corophium volutator* on the abundance of benthic diatoms, bacteria and sediment stability in two estuaries in southeastern England. *Marine Biology Progress Series*, **114**, 109–15.

Gerrard, A.J. (1992) *Soil Geomorphology; an Integration of Geomorphology and Pedology*, Chapman & Hall, London.

Gibbons, R. (1994) Newborough Warren NNR, Anglesey. *British Wildlife*, **6**, 109–11.

Gillham, M.E. (1977) *The Natural History of Gower*, D. Brown and Sons, Cowbridge, South Wales.

Gillham, M.E. (1982) *Swansea Bay's Green Mantle*, D. Brown and Sons, Cowbridge, South Wales.

Gimingham, C.H. (1948) The role of *Barbula fallax* Hedw. and *Bryum pendulum* Schp. in sand-dune fixation. *Transactions of the British Bryological Society*, **1**, 70–2.

Gimingham, C.H. (1951a) The use of life form and growth form in the analysis of community structure, as illustrated by a comparison of two dune communities. *Journal of Ecology*, **39**, 396–406.

Gimingham, C.H. (1951b) Contributions to the maritime ecology of St. Cyrus, Kincardineshire. II. The sand-dunes. *Transactions of the Botanical Society of Edinburgh*, **35**, 387–414.

Gimingham, C.H. (1964) Maritime and sub-maritime communities, in *The Vegetation of Scotland* (ed. J.H. Burnett), Oliver and Boyd, Edinburgh, pp. 67–142.

Gimingham, C.H. and Birse, E.M. (1957) Ecological studies on growth-form in Bryophytes. I. Correlations between growth-form and habitat. *Journal of Ecology*, **45**, 533–45.

Gimingham, C.H., Gemmell, A.R. and Greig-Smith, P. (1949) The vegetation of a sand-dune system in the Outer Hebrides. *Transactions of the Botanical Society of Edinburgh*, **35**, 82–96.

Ginsberg, R.N., Isham, L.B., Bein, S.J. *et al.* (1954) *Laminated algal sediments of south Florida and their recognition in the fossil record*, Marine Laboratory, University of Miami, Coral Gables, Florida, report 54–21.

Gleason, H.A. (1917) The structure and development of the plant association. *Bulletin of the Torrey Botanical Club*, **43**, 463–81.

Gleason, H.A. (1926) The individualistic concept of the plant association. *Bulletin of Torrey Botanical Club*, **53**, 7–26.

Glenn, E.P. (1987) Relationship between cation accumulation and water content of salt-tolerant grasses and a sedge. *Plant, Cell and Environment*, **10**, 205–12.

Glenn, E.P. and O'Leary, J.W. (1984) Relationship between salt accumulation and water content of dicotyledonous halophytes. *Plant, Cell and Environment*, **7**, 253–61.

Godwin, H. (1956) *The History of the British Flora*, Cambridge University Press, Cambridge.

Godwin, H. (1978) *Fenland: its Ancient Past and Uncertain Future*, Cambridge University Press, Cambridge.

Good, R. (1935) Contributions towards a survey of the plants and animals of South Haven Peninsula, Studland Heath, Dorset. II. General ecology of the flowering plants and ferns. *Journal of Ecology*, **23**, 361–405.

Goodman, P.J. (1969) *Spartina* Schreb. genus: Biological Flora of the British Isles. *Journal of Ecology*, **57**, 285–7.

Goodman, P.J. and Williams, W.T. (1961) Investigations into 'die-back' in *Spartina townsendii* agg. III. Physiological correlates of 'die-back'. *Journal of Ecology*, **49**, 391–8.

Goodman, P.J., Braybrooks, E.M. and Lambert, J.M. (1959) Investigations into 'die-back' in *Spartina townsendii* agg. I. The present status of *Spartina townsendii* in Britain. *Journal of Ecology*, **47**, 651–77.

Goodman P.J., Braybrooks, E.M., Marchant, C.M. and Lambert, J.M. (1969) *Spartina* × *townsendii* H. and J. Groves *sensu lato*: Biological Flora of the British Isles. *Journal of Ecology*, **57**, 298–313.

Gorham, E. (1958) Soluble salts in dune sands from Blakeney Point in Norfolk. *Journal of Ecology*, **46**, 373–9.

Gorham, J., Hughes, L. and Wyn Jones, R.G. (1980) Chemical composition of salt-marsh plants from Ynys Mon (Anglesey): the concept of physiotypes. *Plant, Cell and Environment*, **3**, 309–18.

Gorman, M. and Raffaelli, D. (1983) Classic sites. The Ythan Estuary. *Biologist*, **40**, 10–13.

Goss-Custard, J.D. (1969) The winter feeding ecology of the redshank, *Tringa totanus*. *Ibis*, **111**, 338–56.

Goss-Custard, J.D. (1975) Beach Feast. *Birds*, **September/October**, 23–6.

Goss-Custard, J.D. (1977) The Ecology of the Wash. III. Density-related behaviour and the possible effects of a loss of feeding grounds on wading birds (Charadrii). *Journal of Applied Ecology*, **14**, 721–40.

Goss-Custard, J.D. (1980) Competition for food and interference among waders. *Ardea*, **68**, 31–52.

Goss-Custard, J.D. and Moser, M.E. (1990) Changes in the numbers of dunlin (*Calidris alpina*) in British estuaries in relation to changes in the abundance of *Spartina*, in Spartina anglica – *a research review* (eds A.J. Gray and P.E.M. Benham), HMSO, London, pp. 69–71.

Gray, A.J. (1971) *Variation in* Aster tripolium *L. with particular reference to some British populations*. PhD Thesis, University of Keele.

Gray, A.J. (1972) The ecology of Morecambe Bay. V. The salt marshes of Morecambe Bay. *Journal of Applied Ecology*, **9**, 207–20.

Gray, A.J. (1985) Adaptation in perennial coastal plants – with particular reference to heritable variation in *Puccinellia maritima* and *Ammophila arenaria*. *Vegetatio*, **61**, 179–88.

Gray, A.J. (1992) Saltmarsh plant ecology: zonation and succession revisited, in *Saltmarshes: Morphodynamics, Conservation and Engineering Significance* (eds J.R.L. Allen and K. Pye), Cambridge University Press, Cambridge, pp. 63–79.

Gray, A.J. and Adam, P. (1974) The reclamation history of Morecambe Bay. *Nature in Lancashire*, **4**, 13–20.

Gray, A.J. and Benham, P.E.M. (1990) Spartina anglica – *a Research Review*. HMSO, London.

Gray, A.J. and Bunce, R.G.H. (1972) The ecology of Morecambe Bay. VI. Soils and vegetation of the salt marshes: a multivariate approach. *Journal of Applied Ecology*, **9**, 221–34.

Gray, A.J. and Scott, R. (1977a) *Puccinellia maritima* (Huds.) Parl.: Biological Flora of the British Isles. *Journal of Ecology*, **65**, 699–716.

Gray, A.J. and Scott, R. (1977b) The ecology of Morecambe Bay. VII. The distribution of *Puccinellia maritima*, *Festuca rubra* and *Agrostis stolonifera* in the salt marshes. *Journal of Applied Ecology*, **14**, 229–41.

Gray, A.J., Benham, P.E.M. and Raybould, A.F. (1990) *Spartina anglica* – the evolutionary and ecological background, in Spartina anglica – *a Research Review*. HMSO, London, pp. 5–10.

Gray, A.J., Drury, M. and Raybould, A.F. (1990) *Spartina* and the ergot fungus *Claviceps purpurea* – a singular contest?, in *Pests, Pathogens and Plant Communities* (eds Burdon, J.J. and Leather, S.R.), Blackwell Scientific Publications, Oxford, pp. 63–79.

Gray, A.J., Marshall, D.F. and Raybould, A.F. (1991) A century of evolution in *Spartina anglica*. *Advances in Ecological Research*, **21**, 1–62.

Gray, A.J., Warmen, E.A., Clarke, R.T. and Johnson, P.J. (1995) The niche of *Spartina anglica* on a changing coastline. *Coastal Zone Topics: Process, Ecology and Management*, **1**, 29–34.

Green, J. (1968) *The Biology of Estuarine Animals*, Sidgwick and Jackson, London.

Green, R.D. (1968) *Soils of Romney Marsh*, Bulletin of the Soil Survey of Great Britain, 4, Harpenden.

Greenhalgh, M.E. (1975) The breeding birds of the Ribble Estuary saltmarshes. *Nature in Lancashire*, **5**, 11–19.

Greenly, E. (1919) *The Geology of Anglesey*, 2 vols. Memoirs of the Geological Survey, HMSO, London.

Greenway, H. and Munns, R. (1980) Mechanisms of salt tolerance in nonhalophytes. *Annual Review of Plant Physiology*, **31**, 149–90.

Greig-Smith, P. (1961) Data on pattern within plant communities. II. *Ammophila arenaria* (L.) Link. *Journal of Ecology*, **49**, 703–8.

Greipsson, S. and Davy, A.J. (1994a) Germination of *Leymus arenarius* and its significance for land reclamation in Iceland. *Annals of Botany*, **73**, 393–401.

Greipsson, S. and Davy, A.J. (1994b) *Leymus arenarius*. Characteristics and uses of a dune-building grass. *Busivindi*, **8**, 41–50.

Greipsson, S. and Davy, A.J. (1995) Seed mass and germination behaviour in populations of the dune-building grass *Leymus arenarius*. *Annals of Botany*, **76**, 493–501.

Grime, J.P. (1974) Vegetation classification by reference to strategies. *Nature*, **250**, 26–31.

Grime, J.P. (1979) *Plant Strategies and Vegetation Processes*, Wiley, Chichester.

Grime, J.P., Hodgson, J.G. and Hunt, R. (1988) *Comparative Plant Ecology*, Unwin Hyman, London.

Gulick, P.J. and Dvorak, J. (1992) Coordinate gene response to salt stress in *Lophopyrum elongatum*. *Plant Physiology*, **100**, 1384–8.

Hajibagheri, M.A., Hall, J.L. and Flowers, T.J. (1984) Stereological analysis of leaf cells of the halophyte *Suaeda maritima* (L.) Dum. *Journal of Experimental Botany*, **35**, 1547–57.

Hajibagheri, M.A., Yeo, A.R. and Flowers, T.J. (1985) Salt tolerance in *Suaeda maritima* (L.) Dum. Fine structure and ion concentrations in the apical region of roots. *New Phytologist*, **99**, 331–43.

Hale, W.G. (1980) *Waders*, Collins, London.

Harley, J.L. and Harley, E.L. (1987) A check-list of mycorrhiza in the British flora. *New Phytologist (Supplement)*, **105**, 1–102.

Harmes, P.A. and Spiers, A. (1993) *Polygonum maritimum* L. in East Sussex (v.c. 14). *Watsonia*, **19**, 271–3.

Harper, J.L. (1977) *Population Biology of Plants*, Academic Press, London.

Harris, D. and Davy, A.J. (1986a) Strandline colonization by *Elymus farctus* in relation to sand mobility and rabbit grazing. *Journal of Ecology*, **74**, 1045–56.

Harris, D. and Davy, A.J. (1986b) Regenerative potential of *Elymus farctus* from rhizome fragments and seed. *Journal of Ecology*, **74**, 1057–67.

Harris, D. and Davy, A.J. (1987) Seedling growth in *Elymus farctus* after episodes of burial with sand. *Annals of Botany*, **60**, 587–93.

Harvey, J.W., Germann, P.F. and Odum, W.E. (1987) Geomorphological control of subsurface hydrology in the creekbank zone of tidal marshes. *Estuarine, Coastal and Shelf Science*, **25**, 677–91.

Haslam, S.M. (1972) *Phragmites communis* Trin.: Biological Flora of the British Isles. *Journal of Ecology*, **60**, 585–610.

Hewett, D.G. (1985) Grazing and mowing as management tools on dunes. *Vegetatio*, **62**, 441–7.

Hicks, S.D., Debaugh, H.A. and Hickman, L.E. (1983) Sea level variations for the United States 1855–1980. National Oceanic and Atmospheric Administration, Rockeville, Maryland, USA.

Hik, D.S. and Jefferies, R.L. (1990) Increases in the net above-ground primary production of a salt-marsh forage grass: a test of the predictions of herbivore optimization model. *Journal of Ecology*, **78**, 180–95.

Hik, D.S., Sadul, H.A. and Jefferies, R.L. (1991) Effects of the timing of multiple grazings by geese on net above-ground primary production of swards of *Puccinellia phryganodes*. *Journal of Ecology*, **79**, 715–30.

Hik, D.S., Jefferies, R.L. and Sinclair, A.R.E. (1992) Foraging by geese, isostatic uplift and asymmetry in the development of salt-marsh plant communities. *Journal of Ecology*, **80**, 395–406.

Hill, A.E. and Hill. B.S. (1973) The *Limonium* salt gland: a biophysical and structural study. *International Review of Cytology*, **35**, 299–319.

Hill, D. and Makepeace, P. (1989) Population trends in bird species at Dungeness, Kent. *Botanical Journal of the Linnean Society*, **101**, 137–51.

Hill, M.O., Bunce, R.G.H. and Shaw, M.W. (1975) Indicator species analysis: a divisive polythetic method of classification and its application to a survey of native pinewoods in Scotland. *Journal of Ecology*, **63**, 597–613.

Hill, T.G. and Hanley, J.A. (1914) The structure and water-content of shingle beaches. *Journal of Ecology*, **2**, 21–38.

Hillen, R. and Roelse, P. (1995) Dynamic preservation of the coastline in the Netherlands. *Journal of Coastal Conservation*, **1**, 17–28.

Hobbs, R.J., Gimingham, C.H. and Band, W.T. (1983) The effects of planting technique on the growth of *Ammophila arenaria* (L.) Link and *Leymus arenarius* (L.) Hochst. *Journal of Applied Ecology*, **20**, 659–72.

Hodgkin, S.E. (1984) Scrub encroachment and its effects on soil fertility on Newborough Warren, Anglesey, Wales. *Biological Conservation*, **29**, 99–119.

Hodson, M.J., Smith, M.M., Wainwright, S.J. and Öpik, H. (1981) Cation cotolerance in a salt-tolerant clone of *Agrostis stolonifera*. *New Phytologist*, **90**, 253–61.

Hollis, T., Thomas, D. and Heard, S. (1990) *The Effects of Sea level Rise on Sites of Conservation Value in Britain and North West Europe*, University College London and World Wildlife Fund.

Holmes, A. (1965) *Principles of Physical Geology*, 2nd edn, Nelson, London.

Hope-Simpson, J.F. and Yemm, E. W. (1979) Braunton Burrows: developing vegetation in dune slacks, in *Ecological Processes in Coastal Environments* (eds R.L. Jefferies and A.J. Davy), Blackwell Scientific Publications, Oxford, pp. 113–27.

Houghton, J.T., Jenkins, G.J. and Ephraums, J.J. (1990) *Climatic Change: The IPCC Scientific Assessment*, Cambridge University Press, Cambridge.

Houghton, J.T., Callander, B.A. and Varney, S.K. (1992) *Climatic Change 1992. The Supplementary Report to the IPCC Scientific Assessment*, Cambridge University Press, Cambridge.

Howells, G. (1988) The Conwy crossing: conservation strategies. *Bulletin of the Estuarine and Brackish Water Sciences Association*, **50**, 24–8.

Hoyt, J. (1967) Barrier islands formation. *Bulletin Geological Society of America*, **78**(9), 1125–36.

Hubbard, J.C.E. (1970a) Effects of cutting and seed production in *Spartina anglica*. *Journal of Ecology*, **58**, 329–34.

Hubbard, J.C.E. (1970b) The shingle vegetation of southern England: a general survey of Dungeness, Kent and Sussex. *Journal of Ecology*, **58**, 713–22.

Huggett, D. (1995) The role of the Birds Directive and the Habitats and Species Directive in delivering integrated coastal zone planning and management, in *Directions in European Coastal Management* (eds M.G. Healy and J.P. Doody), Samara Publishing, Cardigan, pp. 9–18.

Hughes, R.G.(1998) Interaction of sediment macrofauna and colonisation of saltmarsh plants, in *British Saltmarshes* (eds T. Harris and B. Sherwood), Linnean Society/Samara Publishing, Cardigan.

Huiskes, A.H.L. (1977) *The population dynamics of* Ammophila arenaria (*L.*) *Link*. PhD Thesis, University of Wales.

Huiskes, A.H.L. (1979) *Ammophila arenaria* (L.) Link: Biological Flora of the British Isles. *Journal of Ecology*, **67**, 363–82.

Huiskes, A.H.L., Koutstaal, B.P., Herman, P.M.J., Beeftink, W.G., Makusse, M.M. and De Munck, W. (1995) Seed dispersal of halophytes in tidal salt marshes. *Journal of Ecology*, **83**, 559–67.

Hulme, B.A. (1957) *Studies on some British species of* Atriplex *L*. PhD Thesis, University of Edinburgh.

Humphreys, M.O. (1982) The genetic basis of tolerance to salt spray in populations of *Festuca rubra* L. *New Phytologist*, **91**, 287–96.

Hussey, A. and Long, S.P. (1982) Seasonal changes in weight of above- and below-ground vegetation and dead plant material in a salt marsh at Colne Point, Essex. *Journal of Ecology*, **70**, 757–71.

Hutchings, M.J. and Russell, P.J (1989) The seed regeneration dynamics of an emergent salt marsh. *Journal of Ecology*, **77**, 615–37.

Hutchinson, G.E. (1957) Concluding remarks. *Cold Spring Harbor Symposia on Quantitative Biology*, **22**, 415–27.

Hydraulics Research (1980) *Design of Seawalls allowing for Wave Overtopping*, Report No. EX 924, Hydraulics Research Station, Wallingford, UK.

Hydraulics Research (1988) *Review of the use of Saltings in Coastal Defence*, Report No. SR 170, Hydraulics Research Station, Wallingford, UK.

Ingold, A. and Havill, D.C. (1984) The influence of sulphide on the distribution of higher plants in salt marshes. *Journal of Ecology*, **72**, 1043–54.

Ingram, H.A.P., Barclay, A.M., Coupar, A.M., Glover, J.G., Lynch, B.M. and Sprent, J.I. (1980) *Phragmites* performance in reedbeds in the Tay Estuary. *Proceedings of the Royal Society of Edinburgh*, **75B**, 89–107.

Inman, D.L., Ewing, G.C. and Corliss, J.B. (1966) Coastal sand dunes of Guerrero Negro, Baja Calif., Mexico. *Bulletin Geological Society of America*, **77**(8), 787–802.

Inns, H. (1995) Reptiles and amphibians. *British Wildlife*, **6**, 183.

Jackson, D., Long, S.P. and Mason, C.F. (1986) Net primary production, decomposition and export of *Spartina anglica* on a Suffolk salt-marsh. *Journal of Ecology*, **74**, 647–62.

Jackson, H.C. (1979) The decline of the sand lizard, *Lacerta agilis* L., population on the sand dunes of the Merseyside coast, England. *Biological Conservation*, **16**, 177–91.

Jackson, M.B. (1985) Ethylene and responses of plants to soil waterlogging and submergence. *Annual Review of Plant Physiology*, **36**, 145–74.

James, P.A. and Wharfe, A.J. (1984) The chemistry of precipitation in a coastal locality in North-West England. *Catena*, **11**, 219–27.

James, P.A. and Wharfe, A.J. (1989) Timescales of soil development in a coastal sand dune system, Ainsdale, North-West England, in *Perspectives in*

Coastal Dune Management (eds F. van der Meulen, P.D. Jungerius and J. Visser), SPB Academic Publishing, The Hague, pp. 287–95.

James, P.A., Wharfe, A.J., Pegg, R.K. and Clarke, D. (1986) A cation budget analysis for coastal dune system in North-West England. *Catena*, **13**, 1–10.

Jefferies, R.L. (1976) The north Norfolk coast, in *Nature in Norfolk*, Jarrold and Sons, Norwich, pp. 130–8.

Jefferies, R.L. (1977) The vegetation of salt marshes at some coastal sites in Arctic North America. *Journal of Ecology*, **65**, 661–72.

Jefferies, R.L., Davy, A.J. and Rudmik, T. (1979) The growth strategies of coastal halophytes, in *Ecological Processes in Coastal Environments* (eds R.L. Jefferies and A.J. Davy), Blackwell Scientific Publications, Oxford, pp. 243–68.

Jefferies, R.L., Jensen, A. and Abraham, K.F. (1979) Vegetational development and the effect of geese on vegetation at La Pérouse Bay, Manitoba. *Canadian Journal of Botany*, **57**, 1439–50.

Jefferies, R.L., Davy, A.J. and Rudmik, T (1981) Population biology of the salt marsh annual *Salicornia europaea* agg. *Journal of Ecology*, **69**, 17–31.

Jefferies, R.L., Jensen, A. and Bazely, D. (1983) The biology of the annual *Salicornia europaea* agg. at the limits of its range in Hudson Bay. *Canadian Journal of Botany*, **61**, 762–73.

Jefferies, R.L., Klein, D.R. and Shaver, G.R. (1994) Vertebrate herbivores and northern plant communities: reciprocal influences and responses. *Oikos*, **71**, 193–206.

Jefferies, R.L., Gadallah, F.L., Srivastava, D.S. and Wilson, D.J. (1995) Desertification and trophic cascades in arctic coastal ecosystems: a potential climatic change scenario?, in *Global Change and Arctic Terrestrial Ecosystems. Ecosystems Research Report 10, EUR 15519 EN* (eds T.V. Callagan *et al.*), European Commission, Brussels, pp. 201–6.

Jennings, D.H. (1976) The effect of sodium chloride on higher plants. *Biological Reviews*, **51**, 453–86.

Jensen, A. (1985) The effect of cattle and sheep grazing in salt-marsh vegetation at Skallingen, Denmark. *Vegetatio*, **60**, 37–48.

Johnson, D.A. and Tieszen, L.L. (1976) Aboveground biomass allocation, leaf growth and photosynthesis patterns in tundra plant forms in Arctic Alaska. *Oecologia*, *(Berlin)*, **24**, 159–73.

Johnson, D.S. and York, H.H. (1915) The relation of plants to tide levels. *Carnegie Institute Publication* 206.

Jones, A.M. (1990) *An assessment of change in the vegetation of the Cefni Salt Marsh 1966–1990*, MSc Thesis, University College of North Wales, Bangor.

Jones, H.G. (1983) *Plants and Microclimate*, Cambridge University Press, Cambridge.

Jones, K. (1974) Nitrogen fixation in a salt marsh. *Journal of Ecology*, **62**, 553–65.

Jones, P.S. and Etherington, J.R. (1989) Ecological and physiological studies of sand dune slack vegetation, Kenfig Pool and Dunes Local Nature Reserve, Mid-Glamorgan, Wales, UK, in *Perspectives in Coastal Dune Management* (eds F. van der Meulen, P.D. Jungerius and J. Visser), SPB Academic Publishing, The Hague, pp. 297–303.

Jones, P.S., Kay, Q.O.N. and Jones, A. (1995) The decline of rare plant species and community types in the sand dune systems of South Wales, in *Directions in European Coastal Management* (eds M.G. Healy and J.P. Doody), Samara Publishing, Cardigan, pp. 547–55.

Jones, V. and Richards, P.W. (1954) *Juncus acutus* L.: Biological Flora of the British Isles. *Journal of Ecology*, **42**, 639–50.

Jumars, P.A. (1993) Gourmands of mud: diet selection in marine deposit feeders, in *Diet Selection: an Inter-disciplinary Approach to Foraging Behaviour* (ed. R.N. Hughes), Blackwell Scientific Publications, Oxford, pp. 124–56.

Jumars, P.A., Newell, R.C., Angel, M.V., Fowler, S.W., Poulet, S.A., Rowe, G.T. and Smetacek, V. (1984) Detritivory, in *Flows of Material and Energy in Marine Ecosystems* (ed. M.J.R. Fasham), Plenum, New York, pp. 685–93.

Jungerius, P.D., Koehler, H., Kooijman, A.M., Mucher H.J. and Graefe, U. (1995) Response of vegetation and soil ecosystem to mowing and sod removal in the coastal dunes 'Zwanenwater', the Netherlands. *Journal of Coastal Conservation*, **1**, 3–16.

Kachi, N. and Hirose, T. (1983) Limiting nutrients for plant growth in coastal sand dune soils. *Journal of Ecology*, **71**, 937–44.

Keddy, P.A. (1981) Experimental demography of the sand-dune annual, *Cakile edentula*, growing along an environmental gradient in Nova Scotia. *Journal of Ecology*, **69**, 615–30.

Keddy, P.A. (1982) Population ecology on an environmental gradient: *Cakile edentula* on a sand dune. *Oecologia*, **52**, 348–55.

Kerr, R.A. (1981) Whither the shoreline. *Science*, **214**, 428.

Kershaw, K.A. (1976) The vegetational zonation of the East Pen Island salt marshes, Hudson Bay. *Canadian Journal of Botany*, **54**, 5–13.

Kessel, R.H. and Smith, J.S. (1978) Tidal creek and pan formation in intertidal marshes, Nigg Bay, Scotland. *Scottish Geographical Magazine*, **94**, 159–68.

King, G.M., Klug, M.J., Wiegert, R.G. and Chalmers, A.G. (1982) Relations of soil water movement and sulphide concentration to *Spartina alterniflora* production in a Georgia saltmarsh. *Science, New York*, **218**, 61–3.

Koehler, H., Munderloh, E. and Hofmann, S. (1995) Soil microarthropods (Acari, Collembola) from beach and dune: characteristics and eco-system context. *Journal of Coastal Conservation*, **1**, 77–86.

Koutstaal, B.P., Markusse, M.M. and De Munck, W. (1987) Aspects of seed dispersal by tidal movements, in *Vegetation Between Land and Sea* (eds A.H.L. Huiskes, C.W.P.M. Blom and J. Rozema), Junk, Dordrecht, pp. 226–33.

Krebs, C.J. (1985) *Ecology: the Experimental Analysis of Distribution and Abundance*, 3rd edn, Harper and Row, New York.

Krumbein, W.C. and Slack, H.A. (1956) The relative efficiency of beach sampling methods. *Technical Memoranda Beach Erosion Board, US*, **90**, 1–34.

Kuipers, B.R., De Wilde, P.A.W.J. and Creutberg, F. (1981) Energy flow in a tidal flat ecosystem. *Marine Ecology, Progress Series*, **5**, 215–21.

Ladle, M. (ed.) (1981) *The Fleet and Chesil Beach: Structure and Biology of a Unique Coastal Feature*, Fleet Study Group, Dorchester, Dorset.

Laing, C. (1954) *The ecological life history of the marram grass community on Lake Michigan dunes*, PhD Thesis, University of Chicago.

Landsberg, S.Y. (1955) *The morphology and vegetation of the sands of Forvie*, PhD Thesis, University of Aberdeen.

Laundon, J.R. (1989) Lichens at Dungeness. *Botanical Journal of the Linnean Society*, **101**, 103–9.

Lee, J.E. and Ignaciuk, R. (1985) The physiological ecology of strandline plants. *Vegetatio*, **62**, 319–26.

Lemon.E.R. and Erickson, A.E. (1955) Principle of the platinum microelectrode as a method of characterising soil aeration. *Soil Science*, **79**, 383–92.

Lewis, W.V. (1932) The formation of the Dungeness foreland. *Geographical Journal*, **80**, 309–24.

Liddle, M.J. (1975a) A selective review of the ecological effects of human trampling on natural ecosystems. *Biological Conservation*, **7**, 17–36.

Liddle, M.J. (1975b) A theoretical relationship between the primary productivity of vegetation and its ability to tolerate trampling. *Biological Conservation*, **8**, 251–5.

Liddle, M.J. and Greig-Smith, P. (1975a) A survey of tracks and paths in a sand dune ecosystem. I. Soils. *Journal of Applied Ecology*, **12**, 893–908.

Liddle, M.J. and Greig-Smith, P. (1975b) A survey of tracks and paths in a sand dune ecosystem. II. Vegetation. *Journal of Applied Ecology*, **12**, 909–30.

Liddle, M.J. and Moore, K.G. (1974) The microclimate of sand dune tracks: the relative contribution of vegetation removal and soil compression. *Journal of Applied Ecology*, **11**, 1057–68.

Lindeman, R.L. (1942) The trophic–dynamic aspect of ecology. *Ecology*, **23**, 399–418.

Little, L.R. and Maun, M.A. (1996) The '*Ammophila* problem' revisited: a role for mycorrhizal fungi. *Journal of Ecology*, **84**, 1–7.

Llewellyn, P.J. and Shackley, S.E. (1996) The effects of mechanical beach-cleaning on invertebrate populations. *British Wildlife*, **7**, 147–55.

Lockwood, A.P.M. and Inman, C.B.E. (1979) Ecophysiological responses of *Gammarus duebeni* to salinity fluctuations, in *Ecological Processes in Coastal Environments* (eds R.L. Jefferies and A.J. Davy), Blackwell Scientific Publications, Oxford, pp. 269–84.

Loffler, M. and Coosen, J. (1995) Ecological impact of sand replenishment, in *Directions in European Coastal Management* (eds M.G. Healy and J.P. Doody), Samara Publishing, Cardigan, pp. 291–9.

Long, S.P. (1990) The primary productivity of *Puccinellia maritima* and *Spartina anglica*: a simple predictive model of response to climatic change, in *Expected Effects of Climatic Change on Marine Coastal Ecosystems* (eds J.J. Beukema, W.J. Wolff and J.J.W.N. Brouns), Kluwer, Dordrecht, pp. 33–9.

Long, S.P. and Mason, C.F. (1983) *Saltmarsh Ecology*, Blackie, Glasgow.

Long, S.P. and Woolhouse, H.W. (1979) Primary production in *Spartina* marshes, in *Ecological Processes in Coastal Environments* (eds R.L. Jefferies and A.J. Davy), Blackwell Scientific Publications, Oxford, pp. 333–52.

Lonsdale, W.M. and Watkinson, A.R. (1983) Plant geometry and self-thinning. *Journal of Ecology*, **71**, 285–97.

Looney, P.B and Gibson, D.J. (1995) The relationship between the soil seed bank and aboveground vegetation of a coastal barrier island. *Journal of Vegetation Science*, **6**, 825–36.

Lopez, G.R. and Levington, J.S. (1987) Ecology of deposit-feeding animals in marine sediments. *Quarterly Review of Biology*, **62**, 235–60.

Lopez, G.R., Taghon, G.L. and Levington, J.S. (eds) (1989) *Ecology of Marine Deposit Feeders*, Springer Verlag, Berlin.

Luxton, M. (1964) Some aspects of the biology of salt marsh Acarina. *Acaralogia*. C.R. 1er Congrès Int. d'Acaralogie, Fort Collins, Colorado, USA, 1963, pp. 172–82.

MacArthur, R.H. and Wilson, E.O. (1967) *The Theory of Island Biogeography*, Princetown University Press, New Jersey.

Mack, R.N. (1976) Survivorship of *Cerastium atrovirens* at Aberffraw, Anglesey. *Journal of Ecology*, **64**, 309–12.

Mack, R.N. and Harper, J.L. (1977) Interference in dune annuals: spatial pattern and neighbourhood effects. *Journal of Ecology*, **65**, 345–63.

MAFF. *The Saltmarsh Habitat Scheme: Guidelines for Farmers*. HMSO, London.

Magnusson, M. (1982) Composition and succession of lichen communities in an inner coastal dune area in southern Sweden. *Lichenologist*, **14**, 153–63.

Malloch, A.J.C. (1989) Plant communities of the British sand dunes. *Proceedings of the Royal Society of Edinburgh*, **96B**, 53–74.

Marchant, C.J. (1975) *Spartina* Schreb., in *Hybridization and the Flora of the British Isles* (ed. C.A. Stace), Academic Press, London, pp. 586–7.

Marinucci, A.C. (1982) Trophic importance of *Spartina alterniflora* production and decomposition to the marsh–estuarine ecosystem. *Biological Conservation*, **22**, 35–58.

Marples, T.G. (1966) A radionuclide tracer study of Arthropod food chains in a *Spartina* salt marsh ecosystem. *Ecology*, **47**, 270–7.

Marschner, H. (1995) *Mineral Nutrition of Higher Plants*, 2nd edn, Academic Press, London.

Marshall, J.K. (1965) *Corynephorus canescens* (L.) P. Beauv. as a model for the *Ammophila* problem. *Journal of Ecology*, **53**, 447–63.

Martin, W.E. (1959) The vegetation of Island Beach State Park, New Jersey. *Ecological Monographs*, **29**, 1–46.

Maun, M.A. and Lapierre, J. (1984) The effects of burial by sand on *Ammophila breviligulata*. *Journal of Ecology*, **72**, 827–39.

McDonnell, M.J. (1981) Trampling effects on coastal dune vegetation in the Parker River National Wildlife Refuge, Massachusetts, USA. *Biological Conservation*, **21**, 289–301.

McLusky, D.S. (1989) *The Estuarine Ecosystem*, 2nd edn, Blackie, Glasgow.

McMinn, S. (1989) Migration studies at Dungeness Bird Observatory. *Botanical Journal of the Linnean Society*, **101**, 79–89.

Miles, J. (1979) *Vegetation Dynamics*, Chapman & Hall, London.

Miles, J. (1987) Vegetation succession: past and present perceptions, in *Colonization, Succession and Stability* (eds A.J. Gray, M.J. Crawley and P.J. Edwards), Blackwell Scientific Publications, Oxford, pp. 1–29.

Mitchell, J.F.B., Manabe, S., Meleshko, V. and Tokioka, T. (1990) Equilibrium climate change – and its implications for the future, in *Climate Change: the IPCC Scientific Assessment* (eds J.T. Houghton, G.J. Jenkins and J.J. Ephraums), Cambridge University Press, Cambridge, pp. 131–72.

Monteith, J.L. and Unsworth, M.H. (1990) *Principles of Environmental Physics*, 2nd edn, Edward Arnold, London.

Moon, G.J., Clough, B.F., Peterson, C.A. and Allaway, G.W. (1986) Apoplastic and symplastic pathways in *Avicennia marina* (Forsk.) Vierh. roots revealed by fluorescent tracer dyes. *Australian Journal of Plant Physiology*, **13**, 637–48.

Moore, C.J. and Scott, G.A.M. (1979) The ecology of mosses on a sand dune in Victoria, Australia. *Journal of Bryology*, **10**, 291–311.

Moore, P.D. (1971) Computer analysis of sand dune vegetation in Norfolk, England, and its implications for conservation. *Vegetatio*, **23**, 323–38.

Morton, A.J. (1974) Ecological studies of fixed dune grassland at Newborough Warren, Anglesey. II. Causal factors of the grassland structure. *Journal of Ecology*, **62**, 261–78.

Moss, S.T. (ed.) (1986) *The Biology of Marine Fungi*, Cambridge University Press, Cambridge.

Mudge, G.P. and Allen, D.S. (1980) Wintering seaducks in the Moray and Dornoch Firths. *Wildfowl*, **31**, 123–30.

Newell, R.C. (1965) The role of detritus in the nutrition of two marine deposit feeders, the Prosobranch *Hydrobia ulvae* and the bivalve *Macoma balthica*. *Proceedings of the Zoological Society of London*, **144**, 25–45.

Nichol, E.A.T. (1935) The ecology of a salt marsh. *Journal of the Marine Biological Association*, **20**, 203–61.

Nicholson, I.A. (1952) *A study of* Agropyron junceum *(L.) Beauv. in relation to the stabilization of coastal sand and the development of sand dunes*, MSc Thesis, University of Durham.

Niu, X., Narasimhan, M.L., Salzman, R.A., Bressan, R.A. and Hasegawa, P.M. (1993) NaCl regulation at the plasma membrane H^+ATPase gene expression in a glycophyte and a halophyte. *Plant Physiology*, **103**, 713–18.

Nixon, S.W. (1980) Between coastal marshes and coastal water – a review of twenty years of speculation and research on the role of saltmarshes in estuarine productivity and water chemistry, in *Estuarine and Wetlands Processes with Emphasis on Modeling* (eds P. Hamilton and K.B. MacDonald), Plenum, New York, pp. 437–525.

Noble, J.C. (1976) *The population biology of rhizomatous plants*, PhD Thesis, University of Wales.

Noble, J.C. (1982) *Carex arenaria* L.: Biological Flora of the British Isles. *Journal of Ecology*, **70**, 867–86.

Noble, S.M., Davy, A.J. and Oliver, R.P. (1992) Ribosomal DNA variation and population differentiation in *Salicornia* L. *New Phytologist*, **122**, 553–65.

Nordstrom, K.F. and Jackson, N.L. (1995) Temporal scales of landscape change following storms on a human-altered coast. *Journal of Coastal Conservation*, **1**, 51–62.

Nordstrom, K.F., Psuty, N. and Carter, R.W.G. (eds) (1990) *Coastal Dunes: Form and Process*, Wiley, Chichester.

North, S. (1981) *Sands of Forvie and Ythan Estuary National Nature Reserve*, Nature Conservancy Council.

NRA (1991) *Grey tide against green marsh – sea defences in East Anglia*. Synopsis of Proceedings of Conference held at Snape on 1st November 1991, National Rivers Authority, UK.

Nuttle, W.K. and Hemond, H.F. (1988) Salt marsh hydrology: implications for biogeochemical fluxes to the atmosphere and estuaries. *Global Biogeochemical Cycles*, **2**, 91–114.

Nuttle, W.K. and Portnoy, J.W. (1992) Effect of rising sea level on runoff and groundwater discharge to coastal ecosystems. *Estuarine, Coastal and Shelf Science*, **34**, 203–12.

Nyman, J.A., Delaune, R.D. and Patrick, W.H. (1990) Wetland soil formation in the rapidly subsiding Mississippi River Deltaic Plain: mineral and organic matter relationships. *Estuarine, Coastal and Shelf Science*, **31**, 57–69.

O'Connor, R.J. (1981) Patterns of shorebird feeding, in *Estuary Birds of Britain and Ireland* (ed. A.J. Prater), Poyser, Calton, Waterhouses, Staffordshire, UK, pp. 34–50.

Odum, E.P. (1969) The strategy of ecosystem development. *Science*, **165**, 262–70.

Odum, W.E., Fisher, J.S. and Pickral, J.C. (1979) Factors controlling the flux of particulate organic carbon from estuarine wetlands, in *Ecological Processes in Coastal and Marine Systems* (ed. D.J. Livingston), Plenum, New York, pp. 69–80.

Olesen, B. and Sand-Jensen, K. (1994) Demography of shallow eelgrass (*Zostera marina*) populations – shoot dynamics and biomass development. *Journal of Ecology*, **82**, 379–90.

Olff, H., Huisman, J. and van Tooren, B.F. (1993) Species dynamics and nutrient accumulation during early primary succession in coastal sand dunes. *Journal of Ecology*, **81**, 693–706.

Oliver, F.W. (1912) The shingle beach as a plant habitat. *New Phytologist*, **11**, 73–99.

Olson, J.S. (1958) Lake Michigan dune development. I. Wind velocity profiles. *Journal of Geology*, **66**, 254–63.

Olson, J.S. and van der Maarel, E. (1989) Coastal dunes in Europe: a global view, in *Perspectives in Coastal Dune Management* (eds F. van der Meulen, P.D. Jungerius and J. Visser), SPB Academic Publishing, The Hague, pp. 3–32.

Olsson-Seffer, P. (1909) Hydrodynamic factors influencing plant-life on sandy sea shores. *New Phytologist*, **8**, 37–51.

Oosterveld, P. (1985) Grazing in dune areas: the objectives of nature conservation and the aims of research for nature conservation management, in *Sand Dunes and their Management* (ed. J.P. Doody), Nature Conservancy Council, Peterborough, pp. 187–203.

Oosting, H.J. (1956) *The Study of Plant Communities*, 2nd edn, W.H. Freeman, London.

Orians, G.H. (1975) Diversity, stability and maturity in natural ecosystems, in *Unifying Concepts in Ecology* (eds W.H. van Dobben and R.H. Lowe-McConnell), Junk B.V. Publishers, The Hague, pp. 139–50.

Otte, M.L. (1991) Contamination of coastal wetlands with heavy metals: factors affecting uptake of heavy metals by salt marsh plants, in *Ecological Responses to Environmental Stresses* (eds J. Rozema and J.A.C. Verkleij), Kluwer Academic Publishers, Dordrecht, The Netherlands, pp. 126–33.

Owen, M., Black, J.M., Agger, M.K. and Campbell, C.R.G. (1987) The use of the Solway Firth, Britain, by Barnacle Geese *Branta leucopsis* Bechst. in relation to refuge establishment and increases in numbers. *Biological Conservation*, **39**, 63–81.

Packham, J.R. (1971) Coping with oil spills. *Marine Technology*, **2**, 93–102.

Packham, J.R. and Harding, D.J.L.(1982) *Ecology of Woodland Processes*, Edward Arnold, London.

Packham, J.R. and Liddle, M.J. (1970) The Cefni salt marsh, Anglesey, and its recent development. *Field Studies*, **3**, 331–56.

Packham, J.R. and Willis, A.J. (1995) Plant growth form, substrate properties and sediment stabilization in coastal ecosystems, in *Directions in European Coastal Management* (eds M.G. Healy and J.P. Doody), Samara Publishing, Cardigan, pp. 351–9.

Packham, J.R., Willis, A.J. and Poel, L.W. (1966) Seasonal variation of some soil features and the ecology of Kennel Field, Warwickshire. *Journal of Ecology*, **54**, 383–401.

Packham, J.R., Harding, D.J.L., Hilton, G.M. and Stuttard, R.A. (1992) *Functional Ecology of Woodlands and Forests*, Chapman & Hall, London.

Packham, J.R., Harmes, P.A. and Spiers, A. (1995) Development of a shingle community related to a specific sea defence structure, in *Directions in European Coastal Management* (eds M.G. Healy and J.P. Doody), Samara Publishing, Cardigan, pp. 369–71.

Page, R.R., da Vinha, S.G. and Agnew, A.D.Q. (1985) The reaction of some sand-dune plant species to experimentally imposed environmental change: a reductionist approach to stability. *Vegetatio*, **61**, 105–14.

Pakeman, R.J. and Lee, J.A. (1991a) The ecology of the strandline annuals *Cakile maritima* and *Salsola kali*. I. Environmental factors affecting plant performance. *Journal of Ecology*, **79**, 143–53.

Pakeman, R.J. and Lee, J.A. (1991b) The ecology of the strandline annuals *Cakile maritima* and *Salsola kali*. II. The role of nitrogen in controlling plant performance. *Journal of Ecology*, **79**, 155–65.

Patton, D. and Stewart, E.J.A. (1917) The flora of the Culbin Sands. *Transactions of the Botanical Society of Edinburgh*, **26**, 345–74.

Paviour-Smith, K. (1956) The biotic community of a salt meadow in New Zealand. *Transactions of the Royal Society of New Zealand*, **83**, 525–54.

Pearson, M.C. and Rogers, J.A. (1962) *Hippophaë rhamnoides* L.: Biological Flora of the British Isles. *Journal of Ecology*, **50**, 501–13.

Pemadasa, M.A. (1981) Cyclic change and pattern in an *Anthrocnemum* community in Sri Lanka. *Journal of Ecology*, **69**, 565–74.

Pemadasa, M.A. and Lovell, P.H. (1974a) Interference in populations of some dune annuals. *Journal of Ecology*, **62**, 855–68.

Pemadasa, M.A. and Lovell, P.H. (1974b) Factors controlling the flowering time of some dune annuals. *Journal of Ecology*, **62**, 869–80.

Pemadasa, M.A. and Lovell, P.H. (1975) Factors controlling germination of some dune annuals. *Journal of Ecology*, **63**, 41–59.

Pemadasa, M.A. and Lovell, P.H. (1976) Effects of the timing of the life-cycle on the vegetative growth of some dune annuals. *Journal of Ecology*, **64**, 213–22.

Pemadasa, M.A., Greig-Smith, P. and Lovell, P.H. (1974) A quantitative description of the distribution of annuals in the dune system at Aberffraw, Anglesey. *Journal of Ecology*, **62**, 379–402.

Peterken, G.F. and Hubbard, J.C.E. (1972) The shingle vegetation of Southern England: the holly wood on Holmstone Beach, Dungeness. *Journal of Ecology*, **60**, 547–72.

Pethick, J.S. (1974) The distribution of salt pans on tidal salt marshes. *Journal of Biogeography*, **1**, 57–62.

Pethick, J.S. (1981) Long-term accretion rates on tidal salt marshes. *Journal of Sedimentary Petrology*, **51**, 571–7.

Pethick, J.S. (1984) *An Introduction to Coastal Geomorphology*, Edward Arnold, London.

Pethick, J.S. (1992) Saltmarsh geomorphology, in *Saltmarshes: Morphodynamics, Conservation and Engineering Significance* (eds J.R.L. Allen and K. Pye), Cambridge University Press, Cambridge, pp. 41–62.

Pethick, J.S. (1996) The sustainable use of coasts: Monitoring, Modelling and Management, in *Studies in European Coastal Management* (eds P.S. Jones, M.G. Healy and A.T. Williams), Samara Publishing Ltd, Cardigan, pp. 83–92.

Pigott, C.D. (1969) Influence of mineral nutrition on the zonation of flowering plants in coastal saltmarshes, in *Ecological Aspects of the Mineral Nutrition of Plants* (ed. I.H. Rorison), Blackwell Scientific Publications, Oxford, pp. 25–35.

Pizzey, J.M. (1975) Assessment of dune stabilisation at Camber, Sussex, using air photographs. *Biological Conservation*, **7**, 275–88.

Pomeroy, L.R. and Wiegert, R.G. (eds) (1981) *The Ecology of a Salt Marsh*, Springer, New York.

Pons, L.J. and Zonneveld, I.S. (1965) *Soil ripening and soil classification; initial soil formation of alluvial deposits with a classification of the resulting soils.* Wageningen International Institute for Land Reclamation and Improvement, Wageningen.

Proctor, M.C.F. (1980) Vegetation and environment in the Exe Estuary, in *Essays on the Exe Estuary* (ed. G.T. Boalch). Devonshire Association for the Advancement of Science, Literature and Art. Special Volume No. 2, pp. 117–34.

Proudfoot, A.M. (1993) *Relationships between* Coleophora atriplicis *and its host plants on a salt marsh*, PhD Thesis, University of East Anglia.

Pugh, G.J.F. (1979) The distribution of fungi in coastal regions, *in Ecological Processes in Coastal*

Environments (eds R.L. Jefferies and A.J. Davy), Blackwell Scientific Publications, Oxford, pp. 415–27.

Pye, K. (1983) Coastal dunes. *Processes in Physical Geography*, **7**, 531–57.

Pye, K. (1992) Saltmarshes on the barrier coastline of North Norfolk, eastern England, in *Saltmarshes: Morphodynamics, Conservation and Engineering Significance* (eds J.R.L. Allen and K. Pye), Cambridge University Press, Cambridge, pp. 148–78.

Pye, K. and Crooks, S. (1996) Geochemical controls on the moisture content and shear strength of saltmarsh sediments, in *Saltmarsh Management for Flood Defence* (eds Sir William Halcrow and Partners Ltd), NRA, Bristol.

Pye, K. and French, P.W. (1993a) *Targets of Coastal Habitat Re-creation*, English Nature, Peterborough.

Pye, K. and French, P.W. (1993b) *Erosion and Accretion Processes on British Saltmarshes*. Vol. 1. *Introduction: Saltmarsh Processes and Morphology*, Cambridge Environmental Research Consultants, Cambridge.

Pye, K. and French, P.W. (1993c) *Erosion and Accretion Processes on British Saltmarshes*. Vol. 5. *Management of Saltmarshes in the Context of Flood Defence and Coastal Protection*, Cambridge Environmental Research Consultants, Cambridge.

Pye, K. and Tsoar, H. (1990) *Aeolian Sand and Sand Dunes*, Unwin Hyman, London.

Quigley, M.B. (ed.) (1991) *A Guide to the Sand Dunes of Ireland*, EUCC, Dublin.

Race, M.S. and Christie, D.R. (1982) Coastal zone development: mitigation, marsh creation and decision-making. *Environmental Management*, **6**, 317–28.

Radley, G.P. (1994) *Sand Dune Vegetation Survey of Great Britain. Part 1: England*, English Nature, Peterborough.

Randall, R.E. (1973) Shingle Street, Suffolk: an analysis of a geomorphic cycle. *Bulletin of the Geological Society of Norfolk*, **24**, 15–35.

Randall, R.E. (1989) Shingle habitats in the British Isles. *Botanical Journal of the Linnean Society*, **101**, 3–18.

Randall, R.E. (1992) The shingle vegetation of the coastline of New Zealand: Nelson Boulder Bank and Kaitorete Spit. *New Zealand Journal of Geography*, **93**, 11–19.

Randall, R.E. and Doody, J.P. (1995) Habitat inventories and the European Habitats Directive: the example of shingle beaches, in *Directions in European Coastal Management* (eds M.G. Healy and J.P. Doody), Samara Publishing, Cardigan, pp. 19–36.

Rankin, S.C. (1986) *The ecology of the Mersey Estuary and likely effects of the proposed Mersey barrage, with special reference to the bird populations of the area.* MSc Thesis, University of Manchester.

Ranwell, D.S. (1958) Movement of vegetated sand dunes at Newborough Warren, Anglesey. *Journal of Ecology*, **46**, 83–100.

Ranwell, D.S. (1959) Newborough Warren, Anglesey. I. The dune system and dune slack habitat. *Journal of Ecology*, **47**, 571–601.

Ranwell, D.S. (1960a) Newborough Warren, Anglesey. II. Plant associes and succession cycles of the sand dune and dune slack vegetation. *Journal of Ecology*, **48**, 117–41.

Ranwell, D.S. (1960b) Newborough Warren, Anglesey. III. Changes in the vegetation on parts of the dune system after loss of rabbits by myxomatosis. *Journal of Ecology*, **48**, 385–95.

Ranwell, D.S. (1961) *Spartina* salt marshes in southern England. I. The effects of sheep grazing at the upper limits of *Spartina* marsh in Bridgwater Bay. *Journal of Ecology*, **49**, 325–40.

Ranwell, D.S. (1964) *Spartina* salt marshes in southern England. III. Rates of establishment, succession and nutrient supply at Bridgwater Bay, Somerset. *Journal of Ecology*, **52**, 95–105.

Ranwell, D.S. (1967) World resources of *Spartina townsendii* (*sensu lato*) and economic use of *Spartina* marshland. *Journal of Applied Ecology*, **4**, 239–56.

Ranwell, D.S. (1972) *Ecology of Salt Marshes and Sand Dunes*, Chapman & Hall, London.

Ranwell, D.S. (1974) The salt marsh to tidal woodland transition. *Hydrobiological Bulletin (Amsterdam)*, **8**, 139–51.

Ranwell, D.S. (1981) Introduced coastal plants and rare species in Britain, in *The Biological Aspects of Rare Plant Conservation* (ed. H. Synge), John Wiley, Chichester, pp. 413–19.

Ranwell, D.S. and Boar, R. (1986) *Coast Dune Management Guide*, Institute of Terrestrial Ecology, Huntingdon, UK.

Ranwell, D.S. and Downing, B.M. (1959) Brent goose (*Branta bernicla* L.) winter feeding pattern and *Zostera* resources at Scolt Head Island, Norfolk. *Animal Behaviour*, **7**, 42–56.

Ranwell, D.S., Bird, E.C.F., Hubbard, J.C.E. and Stebbings, R.E. (1964) *Spartina* salt marshes in southern England. V. Tidal submergence and chlorinity in Poole Harbour. *Journal of Ecology*, **52**, 627–41.

Raunkiaer, C. (1934) *The Life Forms of Plants and Statistical Plant Geography*, Clarendon Press, Oxford.

Read, D.J. (1989) Mycorrhizas and nutrient cycling in sand dune ecosystems. *Proceedings of the Royal Society of Edinburgh*, **96B**, 89–110.

Reading C.J. and S. McGrorty (1978) Seasonal variations in the burying depth of *Macoma balthica* (L.) and its accessibility to wading birds. *Estuarine and Coastal Marine Science*, **6**, 135–44.

Redfield, A.C. (1972) Development of a New England salt marsh. *Ecological Monographs*, **42**, 201–37.

Reed, D.J. and Cahoon, D.R. (1993) Marsh submergence vs. marsh accretion: interpreting accretion deficit data in coastal Louisiana. *Coastal Zone '93, Proceedings 8th Symposium on Coastal and Ocean Management*, pp. 243–57.

Reed, R.H. (1984) Use and abuse of osmo-terminology. *Plant, Cell and Environment*, **7**, 165–70.

Reimold, R.J. (1977) Mangals and salt marshes of eastern United States, in *Wet Coastal Ecosystems* (ed. V.J. Chapman), Elsevier, Amsterdam, pp. 157–66.

Rhodes, D. and Hanson, A.D. (1993) Quaternary ammonium and tertiary sulfonium compounds in higher plants. *Annual Review of Plant Physiology and Plant Molecular Biology*, **44**, 357–84.

Richards, P.W. (1929) Notes on the ecology of the bryophytes and lichens at Blakeney Point, Norfolk. *Journal of Ecology*, **17**, 127–40.

Riley, H. and Ferry, B. (1989) *Dungeness Bibliography*, Nature Conservancy Council, Peterborough.

Ritchie, W. (1995) Maritime oil spills – environmental lessons and experiences with special reference to low-risk coastlines. *Journal of Coastal Conservation*, **1**, 63–76.

Ritchie, W. and Mather, A.S.(1971) Conservation and use; case-study of beaches of Sutherland, Scotland. *Biological Conservation*, **3**, 199–207.

Roberts, B.A. and Robertson, A. (1986) Salt marshes of Atlantic Canada: their ecology and distribution. *Canadian Journal of Botany*, **64**, 455–67.

Robertson, D.A. (1955) *The ecology of the sand dune vegetation of Ross Links, Northumberland, with special reference to secondary succession in the blowouts*. PhD Thesis, University of Durham.

Rodwell, J.S. (1998) *British Plant Communities. 5. Maritime Vegetation and Communities of Open Habitats*, Cambridge University Press, Cambridge.

Round, F.E. (1958) Observations on the diatom flora of Braunton Burrows, North Devon. *Hydrobiologia*, **11**(2), 119–27.

Round, F.E. (1959) A note on the diatom flora of the Harlech sand dunes. *Journal of the Royal Microscopical Society*, **77**, 130–5.

Rozema, J. and Leendertse, P.C. (1991) Natural and man-made stresses in coastal wetlands, in *Ecological Responses to Environmental Stresses* (eds J. Rozema and J.A.C. Verkleij), Kluwer Academic Publishers, Dordrecht, The Netherlands, pp. 92–101.

Rozema, J. and Spruyt, T.J.M. (1985) Ecophysiology of a calcicolous and calcifuge species of the genus *Cladonia*. *Vegetatio*, **61**, 561–3.

Rozema, J., Arp, W., van Esbroek, M., Broekman, R., Punte, H. and Schat, H. (1986) Vesicular arbuscular mycorrhiza in salt marsh plants in response to soil salinity and flooding and the significance to the water relations, in *Physiological and Genetical Aspects of Mycorrhizae* (eds V. Gianinazzi-Pearson and S. Gianinazzi), INRA, Dijon, pp. 657–60.

Rozema, J., Arp, W., van Diggelen, J., Kok, E. and Letschert, J. (1987) An ecophysiological comparison of measurements of the diurnal rhythm of the leaf elongation and changes of the leaf thickness of salt-resistant Dicotyledonae and Monocotyledonae. *Journal of Experimental Botany*, **38**, 442–3.

Rozema, J., Lenssen, G.M., Broekman, R.A. and Arp, W.P. (1990) Effects of atmospheric carbon dioxide enrichment on salt marsh plants, in *Expected effects of climatic change on marine coastal ecosystems* (eds J.J. Beukema, W.J. Wolff and J.J.W.M. Brouns), Kluwer Academic Press, Dordrecht, pp. 49–54.

Rozema, J., Dorel, F., Janissen, R., Lenssen, G.M., Broekman, R.A., Arp, W.P. and Drake, B.G. (1991) Effect of elevated atmospheric CO_2 on growth, photosynthesis and water relations of salt marsh grass species. *Aquatic Botany*, **39**, 45–55.

Rudeforth, C.C., Hartnup, R., Lea, J.W., Thompson, T.R.E. and Wright, P.S. (1984) *Soils and their use in Wales*, Soil Survey, Harpenden.

Russell, P.J., Flowers, T.J. and Hutchings, M.J. (1985) Comparison of niche breadths and overlaps of halophytes on salt marshes of differing diversity. *Vegetatio*, **61**, 171–8.

Rutter, A.J. (1981) Concluding remarks, in *Plants and their Atmospheric Environment* (eds J. Grace, E.D. Ford and P.G. Jarvis), Blackwell Scientific Publications, Oxford, pp. 403–11.

Sackett, R. (1983) *Edge of the Sea*, Time-Life Books, Amsterdam.

Salinas, L.M., Delaune, R.D. and Patrick, W.H. (1986) Changes occurring along a rapidly submerging coastal area: Louisiana, USA. *Journal of Coastal Research*, **2**, 269–84.

Salisbury, E.J. (1922) The soils of Blakeney Point: a study of soil reaction and succession in relation to the plant covering. *Annals of Botany, London*, **36**, 391–431.

Salisbury, E.J. (1942) *The Reproductive Capacity of Plants*, Bell & Sons, London.

Salisbury, E.J. (1952) *Downs and Dunes*, Bell, London.

Salisbury, F.B. and Ross, C.W. (1992) *Plant Physiology*, 4th edn, Wadsworth, Belmont, California.

Sargent, D.M. (1981) The development of a viable method of stream flow measurement using the integrating float technique. *Proceedings of the Institution of Civil Engineers*, **71**(2), 1–15.

Savidge, J.F. (1976) The sand-dune flora, in *Ynyslas Nature Reserve Handbook* (ed. E.E. Watkin), Nature Conservancy Council and University College of Wales, Aberystwyth, pp. 37–66.

Schat, H. and Scholten, M. (1985) Comparative population ecology of dune slack species: the relation between population stability and germination behaviour in brackish environments. *Vegetatio*, **61**, 189–95.

Scholten, M.C.T. and Leendertse, P.C. (1991) The impact of oil pollution on salt marsh vegetation, in *Ecological Responses to Environmental Stresses* (eds J. Rozema and J.A.C. Verkleij), Kluwer Academic Publishers, Dordrecht, pp. 184–90.

Schroeder, P.M.. Dolan, R. and Hayden, B.P. (1976) Vegetation changes associated with barrier–dune construction on the Outer Banks of North Carolina. *Environmental Management*, **1**, 105–14.

Schwarz, M.L. (ed.) (1972) *Spits and Bars*, Dowden, Hutchinson and Ross, Stroudsberg, Pennsylvania.

Scott, G.A.M. (1963a) The ecology of shingle beach plants. *Journal of Ecology*, **51**, 517–27.

Scott, G.A.M. (1963b) *Mertensia maritima* (L.) S.F. Gray: Biological Flora of the British Isles. *Journal of Ecology*, **51**, 733–42.

Scott, G.A.M. (1963c) *Glaucium flavum* Crantz: Biological Flora of the British Isles. *Journal of Ecology*, **51**, 743–54.

Scott, G.A.M. (1965) The shingle succession at Dungeness. *Journal of Ecology*, **53**, 21–31.

Scott, G.A.M. and Randall, R.E. (1976) *Crambe maritima* L.: Biological Flora of the British Isles. *Journal of Ecology*, **64**, 1077–91.

Scott, R., Callaghan, T.V. and Lawson, G.J. (1990) *Spartina* as a biofuel, in Spartina anglica – *a Research Review* (eds A.J. Gray and P.E.M. Benham), HMSO, London.

Scudlark, J.R. and Church, T.M. (1989) The sedimentary flux of nutrients in a Delaware saltmarsh site – a geochemical perspective. *Biogeochemistry*, **7**(1), 55–75.

Seago, M.J. (1989) Birds of Blakeney Point and Scolt Head Island, in *Blakeney Point and Scolt Head Island* (eds H. Allison and J.P. Morley), National Trust, Norfolk, pp. 87–108.

Seliger, U. (ed.) (1992) *Coastal Plant Communities of Latin America*, Academic Press, San Diego.

Sevink, J. (1991) Soil development in the coastal dunes and its relation to climate. *Landscape Ecology*, **6**, 49–56.

Sheldon, S. (1995) Reptiles and Amphibians, in *Ecology, Management and History of the Wyre Forest* (eds J.R. Packham and D.J.L. Harding), Wolverhampton Woodland Research Group, Wolverhampton, pp. 96–107.

Shennan, I. (1986) Flandrian sea-level changes in the Fenland. II: Tendencies of sea-level movement, altitudinal changes, and local and regional factors. *Journal of Quaternary Science*, **1**, 155–79.

Sherman, D.J. and Bauer, B.O. (1993) Dynamics of beach-dune systems. *Progress in Physical Geography*, **17**, 413–47.

Shomer-Ilan, A. and Waisel, Y. (1986) Effects of stabilizing solutes on salt activation of phosphoenolpyruvate carboxylase from various plant sources. *Physiologia Plantarum*, **67**, 408–14.

Shomer-Ilan, A., Moualem-Beno, D. and Waisel, Y. (1985) Effects of NaCl on the properties of phosphoenolpyruvate carboxylase from *Suaeda monoica* and *Chloris gayana*. *Physiologia Plantarum*, **65**, 72–8.

Slavin, P. and Shisler, J.K. (1983) Avian utilisation of a tidally restored salt hay farm. *Biological Conservation*, **26**, 271–85.

Smith, P.H. and Payne, K.R. (1980) A survey of natterjack toad *Bufo calamita* distribution and breeding success in the North Merseyside sand-dune system, England. *Biological Conservation*, **19**, 27–39.

Smith, P.S. (1975) *A study of the winter feeding ecology and behaviour of the Bar-tailed Godwit* (Limosa lapponica). PhD Thesis, University of Durham.

Sneddon,P. and Randall, R.E. (1993a) *Coastal Vegetated Shingle Structures of Great Britain: main report*. (1993b) *Appendix 1. Shingle sites in Wales*. (1994a) *Appendix 2. Shingle sites in Scotland*. (1994b) *Appendix 3. Shingle sites in England*. JNCC, Peterborough.

Snow, A.A. and Vince, S.W. (1984) Plant zonation in an Alaskan salt marsh. II. An experimental study of the role of edaphic conditions. *Journal of Ecology*, **72**, 669–84.

Southwood, T.R.E. (1977) Habitat, the templet for ecological studies? *Journal of Animal Ecology*, **46**, 337–65.

Sparling, J.H. (1968) *Schoenus nigricans* L.: Biological Flora of the British Isles. *Journal of Ecology*, **56**, 883–99.

Srivastava, D.S. and Jefferies, R.L. (1995) The effect of salinity on the leaf and shoot demography of two arctic forage species. *Journal of Ecology*, **83**, 421–30.

Stace, C.A. (1991) *New Flora of the British Isles*, Cambridge University Press, Cambridge.

Stebbins, G.L. (1957) Self-fertilization and population variability in the higher plants. *American Naturalist*, **9**, 337–54.

Steers, J.A. (1926) Orford Ness: a study in coastal physiography. *Proceedings of the Geological Association*, **37**, 306–25.

Steers, J.A. (ed.) (1960) *Scolt Head Island*, 2nd edn, W. Heffer, Cambridge.

Steers, J.A. (1969) *The Sea Coast*, 4th edn, Collins, London.

Stelzer, R. and Läuchli, A. (1977) Salt- and flooding tolerance of *Puccinellia peisonis*. II. Structural differentiation of the root in relation to function. *Zeitschrift fur Pflanzenphysiologie*, **84**, 95–108.

Sterling, P.H. and Speight, M.R. (1989) Comparative mortalities of the brown-tail moth, *Euproctis chrysorrhoea* (L.) (Lepidoptera: Lymantriidae), in south-east England. *Botanical Journal of the Linnean Society*, **101**, 69–78.

Stevenson, J.C., Ward, L.G. and Kearney, M.S. (1986) Vertical accretion in marshes with varying rates of sea level rise, in *Estuarine Variability* (ed. D.A. Wolfe), Academic Press, London, pp. 241–59.

Stewart, A., Pearson, D.A. and Preston, C.D. (eds) (1994) *Scarce Plants in Britain*, JNCC, Peterborough.

Stewart, E.J.A. and Patton, D. (1927) Additional notes on the flora of the Culbin Sands. *Transactions of the Botanical Society of Edinburgh*, **29**, 27–40.

Stewart, W.D.P. (1967) Transfer of biologically fixed nitrogen in a sand dune slack system. *Nature*, 214, 603–4.

Stewart, A., Pearman, D.A. and Preston, C.D. (1994) *Scarce Plants in Britain*, JNCC, Peterborough, UK.

Stroud, L.M. (1976) *Net primary production of belowground material and carbohydrate patterns in two height forms of* Spartina alterniflora *in two North Carolina marshes*, PhD Thesis, North Carolina State University, Raleigh.

St Omer, L., Horvath, S.M. and Setaro, F. (1983) Salt regulation and leaf senescence in aging leaves of *Jaumea carnosa* (Less.) Gray (Asteraceae), a salt marsh species exposed to NaCl stress. *American Journal of Botany*, **70**(3), 363–8.

Strahler, A.N. (1975) *Physical Geography*, 4th edn, John Wiley, New York.

Stumpf, R.P. (1983) The process of sedimentation on the surface of a saltmarsh. *Estuarine and Coastal Shelf Science*, **17**, 495–508.

Sutherland, W.J. (1995) Habitat Management News. *British Wildlife*, **6**, 177–8.

Tansley, A.G. (1949) *The British Islands and their Vegetation*, Cambridge University Press, Cambridge.

Tarczynski, M.C., Jensen, R.G. and Bohnert, H.J. (1993) Stress protection of transgenic tobacco by production of the osmolyte mannitol. *Science*, **259**, 508–10.

Taschereau, P.M. (1985) Taxonomy of *Atriplex* species indigenous to the British Isles. *Watsonia*, **15**, 183–209.

Taylor, F.J. (1956) *Carex flacca* Schreb.: Biological Flora of the British Isles. *Journal of Ecology*, **44**, 281–90.

Teal, J.M. (1962) Energy flow in the saltmarsh ecosystem of Georgia. *Ecology*, **43**, 614–24.

Thom, B.G. (1967) Mangrove ecology and deltaic geomorphology: Tabasco, Mexico. *Journal of Ecology*, **55**, 301–43.

Thomas, G.J., Glover, J. and Makepeace, P. (1989) Management plan for the Royal Society for the Protection of Birds reserve at Dungeness, Kent. *Botanical Journal of the Linnean Society*, **101**, 153–61.

Thompson, H.S. (1922) Changes in the coast vegetation near Berrow, Somerset. *Journal of Ecology*, **10**, 53–61.

Tilbrook, P.J. (1986) Nature conservation in the Moray Firth. *Proceedings of the Royal Society of Edinburgh*, **91B**, 169–91.

Toft, A.R. (1995) Management guidelines, in *Guide to the Understanding and Management of Saltmarshes*, National Rivers Authority, Bristol.

Toft, A.R. and Townend, I.H. (1991) *Saltings as a Sea Defence*, Sir William Halcrow and Partners, R & D Note 29, National Rivers Authority, Bristol.

Tomlinson, P.B. (1994) *The Botany of Mangroves*, Cambridge University Press, Cambridge.

Tooley, M.J. (1992) Recent sea-level changes, in *Saltmarshes; Morphodynamics, Conservation and Engineering Significance* (eds J.R.L. Allen and K. Pye), Cambridge University Press, Cambridge, pp. 19–39.

Tooley, M.J. (1998) Saltmarshes and sea level changes: a Holocene perspective, in *British Saltmarshes* (eds T. Harris and B. Sherwood), Linnean Society/Samara Publishing, Cardigan.

Tooley, M.J. and Jelgersma, S. (eds) (1993) *Impacts of Sea-Level Rise on European Coastal Lowlands*, Blackwell Scientific Publications, Oxford.

Treherne, J.E. and Foster, W.A. (1979) Adaptive strategies of air-breathing arthropods from marine salt marshes, in *Ecological Processes in Coastal Environments* (eds R.L. Jefferies and A.J. Davy), Blackwell Scientific Publications, Oxford, pp. 165–75.

Tsoar, H. and Zohar, Y. (1985) Desert dune sand and its potential in modern agricultural development, in *Desert Development* (ed. Y. Gradus), Reidel, Dordrecht, pp. 184–200.

Tubbs, C.R. (1995a) Sea level change and estuaries. *British Wildlife*, **6**, 168–76.

Tubbs, C.R. (1995b) The meadows in the sea. *British Wildlife*, **6**, 351–5.

Tunnicliffe, C.F. (1952) *Shorelands Summer Diary*, Collins, London.

Tutin, T.G. (1942) *Zostera marina* L.: Biological Flora of the British Isles. *Journal of Ecology*, **30**, 217–24.

Underwood, G.J.C. and Patterson, D.M. (1993) Seasonal changes in diatom biomass, sediment stability and biogenic stabilization in the Severn Estuary. *Journal of the Marine Biological Association of the United Kingdom*, **73**, 871–87.

Underwood, G.J.C., Paterson, D.M. and Parkes, R.J. (1995) The measurement of microbial carbohydrate exopolymers from intertidal sediments. *Limnology and Oceanography*, **40**, 1243–53.

Ungar, I.A. and Woodell, S.R.J. (1993) The relationship between the seed bank and species composition of plant communities in two British salt marshes. *Journal of Vegetation Science*, **4**, 531–6.

Valiela, I. and Teal, J.M. (1979a) The nitrogen budget of a salt marsh ecosystem. *Nature*, **280**, 652–6.

Valiela, I. and Teal, J.M. (1979b) Inputs, outputs and interconversions of nitrogen in a salt marsh ecosystem, in *Ecological Processes in Coastal Environments* (eds R.L. Jefferies and A.J. Davy), Blackwell Scientific Publications, Oxford, pp. 399–414.

van Beckhoven, K. (1992) Effects of groundwater manipulation on soil processes and vegetation in wet dune slacks, in *Coastal Dunes* (eds R.W.G. Carter, T.G.F. Curtis and M.J. Sheehy-Skeffington), A.A. Balkema, Rotterdam, pp. 251–63.

van der Maarel, E. (1979) Environmental management of coastal dunes in the Netherlands, in *Ecological Processes in Coastal Environments* (eds R.L. Jefferies and A.J. Davy), Blackwell Scientific Publications, Oxford, pp. 543–70.

van der Maarel, E. (1981) Fluctuations in a coastal dune grassland due to fluctuations in rainfall: experimental evidence. *Vegetatio*, **47**, 259–65.

van der Maarel, E. (ed.) (1993) *Dry Coastal Ecosystems: Polar Regions and Europe. Ecosystems of the World*, Vol. 2A, Elsevier, Amsterdam.

van der Meulen, F. (1982) Vegetation changes and water catchment in a Dutch coastal dune area. *Biological Conservation*, **24**, 305–16.

van der Meulen, F., Witter, J.V. and Ritchie, W. (eds) (1991) *Impact of Climatic Change on Coastal Dune Landscapes of Europe*, SPB Academic Publishing, The Hague.

van der Putten, W.H. and Peters, B.A.M. (1995) Possibilities for management of coastal foredunes with deteriorated stands of *Ammophila arenaria* (marram grass). *Journal of Coastal Conservation*, **1**, 29–39.

van der Putten, W.H., van Dijk, C. and Troelsta, S.R. (1988) Biotic soil factors affecting the growth and development of *Ammophila arenaria*. *Oecologia*, **76**, 313–20.

van der Putten, W.H., van Dijk, C. and Peters, B.A.M. (1993) Plant specific soil borne diseases contribute to the succession in fore dune vegetation. *Nature*, **362**, 53–5.

van der Sluijs, P. (1970) Decalcification of marine clay soils connected with decalcification during silting. *Geoderma*, **4**, 209–27.

van der Werf, S. (1974) Infiltratie: met water meer plant, in *Meijendel, Duin-Water-Leven* (ed. Croin Michielson), W. van Hoeve, The Hague.

van de Staaij, J.W.M., Rozema, J. and Stroetenga, M. (1990) Expected changes in Dutch coastal vegetation resulting from enhanced levels of solar UV-B, in *Expected Effects of Climatic Change on Marine Coastal Ecosystems* (eds J.J. Beukema, W.J. Wolff and J.J.W.M. Brouns), Kluwer Academic Publishers, Dordrecht, pp. 211–17.

van de Staaij, J.W.M., Lenssen, G.M., Stroetenga, M. and Rozema, J. (1993) The combined effects of elevated CO_2 levels and UV-B radiation on growth characteristics of *Elymus athericus* (= *E. pycnanthus*). *Vegetatio*, **104/105**, 433–9.

van Dijk, H.W.J. (1985) The impact of artificial dune infiltration on the nutrient content of ground and surface water. *Biological Conservation*, **34**, 149–67.

van Dijk, H.W.J. (1989) Ecological impact of drinking-water production in Dutch coastal dunes, in *Perspectives in Coastal Dune Management* (eds F. van der Meulen, P.D. Jungerius and J. Visser), SPB Academic Publishing, The Hague, pp. 163–82.

van Dijk, H.W.J. and Groot, W.T. de (1987) Eutrophication of coastal dunes by artificial infiltration. *Water Research*, **21**, 11–18.

Vogl, R.G. (1966) Salt-marsh vegetation of Upper Newport Bay, California. *Ecology*, **47**, 80–7.

Vose, P.B., Powell, H.G. and Spence, J.B. (1957) Machair grazings of Tiree. *Transactions of the Botanical Society of Edinburgh*, **37**, 89–110.

Wallén, B. (1980) Changes in structure and function of *Ammophila* during primary succession. *Oikos*, **34**, 227–38.

Walmsley, C.A. and Davy, A.J. (1997a) Germination characteristics of shingle beach species, effects of seed ageing and their implications for vegetation restoration. *Journal of Applied Ecology*, **34**, 131–42.

Walmsley, C.A. and Davy, A.J. (1997b) The restoration of coastal shingle vegetation: effects of substrate composition on the establishment of seedlings. *Journal of Applied Ecology*, **34**, 143–53.

Walmsley, C.A. and Davy, A.J. (1997c) The restoration of coastal shingle vegetation: effects of substrate composition on the establishment of container-grown plants. *Journal of Applied Ecology*, **34**, 154–65.

Walmsley, J.G. (1995) A practical approach to wildlife management in Mediterranean salinas. *Coastline*, **4**(1), 21–5.

Walter, H. (1971) *Ecology of Tropical and Subtropical Vegetation*, Oliver and Boyd, Edinburgh.

Warrick, R.A. and Oerlemans, J. (1990) Sea level rise, in *Climate Change: The IPCC Scientific Assessment* (eds J.T. Houghton, G.J. Jenkins and J.J. Ephraums), Cambridge University Press, Cambridge, pp. 257–81.

Waters, S.J.P. (1989) The Darss Peninsula – a comparison and a contrast with Dungeness. *Botanical Journal of the Linnean Society*, **101**, 178–9.

Waters, S.J.P. and Ferry, B.W. (1989) Shingle-based wetlands at Dungeness. *Botanical Journal of the Linnean Society*, **101**, 59–67.

Watkin, E.E. (ed.) (1976) *Ynyslas Nature Reserve Handbook*, Nature Conservancy Council and University College of Wales, Aberystwyth.

Watkinson, A.R. (1978a) The demography of a sand dune annual: *Vulpia fasciculata*. II. The dynamics of seed populations. *Journal of Ecology*, **66**, 35–44.

Watkinson, A.R. (1978b) The demography of a sand dune annual: *Vulpia fasciculata*. III. The dispersal of seeds. *Journal of Ecology*, **66**, 483–98.

Watkinson, A.R. (1978c) *Vulpia fasciculata* (Forskal) Samp.: Biological Flora of the British Isles. *Journal of Ecology*, **66**, 1033–49.

Watkinson, A.R. (1990) The population dynamics of *Vulpia fasciculata*: a nine-year study. *Journal of Ecology*, **78**, 196–209.

Watkinson, A.R. and Davy, A.J. (1985) Population biology of salt marsh and sand dune annuals. *Vegetatio*, **62**, 487–97.

Watkinson, A.R. and Harper, J.L. (1978) The demography of a sand dune annual: *Vulpia fasciculata*. I. The natural regulation of populations. *Journal of Ecology*, **66**, 15–33.

Watkinson, A.R., Huiskes, A.H.L. and Noble, J.C. (1979) The demography of sand dune species with contrasting life cycles, in *Ecological Processes in Coastal Environments* (eds R.L. Jefferies and A.J. Davy), Blackwell Scientific Publications, Oxford, pp. 95–112.

Watling, R. and Rotheroe, M. (1989) Macrofungi of sand dunes. *Proceedings of the Royal Society of Edinburgh*, **96B**, 111–26.

Watson, E.V. (1981) *British Mosses and Liverworts*, 3rd edn, Cambridge University Press, Cambridge.

Watson, R.T., Rodhe, H., Oeschger, H. and Siegenthaler, U. (1990) Greenhouse gases and aerosols, in *Climate Change: The IPCC Scientific Assessment* (eds J.T. Houghton, G.J. Jenkins and J.J. Ephraums), Cambridge University Press, Cambridge, pp. 1–40.

Watson, W. (1918) Cryptogamic vegetation on the sand-dunes of the west coast of England. *Journal of Ecology*, **6**, 126–43.

Watson, W. (1922) List of lichens, etc. from Chesil Beach. *Journal of Ecology*, **10**, 255–6.

Watt, A.S. (1947) Pattern and process in the plant community. *Journal of Ecology*, **35**, 1–22.

Watt, A.S. (1957) The effect of excluding rabbits from grassland B (Mesobrometum) in Breckland. *Journal of Ecology*, **45**, 861–78.

Welburn, A. (1994) *Air Pollution and Climate Change*, 2nd edn, Longman, London.

Welch, R.C. (1989) Invertebrates of Scottish sand dunes. *Proceedings of the Royal Society of Edinburgh*, **96B**, 267–87.

West, R.C. (1977) Tidal salt-marsh and mangal formations of middle and south America, in *Wet Coastal Ecosystems* (ed. V.J. Chapman), Elsevier, Amsterdam, pp. 193–213.

Westhoff, V. (1985) Nature management in coastal areas of Western Europe. *Vegetatio*, **62**, 523–32.

Wheeler, A.J. (1994) *Stratigraphy and palaeoenvironments of late Holocene sediments in north-central Fenland, UK*. DPhil. Thesis, University of Cambridge.

Wheeler, A.J. (1995) Salt-marsh development from fen: analysis of Late Holocene deposits from north-central Fenland, UK. *Quaternary International*, **26**, 139–45.

Whiting, G.W., McKellar, H.N., Kjerfve, B. and Spurrier, J.D. (1985) Sampling and computational design of nutrient flux from a southeastern U.S. saltmarsh. *Estuarine, Coastal and Shelf Science*, **21**, 273–86.

Wiedemann, A.M. (1984) The ecology of Pacific Northwest coastal sand dunes: a community profile. *U.S. Fisheries and Wildlife Service*, FWS/OBS-84/04, pp. 130.

Wiegert, R.G. (1979) Ecological processes characteristic of coastal *Spartina* marshes of the south-eastern USA, in *Ecological Processes in Coastal Environments* (eds R.L. Jefferies and A.J. Davy), Blackwell Scientific Publications, Oxford, pp. 467–90.

Wiegert, R.G. and Wetzel, R.L. (1979) Simulation experiments with a fourteen-compartment model of a *Spartina* salt marsh, in *Marsh–Estuarine Systems Simulation* (ed. R.F. Dame), University of South Carolina Press, Columbia.

Wiehe, P.O. (1935) A quantitative study of the influence of tide upon populations of *Salicornia europaea*. *Journal of Ecology*, **23**, 323–33.

Wilcock, F.A. and Carter, R.W.G. (1977) An environmental approach to the restoration of badly eroded sand dunes. *Biological Conservation*, **11**, 279–91.

Wilkin, P.J. (1989) The medicinal leech, *Hirudo medicinalis* (L.) (Hirudinea: Gnathobdellae), at Dungeness, Kent. *Botanical Journal of the Linnean Society*, **101**, 45–57.

Wilkins, D.A. (1978) The measurement of tolerance to edaphic factors by means of root growth. *New Phytologist*, **80**, 623–33.

Williams, P.H. (1989) Why are there so many species of bumble bees at Dungeness? *Botanical Journal of the Linnean Society*, **101**, 31–44.

Williams, T.D., Cooch, E.G., Jefferies, R.L. and Cooke, F. (1993) Environmental degradation, food limitation and reproductive output: juvenile survival in lesser snow geese. *Journal of Ecology*, **62**, 766–77.

Willis, A.J. (1963) Braunton Burrows: the effects on the vegetation of the addition of mineral nutrients to the dune soils. *Journal of Ecology*, **51**, 353–74.

Willis, A.J. (1964) Investigations on the physiological ecology of *Tortula ruraliformis*. *Transactions of the British Bryological Society*, **4**, 668–83.

Willis, A.J. (1965) The influence of mineral nutrients on the growth of *Ammophila arenaria*. *Journal of Ecology*, **53**, 735–45.

Willis, A.J. (1985a) Plant diversity and change in a species-rich dune system. *Transactions of the Botanical Society of Edinburgh*, **44**, 291–308.

Willis, A.J. (1985b) Dune water and nutrient regimes – their ecological relevance, in *Sand Dunes and their Management* (ed. J.P. Doody), Nature Conservancy Council, Peterborough, pp. 159–74.

Willis, A.J. (1989) Coastal sand dunes as biological systems. *Proceedings of the Royal Society of Edinburgh*, **96B**, 17–36.

Willis, A.J. (1990) The development and vegetational history of Berrow salt marsh. *Proceedings of the Bristol Naturalists' Society*, **50**, 57–73.

Willis, A.J. (1998) The changing structure and vegetational history of the 85-year-old saltmarsh at Berrow, North Somerset, in *British Saltmarshes* (eds T. Harris and B. Sherwood), Linnean Society/Samara Publishing, Cardigan.

Willis, A.J. and Davies, E.W. (1960) *Juncus subulatus* Forsk. in the British Isles. *Watsonia*, **4**, 211–17.

Willis, A.J. and Jefferies, R.L. (1963) Investigations on the water relations of sand-dune plants under natural conditions, in *The Water Relations of Plants* (eds A.J. Rutter and F.H. Whitehead), Blackwell Scientific Publications, Oxford, pp. 168–89.

Willis, A.J. and Yemm, E.W. (1961) Braunton Burrows: mineral nutrient status of the dune soils. *Journal of Ecology*, **49**, 377–90.

Willis, A.J., Folkes, B.F., Hope-Simpson, J.F. and Yemm, E.W. (1959a) Braunton Burrows: the dune system and its vegetation. Part I. *Journal of Ecology*, **47**, 1–24.

Willis, A.J., Folkes, B.F., Hope-Simpson, J.F. and Yemm, E.W. (1959b) Braunton Burrows: the dune system and its vegetation. Part II. *Journal of Ecology*, **47**, 249–88.

Wilson, J.B., Hubbard, J.C.E. and Rapson, G.L. (1988) A comparison of realised niche relations of species in New Zealand and Britain. *Oecologia*, **76**, 106–10.

Wilson, K. (1960) The time factor in the development of dune soils at South Haven Peninsula, Dorset. *Journal of Ecology*, **48**, 341–59.

Wilson, P. (1991) Buried soils and coastal aeolian sands at Portstewart, Co. Londonderry, Northern Ireland. *Scottish Geographical Magazine*, **107**, 198–222.

Wilson, P. (1992) Trends and timescales in soil development on coastal dunes in the north of Ireland, in *Coastal Dunes* (eds R.W.G. Carter, T.G.F. Curtis and M.J. Sheehy-Skeffington), Balkema, Rotterdam, pp. 153–62.

Winter, E. (1982) Salt tolerance of *Trifolium alexandrinum* L. II. Ion balance in relation to salt tolerance. *Australian Journal of Plant Physiology*, **9**, 227–37.

Wolaver, T.G. and Spurrier, J.D. (1988) The exchange of phosphorus between a euhaline vegetated marsh and the adjacent tidal creek. *Estuarine, Coastal and Shelf Science*, **26**, 203–14.

Wolaver, T.G., Zieman, J.C. Wetzel, R. and Webb, K.I. (1983) Tidal exchange of nitrogen and phosphorus between a mesohaline vegetated marsh and the surrounding estuary in the Lower Chesapeake Bay. *Estuarine, Coastal and Shelf Science*, **16**, 321–32.

Woodell, S.R.J. (1974) Anthill vegetation in a Norfolk saltmarsh. *Oecologia (Berlin)*, **16**, 221–5.

Woodell, S.R.J. (1985) Salinity and seed germination patterns in coastal plants. *Vegetatio*, **61**, 223–9.

Woodell, S.R.J. and Dale, A. (1993) *Armeria maritima* (Mill.) Willd.: Biological Flora of the British Isles. *Journal of Ecology*, **81**, 573–88.

Woodwell, G.M., Houghton, R.A., Hall, C.A.S., Whitney, D.E., Moll, R.A. and Juers, D.W. (1979) The Flax Pond ecosystem study: the annual metabolism and nutrient budgets of a salt marsh, in *Ecological Processes in Coastal Environments* (eds R.L. Jefferies and A.J. Davy), Blackwell Scientific Publications, Oxford, pp. 491–511.

Wootton, R.J. and Sinclair, W. (1976) Notes on the fauna of the Ynyslas sand dunes, in *Ynyslas Nature Reserve Handbook* (ed. E.E. Watkin), Nature Conservancy Council and University College of Wales, Aberystwyth, pp. 79–95.

Wyn Jones, R.G. (1980) An assessment of quaternary ammonium and related compounds as osmotic effectors in crop plants, in *Genetic Engineering of Osmoregulation* (eds D.W. Rains, R.C. Valentine and A. Holaender), Plenum Press, New York, pp. 155–70.

Yapp, R.H., Johns, D. and Jones, O.T. (1917) The salt marshes of the Dovey Estuary. II. The salt marshes. *Journal of Ecology*, **5**, 65–103.

Yates, M.G., Goss-Custard, J.D., McGrorty, S., Lakhani, K.H., dit Durell, S.E.A. Le V., Clarke, R.T., Rispin, W.E., Moy. I., Yates, T., Plant, R.A. and Frost, A.J. (1993) Sediment characteristics, invertebrate densities and shorebird densities on the inner banks of the Wash. *Journal of Applied Ecology*, **30**, 599–614.

Yeo, A.R. and Flowers, T.J. (1980) Salt tolerance in the halophyte *Suaeda maritima* (L.) Dum.: Evaluation of the effect of salinity upon growth. *Journal of Experimental Botany*, **31**, 1171–83.

Yeo, A.R. and Flowers, T.J. (1986) Ion transport in *Suaeda maritima*: its relation to growth and implications for the pathway of radial transport of ions across the root. *Journal of Experimental Botany*, **37**, 143–59.

Zaremba, R.E. and Leatherman, S.P. (1986) Vegetative physiographic analysis of a U.S. northern barrier system. *Environmental Geology and Water Sciences*, **8**, 193–207.

Zellmer, I.D., Clauss, M.J., Hik, D.S. and Jefferies, R.L. (1993) Growth responses of arctic graminoids following grazing by captive lesser snow geese. *Oecologia*, **93**, 487–92.

Zhang, J. (1996) Interactive effects of soil nutrients, moisture and sand burial on the development, physiology, biomass and fitness of *Cakile edentula*. *Annals of Botany*, **78**, 591–8.

FURTHER READING

Adam, P. (1990) *Saltmarsh Ecology*, Cambridge University Press, Cambridge.

Allen, J.R.L. and Pye, K. (eds) (1992) *Saltmarshes: Morphodynamics, Conservation and Engineering Significance*, Cambridge University Press, Cambridge.

Boaden, P.J.S. and Seed, R. (1985) *An Introduction to Coastal Ecology*, Blackie, London.

Buck, A.L. (1993) *An Inventory of U.K. Estuaries*, JNCC, Peterborough.

Burd, F.H. (1989) *The Saltmarsh Survey of Great Britain: an Inventory of British Saltmarshes*, Nature Conservancy Council, Peterborough.

Carter, R.W.G. (1988) *Coastal Environments*, Academic Press, London.

Carter, R.W.G., Curtis, T.G.F. and Sheehy-Skeffington, M.J. (eds) (1992) *Coastal Dunes*, Balkema, Rotterdam.

Chapman, V.J. (1976) *Coastal Vegetation*, 2nd edn, Pergamon, Oxford.

Chapman, V.J. (ed.) (1977) *Wet Coastal Ecosystems*, Elsevier, Amsterdam.

Daiber, F.C. (1982) *Animals of the Tidal Marsh*, Van Nostrand Reinhold, New York.

Dargie, T.C.D. (1993) *Sand Dune Vegetation Survey of Great Britain. Part 2: Scotland*, JNCC, Peterborough.

Dargie. T.C.D. (1995) *Sand Dune Vegetation Survey of Great Britain. Part 3: Wales*, JNCC, Peterborough.

Doody, J.P. (ed.) (1985) *Sand Dunes and their Management*, Nature Conservancy Council, Peterborough.

Doody. J.P. (ed.) (1991) *Sand Dune Inventory of Europe*, Joint Nature Conservation Committee, UK and EUCC; Peterborough.

Dring, M.J. (1992) *The Biology of Marine Plants*, Cambridge University Press, Cambridge.

Ferry, B.W., Waters, S.J.P and Jury, S.L. (eds) (1989) Dungeness: the ecology of a shingle beach. *Botanical Journal of the Linnean Society*, **101**, 1–179.

Healy, M.G. and Doody, J.P. (eds) (1995) *Directions in European Coastal Management*, Samara Publishing Ltd, Cardigan.

Holmes, A. (1965) *Principles of Physical Geology*, 2nd edn, Nelson, London.

Jefferies, R.L. and Davy, A.J. (eds) (1979) *Ecological Processes in Coastal Environments*, Blackwell Scientific Publications, Oxford.

Jones, P.S., Healy, M.G. and Williams, A.T. (eds) (1996) *Studies in European Coastal Management*, Samara Publishing Limited, Cardigan.

Lewis, R.R. (1982) *Creation and Restoration of Coastal Plant Communities*, CRC Press, Boca Raton.

Long, S.P. and Mason, C.F. (1983) *Saltmarsh Ecology*, Blackie, Glasgow.

Marschner, H. (1995) *Mineral Nutrition of Higher Plants*, 2nd edn, Academic Press, London.

Pethick, J. (1984) *An Introduction to Coastal Geomorphology*, Edward Arnold, London.

Pomeroy, L.R. and Wiegert, R.G. (eds) (1981) *The Ecology of a Salt Marsh.* Springer, New York.

Proctor, M.C.F. and Yeo, P.F. (1996) *The Natural History of Pollination*, Collins, London.

Quigley, M.B. (ed.) (1991) *A Guide to the Sand Dunes of Ireland*, EUCC; Dublin, Ireland.

Radley, G.P. (1994) *Sand Dune Vegetation Survey of Great Britain. Part 1: England*, English Nature, Peterborough.

Ranwell, D.S.(1972) *Ecology of Saltmarshes and Sand Dunes*, Chapman & Hall, London.

Rodwell, J.S. (1998) *British Plant Communities, 5. Maritime Vegetation and Communities of Open Habitats.* Cambridge University Press, Cambridge, (in press).

Seliger, U. (ed.) (1992) *Coastal Plant Communities of Latin America.* Academic Press, San Diego.

Steers, J.A. (1969) *The Sea Coast*, 4th edn, Collins, London.

Tansley, A.G. (1939) *The British Islands and Their Vegetation*, Cambridge University Press, Cambridge.

Tansley, A.G. (1968) *Britain's Green Mantle*, 2nd edn (revised M.C.F. Proctor), Cambridge University Press, Cambridge.

Tomlinson, P.B. (1994) *The Botany of Mangroves*, Cambridge University Press, Cambridge.

van der Maarel, E. (ed.) (1993) *Dry Coastal Ecosystems*, Vols 1, 2A, 2B, Elsevier, Amsterdam.

van der Meulen,F., Jungerius, P.D. and Visser,J. (eds) (1989) *Perspectives in Coastal Dune Management*, SPB Academic Publishing, The Hague.

van der Meulen, F., Witter, J.V. and Ritchie, W. (eds) (1991) Impact of Climatic Change on the Dune Landscapes of Europe. *Landscape Ecology*, **6**, issue 1/2.

Viles, H. and Spencer, T. (1995) *Coastal Problems: Geomorphology, Ecology and Society at the Coast*, Edward Arnold, London.

INDEX

Bold numbers refer to pages giving a definition of key terms; ***bold italic*** numbers are for Figures.

ENGLISH AND SCIENTIFIC NAMES OF SOME COMMON COASTAL ORGANISMS

Superscript numbers indicate volumes of the *Journal of Ecology* in which Biological Flora accounts occur. Synonyms are given in the index.

Amphipoda (crustacean 'sandhoppers') e.g.
  Talitrus saltator, Corophium volutator
Baltic tellin *Macoma balthica*
Bent, common *Agrostis capillaris*
  creeping *A. stolonifera*
Bird's-foot-trefoil, common *Lotus corniculatus*[74]
Biting stonecrop *Sedum acre*
Black bog-rush *Schoenus nigricans*[56]
Bracken *Pteridium aquilinum*
Canada goose *Branta canadensis*
Cockle, common edible *Cerastoderma edule*
Cord-grass *Spartina*[57]
  Common *S. anglica*
  Prairie *S. pectinata*
  Small *S. maritima*[57]
  Smooth *S. alterniflora*[57]
  Townsend's × *townsendii*[57]
Creeping willow *Salix repens*
Curled dock *Rumex crispus*[52]
Dunlin *Calidris alpina*
Early sand-grass *Mibora minima*
Eelgrass *Zostera marina*[30]
  Dwarf *Z. noltii*
  Narrow-leaved *Z. angustifolia*[30]
Eider *Somateria mollissima*
Fescues
  Dune *Vulpia fasciculata*[66]
  Red *Festuca rubra*
Glassworts, annual *Salicornia*
Glassworts, perennial *Sarcocornia*
Hair-grasses *Aira*
  Early *A. praecox*
  Silver *A. caryophyllea*
Heather *Calluna vulgaris*[48]
Holly *Ilex aquifolium*[55]
Hound's-tongue *Cynoglossum officinale*[78]
Lady's bedstraw *Galium verum*
Laver spire shell *Hydrobia ulvae*
Lesser hawkbit *Leontodon saxatilis*
Lesser snow goose *Anser* (*Chen*) *caerulescens caerulescens*
Lugworm *Arenicola marina*
Lyme-grass *Leymus arenarius*[40]
Marram grass *Ammophila arenaria*[67]
Marsh pennywort *Hydrocotyle vulgaris*
Meadow-grass, smooth *Poa pratensis*
Mouse-ears *Cerastium*
  Common *C. fontanum*
  Sea *C. diffusum*
Mussels
  Common *Mytilus edulis*
  Horse *Modiolus demissus*
Natterjack toad *Bufo calamita*
Oraches *Atriplex*
  Babington's *A. glabriuscula*
  Frosted *A. laciniata*
  Spear-leaved *A. prostrata*
Oysters
  Common European *Ostrea edulis*
  Cup *Crassostrea virginica*
Oystercatcher *Haematopus ostralegus*
Oysterplant *Mertensia maritima*[51]
Peppery furrow shell *Scrobicularia plana*
Plantain *Plantago*
  Buck's-horn *P. coronopus*[41]
  Ribwort *P. lanceolata*[52]
  Sea *P. maritima*
Prickly saltwort *Salsola kali*
Rabbit *Oryctolagus cuniculus*
Ragwort, common *Senecio jacobaea*[45]
Reed, common *Phragmites australis*[60]
Restharrow *Ononis repens*
Rock samphire *Crithmum maritimum*
Rue-leaved saxifrage *Saxifraga tridactylites*

Rushes *Juncus*
  Black needle *J. roemerianus*
  Mud *J. gerardii*
  Sea *J. maritimus*
  Sharp *J. acutus*[42]
Saltmarsh-grass, common *Puccinellia maritima*[65]
Sand couch *Elytrigia juncea*
Sand lizard *Lacerta agilis*
Scurvygrass, common *Cochlearia officinalis*
Sea arrowgrass *Triglochin maritimum*[79]
Sea aster *Aster tripolium*[30]
Sea beet *Beta vulgaris* ssp. *maritima*
Sea blites *Suaeda*
  Annual *S. maritima*[35]
  Shrubby *S. vera* (= *S. fruticosa*)[35]
Sea-buckthorn *Hippophaë rhamnoides*[50]
Sea clubrush *Bolboschoenus maritimus*
Sea couch *Elytrigia atherica*
Sea-heath *Frankenia laevis*[67]
Sea holly *Eryngium maritimum*
Sea kale *Crambe maritima*[64]
Sea-lavenders *Limonium*
  Common *L. vulgare*[55]
  Lax-flowered *L. humile*[55]
  Rock *L. binervosum*
Sea mayweed *Tripleurospermum maritimum*[82]
Sea pea *Lathyrus japonicus*[51]
Sea-purslane *Atriplex portulacoides*[38]
Sea rocket *Cakile maritima*
Sea sandwort *Honckenya peploides*
Sea spurge *Euphorbia paralias*
Sedges *Carex*
  Glaucous *C. flacca*[44]
  Sand *C. arenaria*[70]
Shelduck *Tadorna tadorna*
Sweet vernal grass *Anthoxanthum odoratum*
Terns
  Arctic *Sterna macrura*
  Common *Sterna hirundo*
  Sandwich *Sterna sanvicencis*
Thrift *Armeria maritima*[81]
Whitlowgrass, common *Erophila verna*
Wigeon *Anas penelope*
Yellow horned-poppy *Glaucium flavum*[51]

Avocet (*Recurvirostra avosetta*). (Drawn by P.R. Hobson.)